POLYMER SCIENCE AND TECHNOLOGY

Plastics, Rubbers, Blends and Composites

Third Edition

POLYMER SCIENCE AND TECHNOLOGY

Plastics, Rubbers, Blends and Composites

Third Edition

Premamoy Ghosh
Former Professor and Head,
Department of Polymer Science and Technology,
Calcutta University,
Kolkata

Tata McGraw Hill Education Private Limited
NEW DELHI

McGraw-Hill Offices

New Delhi New York St Louis San Francisco Auckland Bogotá Caracas
Kuala Lumpur Lisbon London Madrid Mexico City Milan Montreal
San Juan Santiago Singapore Sydney Tokyo Toronto

Tata McGraw-Hill

Published by Tata McGraw Hill Education Private Limited,
7 West Patel Nagar, New Delhi 110 008.

ISBN (13): 978-1-25-906166-0
ISBN (10): 1-25-906166-3

Vice President and Managing Director—Asia-Pacific Region: *Ajay Shukla*
Executive Publisher—Professional: *R Chandra Sekhar*
Assistant Sponsoring Editor—Science, Technology and Computing: *Simanta Borah*
Manager—Production: *Sohan Gaur*
Manager—Sales and Marketing: *S Girish*
Deputy Marketing Manager—Science, Technology and Computing: *Rekha Dhyani*
General Manager—Production: *Rajender P Ghansela*
Assistant General Manager—Production: *B L Dogra*

Typeset at Tej Composers, WZ-391, Madipur, New Delhi 110063, and printed at Sai Printo Pack, A-102/4, Okhla Industrial Area, Phase-II, New Delhi-110 020

Cover Printer: Sai Printo Pack

Cover Designer: Kapil Gupta

RZCYCDDZDZLZC

The McGraw-Hill Companies

Preface to the Third Edition

I felt the need for a revised and updated edition as the field of polymer technology has experienced rapid advances.

Many of the chapters have been updated to keep up with the recent advancements in the field. Besides, there are two new chapters in the book. The new chapter titled 'Miscellaneous Polymers and Nanotechnology' covers topics related to dendritic polymers and dendrimers; useful inorganic polymers; power polymers that are being manufactured and used for development of solar photovoltaic devices with emphasis on green and clean energy technology; and nanotechnology including nanocomposites, nanoelectronics and nanobiotechnology. The other new chapter titled 'Polymers in Wastes and Their Environmental Impact'—included towards the end of the book—covers minimization and segregation of industrial and municipal waste; waste recovery, recycle and reuse with stress on green technology and clean energy technology; and waste disposal, with focus on turning waste—particularly polymer-related—to wealth and value-added items.

It is hoped that readers—students, teachers and professionals alike—will find the new edition useful. Efforts have been made to remove some errors that had been inadvertently left out in the second edition.

I would like to thank my wife Suparna for her forbearance, understanding and cooperation.

PREMAMOY GHOSH

Preface to the First Edition

This book has been designed with the idea of blending and integrating basic polymer science and the technologies of plastics and rubbers into a composite structure. The title projects this structural pattern. It is intended for use as a textbook at the graduate and post-graduate levels in Science and Engineering, and it is hoped to be particularly helpful to those studying Material Science and Technology, with special reference to courses in polymers, coating resins, adhesives, plastics and rubbers. The structure and contents of the book have been so organized and framed as to make it useful to professionals in the productive and R & D units of related industries as well.

The overall subject matter has been presented in ten chapters. The structural sequence from Chapter 1 to Chapter 10 is: (i) basic concepts including structure-property relationships, (ii) step-growth polymerization, (iii) chain-growth polymerization, (iv) copolymerization and polymerization techniques, (v) ionic and stereo-regular polymerization, (vi) polymer characteristics and polymer characterization, (vii) polymer rheology and polymer morphology, (viii) plastics materials and related technology, (ix) rubber materials and related technology, and (x) testing and evaluation of plastics and rubbers.

SI units are desired to be increasingly used in science and technology, but the long-used traditional cgs system and fps systems continue to enjoy widespread acceptance, particularly in relation to plastics and rubbers. In this volume, the prevalent use of units in the cgs and fps systems has been followed mostly, to keep in tune with the vast existing literature. A conversion table has been included after the Table of Contents to the book to assist the reader in the conversion of different physical quantities from cgs or fps units to SI units.

In citing references, I have tried to be selective. It was not practically possible to include all possible references. A list of selected books and encyclopedias is given at the end to draw the attention of the reader to some specific and comprehensive treatises on different topics treated in this book.

I am indebted to my predecessors whose published works and contributions have been of great help. I am thankful to my students whose oftrepeated clamour over the years for a comprehensive textbook covering polymer science as well as technology of plastics and rubbers added a genuine purpose to this undertaking. I sincerely acknowledge the help and initiative of Dr. A.S. Bhattacharyya and Mr. S.P. Chatterjee in getting much of the material typed. I am grateful to late Prof S.R. Palit who encouraged me, and whose inspiration and guidance in research and teaching in the early years of my professional career, in a way, induced me to think in terms of a work of the present kind. I am happy that the idea conceived earlier has finally taken the shape of this volume. Direct and indirect assistance from a number of my students and associates has enriched my efforts in moulding this book. It will be a matter of immense pleasure if this book fulfils the objective with which it was conceived and written, finds ready acceptance among students and teachers alike and interests a wide range of readers in the professional field. I have endeavoured to treat and explain different topics in clear terms and in a lucid language as far as possible.

This book is dedicated to my wife and to the memory of my late parents. I am grateful to my wife Suparna for her great forbearance, understanding, moral support and graceful sacrifice, words would be inadequate to measure her cooperation in shaping this volume. I am grateful to my late parents who had always given me encouragement, without which I would have never reached the stage of being able to undertake the writing of this book.

PREMAMOY GHOSH

Contents

Some Physical Quantities and Conversion of Units

Physical quantity		*Typical*	*Symbol*	*SI unit*		*For conversion*
Base quantity	*Derived quantity*	*quantity symbol*	*for customary unit*	*Unit name and symbol*	*Defined in base unit symbols*	*from customary unit to SI unit multiply by*
(1)	*(2)*	*(3)*	*(4)*	*(5)*	*(6)*	*(7)*
Length	–	l	in	metre, m	m	2.54×10^{-2}
			ft	"	"	3.048×10^{-1}
Mass	–	M	lb	kilogram, kg	kg	4.536×10^{-1}
Time	–	t	sec, s	second, s	s	1
Temperature	–	T	C	kelvin, K	K	add 273.16
Amount of substance	–	n	g mole	mole, mol	mol	1
			lb mole	"	"	4.536×10^{2}
	Volume	V	L	metre cubed, m^3	m^3	1×10^{-3}
			ft^3	"	"	2.831×10^{-2}
	Area	A	in^2	metre squared, m^2	m^2	6.452×10^{-4}
			ft^2	"	"	9.290×10^{-2}
	Density	ρ	lb/ft^3	kilogram per metre cubed, kg m^{-3}	kg m^{-3}	1.602×10
			g/cm^3	"	"	1×10^{3}
	Force	F	dyne	newton, N	kg m s^{-2}	1×10^{-5}
			kgf	"	"	9.807
			lbf	"	"	4.448
			tonnef	"	"	9.807×10^{3}
	Pressure	P	$dyne/cm^2$	pascal, Pa (N m^{-2})	kg $m^{-1}s^{-2}$	1×10^{-1}
	(Modulus)	(E, G)	lbf/in^2 or psi	"	"	6.895×10^{3}
			kgf/cm^2	"	"	9.807×10^{4}
			atm	"	"	1.013×10^{5}
	Energy	E	Btu	joule, J (Nm)	kg m^2 s^{-2}	1.055×10^{3}
	(Torque)	(τ)	ft lbf	"	"	1.356
	(Heat)	(Q)	kcal	"	"	4.187×10^{3}
	(Work)	(W)	ft.pdl	"	"	4.214×10^{-2}
			erg (dyne/cm)	"	"	1.000×10^{-7}
	Power	P	Btu/s	watt, W(J.s^{-1})	kg m^2 s^{-3}	1.055×10^{3}
			ft.lbl/s	"	"	4.214×10^{-2}
			kcal/s	"	"	4.187×10^{3}
	Viscosity η		poise	pascal second, Pa s	kg m^{-1} s^{-1}	1×10^{-1}

CHAPTER 1

Basic Concepts of High Polymer Systems

1.1 Introduction and Historical Background

The quest for new materials has been with us from ancient times. Polymers, though introduced in the materials field in a meaningful manner only very recently, occupy a major place and pivotal position in the materials field today. In performance characteristics and application prospects and diversity, they offer novelty and versatility not found in any other kind of materials.

Till about the first two decades of the twentieth century, confusion prevailed over the basic understanding of polymer molecules. They were recognized more as colloids or associated molecules. Attempts to determine their molecular weights using solutions in suitable solvents by cryoscopy often gave confusive, irreproducible, uncertain and sometimes very high values. For starch, natural rubber and cellulose derivatives, molecular weights ranging up to 40,000–50,000 or even more were obtained. Such high values implied existence of very large molecules. But this was not accepted in view of the total lack of structural concept about such large molecules. The chemists continued to favour the concept of large association of much smaller molecules of short chain or cyclic structures.

1.2 Macromolecular Concept

The realization of the macromolecular concept of polymeric materials in 1920s proved to be a vital turning point. Accumulated confused ideas of the earlier decades became meaningful and they were then translated into the production of a large variety of materials of hitherto unknown structures through polymerization and copolymerization of olefinic, diolefinic and vinylic compounds and combinations thereof, and through polycondensation reactions between a large variety of bi- or polyfunctional compounds bearing well characterized chemical functional groups.

The macromolecular concept came into wide acceptance and appreciation soon after it was propounded by Staudinger[1] in 1920. He advanced long-chain structural formulas for polystyrene, natural rubber, etc. Long sequences of simple chemical units linked with each other with covalent bonds came into recognition as the basic structural feature of a polymer molecule. The polystyrene molecule obtained from polymerization of styrene 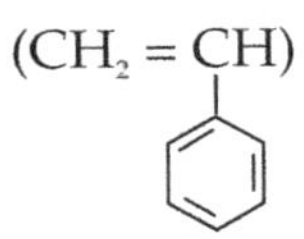 came to be recognized to have the formula — 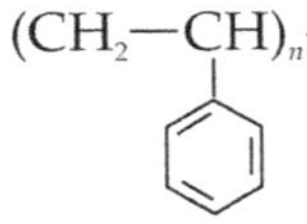 —where n had a large value. Staudinger was not sure about the terminal points or end groups of the long-chain polymer molecules. He was of the view that there was no need to saturate the terminal valences and suggested[2] that no end groups were needed. However, development and elucidation of the chain reaction theory and mechanism of vinyl and related polymerization by Flory[3] and others[4] clearly established that polymer materials consisted of long-chain molecules with well characterized end groups. Another distinctive feature of the same decade was the pioneering work of Carothers[5] in the area of polycondensation and towards the first rational synthesis of linear macromolecules, polyesters and polyamides.

A stage was reached by the middle of the 20th century, when the possibility of attaining almost any conceivable material property of structural importance through polymerization, copolymerization or polycondensation and through modification of available polymers by chemical treatment, reinforcements, compounding or blending, thermal or mechanical treatment, irradiation, etc., appeared to be within our competence. We had reached the era of tailor-made polymers in a comprehensive sense. Material prospect brightened through what may be called "molecular engineering". Nowadays, through proper selection of monomers and their combinations, catalysts and other additives and adopting appropriate polymerization conditions and techniques, experts are able to construct polymer molecules of almost any desired size, shape and complexity and of any desired chemical structure suited to almost any contemplated end-use.

1.3 Structural Features of a Polymer

The word polymer literally means many(*poly*) units(*mer*). A simple chemical unit repeats itself a very large number of times in the structure of a polymer molecule; this unit may consist of a single atom, or more commonly, a small group of atoms linked chemically. An example of the first kind (repeat units consisting of a single atom) is the well-known plastic sulphur, -S-S-S-S-S-S-, in the molecules of which the repeating S atoms are joined together by covalent bonds. This polymeric sulphur is, however, unstable and it slowly changes back to the powdery form of sulphur on keeping at room temperature.

In bulk of the polymers, the repeating unit or simply the repeat unit is constituted of a small group of atoms combined in a specific fashion. The simple molecule from which its polymer is obtained or to which a polymer may be degraded is called the monomer. Table 1.1 lists some polymers and structures of the respective monomers and repeat units.

The number of repeat units in a given polymer molecule is known as its chain length or degree of polymerization(DP). The molecular weight of the polymer is the product of the molecular weight of the repeat unit and DP.

Table 1.1 *Some polymers and respective monomers and repeat units*

Polymer	*Monomer*	*Repeat Unit*
Polyethylene	Ethylene, $CH_2{=}CH_2$	$-CH_2-CH_2-$
Polypropylene	Propylene, $CH_2{=}CH(CH_3)$	$-CH_2-CH(CH_3)-$
Polystyrene	Styrene, $CH_2{=}CH(C_6H_5)$	$-CH_2-CH(C_6H_5)-$
Poly (vinyl chloride)	Vinyl chloride, $CH_2{=}CH(Cl)$	$-CH_2-CH(Cl)-$
Poly (viny acetate)	Vinyl acetate, $CH_2{=}CH(OCOCH_3)$	$-CH_2-CH(OCOCH_3)-$
Polyacrylonitrite	Acrylonitrite, $CH_2{=}CH(CN)$	$-CH_2-CH(CN)-$

(*Contd.*)

Table 1.1 *(Contd.)*

Polymer	*Monomer*	*Repeat Unit*
Poly (acrylic acid)	Acrylic acid, $CH_2{=}CH$ $\vert$ COOH	$-CH_2-CH-$ $\vert$ COOH
Poly (methacrylic acid)	Methacrylic acid, $CH_2{=}C(CH_3)$ $\vert$ COOH	$-CH_2-C(CH_3)-$ $\vert$ COOH
Poly (methyl acrylate)	Methyl acrylate, $CH_2{=}CH$ $\vert$ $COOCH_3$	$-CH_2-CH-$ $\vert$ $COOCH_3$
Poly (methyl methacrylate)	Methyl methacrylate, CH_3 $\vert$ $CH_2{=}C$ $\vert$ $COOCH_3$	CH_3 $\vert$ $-CH_2-C-$ $\vert$ $COOCH_3$
Poly (tetrafluoroethylene)	Tetrafluoroethylene, $CF_2{=}CF_2$	$-CF_2-CF_2-$
Poly (oxymethylene)	Formaldehyde, $CH_2{=}O$	$-CH_2O-$
1,4 Polyisoprene	Isoprene, CH_3 $\vert$ $CH_2{=}C-CH{=}CH_2$	CH_3 $\vert$ $-CH_2-C{=}CH-CH_2-$
Polycaprolactam (Nylon 6)	Caprolactam, ┌ – – – – ┐ $HN-(CH_2)_5-C{=}O$	O $\Vert$ $-HN-(CH_2)_5-C-$

When more than one kind of repeat units are present in a polymer, it is known as a copolymer. Polymers having molecular weight roughly in the range of 1000–20,000 are called low polymers and those having molecular weights higher than 20,000 as high polymers.

1.4 Length to Diameter Ratio

Length to diameter ratio of a polymer molecule is very high compared to that of a simple molecule. If a small molecule such as the ethylene molecule is represented by a dot (·), then the molecule of a polymer, say polyethylene of 1000 DP will be represented by a line (—) formed by addition or sticking together of 1000 dots in a linear fashion. This characteristic difference in length to diameter ratio between a small molecule and a polymer or a high polymer molecule makes all the difference in their physical properties.

Due to their long lengths, molecules in a given sample of a polymer material remain in a state of entanglement, both intermolecular and intramolecular. High melting or softening temperature, high melt or solution viscosity exhibited by polymer materials arise as a consequence of the big size of their molecules and the entanglements of the molecular chains.

1.5 Classification of Polymers

Important classifications of polymers are shown in Table 1.2. The bases of classification and major classes of polymers along with examples are given below:

Table 1.2 *Classification of polymers*

Basis of Classification	*Polymer Types*
Origin	Natural, semisynthetic, synthetic
Thermal response	Thermoplastic, thermosetting
Mode of formation	Addition, condensation
Line structure	Linear, branched, cross-linked
Application and physical properties	Rubbers, plastics, fibres
Tacticity	Isotactic, syndiotactic, atactic
Crystallinity	Non-crystalline (amorphous), semi-crystalline, crystalline

The basis of classification and major classes of polymers along with examples are also given below.

1.5.1 Origin

(a) ***Natural Polymers*** They are available in nature. Examples of such polymers are: natural rubber (1, 4-*cis*-polyisoprene), natural silk, cellulose, starch, proteins, etc.

(b) ***Semisynthetic Polymers*** They are chemically modified natural polymers such as hydrogenated, halogenated or hydro-halogenated natural rubber, cellulosics, i.e., esters and ethers of cellulose such as cellulose nitrate, methyl cellulose, etc.

(c) ***Synthetic Polymers*** They are man-made polymers prepared synthetically. Polyethylene, polystyrene, poly (vinyl chloride), polyesters, phenol-formaldehyde resins, etc., are example of synthetic polymers.

1.5.2 Thermal Response

(a) ***Thermoplastic Polymers*** They can be softened or plasticized repeatedly on application of thermal energy, without much change in properties if treated with certain precautions, e.g., polyolefins, polystyrene, nylons, linear polyesters and polyethers, poly (vinyl chloride), etc. They normally remain soluble and fusible after many cycles of heating and cooling.

(b) ***Thermosetting Polymers*** They can be obtained in soluble and fusible forms in early or intermediate stages of their synthesis, but they get set or cured and become insoluble and infusible when further heated or thermally treated; the curing or setting process involves chemical reactions leading to further growth and cross linking of the polymer chain molecules and producing giant molecules, e.g. phenolic resins, urea/melamine resins, epoxy resins, diene rubbers, unsaturated polyesters, etc.

1.5.3 Mode of Formation

(a) ***Addition Polymers*** They are formed from olefinic, diolefinic, vinylic and related monomers. They all have —C—C— linkages along the main chains of the polymer molecules and usually no other atom appears in the main chain. These polymers are formed by simple additions of monomer molecules to each other in quick succession by a chain mechanism. This is known as addition polymerization or chain-growth polymerization. The examples of such polymers are: polyethylene, polypropylene, polystyrene, polybutadiene, poly (vinyl chloride), etc. (*See also* Chapters 3 and 8).

(b) ***Condensation Polymers*** They are formed from intermolecular reactions between bifunctional or polyfunctional monomer molecules having reactive functional groups such as —OH, —COOH, —NH_2, —NCO, etc. The related functional groups react with each other in a stepwise manner producing condensed structural linkages, usually with the elimination of a simple, small byproduct molecule in each step of reaction. Besides —C—C— linkages, they contain such atoms as O, N, S, etc., at regular intervals in the main chain. The process of their formation is called condensation polymerization or step-growth polymerization. Polyamides, polyesters, polyethers, polyurethanes, phenolics, epoxy resins, etc., are examples of condensation polymers. (*See also* Chapters 2 and 8.)

1.5.4 Line Structure

(a) *Linear Polymers* They can schematically be represented by lines of finite lengths (Fig. 1.1). They are formed from olefinic, vinyl or related polymerization under suitable conditions or by condensation polymerization of bifunctional monomers. Linear polymers such as high density polyethylene, poly(vinyl chloride), polystyrene, nylon 6, etc., are soluble and fusible.

(b) *Branched Polymers* They can be schematically represented by lines of finite lengths with short or long branch structures of repeat units (Fig. 1.1). The branches appear as a consequence of uncontrolled side reactions during polymerization or by design of polymerization. Branched polymers are usually more readily soluble and fusible than linear polymers of comparable chain length or molecular weight.

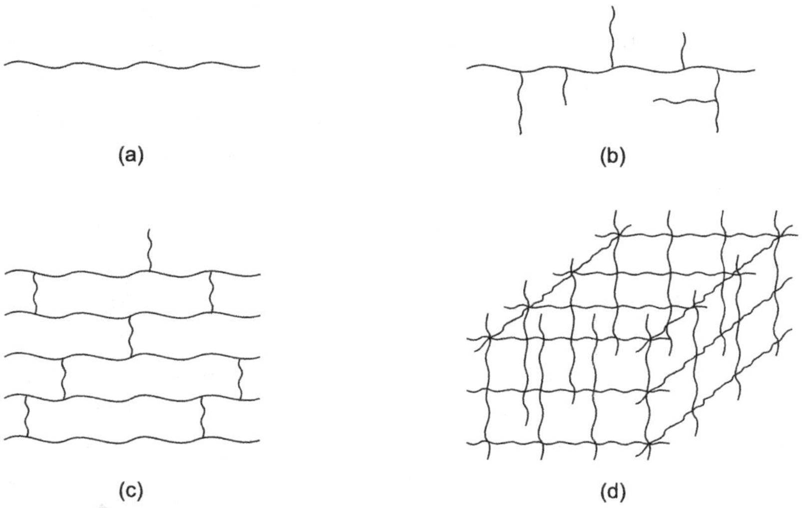

Fig. 1.1 *Schematic representation of (a) linear (b) branched (c) planar network, and (d) space network polymer structures*

(c) Cross-linked or Network Polymers They can be represented by a network structure, planar-network as in graphite or space-network as in diamond (Fig. 1.1).

Cross-linked polymers are insoluble and infusible as the molecules in them are giant molecules, often of unusually high or infinite molecular weight. Depending on the nature and frequency of cross-links, such polymers may show different orders of swelling in solvents. Examples are: phenol-formaldehyde resins, epoxy resins, vulcanized rubber, etc.

1.5.5 Application and Physical Properties

(a) ***Rubbers*** These polymers are characterized by long-range elasticity. On consideration of mechanical strength, they are rather weak, dimensionally unstable and undergo high elongations even on application of low stresses. They exhibit tensile strengths in the range of 300–3000 psi and elongation at break ranging between 300–1000%. Examples are natural rubber and synthetic rubbers. They are also known as elastomers. Technically, rubbers become useful if cross-linked to the desired extent. The cross-linking of rubber is commonly known as vulcanization. The rubbery polymers are characterized by low molar cohesion (< 2 kcal/g mol per 5Å chain length).

(b) ***Plastics*** These are usually much stronger than rubbers. Some of them are hard, horny, rigid, stiff and dimensionally stable, while others may be soft and flexible. Plastics exhibit tensile strength ranging between 4,000–15,000 psi and elongation at break ranging usually from 20 to 200% or even higher. The examples of plastics are: polyethylene, polypropylene, polystyrene, poly(vinyl chloride), nylon polyamides, urea-formaldehyde resins and phenol-formaldehyde resins, polycarbonates, etc. Plastics usually exhibit molar cohesion in the range of 2–5 kcal/g mol per 5Å chain length.

(c) ***Fibres*** They are the strongest of the three different types of polymers mentioned here. Among the natural polymers of industrial importance, the foremost place is occupied by fibres, which may be of either plant or animal origin. The suitability of fibres, natural or synthetic, for use as textile materials is unique and it is associated not only with their high mechanical strength (tensile strength in the range of 20,000–150,000 psi) but also with other useful properties such as warmth or thermal insulation, softness and flexibility. Cotton cellulose, wool, silk, synthetic polyamides and polyester fibres, and acrylic fibres fall in this clan of polymers. They usually exhibit molar cohesion in the range of 4–10 kcal/g mol per 5 Å chain length.

1.5.6 Tacticity

A vinyl polymer, $—(CH_2—CHR)_n—$ is characterized by an asymmetric carbon atom in each of the repeat units of the chain molecule, and a type of stereochemical difference that may arise in the segments of the chain is *d, l* isomerism. In a polymer of this kind, such as polypropylene or polystyrene (Fig. 1.2), every alternate carbon atom indicated by an asterisk along the chain backbone is asymmetric:

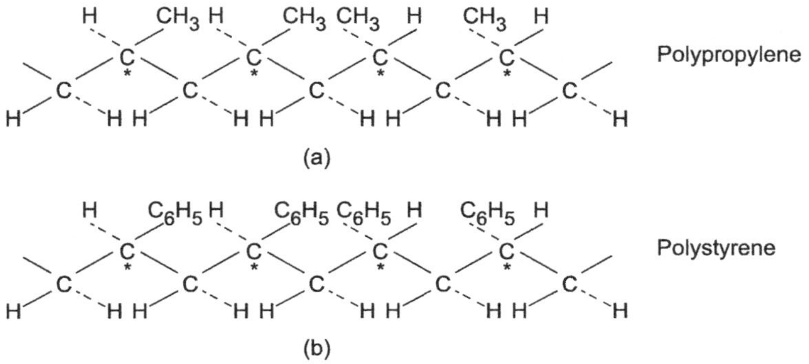

Fig. 1.2 ***Structures showing stereochemically similar and different sequences of repeat units for (a) polypropylene and (b) polystyrene***

Considering head to tail linkage of the repeat units, the first two asymmetric carbon atoms in each case (Fig. 1.2) have a different configuration in comparison with the last two, giving stereochemical sequences represented as *–d–d–l–l*. Depending on the tacticity or spatial disposition of the substituents attached to the asymmetric carbon atoms in the chain, we can have three major and distinctive sequences; (i) isotactic sequence represented as *–d–d–d–d–d–* or *–l–l–l–l–l*, giving similar or identical spatial disposition of the substituent in the repeat units, (ii) syndiotactic sequence represented as *–d–l–d–l–d–l–* where the steric placements of the substituent are such as to give strictly alternating *d* and *l* configurations, and (iii) completely random placements of the *d* and *l* configurations giving what is known as the atactic structure represented as *–d–l–d–d–l–l–l–l–d–l–d–d–*, which ultimately turns out to be a combination of isotactic and syndiotactic sequences of varying lengths. Because of high order of steric regularity in isotactic structures fostering close packing of molecular chains, isotactic polymers are usually highly crystallines, whereas the randomness in placements of *d* and *l* configurations hinders molecular packing and hence development of crystallinity in atactic polymers. Syndiotactic structure, being also regular through alternation of *d* and *l* configurations, shows good tendency to crystallize. Differences in molecular structure due to tacticity difference may be understood from the line structures (Fig. 1.3).

1.5.7 Crystallinity

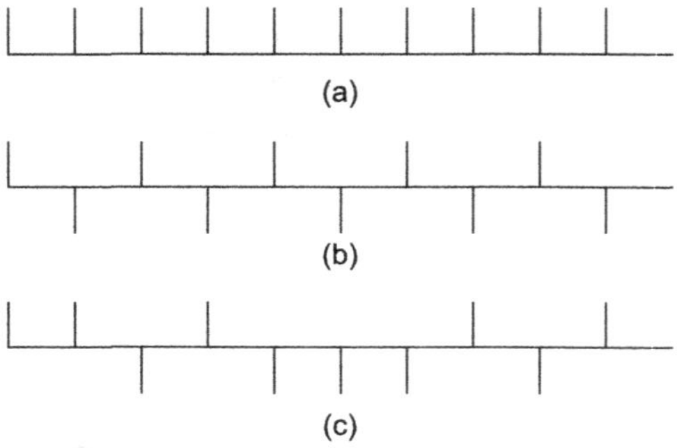

Fig. 1.3 *Line structures for (a) isotactic (b) syndiotactic, and (c) atactic sequences. Small branches or vertical lines indicate the spatial dispositions of the substituent groups (R) attached to the alternate carbon atoms along the length of the vinyl polymer chain, —(CH_2—CHR)$_n$—, represented by the horizontal line*

When cooled from molten state, different polymers exhibit different tendencies to crystallize at different rates depending on many factors including prevailing physical conditions, chemical nature of the polymers, their molecular symmetry and structural regularity or irregularity. Bulky pendent groups or short-chain branches of different lengths hinder molecular packing and inhibit crystallization. The nature of the crystalline state of polymers is not simple and it should not be confused with the regular geometry of the crystal of low molecular weight compounds such as sodium chloride and benzoic acid. There are polymers which are, by and large, amorphous and have very poor tendency to get transformed into ordered or oriented structures on cooling to near or even below room temperature. Natural or synthetic rubbers, and glassy polymers such as polystyrene, acrylate and methacrylate polymers are polymers of this nature. In a crystalline polymer, a given polymer chain exists in or passes through several crystalline and amorphous zones. The crystalline zones are made up of intermolecular/intramolecular alignment or orderly (and hence closely packed) arrangement of molecules or chain segments, and a lack of it leads to formation of amorphous zones. Polymers showing a high extent of crystallinity (>60%) are commonly known as crystalline polymers and those showing a significant but relatively poor degree of crystallinity are termed as semicrystalline polymers. Stress-induced molecular orientation in a polymeric system is schematically represented in Fig. 1.4 (*See also* Secs. 7.21 to 7.29).

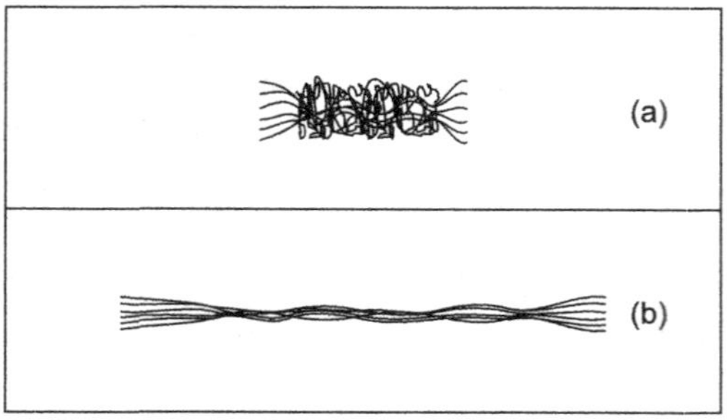

Fig. 1.4 *Schematic representation of (a) unoriented (melt normally cooled), and (b) oriented (cold drawn from melt) structures of polymer molecules, orientation having taken place in the direction of the applied stress*

1.6 Structure–Property Relationship

The wide variety of natural, semisynthetic and synthetic polymers known today exhibit wide diversity of properties. Some are rigid, hard and strong and dimensionally stable, while others are soft, flexible or largely deformable under stress. Some are soluble and fusible while others are more resistant to heat and solvents and may be even insoluble and infusible. All such properties vary from a polymer of one type to a polymer of another type. They may vary even between samples of the same type of polymer depending on how they were prepared and treated thermomechanically before being tested. Depending on the property ranges they exhibit, the polymers are classified as rubbers, plastics and fibres. To find a relation between the structure of a polymer and its physical properties, the factors which largely influence the properties should be primarily considered. These factors are: molecular weight, polarity, crystallinity, molar cohesion, linearity and non-linearity of polymeric chains, thermomechanical history of the polymer and temperature of observation, etc. The simple consideration of chemical structure of the repeat units to predict properties, particularly the mechanical behaviour has certain limitations. This is because the chemical structure of individual molecules or segments thereof contributes partly and often indirectly to mechanical properties and a more consequential and direct role is very often played by the supramolecular structure, i.e. the physical arrangement of the chain molecules with respect to each other, more so for crystalline polymers.

1.6.1 Molecular Weight

The molecular weight and molecular weight distribution in polymer systems play an important role in determining their bulk properties. Higher molecular weight permits greater degree of chain entanglements resulting in higher melting or softening temperature and tensile strength. The general trend of variation of melting (softening) point and tensile strength of a polymer is shown in Fig. 1.5.

1.6.2 Linearity and Non-linearity of Polymer Chains

The properties of a polymer would largely depend on whether its chains are predominantly linear, branched to different extents or cross-linked. Linear polymers are mostly soluble and fusible. Small extents of branching make the otherwise equivalent polymer less resistant to solvents, chemicals and heat, owing to enhanced

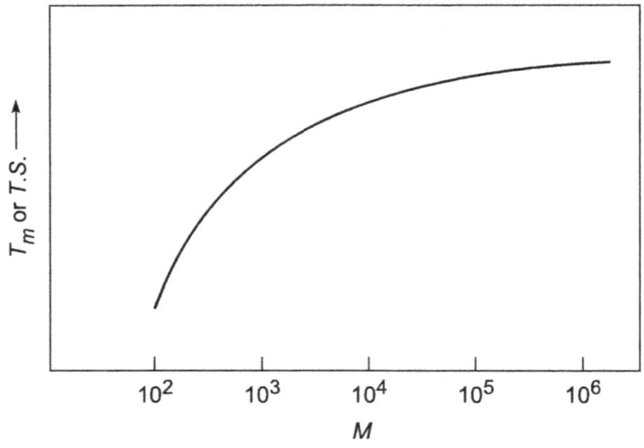

Fig. 1.5 *General trend of variation of melting (softening) temperature, (T_m) and tensile strength (T.S.) of a polymer with variation of molecular weight (M)*

molecular mobility manifested through the branches or pendent groups. High degrees of branching and ultimate cross-linking make the polymer relatively stiff through greater degree of chain entanglements and ultimately forming giant molecules of a network structure, thus restraining large scale molecular mobility or chain slippage and improving dimensional stability. Polymers thereby become less soluble and less fusible, and ultimately, insoluble and infusible. The mechanical strength increases substantially due to crosslinking which is clearly exemplified by the curing or cross-linking of a host of thermosetting polymer systems such as the curing of phenolic and urea resins, vulcanization of rubbers, etc.

Through establishment of cross-links, basic structural changes in the polymers are introduced and consequently, basic changes or improvements in properties are often achieved. By proper design of the process and according to the set objectives, different levels of cross-linking, low or high, may be achieved (Fig. 1.6).

In most commercial cross-linkable polymer systems, the established cross-links are covalent bonds and the process limitation is that the shaping of the polymer to a useful article must be accomplished before completion of the cross-linking. The process is mostly irreversible, and regeneration of the heat softenable uncross-linked structure for reprocessing is seldom possible.

1.6.3 Molar Cohesion, Polarity and Crystallinity

These factors or parameters are interrelated. A chain molecule having a strong polar structure for each repeat unit exerts strong attractive force on molecules around it.

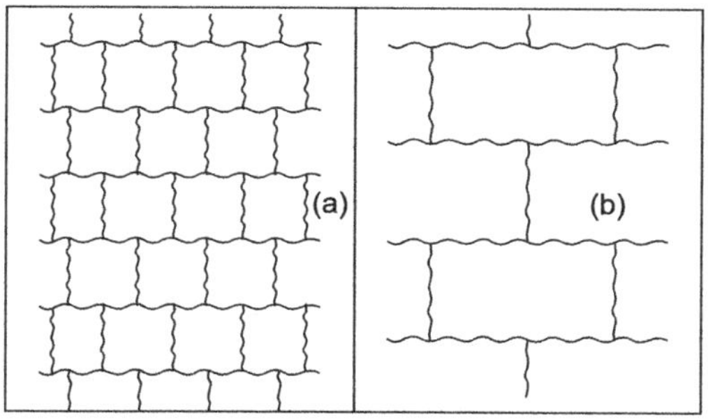

Fig. 1.6 ***Schematic representation of cross-linked polymer structures: (a) high cross-link density, and (b) low cross-link density***

In the absence of such polar structures, the molar cohesion would be weak. Let us consider the two polymers: polyethylene, $—(CH_2—CH_2)_n—$, and nylon 6 polyamide, $[—(CH_2)_5—\overset{\displaystyle O}{\overset{\|}{C}}—\underset{\displaystyle H}{\underset{|}{N}}—]_n$. The methylene links in each polymer are practically non-polar, whereas the interunit amide linkage (—CONH—) in the polyamide is strongly polar. Assuming degree of polymerization such that the chains are of equal lengths, the forces of attraction between the polyamide chains will be very much stronger than those between the polyethylene chains. This is reflected in much higher rigidity, tensile strength, melting point and crystallizing tendencies of the polyamide. Moreover, extensive intermolecular hydrogen bond formation takes place in the polyamide, particularly on stretching it from the melt condition as it is cooled, resulting in the formation of the polyamide fibre.

Again, though polyethylene has a weak molar cohesion which is of the same order as for the various rubbers, natural and synthetic, it does not really exhibit rubbery properties: instead, it is one of the most useful plastics known. Its improved physical (mechanical) properties are basically due to its very simple chemical structure. The architectural symmetry of its molecules facilitates close packing of the chains into a lattice structure, inducing crystallinity and strength. Besides the architectural symmetry, recurrence symmetry is also important in determining the total forces of attraction between chains, particularly when oriented. This is very often illustrated by copolymerization of two monomers leading to loss of recurrence symmetry, and also by chain branching. Thus, polyethylene prepared by the high-pressure method (free radical mechanism) and the one by the low-pressure method (anionic co-ordination mechanism) greatly differ in their properties; the low-pressure polyethylene being higher melting, its density and rigidity being higher and this polymer having higher degree of crystallinity. The reason is that the low-pressure polyethylene is predominantly linear and hence it has high degree of molecular symmetry, while the high pressure polyethylene is fairly branched, thus poorer in molecular symmetry.

The influence of molecular symmetry on polymer properties is also exemplified by the chemical modification of cellulose. Cellulose, originally an infusible fibre of good strength, resistant to solvents and having a good degree of crystallinity due to extensive intermolecular hydrogen bonds through the hydroxyl groups, is easily transformed into useful plastic products with good solubility in selected solvents and more or less well defined softening or melting points on progressive substitution of OH groups by esterification and etherification. As a consequence of these reactions, the high architectural and recurrence symmetry of cellulose chain molecules are progressively lost and hydrogen bonds are destroyed considerably. The completely modified cellulose, giving a degree of substitution equal to three, would normally have better strength characteristics than the incompletely modified products, because by complete modification the overall molecular symmetry is regained.

It is apparently surprising that the melting points of most linear aliphatic polyesters $\text{HO}—[(\text{CH}_2)_x—\text{O}\underset{\underset{\text{O}}{\|}}{\text{C}}—(\text{CH}_2)_y—\underset{\underset{\text{O}}{\|}}{\text{C}}\text{O}]_n—\text{H}$ having polar $—\overset{\overset{\text{O}}{\|}}{\text{C}}—\text{O}—$ interunit linkages lie well below that of polyethylene[6] (Fig. 1.7). This may be partly explained considering that the contributions of the polar ($—\overset{\overset{\text{O}}{\|}}{\text{C}}—\text{O}—$) groups in enhancing molar cohesion are far outweighed by the flexibility effects imparted to the chains through the oxygen linkages in the chain backbone. The increase in chain flexibility and hence lowering of melting point by oxygen linkages in the main chain is also illustrated by a much lower melting point (nearly 150°C) for the polyurethane $[—(\text{CH}_2)_5—\underset{\underset{\text{H}}{|}}{\text{N}}—\underset{\underset{\text{O}}{\|}}{\text{C}}—\text{O}—]_n$ than that for the analogous polyamide $[—(\text{CH}_2)_5—\underset{\underset{\text{H}}{|}}{\text{N}}—\underset{\underset{\text{O}}{\|}}{\text{C}}—]_n$ having melting point of 215°C.

Incorporation of rigid bulky groups such as an aromatic ring in the chain makes the polymer more stiff and hence higher melting.[7] Thus, while linear aliphatic

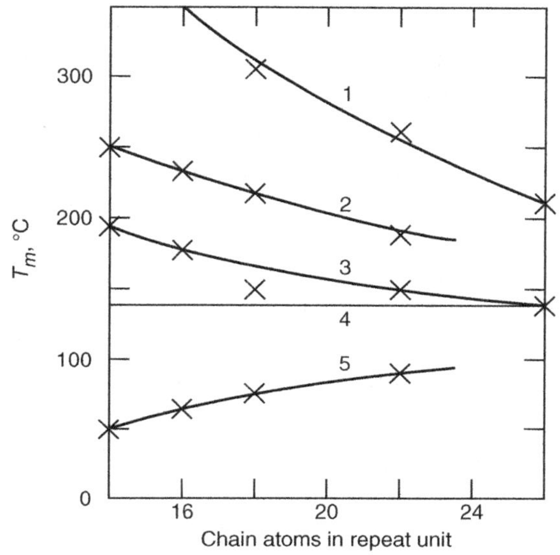

Fig. 1.7 ***Trend of change of melting point (T_m) with variation of number of chain carbon atoms in repeat units in several linear (aliphatic) condensation polymers, viewed against the melting point of linear polyethylene (Curve-4): Curves 1, 2, 3 and 5 are for polyureas, polyamides, polyurethanes and polyesters respectively. The melting points approach that of polyethylene[6] as spacing between the polar interunit linkages is increased (Hill & Walker, 1948; Courtesy, Wiley-Interscience, New York)***

polyesters are highly flexible and low melting polymers (Fig. 1.7), the aromatic linear polyester, poly(ethylene terephthalate), H—[OOC—C_6H_4—COO—$(CH_2)_2]_n$—OH is a relatively rigid polymer of high melting point (>250°C) and it can be readily drawn into a fibre of high strength, crystallinity and resistance to solvents and weathering conditions. Its improved physical properties are not due to enhanced molar cohesion, since an aromatic ring has nearly the same molar cohesion as six linear methylene groups, but are mostly due to the bulk and rigidity of the aromatic ring in each repeat unit, resulting in great hindrance to rotation and thus, reducing chain flexibility and at the same time enhancing chain stiffness enormously. Symmetrical structure of the terephthalic acid units also permits a close fitting of the chain molecules into a crystal lattice. On the other hand aromatic bifunctional compounds having functional groups in the *ortho* and *meta* positions yield similar polymers with inferior physical and mechanical properties mainly due to their poor molecular symmetry, resulting in their inability to fit into a lattice structure as closely as for polymers from the corresponding aromatic bifunctional compounds with functionalities in the *para* positions.

The intermolecular forces are opposed by thermal agitation. The effect of chain length is mainly extensive, i.e. total force of intermolecular attraction increases owing to greater molecular surface while the force per unit length or area remains constant. By contrast, polarity is intensive; more polar the structure, greater the attractive force per unit length or area. It is important to note that the effects of these factors are relatively short range and hence, minor or subtle variations in chemical structure may notably affect the physical behaviour of polymers. Structure–property relationship highlighting differences effected by aliphatic and aromatic structural units and the effects of flexible (—O—, —S—) and polar (—SO_2—) interunit linkages can be further appreciated by examining Table 1.3.

The polymer properties are also highly dependent on the spatial disposition of the side groups or side chains. The stereoregular (isotactic) polymers are of high degree of crystallinity and they exhibit higher strength and melting point than those exhibited by the corresponding random (atactic) polymers. The lengthening of the substituent group of the higher isotactic poly(α-olefin) homologues decreases the melting point up to polyheptene; from this point the melting point curve follows an increasing trend (Fig. 1.8). The initial falling trend in melting point is due to repulsion of the bulky side groups in the crystalline region. The

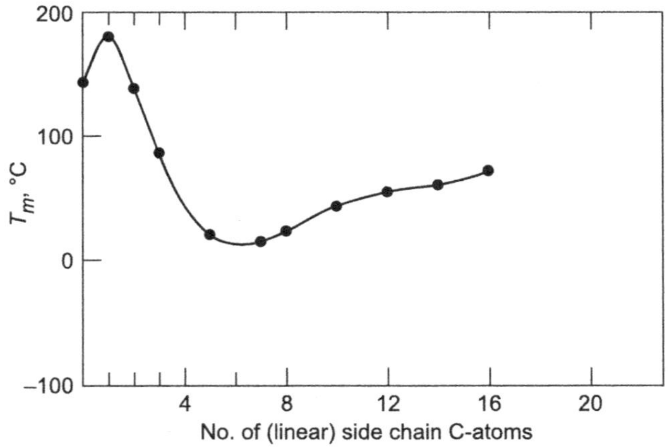

Fig. 1.8 ***Effect of variation of branch (substituent) length on the melting point of isotactic poly(α–olefins), (—CH_2—CHR)$_n$—***

Table 1.3 *Thermal properties of some aliphatic and aromatic polymers (thermoplastics)*

Polymer	*Repeat Unit*	T_g, °C	T_m, °C	*Remarks*
Polyethylene (linear), PE	$—CH_2—CH_2—$	–115, –60	137	General purpose, low modulus polymer with good degree of toughness; excellent electrical insulator
Polystyrene, PS	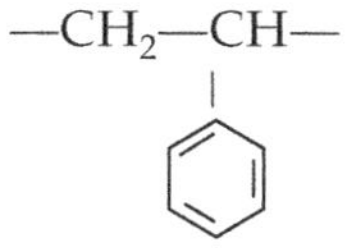	80 to 100	240	Rigid, brittle transparent polymer; excellent insulator
Poly(vinyl naphthalene), PVN	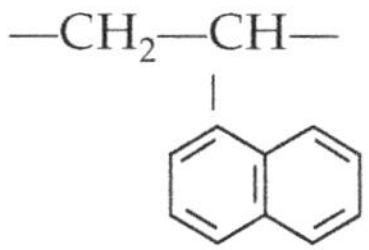	150	360	Highly, rigid, brittle polymer; excellent insulator
Poly(*para*-phenylene), PPP		—	530	Tough, thermostable, intractable polymer with semiconducting or even good conducting (when doped) properties with electrical conductivity in the range 10^{-18} to 10^2 $(\Omega\ cm)^{-1}$
Poly(tetramethylene oxide), PTMO	$—(CH_2)_4—O—$	–80	60	Weak, flexible polymer
Poly (phenylene ether), PPE	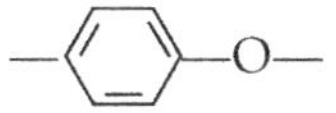	80	300	A good engineering thermoplastic, having high strength and toughness
Poly(phenylene oxide), PPO	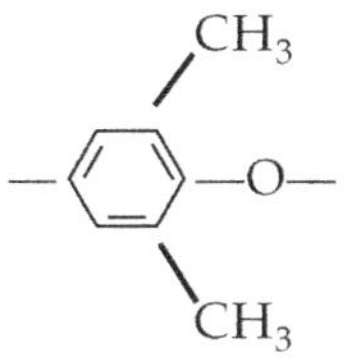	209	—	Very low water absorption; more resistant (chemically, mechanically and thermally) than PPE

(*Contd.*)

Table 1.3 (Contd.)

Polymer	*Repeat Unit*	T_g, °C	T_m, °C	*Remarks*
Poly (tetramethylene sulphide), PTMS	$—(CH_2)_4—S—$	—	70	Weak flexible polymer
Poly(phenylene sulphide), PPS	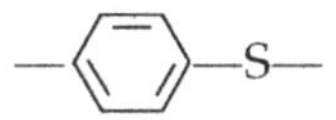	90	295	Strong, tough, thermally resistant polymer. Normally a good insulator, but when properly doped, say with 2–5% AsF_5, it assumes a semiconducting character
Poly(tetramethylene sulphone), PTMSO	$—(CH_2)_4—SO_2—$	—	270	Presence of strongly polar sulphone ($—SO_2—$) links makes the polymer mechanically strong and thermally more stable, as can be realized by comparing PTMSO with PTMS or PTMO
Poly(phenylene sulphone), PPSO	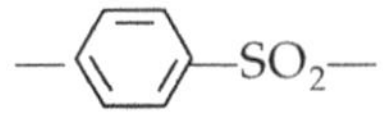	350	525	Mechanically very strong and thermally stable up to 500 °C; high-performance plastic
Poly(ether sulphone), PES	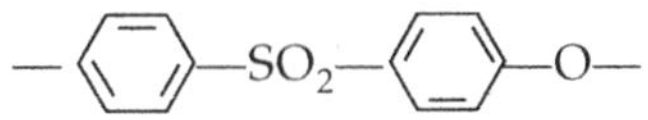	230	—	Thermostable, high performance transparent thermoplastic with improved processability in comparison with PPSO
Poly(ether ketone), PEK	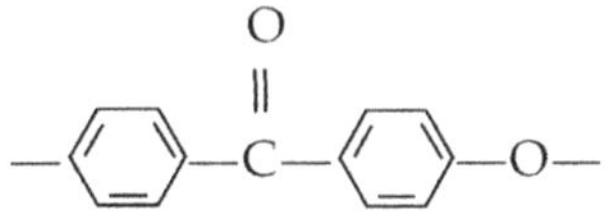	155	365	High performance thermoplastic with excellent environmental stress cracking resistance; continuous use temperature nearly 275°C
Poly(ether ether ketone) PEEK	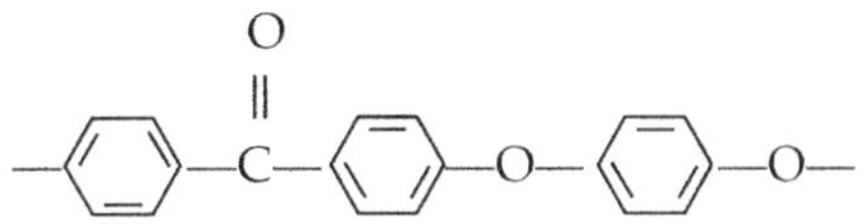	140	343	High performance, chemical resistant thermoplastic, showing a high continuous use temperature (250°C)

Rigid, bulky aromatic ring substituent makes a vinyl polymer more rigid and brittle. When aromatic rings constitute the chain backbone, the polymer becomes rigid, stiff, strong and tough. —S— and —O— inter unit linkages make the polymer weak, relatively soft and flexible, while highly polar $—SO_2—$ linkages along the chain backbone make the polymer more rigid, strong and thermostable.

upward trend in melting point is probably a result of developing trends in crystallinity involving the relatively long flexible side groups, i.e. a case of side chain crystallinity. Further, dependence of physical properties on structural isomerism is vividly reflected in the well-known examples of 1,4-*cis*-polyisoprene (natural rubber) and 1,4-*trans*-polyisoprene (*gutta percha*) (*See also* Sec. 5.9); the former, being normally devoid of or poorer in crystallinity, is more flexible and behaves as a rubber, while the latter is a good plastic having good order of crystallinity in the normal condition.

1.6.4 Effect of Temperature

Roughly speaking, all linear amorphous polymers can behave as Hookian elastic (glassy) materials, highly elastic (rubbery) substances or viscous melts according to the prevailing temperature and time-scale of experiments. Different property profiles for the same polymer at different temperatures are related to variation in the physical structures or arrangements of the chain molecules as a consequence of different types and degrees of deformation.

1.6.5 Survey of Deformation Behaviour in the Amorphous State

An idealized graph of log (shear) modulus vs. temperature is shown in Fig. 1.9. The modulus curve usually shows a number of transitions each of which is connected with the gradual development of a specific kind of molecular movement.

At a temperature below the corresponding transition temperature, the molecular processes in question are frozen in. With rise in temperature, and in the vicinity of a transition point, an additional molecular movement begins to contribute to the deformation mechanism and hence lowers the resistance to deformation, i.e. the modulus.

At very low temperatures, the only deformation occurring is Hookian elastic deformation which is time-independent and mechanically reversible. It takes place almost instantaneously and the deformation recovers instantaneously and completely on removal of the force or stress applied. The deformation is also thermodynamically reversible, as no energy is dissipated during the stress–strain cycle. The Hookian deformation is attributed to the small changes in valence angles and valence distances (bend or stretch of bonds) that take place against the strong binding forces of the homopolar bonds (dissociation energy, 30–200 kcal/mol), (Table 1.4), and therefore, it is associated with a high modulus in the region of 10^{10} dynes/cm^2 or higher. At sufficiently high temperatures, flow which is totally irreversible, both mechanically and thermodynamically, takes place. The deformation energy is dissipated as heat in the viscous melt and little configurational free

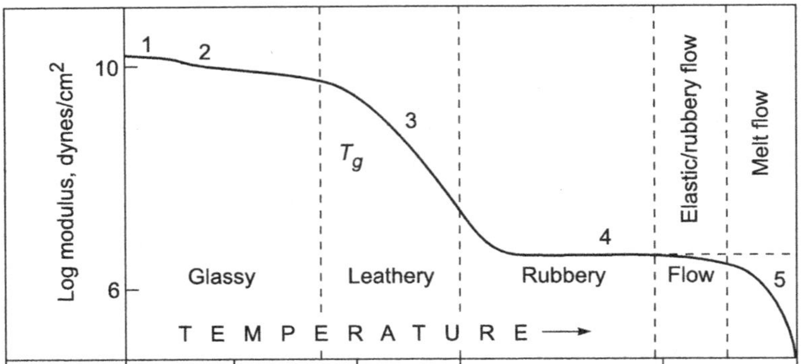

1 Defor-mation	1 Hookian glassy	2 Secondary transition	3 Primary transition	4 Highly elastic rubbery	5 Flow
Molecular motion	Bend & stretch of bonds	Excitation of internal rotation			
		Side groups	Main chain, gradual	Main chain (segmental) large-scale mobility	Chain slippage

Fig. 1.9 *Temperature dependence of log (shear) modulus in polymer system showing molecular mechanism of the deformations taking place at different points*

Table 1.4 *Characteristics of typical primary valence bonds*

Nature/Type of Bond	*Average Bond Distance, Å*	*Average Dissociation Energy, kcal/mol*
C—H	1.10	98
C—C	1.54	80
C=C	1.34	145
C—O	1.46	85
C=O	1.21	180
C—N	1.47	75
C≡N	1.15	215
C—Cl	1.77	80
C—F	1.32–1.40	100–125
O—H	0.96	110
N—H	1.01	92
O—O	1.32	34

energy is left after the deformation and hence there is no recovery. A polymer melt may be viewed as a cluster of randomly coiled or intercoiled molecules with very limited or no restriction on rotation of molecular segments due to closely neighbouring molecules. The deformation mechanism involved at this stage is slippage of molecular chains relative to each other but at the same time large scale displacement of chain segments normal to their chain axes is restricted, thereby giving rise to high melt viscosity. Cross-linked polymers are not expected to exhibit the flow phenomenon.

1.6.6 Transitions and Rubbery Region

The intermediate regions present some interesting deformation behaviours. The different transitions, shown schematically in Fig. 1.9, are manifestations of viscoelastic deformations which are time-dependent in nature, reaching at the respective point an equilibrium value after a certain time and recovering after a time of the same magnitude. The deformation in these regions is often mechanically reversible, but owing to time-dependency, thermodynamically irreversible. A part of the deformation energy is stored as free energy which takes care of mechanical reversibility and the remaining part of the deformation energy is dissipated as heat (Sec. 7.18).

The transitions schematically shown as primary and secondary transitions may be close or widely separated depending on the nature and molecular weight of the polymer. The secondary transition(s) arise as a consequence of inducement of thermal excitations involving rotation of the side groups of the polymeric chains and it generally leads to relatively small decrease in modulus. Although, inception of rotation of small segments of the main chains is also attributed to such transitions but, by and large, the rotation about the bonds of the backbone chain is still strongly hindered and the molecular chains as a whole remain stiff.

The primary transition corresponds to what is commonly known as the glass transition temperature (T_g), and this transition arises as a consequence of excitation of rotation of segments of the main chains. Almost all polymers show a primary transition or a characteristic glass transition, but many of them may not exhibit a distinct secondary transition (*See also* Sec. 7.20). Through the glass transition or primary transition, the polymers suffer a sharp and large change in modulus.

1.6.7 Property Demand and Polymer End-Use

A schematic, though rough and approximate phase diagram for polymers is given in Fig. 1.10. It shows (i) how the melting temperature (T_m) and the glass transition temperature (T_g) vary and tend to level off with increase in molecular weight (*See*

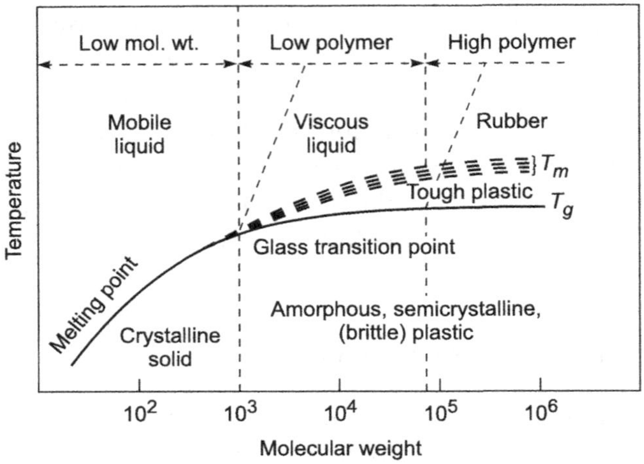

Fig. 1.10 *A schematic phase diagram showing variation of melting point (T_m), glass transition point (T_g) and consistency or viscosity of melt with variation of molecular weight*

also Fig. 1.5), (ii) how viscosity of melts varies with extension of chain lengths and (iii) the different regions or zones over which properties typical of plastics, rubbers, viscous fluids, etc. may be found.

(a) *Elastomers or Rubbers* It is apparent from the diagram that while useful plastic properties may be exhibited by polymers of wide molecular weight range (low polymer and high polymer), good useful rubbery properties are exhibited by polymers which are essentially high polymeric in nature. For revelation of rubberiness, the polymers must be far above their T_g and for practical advantage and common use, T_g for elastomers or rubbers should be in the range – 40 to – 80°C. This is very commonly attained in systems of low cohesive energy density (Table 1.5), and poor molecular symmetry resulting in amorphousness, at least in the unstrained state and permitting enough freedom of molecular motion so that deformation of high magnitude takes place rapidly. Although, these requirements and features for elastomers imply high local or segmental mobility, the gross or full-scale mobility as in the flow region must be low in practical elastomers. Restricted chain slippage must be assured in view of property demand in the form of regain of original shape and dimension on release of stress. This restriction is technically achieved through introduction of widely spaced primary valence cross-links in the chain molecular system such that on application of forces of extension, large deformations can take place without rupture of primary bonds.

(b) *Fibres* For a typical fibre, showing high tensile strength and modulus, the polymer must normally possess a combination of high molecular symmetry and high cohesive energy density manifested through polar structures in the repeat units fostering high order of permanent crystallinity (in the useful temperature zone) on cold drawing. The crystalline melting point, T_m (without decomposition) must normally be above 200°C and below 300°C to make it suitable for hot pressing or ironing and for its trouble-free spinning into a fibre. Resistance to common solvents is an advantage keeping dry cleaning in view. Again, the molecular weight should not be too high for synthetic fibre-forming polymers to ensure speedy and trouble-free spinning from melts or solutions and at the same time it should be high enough to ensure full development of tensile strength and related properties. Extensive intermolecular hydrogen bond formation is a key factor in synthetic fibre technology.

(c) *Plastics* In general, the properties of plastics are more or less intermediate between those of fibres and elastomers, with good amount of overlap on one or the other side. Plastics are put to use in a wide range of applications and hence a wide variety of property combinations is associated with them. Consequently, a wide variety of structures may adequately describe them. Some polymers having high cohesive energy density such as the nylon polyamides, behave as plastics when used without orientation of the molecules by stretching. But on orientation of the molecules by stretching or cold drawing, major changes in the structure of their physical agglomerates take place leading to high order of crystallization and the respective polymers then behave as excellent fibres.

The difference between rubbers, plastics and fibres is not really basic or very intrinsic, it is rather a matter of degree. Small or minor variations in chemical structure or physical conditioning often bring about significant property variations and may transform a rubber to a resin or a plastic or vice-versa and a plastic to a fibre or vice versa. Thus, (i) moderate to extensive cyclization, hydro-halogenation or halogenation of natural rubber, transforms it into a resinous product; (ii) limited chlorosulphonation of polyethylene introducing Cl atoms and —SO_2Cl groups in the chain molecule through simultaneous treatment with chlorine and sulphur dioxide gas under pressure breaks its molecular symmetry and transforms this general purpose plastic into a synthetic rubber despite enhancement of polarity; blending or mixing with a liquid plasticizer readily transforms the hard rigid plastic poly(vinyl chloride), PVC, into a leathery or rubbery product; (iii) limited N-alkylation of nylon polyamides makes them stretchable even when made into a fibre, and highly N-alkylated polyamides of comparable chain length become very soft or even liquids. Cellulose, an insoluble and infusible fibre, is readily transformed into good plastics on esterification or etherification. N-alkylation for the aliphatic polyamides and substitution of OH groups in cellulose reduce intermolecular hydrogen bond formation, resulting in weakening of the polymer systems.

1.6.8 Nature of Intermolecular Forces

Between molecules or chain segments not connected by primary (covalent) bonds, the so-called secondary bonds or Van der Waals forces give rise to attractive forces. The energy of the intermolecular attractive forces varies as the inverse sixth power of the intermolecular separation. It is nearly two orders of magnitude weaker than the covalent bond. The cohesive energy amounts to the total energy required to remove a molecule in a liquid to a position far from its neighbours. The cohesive energy per unit volume, widely referred to as the cohesive energy density for a few commonly used polymers is listed in Table 1.5 which clearly shows how variation in this parameter depends on molecular structure.

Among the different types of secondary valence forces known, one type arises from polarity of the molecules and it is manifested through *interaction of dipoles*. The magnitude of such interaction energy is dependent on the mutual alignment of the interacting dipoles. As molecular alignments are opposed by thermal agitation, the dipole forces are very much dependent on temperature.

Table 1.5 *Cohesive energy densities of some linear polymers*[8,9]

Polymer	*Repeat Unit*	*Cohesive Energy Density, cal/cm*^{3}
Polyethylene	—CH_2—CH_2—	56–64
Polystyrene	—CH_2—$CH(C_6H_5)$—	75–88
Polyisobutylene	—CH_2—$C(CH_3)_2$—	56–64
Polyisoprene (natural rubber)	—CH_2—$C(CH_3)$=CH—CH_2—	60–67
Poly(vinyl acetate)	—CH_2—$CH(OCOCH_3)$—	82–94
Poly(vinyl chloride)	—CH_2—CHCl—	85–95
Poly(methy1 methacrylate)	—CH_2—$C(CH_3)$ $(COOCH_3)$—	78–85
Poly(ethylene terephthalate)	—$OCH_2CH_2OOCC_6H_4CO$—	90–115
Poly(hexamethylene adipamide, nylon–66	—$HN(CH_2)_6NHCO(CH_2)_4CO$—	180–220
Polyacrylonitrile	—CH_2—CH(CN)—	150–230

Another kind of intermolecular interaction is that induced by polar molecules or segments through the dipoles on surrounding molecules or segments that do not have permanent dipoles.

The secondary bond forces consequent to *interactions between permanent dipoles and induced dipoles* are known as the *induction forces*.

The cohesive energy corresponding to the induction force is small and it does not depend on temperature.

There are also *dispersion forces* which are not dependent on polarity or permanent dipoles but which are consequences of different instantaneous configurations of electrons and nuclei of all kinds of molecules, resulting in *time-varying dipole moments* that average out to zero. The dispersion forces exist in all molecular systems, polar or non-polar and constitute a major part of the cohesive forces except in systems having strong dipoles. In non-polar materials the molar cohesion is due only to the dispersion forces which are also independent of temperature. The energy of all the above secondary valence forces usually ranges between 2–5 kcal/mol.

A fourth but an important and major kind of intermolecular attraction, particularly in polar molecules containing fluorine, nitrogen, oxygen and occasionally chlorine atoms is due to hydrogen bonding. The energy of *hydrogen bonds* usually ranges between 5–10 kcal/mol and its overall cumulative effect in polymer material systems is very pronounced and sometimes overwhelming, so much so that suitable polymers having frequent and extensive hydrogen bonds along the chains, such as cellulose, polyamides, polyacrylonitrile, etc., are inherently crystalline and resistant to heat and solvents (*see* the idealized structure of pairs of molecules in each case in Fig. 1.11). The exact nature of the hydrogen bond depends on the chemical nature and environment of the concerned polar groups and it is considered, in most systems, electrostatic or ionic in character. The average attractive force between sections of polymer molecules in an oriented domain or crystal lattice is higher than that between sections that are not in a lattice. Crystallites are usually 10–15% more dense than the supercooled melt at the same temperature.

References

1. Staudinger, H., *Ber.*, **53B** (1920), 1073.
2. Staudinger, H., *Helv. Chim. Acta*, **8** (1925), 67.
3. Flory, P.J., *J. Am. Chem. Soc.*, **59** (1937), 241.
4. Price, C.C., R.W. Kell and E. Krebs, *J. Am. Chem. Soc.*, **64** (1972), 1103.
5. Carothers, W.H., *J. Am. Chem. Soc.*, **51** (1929), 2548. *Chem. Rev.*, **8** (1931), 353.
6. Hill, R. and E.E. Walker, *J. Polym. Sci.*, **3** (1948), 609.
7. Edgar, O.W. and R. Hill, *J. Polym. Sci.*, **8** (1952), 1.
8. Small, P.A., *J Appl. Chem.*, **3** (1951), 71.
9. Burrell, H., *Polymer Handbook*, J. Brandrup and E.H. Immergut Eds. Wiley-Interscience, New York, 1975.

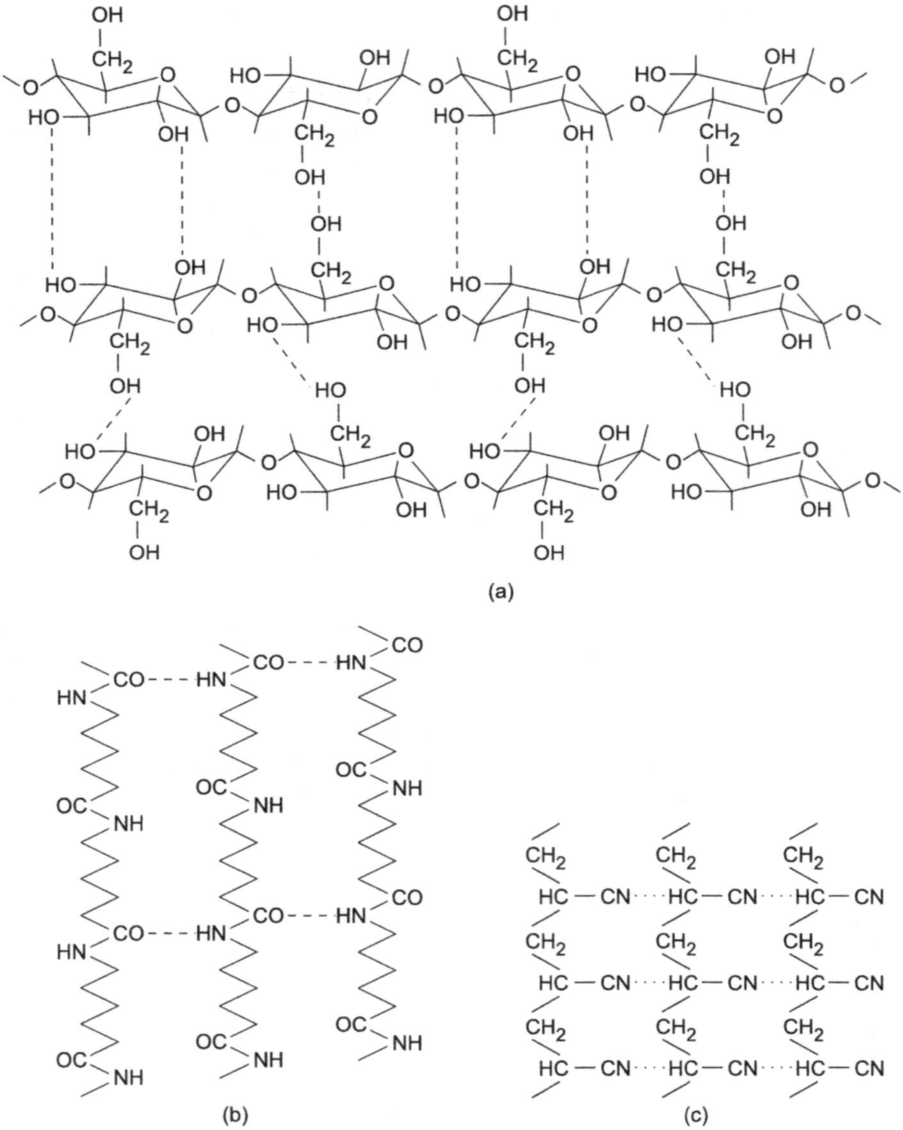

Fig. 1.11 *Idealized representation of segments of chain molecules of (a) cellulose fibre (b) (melt drawn) oriented nylon 6 polyamide fibre, and (c) oriented polyacrylonitrile fibre, showing formation of intermolecular hydrogen bonds in each case*

CHAPTER 2

Condensation Polymerization or Step-Growth Polymerization

2.1 Functionality Principle

Polymerization may be generally defined as intermolecular reaction between bifunctional or polyfunctional compounds (covering olefins and related compounds) avoiding formation of ring or cyclic structures and in a manner that makes the process functionally capable of proceeding to infinity. Functional groups or atoms, through disappearance of which polymerizations of different kinds take place are: reactive hydrogen atom (—H), hydroxyl group (—OH), carboxyl group (—COOH), amino group (—NH_2), halogen atom (—Cl, —Br), isocyanate group (—NCO), aldehyde group (—CHO), double bond (C=C), etc.

According to the functionality principle and concept, for polymerization to take place, each of the reacting molecules must be at least bifunctional, giving average functionality for the system as 2. For reacting compounds containing two different kinds of functional groups, the average functionality is calculated on the basis of stoichiometric equivalence of the two kinds of functional groups. Thus, for reaction between a dicarboxylic acid and a trihydric alcohol, to strike stoichiometric equivalence of —COOH and —OH groups, the acid and the alcohol should be taken in a molar ratio 3 : 2, which would express the average functionality for the system as 12/5, i.e., 2.4. On this basis, any system consisting of a polyfunctional compound (functionality 3 or more) bearing a given kind of functional group (say, —OH) and a related monofunctional compound bearing a different kind of functional group (say, —COOH) would always give an average functionality less than 2. So, for any system, if one type of the reacting molecules is monofunctional, then the system would fail to produce polymer molecules. On consideration of functionality, a double bond (C=C) is bifunctional, as on thermal or photoactivation, the π bond may open up to yield a diradical intermediate.

2.2 Types of Polymerization

It has been suggested in Chapter 1 that polymerization or growth of the polymer usually takes place by two basically different mechanisms. Depending on the mode of formation, one kind of polymerization is called condensation polymerization or step-growth polymerization, and the other kind is known as addition polymerization or chain-growth polymerization.[1] The physicochemical aspects of condensation polymerization will be the subject matter of this chapter.

2.3 Basic Characteristics of Condensation or Step-Growth Polymerization

Condensation polymerization takes place under suitable conditions allowing bifunctional or polyfunctional molecules to condense intermolecularly in a manner that the growth of the molecule or chain extension takes place in controlled, distinguishable steps and the process is normally associated with the formation of condensed interunit chemical links and liberation of a small byproduct such as H_2O, HCl, etc, at each step of reaction.[2] Formation of a polyester by condensation polymerization, or what may be alternatively termed as a polycondensation process between bifunctional monomers eliminating water as the byproduct of condensation, may be cited as a typical example.

2.3.1 Formation of a Polyester

Ist step

$$\underset{\text{(monomers)}}{\text{HOROH + HOOCR'COOH}} \rightarrow \underset{\text{1st step reaction product}}{\text{HOROOCR'COOH}} + H_2O \tag{2.1}$$

byproduct of
condensation

2nd step

$$\text{HOROOCR'COOH} + \text{HOROH} \rightarrow \text{HOROOCR'COOROH} + H_2O \quad (2.2)$$

(HOROOCR'COOROH: 2nd step reaction product; H_2O: byproduct of condensation)

or

$$\text{HOOCR'COOH} + \text{HOROOCR'COOH} \rightarrow \text{HOOCR'COOROOCR'COOH} + H_2O \quad (2.3)$$

The condensation process beyond the first step need not necessarily involve reaction with one of the monomers. Condensation products of all the successive steps retain the bifunctional character, and each bifunctional species formed and remaining into the system at any point of time or extent or reaction is capable of reacting with any other bifunctional species including another of its own size or an unreacted monomer molecule, and the elimination of a molecule of H_2O as a byproduct in each step remains a common feature. One more step of condensation results in the loss of one more bifunctional species and as a consequence two functional groups, one of each kind, are lost in each step. The formation of a long-chain molecule of degree of polymerization n, where n has a large value, may be expressed as follows, combining $(2n - 1)$ successive steps of condensation:

$$n\ \text{HOROH} + n\ \text{HOOCR'COOH} \rightarrow \text{HO—[—ROOCR'COO—]}_n\text{—H} + (2n - 1)\ H_2O \quad (2.4)$$

The chain-growth process at any point of time may be represented by a general equation as follows:

$$\Sigma M_n + \Sigma M_m \rightarrow \Sigma M_{n+m} + H_2O \quad (2.5)$$

where n and m may have any integral value between 1 and α. For the monomer, the value of n or m is 1.

For a hydroxy acid, the reaction would be written as:

$$n\ \text{HORCOOH} \rightarrow \text{H—[—ORCO]}_n\text{—OH} + (n - 1)\ H_2O \quad (2.6)$$

Examples of some step-growth polymerization are given in Table 2.1.

Table 2.1 *Examples of some step-growth polymers*

Polymer type (1)	*General reaction (showing interunit linkage formed and byproduct eliminated, etc.)* (2)	*Example (giving polymer structure)* (3)
Polyester	(a) Esterification (acid/alcohol)	
	$—COOH + HO— \rightleftharpoons —C(=O)—O— + H_2O$	(i) $nHORCOOH \rightleftharpoons H—[—ORCO—]_n—OH + (n-1)H_2O$ (ii) $nHOOCRCOOH + nHOR'OH \rightleftharpoons HO—[—OCRCOOR'O—]_n—H + (2n-1)H_2O$
	(b) Ester interchange (ester/alcohol)	
	$—C(=O)—OR_1 + HOR_3 \rightleftharpoons —C(=O)—OR_3 + R_1OH$	(i) $n\,R_1O—C(=O)—R_2—C(=O)—OR_1 + nHOR_3OH \rightleftharpoons$ $R_1—[—O—C(=O)—R_2—C(=O)—OR_3—]_n—OH + (2n-1)R_1OH$
	(c) Esterification (acid chloride/alcohol)	
	$—C(=O)—Cl + HO— \rightleftharpoons —C(=O)—O— + HCl$	(i) $nClOCRCOCl + nHOR'OH \rightleftharpoons Cl—[—OCRCOOR'O—]_n—H + (2n-1)HCl$
	(d) Ring opening polymerization of Lactone	
	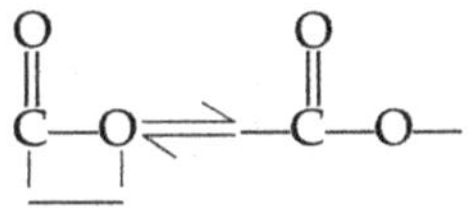	(i) 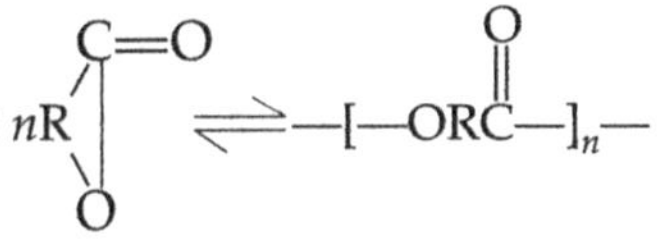

Contd.

Table 2.1 (Contd.)

(1)	(2)	(3)
Polyamide	(a) Amidation (amine/acid) $—NH_2 + HOOC— \rightleftharpoons —NH\overset{O}{\overset{\Vert}{C}}— + H_2O$	(i) $nH_2NRCOOH \rightleftharpoons H—[—HNR\overset{O}{\overset{\Vert}{C}}—]_n—OH + (n-1)H_2O$ (ii) $nH_2NRNH_2 + nHOOCR'COOH \rightleftharpoons$ $H—[—HNRNHCOR'CO—]_n—OH + (2n-1)H_2O$
	(b) Amidation (amine/acid chloride) $—NH_2 + Cl—\overset{O}{\overset{\Vert}{C}}— \rightleftharpoons —\underset{H}{\underset{\vert}{N}}—\overset{O}{\overset{\Vert}{C}}— + HCl$	(i) $nH_2NRNH_2 + nClOCR'COCl \rightleftharpoons$ $H—[—NHRNHCOR'CO—]_n—Cl + (2n-1)HCl$
	(c) Ring opening polymerization of lactam $—NH\overset{O}{\overset{\Vert}{C}}$ (ring) $\rightleftharpoons —\underset{H}{\underset{\vert}{N}}—\overset{O}{\overset{\Vert}{C}}—$	(i) $nR\langle NH, C{=}O \rangle$ (lactam ring) $\rightleftharpoons —[—NHR\overset{O}{\overset{\Vert}{C}}—]_n$
Polyurethane	Formation of urethane structure $—N{=}C{=}O + HO— \rightleftharpoons —NH—\overset{O}{\overset{\Vert}{C}}—O—$	$nOCNRNCO + nHOR'OH \rightleftharpoons OCNR$ $—[—HN—\overset{O}{\overset{\Vert}{C}}—OR'O—\overset{O}{\overset{\Vert}{C}}—NHR—]_{n-1}—HN\overset{O}{\overset{\Vert}{C}}—OR'OH$ (no byproduct elimination)
Polyurea	Formation of urea structure $—N{=}C{=}O + H_2N— \rightleftharpoons —NH\overset{O}{\overset{\Vert}{C}}—NH—$	(i) $nOCNRNCO + nH_2NR'NH_2 \rightleftharpoons$ $OCN—[—RNH\overset{O}{\overset{\Vert}{C}}—HNR'NH\overset{O}{\overset{\Vert}{C}}NH]_{n-1}—RNH\overset{O}{\overset{\Vert}{C}}NHR'NH_2$ (no byproduct elimination)

Contd.

Table 2.1 (Contd.)

(1)	(2)	(3)
Polyacetal (acetal resins) (polyether)	Formation of acetal or ether structure 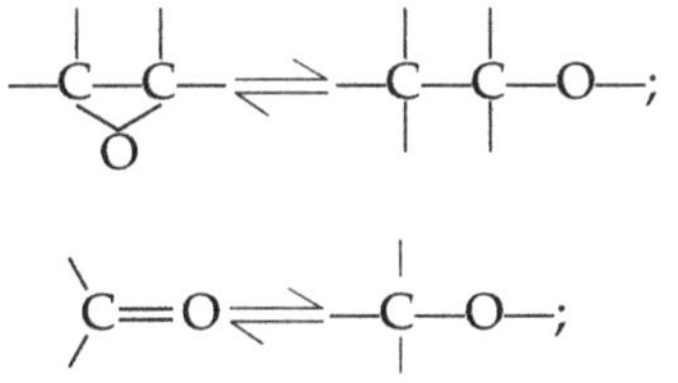	(i) $nR-\underset{\backslash O /}{CH-CH_2} \rightleftharpoons -[CH_2-\underset{R}{\underset{\vert}{CH}}-O]_n-$ (ii) $nH-\underset{H}{\underset{\vert}{C}}=O \rightleftharpoons -[CH_2-O]_n-$ (iii) $nH-\underset{CH_3}{\underset{\vert}{C}}=O \rightleftharpoons -[\underset{CH_3}{\underset{\vert}{CH}}-O-]_n-$ (no byproduct elimination)
Hydrocarbon polymer	Formation of hydrocarbon structure $-R-Br + 2Na + Br-R- \rightleftharpoons -R-R- + 2NaBr$;	$nBrRBr + 2nNa \rightleftharpoons -[-R-]_n + 2nNaBr$
Arylene polyether	Oxidative coupling (phenolic OH/active hydrogen of phenol) $-OH + \frac{1}{2}O_2 + H- \rightleftharpoons -O- + H_2O$	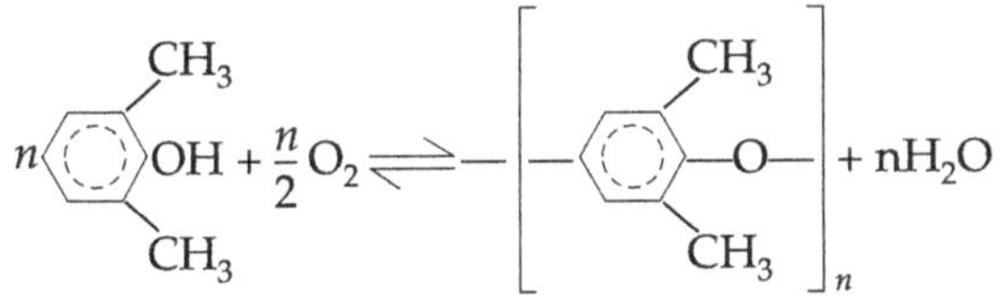
Polyarylenes [e.g. poly (*p*-xylylene)	Pyrolytic oxidative coupling: (Pyrolytic dehydrogenation and spontaneous polymerization forming alkylene linkages)	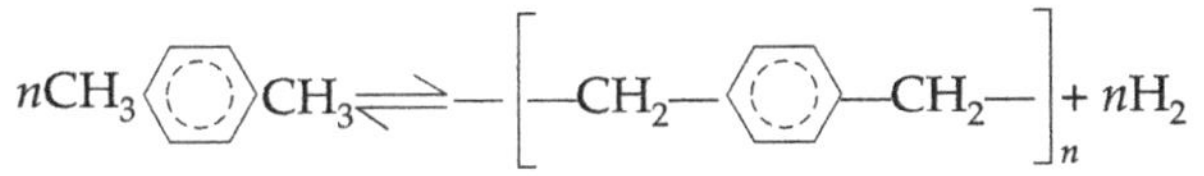
Phenolic or amino resins	Methylolation and subsequent condensation	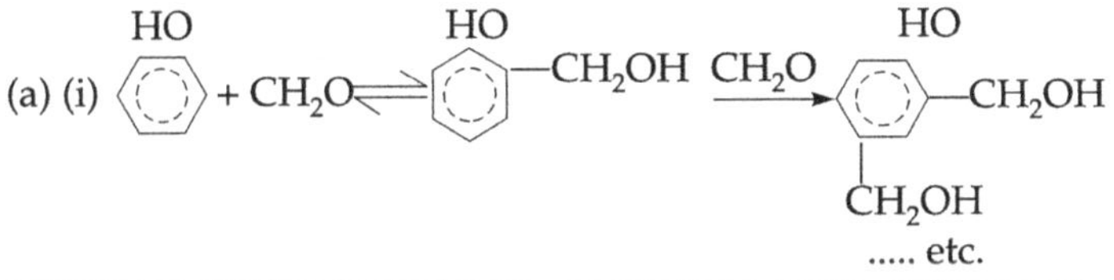

Contd.

Table 2.1 *(Contd.)*

(1)	(2)	(3)
	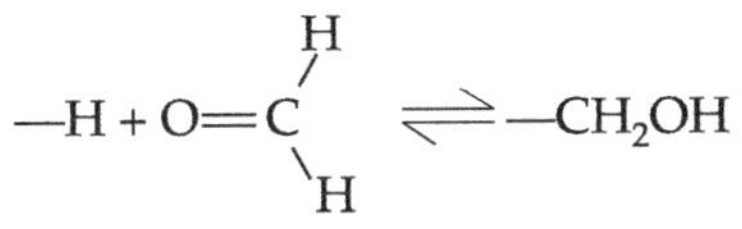	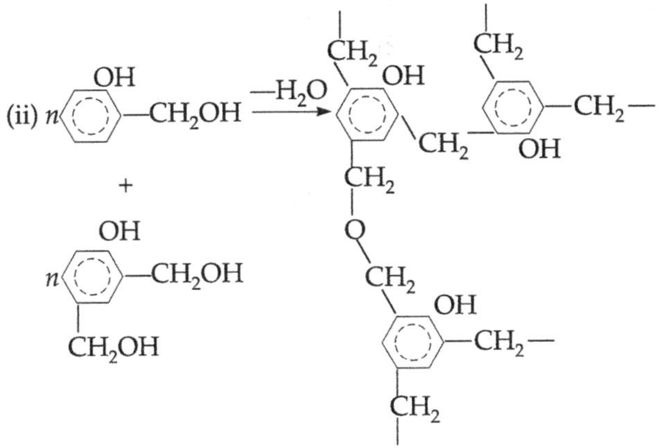
		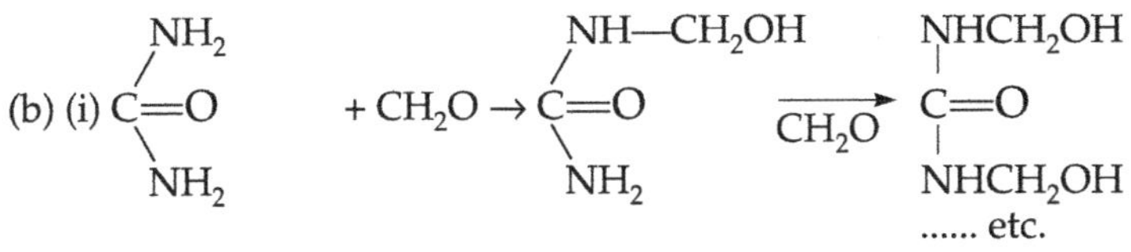
		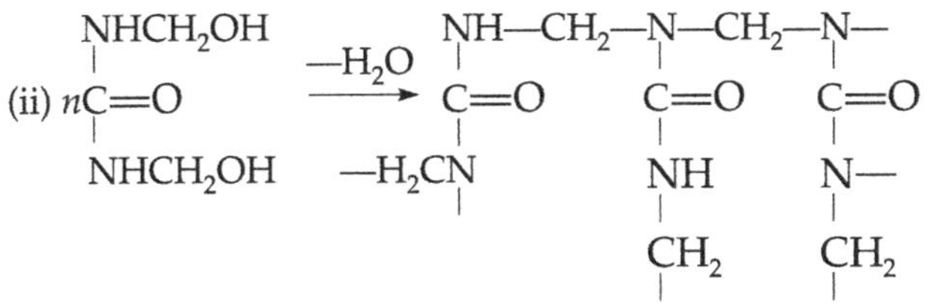

2.4 Relationship between Average Functionality, Extent of Reaction and Degree of Polymerization

W.H. Carothers derived a general expression, relating average functionality (f), extent of reaction (p) and average degree of polymerization ($\overline{X}_n$), for polycondensation reactions carried out for a time period t. The expression is commonly known as Carothers' Equation.

Let N_0 be the total number of reacting molecules initially present giving an average functionality for the system as f. The related functional groups (say, —OH and —COOH) are also considered to be present in stoichiometric equivalence. Let N be the number of molecules present at time t when the extent of reaction is p.

Then, number of molecules lost during the process over the time period t is $(N_0 - N)$; for each molecule lost, the number of functional groups lost is 2 (one of each kind) and hence the total number of functional groups lost is $2(N_0 - N)$ against the initial total number of $N_0 f$ functional groups.

Hence, the extent of reaction $p = \dfrac{2(N_0 - N)}{N_0 f}$

or

$$p = \frac{2}{f}\left(1 - \frac{N}{N_0}\right) \tag{2.7}$$

The average degree of polymerization, $\overline{X}_n$, is defined as the average number of structural units per polymer molecule which can simply be expressed as:

$$\overline{X}_n = \frac{N_0}{N} \tag{2.8}$$

combining Eqs. (2.7) and (2.8), one obtains

$$p = \frac{2}{f}\left(1 - \frac{1}{\overline{X}_n}\right) \tag{2.9}$$

Equation (2.9), known as Carothers' Equation is very important for the understanding and control of the growth of polymers through polycondensation reactions. The critical extent of reaction p_c at the gel point (for $\overline{X}_n \to \alpha$) is given by:

$$p_c = \frac{2}{f} \tag{2.9a}$$

2.5 Bifunctional System

For a bifunctional system, $f = 2$, Eq. (2.9) reduces to:

$$p = \left(1 - \frac{1}{\overline{X}_n}\right) \tag{2.10}$$

for $p = 1$, $\overline{X}_n = \alpha$, which means that for complete reaction, the average degree of polymerization will be infinity; but short of complete reaction, the degree of polymerization is always a finite quantity. Polymer obtained in a bifunctional system is, under all practical situations, linear and hence soluble and fusible. The upper theoretical limit in the degree of polymerization is infinity and this can be attained only in the hypothetical condition of completion of reaction between the functional groups ($p = 1$). For bifunctional systems, Eq. (2.10) may be rearranged to express $\overline{X}_n$ as

$$\overline{X}_n = \frac{1}{(1-p)} \tag{2.11}$$

If the mean molecular weight of the repeat unit in the polymer molecules is M_0, then the average molecular weight $\overline{M}_n$ of the polymer system is expressed as

$$\overline{M}_n = \overline{X}_n \cdot M_0 = \frac{M_0}{(1-p)} \tag{2.12}$$

2.6 Polyfunctional System

A typical polyfunctional system in condensation polymerization is exemplified by the reaction between phthalic anhydride or phthalic acid and glycerol, i.e. by allowing a bifunctional acid to react with a trihydric alcohol. As discussed before, this system is characterized by an average functionality of 2.4.

For this system, Eq. (2.9) will assume the form:

$$p = \frac{2}{2.4}\left(1 - \frac{1}{\overline{X}_n}\right) \tag{2.13}$$

We may now easily find out the critical extent of reaction, p_c at which gelation or cross-linking would commence, i.e. at which point $\overline{X}_n$ would first assume a value of infinity; the calculated value of p_c, according to Eq. (2.9a), will be $2/2.4 = 0.83$.

Thus, it is clearly seen that at the point of a theoretical extent of reaction $p = 0.83$, i.e. when the reaction is 83% complete, the polymer being formed in the polycondensation system gets cross-linked and, hence, turns into an insoluble, infusible gelled mass. Technically, such a condensation polymerization cannot and should not be allowed to proceed beyond this point (83% completion), or otherwise the reaction mass would turn into an unusable product. In fact, it is advisable to arrest the reaction nearly 10–15% earlier in p to avoid damage of the reaction kettle and to collect a good quality product which would be useful and which could be easily manipulated in subsequent steps.

The usefulness of Carothers' Equation lies in the linking of p, f and $\overline{X}_n$ in one expression, and in the fact that it can be used to predict gelation in polyfunctional systems. The limitation of the equation is that practically gelation starts even at earlier stages than predicted by the equation because of uncontrolled local conditions in the hot viscous reaction mass. Moreover, this equation is applicable to polycondensation reactions only, and it loses relevance in the context of addition or chain growth polymerization.

2.7 Molecular Weight Control: Quantitative Effect of Stoichiometric Imbalance on Maximum Attainable Molecular Weight

Condensation polymerization being an equilibrium step process and endothermic in nature in most cases, control of molecular weight can be readily achieved by allowing the reaction to die down at a desired point by withdrawing heat supply or by cooling and further, by allowing the byproduct of condensation to accumulate in the medium. But in any event, the intermediate product may not be adequately stable toward further heating under a favourable condition with respect to further chain extension and gain in molecular weight. Control of molecular weight can be more precisely attained by striking a calculated stoichiometric imbalance of the two types of reacting functional groups. This may be clearly illustrated considering a bifunctional system having monomers of the A–A and B–B types and giving the numbers of A and B functional groups as N_A and N_B respectively, such that $(N_A/N_B) = r$, where r is either equal to 1 or less than 1, i.e. B groups are present in the system in molar excess, if there is stoichiometric imbalance. The total number of molecules initially present, N_0 is then given by $(N_A + N_B)/2$, or $N_A(1 + 1/r)/2$.

Let us now find out the total number of molecules N left into the system after a specific time period of reaction when extent of reaction for A groups is p, i.e. p

is the fraction of A groups which have reacted. The fraction of B groups which have reacted by that time is then given by rp.

The total number of unreacted A and B groups are then expressed as $(N_A - N_A p)$, i.e. $N_A(1 - p)$ and $N_B - N_B rp$, i.e. $N_B(1 - rp)$ respectively. The total number of polymer chain ends is thus, given by $[N_A(1 - p) + N_B(1 - rp)]$, i.e. $[N_A(1 - p) + N_A(1 - rp)/r]$. The value of N, being one half this number, is given by $N = N_A[(1 - p) + (1 - rp)/r]/2$.

Then,
$$\overline{X}_n = \frac{N_0}{N} = \frac{N_A(1+1/r)/2}{N_A[(1-p)+(1-rp)/r]/2}$$

or
$$\overline{X}_n = \frac{1+r}{1+r-2rp} \tag{2.14}$$

and
$$\text{Lt } p \to 1,\ \overline{X}_n = \frac{1+r}{1-r} \tag{2.15}$$

For equimolar proportions of A and B groups, i.e. for $r = 1.0$ $\overline{X}_n = 1/(1 - p)$, which is the same as given by Eq. (2.11). The maximum theoretically attainable degree of polymerization for $r = 1.0$ is α. But for $r \neq 1.0$, and with only, say one mole per cent excess of B groups taken into the system (by taking excess of the related bifunctional compound in calculated amount), and taking the help of Eq. (2.15), the maximum attainable $\overline{X}_n$ becomes:

$$\overline{X}_n = \frac{1+(100/101)}{1-(100/101)} = 201$$

This consideration clearly establishes the importance of stoichiometric equivalence or balance of the reacting functional groups (A and B groups) for attaining high degrees of polymerization, or of introduction of stoichiometric imbalance in order to limit the maximum attainable degree of polymerization to a desired lower level. Figure 2.1 shows how $\overline{X}_n$ varies with variation in stoichiometric imbalance, i.e., variation in the value of r.

Considering a bifunctional system and Eqs. (2.11) and (2.15) applicable to it, it is clearly realized that the step-growth polymerization must always be carried out to very high conversions, at least to 98%, since a degree of polymerization >50 is normally required for a polymer to be widely useful. Higher conversions and lesser deviation from stoichiometric balance are required for higher degrees of polymerization. The interdependence of $\overline{X}_n$ on p and r is given by Eq. (2.14) and Fig. 2.1. Closer adherence to stoichiometric equivalence is required for obtaining

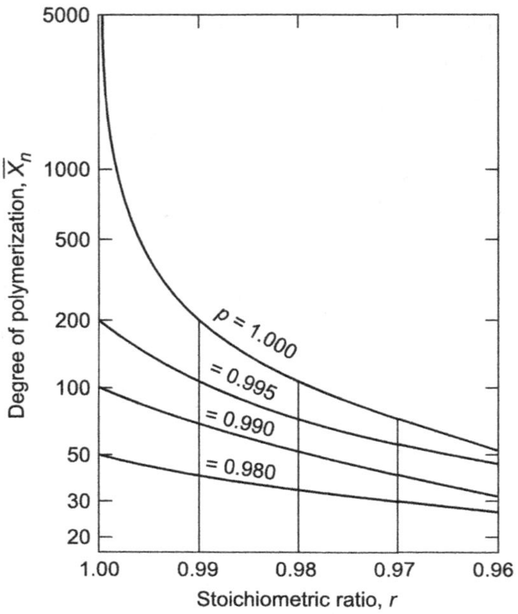

Fig. 2.1 *Trend of variation of number average degree of polymerization $\overline{X}_n$ with stoichiometric ratio r for different extents of reaction, p in the polycondensation involving A–A and B–B type monomers*

higher degrees of polymerization. The effect of stoichiometric imbalance is much more pronounced in the very high conversion zone (>98%) than in the lower conversion zones.

2.8 Kinetics of Step-Growth Polymerization

A powerful tool for a good understanding of any chemical reaction is the study of its kinetics. Kinetic analysis of a reaction helps in the understanding of the mechanism of the reaction.

2.8.1 Principle of Equal Reactivity of Functional Groups

Equation (2.5) describes a characteristic feature of step polymerization. Any two species in the reaction mixture can react with each other. This feature alone complicates kinetic analysis of the overall process, as it involves innumerable separate reactions. However, the kinetic analysis is largely simplified if it is assumed that the reactivities of both functional groups of a bifunctional monomer, e.g. both the carboxyl groups of a dicarboxylic acid, are just the same and that the reactivity of a functional group is independent of the size of the molecule to which it is attached. This means that the reactivity of the specific functional group is independent of the values of n and m. Fortunately, this assumption is more or less valid for all monomers and all larger species that are involved in polycondensation.

The independence of the reactivity of a specific functional group on molecular size is also generally substantiated by the closeness of the rate constants for reactions of a homologous series of compounds (fatty acid) differing from each other only in molecular (weight) size involving esterification with a specific alcohol (ethyl alcohol) in each case. (Table 2.2.)

$$H(CH_2)_n\text{—}COOH + C_2H_5OH \xrightarrow[\text{catalyst (HCl)}]{\text{mineral acid}} H(CH_2)_n\text{—}COOC_2H_5 + H_2O$$

Table 2.2, however, shows a decrease in reactivity with increase in molecular size (n) over the very (initial) low range of n and the rate constant of the esterification reaction levels off to a limiting (constant) value for $n \geq 3$ and maintains constancy and independence of molecular size. This feature of equal reactivity may be generalized and considered applicable to polymerization reactions as well. In the latter case some complications, not encountered in the study of simple compounds,

Table 2.2 *Esterification rate constants, k at 25°C for homologous series*[3]

Molecular size, n	A, $k \times 10^4$	B, $k \times 10^4$
1	22.1	—
2	15.3	6.0
3	7.5	8.7
4	7.5	8.4
5	7.4	7.8
6	—	7.3
8	7.5	
9	7.4	
11	7.6	
15	7.7	
17	7.7	

All k values are in unit of (gram equivalent/litre). sec^{-1}.
A—Esterification between monocarboxylic acids, $H(CH_2)_nCOOH$ and C_2H_5OH.
B—Esterification between dicarboxylic acids $(CH_2)_n(COOH)_2$ and C_2H_5OH.

may arise, but they may be readily understood and accounted for when considered under proper perspective, i.e. equivalent (molar) concentrations of reacting functional groups, solubility, dilution level, etc. An end (functional) group comprising a terminal reactive centre of a polymer (chain) molecule has a much greater mobility than that of the polymer molecule as a whole. The functional group at the free end would have good degree of mobility in view of the rearrangements that would normally occur in closeby segments of the chain molecule. The collision rate of such reactive functional groups with related functional groups in the neighbourhood will be much the same as for small molecules.

In polymer systems, the medium viscosity, either in melt condition or in solution, is usually much higher than in systems containing only small non-polymeric molecules. Further, the polymer molecules may become insoluble beyond a critical chain length or molecular weight and the low molecular weight byproduct molecules would tend to slowly accumulate in the system causing the equilibrium to shift in favour of the reactants [Eq. (2.5)]. All these physical factors may affect the polymerization rate quite measurably. The rate at which independent functional groups react to produce chain extension through establishment of a fresh condensed interunit linkage is dependent on three different rate processes: (i) the rate at which the groups diffuse into a cage, (ii) the rate at which they diffuse apart, and (iii) the rate at which the reactive groups trapped in the cage react to produce the new condensed interunit linkage. Thus, in a given time interval long enough to allow diffusion of a pair of reacting groups, the number of related functional groups with which a given functional group may undergo collisions will be less for a group linked to a polymer molecule than for one linked to a small molecule. But what is more significant is that the overall frequencies of collision in either case are much the same. A pair of groups with a lower diffusion rate will undergo more total collisions before diffusing apart. The net outcome of these considerations points to the important conclusion that the reactivity of a functional group is independent of the size of the molecule to which it is linked.

2.8.2 Rate of Step-growth Polymerization

(a) ***Self-catalyzed Polyesterification*** With the idea of equal reactivity of specific functional groups independent of molecular size, the consideration of rate of step-growth polymerization becomes easy. The formation of a polyester from a dicarboxylic acid and a diol may again be taken as an example, for simplicity. This reaction is prone to catalysis by mineral acids. But in the absence of an added mineral acid such as H_2SO_4 or HCl, the reactant carboxylic acid itself acts as a catalyst; for each step of reaction between a —COOH group

$$\text{—COOH} + \text{HO—} \xrightarrow[\text{(catalyst)}]{\text{— COOH}} \text{—COO—} + H_2O \tag{2.16}$$

and an alcohol (—OH) group, a second COOH group from the reactants acts as a catalyst. The rate of the reaction can be measured by measuring the rate of disappearance of the carboxyl groups:

$$-\frac{d\,[\text{COOH}]}{dt} = k\,[\text{COOH}]^2\,[\text{OH}] \tag{2.17}$$

Here, k is the rate constant of the reaction given by equation (2.16). Considering equimolar presence of carboxyl and hydroxyl groups, and c giving the concentration of each group at a given point,

$$-\frac{dc}{dt} = kc^3 \tag{2.18}$$

or

$$-\frac{dc}{c^3} = k\,dt \tag{2.19}$$

on integration, Eq. (2.19) gives:

$$2kt = \frac{1}{c^2} - \text{constant} \tag{2.20}$$

If c_0 is the initial concentration of each kind of functional group and p is the extent of reaction, i.e. fraction of a given functional group that has reacted at time t, then

$$c = c_0(1 - p) \tag{2.21}$$

By combining Eqs. (2.21) and (2.20), we get

$$2c_0^2kt = \frac{1}{(1-p)^2} - \text{constant} \tag{2.22}$$

Equation (2.22) prescribes linearity of the plot of $1/(1 - p)^2$ vs time. In the case of self catalyzed polycondensation (polyesterification) of diethylene glycol and adipic acid at 166°C, as studied by Flory,[4] the linearity is found to be strictly followed only over the range of nearly 80–94%. Below or above this range of extent of reaction, the experimental points deviated from the third order plot (Fig. 2.2). Equation (2.22) actually shows a linear relationship between square of the average degree of polymerization, $(\overline{X}_n)^2$ with time of polycondensation.

It may be mentioned here that the initial nonlinearity is not limited to polyesterification; the initial deviation from the third order linear plot is characteristic of esterification in general.

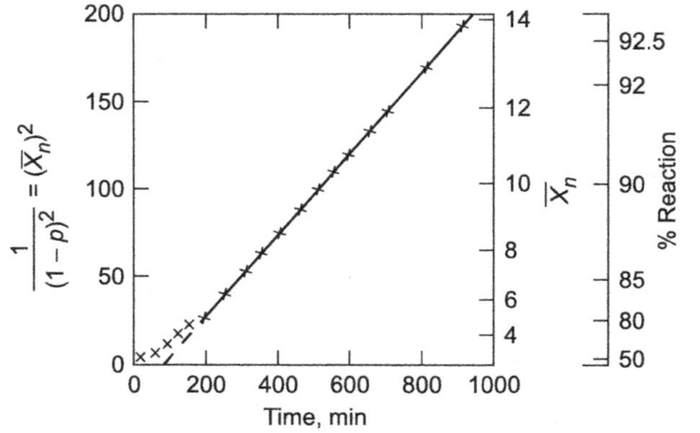

Fig. 2.2 *Plot showing $1/(1-p)^2$ or $(\overline{X}_n)$ vs time for self catalyzed polyesterification of diethylene glycol with adipic acid at 166°C (Flory, 1939;* **Courtesy,** *American Chemical Society, Washington)*

Over the initial 50–80% conversion range, a very significant change takes place in the reaction medium, although the average growth in size of the molecules remains far too low. A large change in the polarity of the reaction mixture, however, takes place due to the disappearance of a large fraction of the polar carboxyl groups resulting in the formation of less polar ester linkages. This definite change in polar nature of the reaction medium has a major role to play in causing the deviation in question. But the exact manner or mechanism in which the rate of esterification is affected as a consequence of this polarity change is not very well understood.

The deviation from linearity in the high conversion region (>94%), however, arises as a consequence of other factors including loss of reactants through degradation or volatilization, or even cyclization and other side reactions. Such losses of reactants are likely to be of much less importance in the initial stages. Difficulty in removal of the byproduct of condensation (water) from the reaction medium due to the very high viscosity it assumes in the later stages, is also an important contributing factor in decreasing the reaction rate with increasing extent of reaction beyond 94% conversion.

(b) ***Polyesterification using Strong Acids as Catalysts*** The direct or self-catalyzed polyesterification is slow and sluggish, considering slow increase in molecular weight or average degree of polymerization and this is understood to be a consequence of the third order kinetics. A better result and a faster rate of chain growth is, however, achieved by adding a small amount of a strong acid such as sulphuric acid, *p*-toluenesulphonic acid, etc. as a catalyst, which remains undepleted in the system with progress of the reaction. The rate of disappearance of COOH groups can then be expressed as

$$-\frac{d[\text{COOH}]}{dt} = k' \, [\text{COOH}] \, [\text{OH}] \tag{2.23}$$

For stoichiometric equivalence of the two functional groups and taking c as the concentration of either group, the Eq. (2.23) may be written as

$$-\frac{dc}{dt} = k'c^2 \tag{2.24}$$

where k' is the overall rate constant. On integration, and making use of Eq. (2.21), the following expressions, relating extent of reaction p or degree or polymerization $\overline{X}_n$ with time of reaction, t, are obtained:

$$c_0 k' t = \frac{1}{(1-p)} - \text{constant} \tag{2.25}$$

or

$$c_0 k' t = \overline{X}_n - \text{constant} \tag{2.26}$$

A plot of $1/(1 - p)$, i.e. $\overline{X}_n$ vs time, would be linear according to Eqs (2.25) and (2.26). Data relating to polymerization of diethylene glycol and adipic acid catalyzed by *p*-toluenesulphonic acid as observed by Flory[4], are shown in Fig. 2.3. The plot shows the experimental variation of $\overline{X}_n$, the average degree of polymerization with time of polycondensation at a given temperature.[5] Comparison of Figs 2.2 and 2.3 clearly shows that the polyesterification process is

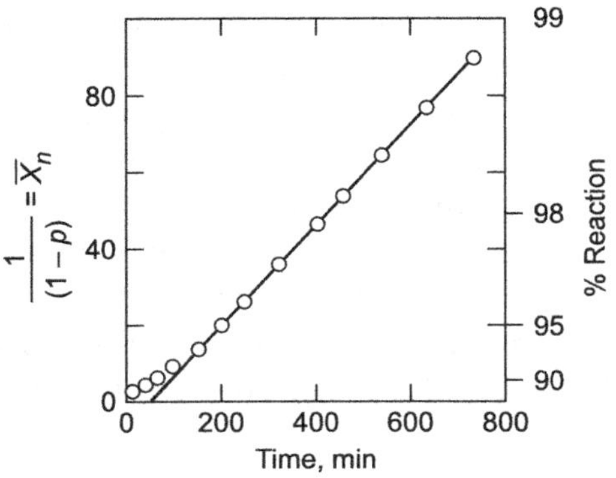

Fig. 2.3 *Plot showing 1/(1 – p) or $\overline{X}_n$ vs time for polycondensation of diethylene glycol and adipic acid at 109°C using 0.4 mole % p-toluene sulphonic acid as the catalyst[4,5] (Flory, 1939; Solomon, 1967;* **Courtesy,** *American Chemical Society, Washington)*

much more readily and economically accomplished when an external strong acid catalyst is used. Not only the linearity of $\overline{X}_n$ vs time plot is very closely maintained, particularly in the more important high conversion zone, but also the rate of increase of $\overline{X}_n$ with time of reaction is much higher. The second order kinetics find good experimental support up to nearly 98.9% conversion, corresponding to $\overline{X}_n$ value of approximately 90 or a molecular weight of nearly 10,000. This approach of kinetic analysis, however, shows that the reactivities of the carboxyl and hydroxyl groups remain practically unchanged despite growth of the chain molecules to a fairly big average molecular size and a big increase in medium viscosity as a consequence.

Step-growth polymerization other than polyesterification may proceed at desired rates in presence of external acidic or basic catalysts or as uncatalyzed reactions. Irrespective of the approach and situation prevailing, the basic and overall kinetics will be much the same as detailed above for polyesterification. For stoichiometric proportions of the two types of reacting functional groups, say A and B groups, Eq. (2.25) would be useful to reasonably describe and analyze the kinetics in the respective systems. Whether or not the reacting monomers are of the A–A and B–B types or simply of the A–B types, the observed kinetics will be the same.

2.9 Distribution of Molecular Weight in (Linear) Bifunctional Polycondensation

Equation (2.5) clearly indicates that random intermolecular reaction between molecules in a bifunctional system would lead to linear chain molecules whose chain lengths admit of wide variation. For both theoretical and practical view points, consideration of an approach to make a quantitative assessment and analysis of the molecular weight distribution is relevant and important at this point. Flory's statistical approach[6] in this context based on equal reactivity of functional groups of a given kind independent of molecular size is essentially and equally applicable to the mixture of A–A and B–B types of monomers with stoichiometric balance of the A and B functional groups and to the simple A–B type of monomers.

2.9.1 Derivation of Distribution Functions

Let the formation of a linear polymer molecule of a degree of polymerization n be considered, starting with the monomer of the formula HORCOOH. The polymer structure may be written as:

$$\underset{1}{\text{HO—R—COO}}\text{—}\underset{2}{\text{R—COO}}\text{—}\underset{3}{\text{R—COO}}\text{—}\ldots\ldots$$

$$\text{—}\underset{(n-2)}{\text{COO—R}}\text{—}\underset{(n-1)}{\text{COO—R}}\text{—}\underset{n}{\text{COOH}}$$

Now, considering reaction of the functional groups, say the COOH groups, and for a degree of polymerization n, for the chain molecule formed, it is necessary that $(n - 1)$ COOH groups had reacted to form $(n - 1)$ interunit –COO– linkages and that one COOH group remained unreacted so as to give the linear chain structure as shown above. The probability that the first or any other subsequent COOH group has reacted up to the point of time t under consideration is given by the extent of reaction p, and the probability that a given COOH group has remained unreacted is then $(1 - p)$. Hence the probability that $(n - 1)$ COOH groups have reacted is given by the product of $(n - 1)$ separate probabilities, i.e. by p^{n-1}. Combining with the probability of one (the nth) carboxyl group remaining unreacted, the probability P_n of formation of the polymer molecule with degree of polymerization n, is given by

$$P_n = p^{n-1}\,(1 - p) \tag{2.27}$$

P_n, in effect, is also the mole fraction or number fraction of molecules containing n structural units in the polymer mixture, $(P_n = N_n/N)$, such that

$$N_n = N\,p^{n-1}(1 - p) \tag{2.28}$$

where N_n is the number of n-mers formed and N is the total number of molecules of all sizes. If the total number of structural units (monomer molecules) initially present is N_0, then $N = N_0(1 - p)$ and Eq. (2.28) assumes the form

$$N_n = N_0(1 - p)^2\,p^{n-1} \tag{2.29}$$

Equation (2.29) gives the number distribution function for a linear step-growth polymerization corresponding to an extent of reaction p.

Further, accounting for the weight loss due to byproduct elimination in each step of reaction, and neglecting weights of the end groups, the weight fraction W_n of the n-mers, i.e. weight fraction of the molecules each containing n structural units is given by $W_n = n\,N_n/N_0$ and with the help of Eq. (2.29), W_n is expressed as:

$$W_n = n(1 - p)^2\,p^{n-1} \tag{2.30}$$

The above equation gives the weight distribution function for a linear step-growth polymerization at the extent of reaction p.

The molecular size distributions given by Eq. (2.29) as derived from Eq. (2.27) and by Eq. (2.30), usually known as the most probable or Flory distribution are shown in graphical plots as P_n vs n and W_n vs n in Figs 2.4 and 2.5 respectively. Comparison and analysis of the two figures gives a better appreciation of the molecular size distribution.[8]

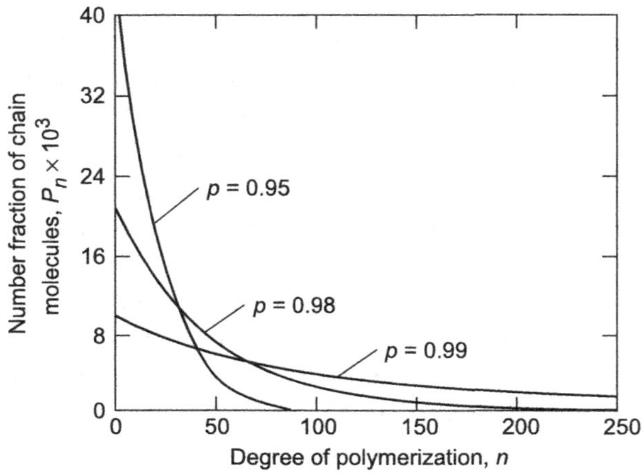

Fig. 2.4 *Number or mole fraction distribution curves (P_n vs n) for several extents of reaction, p for linear step-growth polymerization*[7] *(Flory, 1936;* **Courtesy,** *American Chemical Society, Washington)*

For a given value of p, W_n passes through a maximum and the maximum shifts to a higher value of n as p increases. Further, a distinct broadening of the distribution of n values as p increases is clearly indicated from each figure. Even at high p values, N_n is large for monomers (and for other species where n has a low value) and N_n decreases as n increases, i.e. on a number basis, the fraction of molecules decreases as n increases, irrespective of the value of p, and as p increases, the fraction of molecules with smaller n values decreases and the fraction of molecules with larger n values increases. The steepness of the number fraction distribution curves falls, i.e. the curves tend to flatten as p increases.

2.9.2 The Number Average Degree of Polymerization

The number average degree of polymerization, $\overline{X}_n$, for the mixture of polymers at a given extent of reaction, p, may be expressed as

$$\overline{X}_n = \sum_{n=1}^{N_0} nP_n = \sum_{n=1}^{\infty} n\, p^{n-1}\, (1 - p) \tag{2.31}$$

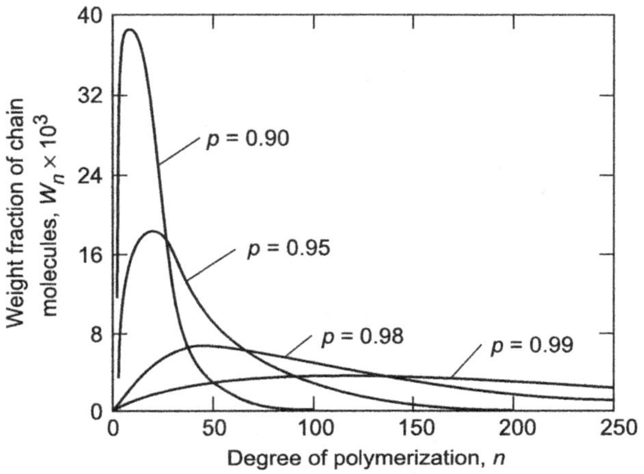

Fig. 2.5 *Weight fraction distribution curves (W_n vs n) for several extents of reaction, p for linear step-growth polymerization*[7] *(Flory, 1936;* **Courtesy,** *American Chemical Society, Washington)*

The upper limit of the second summation is shifted to ∞ from N_0 on general theoretical consideration even though it is of no real practical significance. Considering different values of n and giving a few terms of the summation

$$\bar{X}_n = 1\,p^{\circ}(1 - p) + 2p^1(1 - p) + 3\,p^2(1 - p) + \ldots$$
$$= (1 - p)\,(1 + 2p + 3p^2 + \ldots)$$

The second factor, giving a summation series, can be simplified into the square of a summation series which may be expressed as $(1 + p + p^2 + \ldots)^2$. The series $(1 + p + p^2 + \ldots)$ may be further reduced to $1/(1 - p)$, such that $\bar{X}_n = (1 - p)(1 - p)^{-2}$. Thus, the expression for $\bar{X}_n$ finally takes the form of Eq. (2.11) as derived from Carothers' equation i.e.,

$$\bar{X}_n = 1/(1 - p).$$

2.9.3 Weight Average Degree of Polymerization

The weight average degree of polymerization, $\bar{X}_w$, may be expressed as follows by averaging the contributions of molecules of all n values keeping weight fractions of molecules of each n value in view as weighting factors during the averaging. Thus

$$\overline{X}_w = \frac{\sum_{n=1}^{N_0} nW_n}{\sum_{n=1}^{N_0} W_n} = \frac{\sum_{n=1}^{\alpha} n^2 p^{n-1}(1-p)^2}{\sum_{n=1}^{\alpha} np^{n-1}(1-p)^2} = \frac{\sum_{n=1}^{\alpha} n^2 p^{n-1}}{\sum_{n=1}^{\alpha} np^{n-1}}$$

$$= \frac{1 + 4p + 9p^2 + 16p^3 + 25p^4 + \ldots}{1 + 2p + 3p^2 + 4p^3 + 5p^4 + \ldots}$$

on appropriate simplification by long division, the above equation reduces to:

$$\begin{aligned}\overline{X}_w &= 1 + 2p + 2p^2 + 2p^3 + 2p^4 \ldots \\ &= (2 + 2p + 2p^2 + 2p^3 + 2p^4 \ldots) - 1 \\ &= 2(1 + p + p^2 + p^3 + p^4 + \ldots) - 1\end{aligned}$$

Since $(1 + p + p^2 + \ldots) = 1/(1 - p)$, as mentioned before and as may be verified by long division,

$$\overline{X}_w = \frac{2}{(1-p)} - 1$$

On simplification, the weight average degree of polymerization reduces to the form

$$\overline{X}_w = \frac{(1+p)}{(1-p)} \tag{2.32}$$

The ratio of $\overline{X}_w/\overline{X}_n$ commonly used to represent the breadth of the molecular weight or molecular size distribution will be simply given by $(1 + p)$, such that for $p \rightarrow 1$, $(\overline{X}_w/\overline{X}_n) \rightarrow 2$.

2.10 Multichain Step-Growth Polymers (Polyfunctional Systems)

2.10.1 Branching

A special kind of step-growth polymer may be produced by allowing a bifunctional monomer of the type A–B to polymerize in presence of a small amount of a second monomer A_f bearing f number (>2) of A functional groups per molecule. Even on practical completion of reaction, the resulting polymer will have a branched structure showing f number of chains attached to or growing from a central (A_f) branch unit. Considering f = 3 for the A_f molecule, the polymerization in question will lead to a branched structure of the following kind:

```
A — BA — BA — BA—AB — AB — AB — AB — AB — AB — A
                 |
                 A
                 B
                 |
                 A
                 B
                 |
                 A
                 B
                 |
                 A
```

Branched structure

Formation of network or cross-linked structures cannot take place since no units of the B–B type are present. Two branched polymer molecules of the above type cannot mutually react since all the reactive groups at the end of the branches are A functional groups. For f = 1 and 2, the polymer formed would be linear; for higher values of f, the polymer formed is branched, as stated earlier.

The molecular weight distribution in this class of branched polymerization will be relatively much narrow compared to the linear polymerization by design and as a consequence of the restrictions in chain growth inherent in the process. The size or length of each branch is statistically controlled and determined independent of all other branches, and as such, the likelihood of formation of a molecule with branches, very much different in size from the average, is much less as compared to linear polymerization, more so as the number of branches, i.e. the value of f increases. Thus, the molecular weight distribution gets narrower with increasing f value,[9] as shown in Fig. 2.6. In such cases, the distribution breadth, according to Flory,[10] is expressed as:

$$\frac{\overline{X}_w}{\overline{X}_n} = 1 + \frac{fp}{(fp+1-p)^2} \tag{2.33}$$

for p = 1, Eq. (2.33) reduces to

$$\overline{X}_w/\overline{X}_n = \left(1+\frac{1}{f}\right) \tag{2.34}$$

Thus, the breadth of molecular size or the distribution ratio decreases from 2 for f = 1 to 1.33 for f = 3 and to 1.25 for f = 4, considering p = 1 in each case; f = 1 prescribes a condition for the most probable distribution for linear step-growth polymerization.

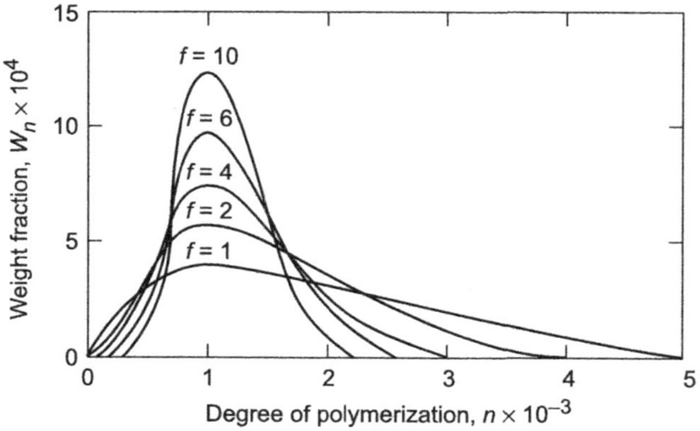

Fig. 2.6 *Molecular weight distribution curves[9] (W_n vs n) for multichain polymers for different values of functionality, f for A_f, $\overline{X}_n$ in each case being identical (Schulz, 1939;* **Courtesy,** *R. Oldenbourg Verlag, Munich)*

2.10.2 Cross-linking

For the formation of cross-linked polymer structure, the necessary condition is to polymerize monomers of the type A–B and $A_f(f > 2)$ in the presence of another monomer of the type B–B. The B–B type monomers now facilitate establishment of cross-links by combining two branches from two different polymers ending in A functional groups, or interaction between two such branches, one ending in an A functional group and the other in a B functional group through the condensation process. A typical cross-linked polymer structure is shown on the next page taking $f = 3$.

Thus, it is clear that cross-linking established by linking of branched segments or units leads to a three dimensional network structure which ultimately gives rise to the phenomenon of gelation of the reaction mixture. The onset of gelation is associated with a steep rise in medium viscosity and formation of a measurable proportion of insoluble mass (gel fraction), while the rest of the mass still retaining solubility (sol fraction). Carrying the polymerization to beyond gel point leads to increase in the gel fraction at the expense of the sol fraction and the reaction mass gets rapidly transformed into a viscoelastic mass from a state of viscous liquid.

2.10.3 Prediction of Gel-point

A statistical approach to derive a useful expression for the prediction of gel point is due to Flory[6] and Stockmayer[11]. For this, the concept of branching coefficient α, defined as the probability that a functional group of a branch unit is linked with

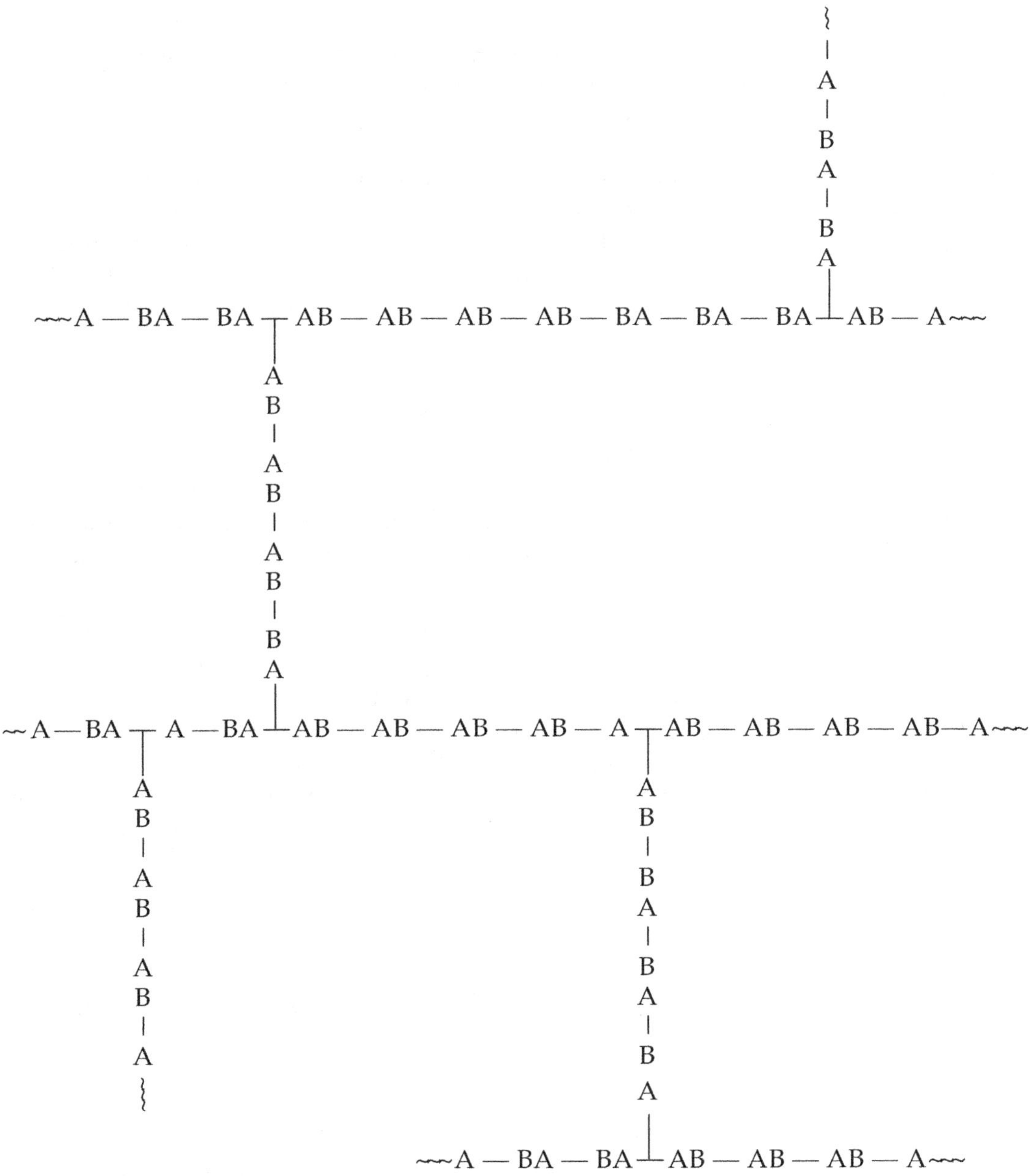

A cross linked structure (f = 3)

another branch unit, is useful. Considering a polycondensation system consisting of monomers of the type A–A, B–B and the branch inducing monomer $A_f (f > 2)$, structures of chain segments that would result will be of the type:

$$A_{(f-1)} - A - [B - BA - A]_n - B - BA - A_{(f-1)}$$

where n may have any value. Other chain species likely to be formed into the system may contain branch units at only one or none of the ends. For gelation, the necessary condition is that at least one of the $f - 1$ chain segments projecting out from the branch unit, as shown, gets in turn linked to another branch unit and the probability for this to happen is simply $1/(f - 1)$. It follows then that the critical value of α for gelation is

$$\alpha_c = \frac{1}{(f-1)} \tag{2.35}$$

Here, f represents the functionality of the branch units only and it should not be confused with the average functionality considered in Carothers' Equation. In case more than one type of branch units differing in functionality are used, an average f value of all the multifunctional monomer molecules (each having $f > 2$) is to be used in Eq. (2.35) for evaluation of α_c.

Now, an expression relating the probability α with the extent of reaction may be obtained by determining the probability of getting a chain segment of the type shown. Considering the extent of reaction of A groups as p_A and that of B groups as p_B and the ratio of A groups on branch units to all A groups in the system as ρ, the probability of a B group having reacted with a branch unit is given by $p_B\rho$ and with a (bifunctional) A–A unit by $p_B(1 - \rho)$. So, the probability of obtaining a chain segment of the type shown is given by $p_A[p_B(1 - \rho)p_A]^n \, p_B\rho$ and then summing over all values of n gives

$$\alpha = \frac{p_A p_B \rho}{1 - p_A p_B (1-\rho)} \tag{2.36}$$

Taking the ratio of all A groups to all B groups, $(N_A/N_B) = r$, as defined in Sec. 2.7, p_B may be expressed in terms of p_A by the relation $p_B = rp_A$. Eq. (2.36) will then reduce to:

$$\alpha = \frac{rp_A^2 \rho}{1 - rp_A^2 (1-\rho)} = \frac{p_B^2 \rho}{r - p_B^2 (1-\rho)} \tag{2.37}$$

combining Eqs (2.37) and (2.35), the critical extent of reaction p_c for A groups for gelation to occur (i.e. at the gel point) is

$$p_c = \frac{1}{[r + r\rho(f-2)]^{1/2}} \tag{2.38}$$

For $N_A = N_B$, i.e. $r = 1$, $p_A = p_B = p$, so that

$$\alpha = \frac{p^2 \rho}{1 - p^2 (1 - \rho)} \tag{2.39}$$

For a system containing only B–B and A_f type monomers and no A–A units, and for $N_A \neq N_B$, giving $\rho = 1$ and $r < 1$,

$$\alpha = r p_A^2 = \frac{p_B^2}{r} \tag{2.40}$$

and in that case

$$p_c = \frac{1}{[r + r(f - 2)]^{1/2}} \tag{2.41}$$

For $r = 1$ and also $\rho = 1$, Eqs (2.39) and (2.40) reduce to

$$\alpha = p^2 \tag{2.42}$$

and

$$p_c = \frac{1}{[1 + (f - 2)]^{1/2}} \tag{2.43}$$

In the event of having a special system with $f > 2$ for all the monomers present, the probability that a functional group on a branch unit reacts to generate another branch unit is simply the probability that it has reacted, thus giving $\alpha = p$, and the critical extent of reaction at gel point is given by considering Eq. (2.35),

$$p_c = \frac{1}{(f - 1)} \tag{2.44}$$

2.11 Some Additional Considerations of Non-stoichiometric Reactant Systems

The Carothers' Equation, Eq. (2.9), and the expression for the critical extent of reaction at the gel point ($p_c = 2/f$) are theoretical expressions for stoichiometric proportions of the two reacting functional groups in a given polycondensation reaction. For non-stoichiometric systems, the general applicability of the Carothers' Equation and those that follow from it as well as the functionality principle of condensation polymerization discussed in the beginning of this chapter becomes less meaningful, more so in ternary or higher mixtures of monomers of different functionality, irrespective of whether or not the reacting functional groups are present in stoichiometric proportions.

Let us first consider a specific case of binary reaction system consisting of 1 mole of the trifunctional alcohol, glycerol and 4 moles of the bifunctional acid, phthalic acid, giving the (number) average functionality, f_{av} of the system as 11/5 or 2.2. A high molecular weight for the products is otherwise indicated and the predicted gel point is $p_c = 0.909$. But the indication and prediction are far from reality, because the gross stoichiometric imbalance between the A(—OH) and B(—COOH) functional groups, giving the value of N_A/N_B, i.e., r as 0.375 (*See* Sec. 2.7), precludes formation of polymers and only some very low molecular weight reaction products will be formed, leaving a large portion of the functional groups unreacted.

This necessitates a different approach for defining average functionality in non-stoichiometric mixtures. According to Pinner[12], f_{av} for non-stoichiometric mixture is equated to the ratio of twice the total number of functional groups that are not in excess to the total number of molecules present. The deficient groups hold key to the growth of the molecules and also to cross-linking, if it is at all feasible under the prevailing condition. The functional group present in excess acts against chain extension through blocking effect and thus lowers the effective functionality (f_{av}) of the system. For the specific non-stoichiometric mixture considered above, (1 mole glycerol and 4 moles phthalic acid), the effective f_{av} value calculated on the basis detailed above is 6/5 or 1.2, a low value indeed; and this clearly explains why attempts to produce useful polymers in such grossly non-stoichiometric systems prove abortive.

For a ternary mixture containing N_A moles of a monomer containing A functional groups with a functionality of f_A, N_c moles of another monomer bearing the same A functional groups but having a functionality of f_c and N_B moles of a third monomer bearing B functional groups and having a functionality of f_B, such that the total number of A functional groups in the system from a total of $N_A + N_c$ moles of monomers is less than the number of B groups from N_B moles of the third monomer, f_{av} for this system is then expressed as

$$f_{av} = \frac{2(N_A f_A + N_c f_c)}{(N_A + N_c) + N_B} \tag{2.45}$$

For this system the stoichiometric imbalance r (ratio of the total number of deficient group to the number of the group in stoichiometric excess) is expressed as

$$r = \frac{N_A f_A + N_c f_c}{N_B f_B} \tag{2.46}$$

A special case of interest is provided by having one of the monomers bearing A functional groups to be polyfunctional and the remaining two monomers, one bearing A functional groups and the other with B functional groups to be bifunctional (so that, say $f_c > 2$ and $f_A = f_B = 2$).

For such a system,

$$f_{av} = \frac{4rf_c}{f_c + 2r\rho + rf_c(1-\rho)} \tag{2.47}$$

where ρ is the fraction of A functional groups belonging to molecules characterized by the functionality f_c, i.e., $\rho = [N_c f_c/(N_A f_A + N_c f_c)]$. The critical extent of reaction, p_c, corresponding to gel point may then be readily obtained using Eq. (2.9a).

2.12 Practical Consideration of Gel Points

It would be interesting to examine, analyze and compare experimental results on gelation using the statistical approach and the Carothers' Equation. The Carothers' Equation is used to predict the extent of polycondensation reaction at which the average degree of polymerization $\overline{X}_n$ becomes infinite. Prediction of a quite high value for p_c is inherent in this approach because polymer molecules having degree of polymerization much larger than $\overline{X}_n$ are present and they will reach the gel point earlier than those of the size $\overline{X}_n$. The statistical treatment [Eq. (2.38)] is theoretically more sound and practically more useful, though predicting in many cases lower values of p_c than the experimentally observed values.[13] The difference in the p_c value predicted by the statistical approach is attributed to some wasteful intramolecular cyclization reactions not taken into account in the derivation and also partly in some cases to the limited applicability or failure of the assumption of equal reactivity of all functional groups of the same type. When glycerol is used in a polyesterification process, the difference between the calculated and observed values of p_c will be substantially narrowed down, though not totally eliminated, if during calculation the lower reactivity of the secondary hydroxyl group of glycerol is duly accounted for.

In the glycerol-phthalic anhydride polyesterification system, traces of diols and sometimes monocarboxylic acids are used to achieve relatively flexible products of lower cross-link density. For equimolar proportions of glycerol, phthalic anhydride and a monofunctional acid, f_{av} is exactly 2. Carothers' Equation predicts no gelation for such systems in practical sense, but many such systems are actually associated with a gel point, thus revealing further limitations of the Carothers' Equation.

However, for polycondensation system of the above kind, the following equation[14] based on the statistical approach is very useful in predicting gel points:

$$p_c = \left[\frac{\gamma}{2(1-\sigma)}\right]^{1/2} \tag{2.48}$$

where

$$\gamma = \frac{\text{Total equivalents of OH functional groups}}{\text{Total equivalents of COOH functional groups}}$$

$$\sigma = \frac{\text{Equivalents of COOH from monofunctional acid}}{\text{Total equivalents of COOH groups}}$$

2.13 Molecular Weight Distribution in Multifunctional Reactant Systems

Expressions for molecular size distributions for multifunctional reactant systems leading to three dimensional polymers are derived following much the same approach as for linear polymers, though with much more difficulty. For a simple case of reaction involving equivalent amounts of two trifunctional monomers in which all the functional groups in each monomer are equally reactive, the weight-distribution function W_n is given by

$$W_n = \left[\frac{(fn-n)!\,f}{(n-1)!\,(fn-2n+2)!}\right] p^{n-1}\,(1-p)^{fn-2n+2} \tag{2.49}$$

Equation (2.49) is similar in form to Eq. (2.30) derived for linear systems for which the first factor of Eq. (2.49) simply reduces to n. The second term gives the probability of establishment of $(n-1)$ links in an n-mer; the last term incorporates the probability of having $(f-2)\,n+2$ or $fn-2n+2$ unreacted ends. Plots showing variation of the weight fraction W_n with number of units or degree of polymerization for different f values at a fixed p or at various stages of reaction for a trifunctional system ($\alpha = p$) are shown in Figs 2.7 and 2.8, respectively. Figure 2.7 clearly shows that for the stepwise polymerizations, the weight fraction distribution broadens out progressively with increasing functionality at comparable extents of reaction. Figure 2.8 reveals a broadening out of the distribution with increasing value of p in the trifunctional case[2].

Plot in Fig. 2.9 shows how the weight fraction of the various molecular species varies with α for $\alpha = p$ in a trifunctional step growth polymerization. The trend of

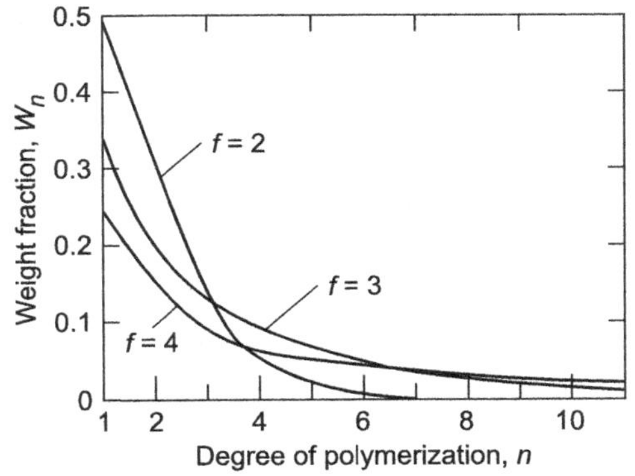

Fig. 2.7 *Molecular weight distribution curves,[15] W_n vs n for step-growth polymerization corresponding to different values of f and a fixed value of p(p = 0.3) (Mark & Tobolsky, 1950;* **Courtesy,** *Wiley-Interscience, New York)*

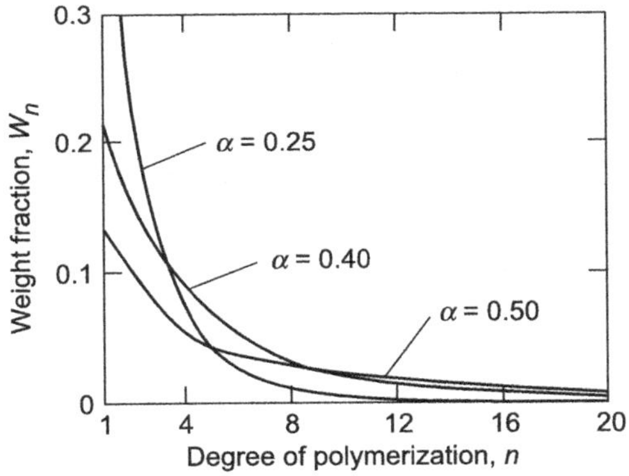

Fig. 2.8 *Molecular weight distribution curves, W_n vs n for step-growth polymerization with trifunctional reactants at different stages of reaction[2], given by $\alpha = p$ (Flory, 1946;* **Courtesy,** *American Chemical Society, Washington)*

change in the weight fraction of the gel (W_{gel}) is also shown in the plot. Unlike the case of linear polycondensation (Fig. 2.5), the branching/gelling system involving trifunctional monomers (Figs 2.8 and 2.9) shows that the weight fraction of monomer is always greater than that of any one of the other species, the weight fraction of the bigger species being progressively lower. With progress of reaction, bigger species are formed at the expense of smaller ones with a maximum being reached in each case at a p value less than 0.5. The maximum, however, shifts to higher p values for the bigger species. The distribution initially broadens with p and

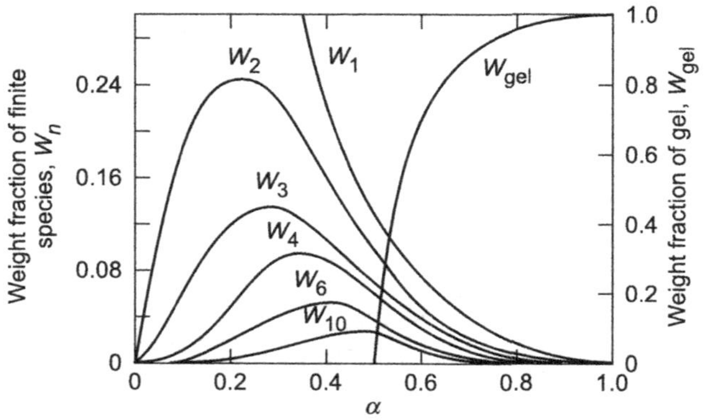

Fig. 2.9 *Variation of weight fractions of different finite species and of the gel species with variation of the branching coefficient, α for trifunctional polycondensation[2], where α = p (Flory, 1946;* **Courtesy,** *American Chemical Society, Washington)*

maximum heterogeneity is reached at the gel point, i.e. the point of first appearance of infinite network ($\alpha = 0.5$). Beyond the intersection of W_1 and W_{gel} curves, the gel formed comprises the most abundant species on weight basis. At $\alpha = p = 1$, only gel is present, i.e. the whole system is converted into an insoluble, infusible mass of one giant molecule and $\overline{X}_w/\overline{X}_n$ becomes equal to unity. At the gel point ($\alpha = p = 0.5$), however, the breadth of the distribution $\overline{X}_w/\overline{X}_n$ assumes an excessively high value, since $\overline{X}_w$ becomes infinite, while $\overline{X}_n$ has a finite value of only four (Fig. 2.10). The general expression for $\overline{X}_n$ and $\overline{X}_w$ for the non-linear multifunctional systems characterized by $\alpha = p$, are given by:

$$\overline{X}_n = \frac{1}{1-(\alpha f/2)} \tag{2.50}$$

$$\overline{X}_w = \frac{(1+\alpha)}{1-(f-1)\alpha} \tag{2.51}$$

2.14 Interfacial Polymerization

Polycondensation reactions such as the amidation between diacids and diamines have been studied widely in melt condition (in bulk), in solution and even in the solid state. Stoichiometric balance between the reactive functional groups is easily obtained by precipitating the 1 : 1 ammonium salt in alcoholic medium.

$$H_2N{-}R{-}NH_2 + HOOC{-}R'{-}COOH \rightarrow \begin{matrix} H_3\overset{+}{N}{-}R{-}\overset{+}{N}H_3 \\ {}^{-}OOC{-}R'{-}COO^{-} \end{matrix}$$

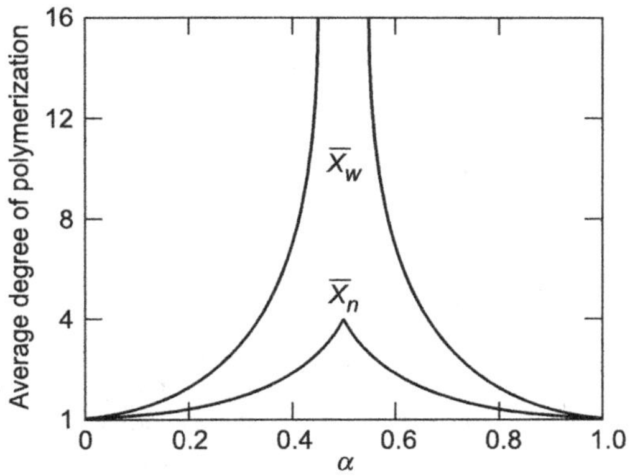

Fig. 2.10 *Variation of number and weight average degrees of polymerization ($\bar{X}_n$ and $\bar{X}_w$) as a function of the branching coefficient, α for a trifunctional polycondensation reaction*[2] *(Flory, 1946;* **Courtesy,** *American Chemical Society, Washington)*

The ammonium salt is commonly called the nylon salt. The salt/polymer equilibrium is favourable more toward polymer for polyamide formation than for the formation of polyesters by ester-interchange step polymerization, such as in the case involving dimethyl terephthalate and ethylene glycol. These reactions are usually carried out at high temperatures (>200°C) in two successive stages, each culminating into a melt polymerization system in the final stage.

Formation of polyesters and polyamides can be accomplished at much lower temperatures, often under the room temperature conditions starting with the more reactive acid chlorides (Table 2.1) and employing a solution polymerization technique. A special technique known as interfacial polymerization has been developed for the formation of polyamides, polyesters, polycarbonates and polyurethanes. The technique is so designed as to permit polymerization at the interface between two immiscible solutions one of which contains one of the dissolved monomers while the second monomer is dissolved in the other.

By the interfacial technique, polyamidation is usually performed at or near room temperature using a set up as given in Fig. 2.11, showing formation of the polyamide at the interface between a layer of aqueous solution of a diamine over a solution of the diacid chloride in an organic solvent layered below. The reactant monomers diffuse to and transform into the polymer at the interface.[16] The polymer gets precipitated which can be taken out or withdrawn from the interface in the form of a continuous filament or film. The rate of polycondensation is determined by the rates of diffusion of the reacting monomers in the two phases to the interface. The

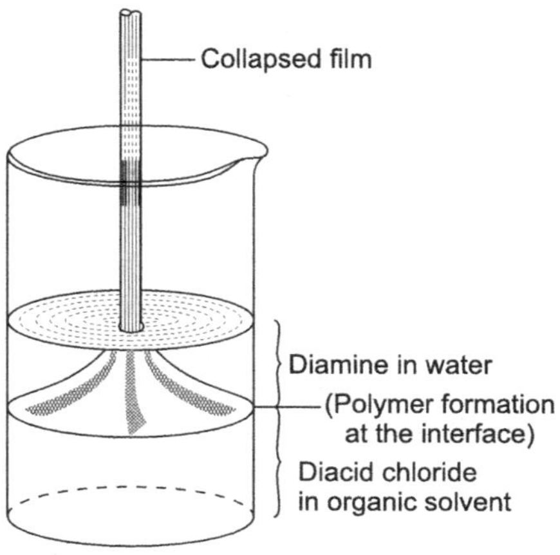

Fig. 2.11 *Interfacial polymerization*

monomers diffusing to the interface, however, react with the polymer chain ends. Thus, the mechanism involved in interfacial polycondensation is different from the usual step-growth polymerization in that the new reactants taking part late in the process contribute to chain extension by reacting with existing chains rather than interacting to form fresh chains. A much higher molecular weight usually results in the interfacial process as a consequence of this difference.

The byproduct of condensation, HCl (using diacid chlorides in the organic layer) is usually neutralized by dissolving an inorganic base in the aqueous layer in order to prevent lowering of reaction rate due to formation of unreactive amine hydrochloride salt. The likelihood of hydrolysis of the acid chloride to the much less reactive or unreactive acid in the presence of an aqueous inorganic (strong) base is avoided or kept to a minimum in view of very low water solubility of the acid chloride and also in view of high rates of polymer formation at the interface thus preventing or putting hindrance to the diffusion of the acid chloride into the aqueous layer through the polymer layer at the interface.

The selection of organic solvent plays a significant role in determining the properties of the polymer formed, particularly the molecular weight. Premature precipitation due to poor solvent action of a selected solvent limits molecular weight of the product and prevents formation of products of high degree of polymerization. The biphasic system and the nonsolvent action of water in interfacial polycondensation lead to molecular weight distributions[17, 18] far different from the most probable distributions [Eq. (2.30)] derived by Flory[2].

From many considerations, the biphasic interfacial polycondensations are more advantageous than the polycondensation done in homogeneous solution or under melt conditions. The impurity of the reactants is of much less consequence as their adverse effect is insignificant or of much lower dimension at the much lower temperature of polymerization for the interfacial system. Stoichiometric equivalence of the reactants in the two phases is not a prerequisite as stoichiometry is automatically struck at the interface where only the polycondensation proceeds. Attainment of high molecular weight for the product is not dependent on p or the extent of reaction. Low temperature of polymerization makes the process more stable toward occurrence of undesired side reactions. Final products in the form of films, fibres and fibrous particles may be formed by direct polymerization. The major demerit of the interfacial process which stands in the way of its commercial development is the unfavourable economy for the high cost of the acid chloride, handling of large amount of solvents and their recovery, etc.

References

1. Carothers, W.H., *J. Am. Chem. Soc.*, **51** (1929), 2548.
2. Flory, P.J., *Chem. Rev.*, **39** (1946), 137.
3. Bhide, B.V. and J.J. Sudborough, *J. Indian Inst. Sci.*, **8A** (1925), 89.
4. Flory, P.J., *J. Am. Chem. Soc.*, **61** (1939), 3334.
5. Solomon, D.H., *J. Macromal. Sci. Revs. Macromol. Chem.*, **C1**(1) (1967), 179.
6. Flory, P.J., *Principles of Polymer Chemistry*, Cornel University Press, Ithaca, New York, 1953, Ch. 8 and 9.
7. Flory, P.J., *J. Am. Chem. Soc.*, **58** (1936), 1877.
8. Howard, G.J., *Progress in High Polymers*, J.C. Robb and F.W. Peaker, Eds. Iliffe, London, Vol. 1 1961, pp. 185–231.
9. Schulz, G.V., *Z. Physik. Chem.*, **B43** (1939), 25.
10. Shaefgen, J.R. and P.J. Flory, *J. Am. Chem. Soc.*, **70** (1948), 2709.
11. Stockmayer, W.H., *J. Polym. Sci.*, **9** (1952), 69; **11** (1953), 424.
12. Pinner, S.H., *J. Polym. Sci.*, **21** (1956), 153.
13. Flory, P.J., *J. Am. Chem. Soc.*, **63** (1941), 3083.
14. Solomon, D.H., *The Chemistry of Organic Film Formers*, Chapter 3, Wiley-Interscience, New York, 1967.
15. Mark, H. and A.V. Tobolsky, *Physical Chemistry of High Polymer Systems*, Interscience, New York, 1950.
16. Morgan, P.W. and S.L. Kwolek, *J. Chem. Ed.*, **36** (1959), 182, *J. Polym. Sci.*, **40** (1959), 299.
17. Morgan, P.W., *Condensation Polymers by Interfacial and Solution Methods*, Wiley-Interscience, New York, 1975.
18. Korshak, V.V., *Pure and Appl. Chem.*, **12** (1966), 101.

CHAPTER 3

Addition Polymerization or Chain-Growth Polymerization

3.1 General Theory of Addition Polymerization

Addition polymerization is a process by which unsaturated monomers are converted to polymers of high molecular weight, exhibiting the characteristics of a typical chain reaction. A large number of different class of unsaturated monomers, such as ethylene ($CH_2 = CH_2$) the simplest olefin, α-olefins ($CH_2 = CHR$, where R is an alkyl group), vinyl compounds ($CH_2 = CHX$, where X = Cl, Br, I, alkoxy, CN, COOH, COOR, C_6H_5, etc., atoms or groups) and conjugated diolefins ($>C=\overset{|}{C}-\overset{|}{C}=C<$) readily undergo this type of polymerization, mechanistically characterized as chain-growth polymerization or simply chain polymerization.

Addition or chain-growth polymerization, as all other typical chain reactions, are fast reactions typified by three normally distinguishable steps, viz. (i) triggering off or initiation of the chain process, (ii) propagation of the chain or the chain growth process, and (iii) cessation of the chain growth, i.e. the chain termination process.

The initiation of the chain polymerization may be activated and induced by heat, light, high energy radiations or a host of chemical additives commonly known as initiators or catalysts (I) for polymerization. The initiation is usually a direct consequence of generation or introduction of a highly reactive species R* through dissociation or degradation of some monomer molecules (M) under the influence of such physical agencies as heat, light, radiation, etc. or as a consequence of dissociation or decomposition of the initiator or catalyst molecules present as shown below:

$$I \rightarrow R^* \tag{3.1}$$

The reactive species may be a free radical, cation or anion which readily attacks the unsaturated monomer molecule if conditions are favourable, adds itself to the monomer molecule by opening the π bond to form a fresh reactive centre (radical, cation or anion centre), depending on the nature of R* and the prevailing chemical environment. The new reactive centre adds to another monomer molecule, M and the process is repeated in quick succession leading many more monomer molecules to add to the same growing species at its reactive or growth centre to carry on the propagation of the chain process:

$$R^* + M \rightarrow R - M^* \xrightarrow{+M} R - M - M^* \xrightarrow{+M} R - [M]_{n-1} - M^* \tag{3.2}$$

For a vinyl monomer (M, structurally written as $CH_2 = CHX$), the growing chain of size n may be structurally written as R—$[CH_2$—$CHX]_{n-1}$—CH_2—*CHX. The chain growth is terminated at some point by annihilation of the reactive centre by one or more convenient and appropriate mechanism largely dependent on the type of the reactive centre (radical, cation or anion), nature of the monomer, M and the overall environment and condition of the reaction. In the present chapter, the case of free radical polymerization will be considered in detail.

3.2 Overall Scheme of Radical Chain Polymerization

Addition or chain polymerization is also commonly known as vinyl and related polymerization in view of the fact that most of the monomers involved generally contain the chemical group ($CH_2 = CH$—), commonly identified as the vinyl group. In radical or free radical polymerization, the chain propagating species is a free radical and the process is usually initiated using a free radical generating catalyst or initiator in the monomer or the polymerization medium.

3.2.1 Chain Polymerization using a Free Radical Initiator

The polymerization of a vinyl or related monomer M, induced in presence of a free radical initiator, I may be schematically expressed as follows:

Initiator decomposition:

$$I \xrightarrow{k_d} 2R^{\bullet} \tag{3.3}$$

followed by *Chain initiation:*

$$R^{\bullet} + M \xrightarrow{k_i} RM_1^{\bullet} \tag{3.4}$$

Chain propagation:

$$\left.\begin{aligned} RM_1^{\bullet} + M &\xrightarrow{k_p} RM_2^{\bullet} \\ RM_2^{\bullet} + M &\xrightarrow{k_p} RM_3^{\bullet} \\ RM_{n-1}^{\bullet} + M &\xrightarrow{k_p} RM_n^{\bullet} \end{aligned}\right\} \tag{3.5}$$

Chain termination:

$$RM_m^{\bullet} + RM_n^{\bullet} \xrightarrow{k_{tc}} R - M_{m+n} - R \tag{3.6}$$

$$RM_m^{\bullet} + RM_n^{\bullet} \xrightarrow{k_{td}} RM_n + RM_m \tag{3.7}$$

where $R^{\bullet}$ is a free radical generated from the initiator I and $RM_1^{\bullet}$, $RM_2^{\bullet}$, $RM_3^{\bullet}$, etc. are the growing polymer chains, each bearing a free valency or radical centre at the chain-end and RM_n, RM_m, $R - M_{n+m} - R$, etc., are the (dead) polymer molecules obtained.

It is apparent from the above scheme that the initiator fragments in the form of radical $R^{\bullet}$ are trapped as end groups in the polymer structure, sealing one or both ends of the polymer chains.

A common feature of vinyl polymerization is that the active centre of the kinetic chain is retained by a single molecular species throughout the chain growth or chain propagation process. A partially polymerized system would practically consist of full grown polymer molecules of high molecular weight and the unreacted monomer; no species in an intermediate stage of growth can be isolated. The fact is that the polymer molecules formed in the early stages of the reaction are usually comparable in molecular weight or chain length to those formed at a much advanced stage of the process. The time period required for the full growth of an individual polymer molecule is an overwhelmingly minute fraction of the time needed for a measurable overall conversion.

3.3 Comparison between Chain-growth and Step-growth Polymerization

A comparative account of the differences between chain polymerization or addition polymerization on the one hand and step polymerization or condensation polymerization on the other is given in Table 3.1.

Table 3.1 *Comparison of addition polymerization and condensation polymerization*

Addition/Chain-growth Polymerizaion	*Condensation/Step-growth Polymerization*
Proceeds by a chain mechanism characterized by the three distinct steps in the main, viz. chain initiation, chain propagation and chain termination. Under favourable condition, the reaction goes fast almost to completion	Proceeds by an equilibrium step reaction mechanism. The step process is usually much slower than the addition or chain polymerization process
Monomers with carbon–carbon unsaturation undergo addition polymerization	Monomers bearing functional groups of the kind —OH, —COOH, —NH_2, —NCO, etc., undergo condensation polymerization
The growth reaction simply involves addition of monomers with one another in quick succession and the chain carrier is either a free radical, cation (carbonium ion) or an anion (carbanion). No byproduct is eliminated in the process. Growth or propagation involves reaction of only monomer molecules with the active centres	The growth process is characterized by relatively slow intermolecular condensation reaction between related functional groups in successive (distinguishable) steps, normally with the elimination of a small byproduct of condensation such as H_2O, HCl, NH_3, etc., in each step. Any two species of any size, including the monomers may react and contribute to the growth process
Monomer concentration decreases throughout the course of the polymerization process, as the yield of polymer increases. Usually, the conversion of monomer to polymer passes through a steady rate zone at some early stage of the reaction	Monomer disappears much faster in the early stages of the step polymerization
Molecular species in the intermediate stages of growth cannot be isolated. Conversion of monomer to polymer increases with time but the molecular weight of the polymer remains more or less unchanged with progress of reaction	Molecular species in the intermediate stages of growth can be readily isolated. Molecular weight slowly increases throughout the process and it is only at a very high range of conversion that polymer molecules of very high molecular weight are obtained

Contd...

Table 3.1 *(Contd.)*

Addition/Chain-growth Polymerizaion	*Condensation/Step-growth Polymerization*
At any instant of time the reaction mixture consists of full grown polymer molecules, unreacted monomer molecules and a very low concentration (10^{-8}–10^{-3} mol. 1^{-1}), of growing chains or active centres	At any instant of time the reaction mixture consists of molecular species of different sizes in a calculable distribution
Full grown polymers are normally formed at the incidence of the termination reaction, and they are considered as dead polymers, being normally incapable of taking further part in the chain growth process	Molecular species of any size retain the reactive functional groups at their chain ends and hence, they are capable of contributing to further chain growth indefinitely. As such, there can be no termination of the process till it reaches 100% extent of reaction
Usually the backbone of the polymer chains consists of —C—C— linkages and other kinds of atoms such as O, N, S, etc. may appear in the side groups	The condensed interunit links give rise to introduction of hetero atoms such as N, O, S, etc., in the backbone of the polymer chains at regular intervals. A special case of a polymer with alternate C and O atoms in the chain backbone is the polyoxymethylene $—(CH_2—O)_n—$
Addition polymerization is usually done in gas phase, in liquid bulk monomer or under melt condition, or by solution, suspension or emulsion techniques	Condensation polymerization is usually accomplished in bulk under melt condition or in solution and employing the biphasic interfacial polymerization technique
Control of molecular weight is achieved by selecting appropriate initiator and monomer concentrations and temperature of polymerization and for a given set of conditions by using calculated amount of a chain transfer agent	Control of molecular weight is achieved by using a stoichiometric imbalance of the reacting functional groups or a calculated amount of an appropriate monofunctional monomer

3.4 Kinetics of Chain Polymerization Initiated by a Free Radical Initiator

According to the scheme presented in Section 3.2.1, the initiator, I first decomposes into a pair of primary (free) radicals, $R^\bullet$, as in Eq. (3.3) and then each $R^\bullet$ attacks a monomer molecule M to produce a chain radical $RM^\bullet$, as in Eq. (3.4). These two steps are characterized by the rate constant, k_d for the decomposition of the initiator and the rate constant k_i for the chain initiation reaction respectively.

The chain growth or the propagation reaction involves addition of the monomer M to $RM^{\bullet}$ and to the successive radicals as in the series of Eq. (3.5); each successive propagation reaction is characterised by the rate constant of propagation k_p, assuming that the radical reactivity is independent of chain length as long as the end structural segment on which the radical site rests is just the same. All successive propagation steps may then be expressed by the general expression

$$M^{\bullet} + M \xrightarrow{k_p} M^{\bullet} \tag{3.8}$$

Termination of polymerization is bimolecular in nature and it involves reaction between a pair of chain radicals leading to annihilation of two chain radicals in each incidence of the termination reaction and production of what is known as (dead) polymer product: Eqs (3.6) and (3.7), showing the termination reaction by two different modes, represent (i) coupling or mutual combination of two chain radicals characterized by the rate constant k_{tc} and (ii) disproportionation of chain radicals characterized by the rate constant k_{td} respectively. The two different modes of termination may be more clearly expressed as follows so as to clearly indicate differences in the number of species formed and their detail structure.

Termination by combination

$$\begin{aligned} & R \sim\sim CH_2—{}^{\bullet}CHX + XH\dot{C}—CH_2 \sim\sim R \\ & \xrightarrow{k_{tc}} R \sim\sim CH_2—CHX—CHX—CH_2 \sim\sim R \end{aligned} \tag{3.9}$$

Termination by disproportionation

$$\begin{aligned} & R \sim\sim CH_2—{}^{\bullet}CHX + {}^{\bullet}CHX—CH_2 \sim\sim R \\ & \xrightarrow{k_{td}} R \sim\sim CH_2—CH_2X + R \sim\sim CH = CHX \end{aligned} \tag{3.10}$$

For kinetic purposes, the bimolecular mechanism of termination alone is important, and the exact mode (combination or disproportionation) does not make any difference, so that the termination step may be simply expressed as:

$$M^{\bullet} + M^{\bullet} \xrightarrow{k_t} \text{dead polymer product} \tag{3.11}$$

The sizes of the reacting chain radicals are not specified in the above reaction in view of the assumption made about equal reactivity of chain radicals of a specific kind independent of their size. Here, k_t is the overall rate constant of bimolecular termination such that $k_t = (k_{tc} + k_{td})$.

It is clear from Eqs (3.9) and (3.10) that the dead polymer molecule formed by combination of chain radicals would bear two initiator fragments (R) per molecule as end groups, whereas disproportionation would lead to polymer molecules having only one initiator fragment (R) per polymer molecule as end groups, and fifty per cent of the polymer molecules formed by disproportionation would bear an unsaturation each at one of the chain ends.

3.4.1 Rate of Initiator Decomposition and Chain Initiation

Equation (3.3) shows a unimolecular decomposition of the initiator I to produce a pair of initiator or primary radicals. In many instances the decomposition of the lone initiator, I involves homolytic cleavage of the weak bond in the molecules such that the chain initiating radicals are generated in pairs. Hence, R_d, the rate of initiator decomposition may be expressed as:

$$R_d = -\frac{d[\mathrm{I}]}{dt} = k_d\,[\mathrm{I}] \quad (3.12)$$

The rate of radical generation $d[\dot{\mathrm{R}}]/dt$ would then be expressed as:

$$\frac{d[\dot{\mathrm{R}}]}{dt} = 2k_d\,[\mathrm{I}] \quad (3.13)$$

since, for each incidence of decomposition of an initiator molecule, two initiator radicals ($R^\bullet$) are produced. As the radicals are transient species and they give rise to chain initiation as soon as they are generated in the monomer system, the rate of chain initiation, R_i will be the same as the rate of radical generation in the absence of undesirable wastage of the chain initiating radicals, $R^\bullet$ by side reactions. Hence,

$$\mathrm{R_i} = \left(\frac{d[\mathrm{M}^\bullet]}{dt}\right)_i = k_i[\mathrm{R}^\bullet]\,[\mathrm{M}] = 2k_d\,[\mathrm{I}] \quad (3.14)$$

Thus, $R_i = 2k_d$ [I] marks the ideal case of hundred per cent efficiency of chain initiation by the primary radicals. For all practical purposes, however, the expression is modified as:

$$R_i = 2f k_d\,[\mathrm{I}] \quad (3.15)$$

where, the factor f is the initiator efficiency or the efficiency of initiation, representing the fraction of primary radicals ($R^\bullet$) which actually contributes to chain initiation as given by Eq. (3.4). Realistically, the value of f is less than unity due to side reactions leading to measurable wastage of primary radicals. f will be dependent on monomer concentration [M] if the wastage of $R^{\bullet}$ is high due to side reactions and in that case R_i should depend on [M]. However, the general independence of f on [M] has in a large measure simplified the kinetic treatment of vinyl and related polymerizations.

3.4.2 Rate of Chain Termination

The rate of overall chain termination (R_t) may be expressed by Eq. (3.16), considering the termination reaction in a simplified and general form as given by the reaction (3.11). Thus,

$$R_t = -\left(\frac{d[M^{\bullet}]}{dt}\right)_t = 2k_t\,[M^{\bullet}]^2 \tag{3.16}$$

The factor 2 is used for the simple reason that for each incidence of a termination reaction, 2 chain radicals disappear from the scene. It is important to note that as radicals are generated in pairs, they are also destroyed in pairs.

3.4.3 The Overall Rate of Polymerization

(a) ***Rate of Propagation*** Considering the chain propagation reaction in the general form given by the reaction (3.8), the rate of propagation or chain growth, R_p may be expressed as:

$$R_p = k_p\,[M^{\bullet}]\,[M] \tag{3.17}$$

The equation (3.17) is not directly useful for evaluating R_p as it contains the term for the concentration of the transient chain radicals, $[M^{\bullet}]$ and because the radical concentrations are very difficult to measure as they are very low ($\simeq 10^{-8}$ mol l^{-1}). An expression free from the radical concentration term will be useful in that event.

(b) ***Steady-state Assumption*** In order to eliminate the radical concentration term $[M^{\bullet}]$ from Eq. (3.17), it is necessary to find an expression for it in terms of known or measurable parameters. For this, the steady-state assumption is made. According to this assumption, soon after chain initiation, a state is attained when the value of $[M^{\bullet}]$ becomes essentially constant or steady as radicals are formed and destroyed at equal rates, i.e. $R_i = R_t$. On the basis of the steady-state assumption and combining Eqs (3.15) and (3.16), $[M^{\bullet}]$ may be expressed as:

$$[M^{\bullet}] = \left(\frac{fk_d\,[I]}{k_t}\right)^{1/2} \tag{3.18}$$

combining Eqs (3.17) and (3.18), the expression for rate of propagation, R_p reduces to

$$R_p = \frac{k_p}{k_t^{1/2}} \cdot (fk_d)^{1/2}[I]^{1/2}[M] \tag{3.19}$$

(c) ***Overall Rate of Polymerization*** Examination of the overall scheme of polymerization represented by Eqs (3.3) to (3.7) clearly shows that monomer disappearance takes place at chain initiation and chain propagation steps only. Hence, the overall rate of monomer disappearance ($-d[M]/dt$), which is the same as the overall rate of polymerization, may be equated as:

$$-\left(\frac{d[M]}{dt}\right) = R_i + R_p \tag{3.20}$$

As the number of monomer molecules consumed at the initiation step is far less or insignificant compared to the number consumed in the long sequence of reactions in the propagation step, the overall rate of polymerization is, for all practical purposes, simply equated to the rate of propagation, R_p. Therefore,

$$\text{overall rate of polymerization, } \left(-\frac{d[M]}{dt}\right) = R_p$$

$$= \frac{k_p}{k_t^{1/2}} \cdot (fk_d)^{1/2}\,[I]^{1/2}\,[M] \tag{3.21}$$

Equation (3.19) or (3.21) may also be expressed as:

$$R_p = \frac{k_p}{k_t^{1/2}} \cdot \left(\frac{1}{2}R_i\right)^{1/2} [M] \tag{3.22}$$

3.5 Analysis of the Kinetic Expressions

Equation (3.22) relating R_p with R_i is a significant kinetic expression and it clearly shows that the polymerization rate at a given monomer concentration is dependent on the square root of the rate of chain initiation. A twofold increase in R_i does not double R_p ; R_p increases only by a factor of $2^{1/2}$ as a consequence.

Similarly, Eq. (3.19) or (3.21) clearly shows a square root dependence of R_p on initiator concentration [I]. The square root dependence of R_p on [I] is a direct consequence of the bimolecular termination mechanism. R_p is, however, directly proportional to the monomer concentration [M], according to the normal reaction scheme considered here. Normal free radical kinetics prescribe constancy of the ratio ($R_p/[I]^{1/2}[M]$), if f is independent of [M].

Practical evaluation of rates of free radical polymerization using different monomer–initiator systems points to the general validity of the kinetic expressions derived [Eqs (3.21) and (3.22)]. The monomer exponent and initiator exponent may be separately obtained from two sets of experiments at a given temperature, i.e. at fixed initiator and variable monomer concentrations and vice versa, respectively, and the values of the monomer exponent and initiator exponent are given by the slope of the log R_p vs. log [M] plot (Fig. 3.1) and of the log R_p vs. log [I] plot (Fig. 3.2), respectively.[1,2] Good agreement between experimental data and the kinetic expression given by Eq. (3.21) or (3.22) points to the general validity of the assumptions made in deriving the expressions.[3,4]

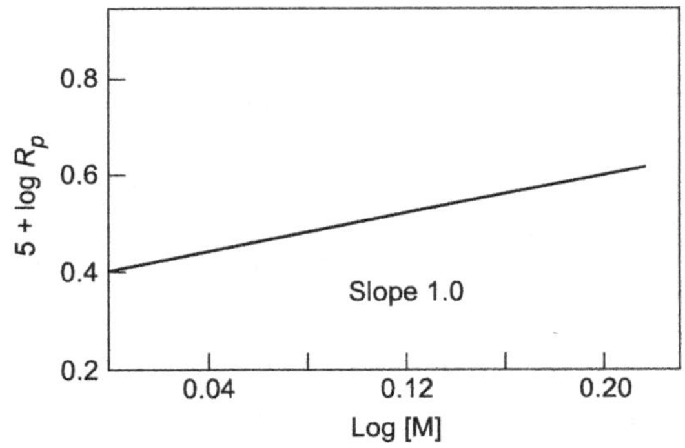

Fig. 3.1 *log—log plot of R_p vs [M] for polymerization of methyl methacrylate using KVO_4 as initiator[1] (25) $\times 10^{-4}$ mol 1^{-1}) in acid aqueous medium at 50°C (Ghosh & Paul, 1983;* Courtesy, *Marcel Dekker, New York)*

3.6 Experimental Approaches for Determination of R_p

The polymerization rate (R_p) can be determined experimentally by following the change in a number of properties of the polymerization system with time or progress of polymerization, such as, density or volume, viscosity, refractive index and loss in monomer concentration or gain in polymer content (using volumetric, gravimetric or spectrophotometric analysis). Most of these methods depend on drawing the monomer–polymer mixture at regular intervals for doing the analysis or characterization. But the density or the volume change due to conversion of

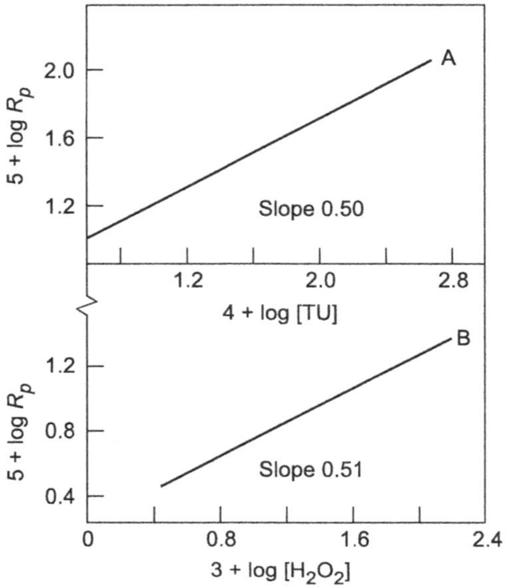

Fig. 3.2 *log–log plot of R_p vs initiator concentration[2] for aqueous polymerization of methyl methacrylate at 35°C using hydrogen peroxide (H_2O_2) and thiourea (TU) combination as the redox initiator[2]: (A)—[H_2O_2] fixed, [TU] varies; B—[TU] fixed, [H_2O_2] varies (Ghosh & Banerjee, 1979;* Courtesy, *Wiley-Interscience, New York)*

monomer to polymer provides adoption of a very simple, accurate and sensitive technique of determining R_p value without disturbing the polymerization system and without the necessity of drawing samples with time. Polymerization is associated with a volume contraction, the density of a polymer being higher than that of the corresponding monomer. The progress of polymerization can be very easily monitored by following the volume contraction in the capillary of a dilatometer used as the reaction vessel. Even a fractional percent polymerization can be accurately measured by the dilatometric technique. Experimental data allow a plot of percent conversion vs. time from which, considering the steady zone of the plot, R_p can be easily calculated. R_p is normally expressed in mol l^{-1} s^{-1}.

3.7 Initiation and Initiator systems

3.7.1 Unimolecular Initiating Systems

Initiator compounds characterized by unimolecular decomposition through dissociation of a relatively weak bond in them leading to generation of a pair of

Table 3.2 *Some initiator systems and radical generation reactions*

Initiator System	*General Mechanism*	*Primary Radical Generation Reaction*
Unimolecular or lone initiator system (peroxy/azo compounds)		
(a) Benzoyl peroxide (Bz_2O_2)	Cleavage of peroxy linkage by thermal or photo activation	$C_6H_5-\overset{O}{\overset{\|}{C}}-O-O-\overset{O}{\overset{\|}{C}}-C_6H_5 \rightarrow 2C_6H_5-\overset{O}{\overset{\|}{C}}-\dot{O} \rightarrow 2\dot{C}_6H_5 + 2CO_2$
(b) Acetyl peroxide (Ac_2O_2)	"	$CH_3-\overset{O}{\overset{\|}{C}}-O-O-\overset{O}{\overset{\|}{C}}-CH_3 \rightarrow 2CH_3-\overset{O}{\overset{\|}{C}}-\dot{O} \rightarrow 2\dot{C}H_3 + 2CO_2$
(c) Di-*t*-butyl peroxide (DTBP)	"	$(CH_3)_3-C-O-O-C(CH_3)_3 \rightarrow 2(CH_3)_3-C-\dot{O} \rightarrow \underset{CH_3\ \ CH_3}{\overset{O}{\overset{\|}{C}}} + \dot{C}H_3$
(d) *t*-Butyl hydroperoxide (TBHP)	"	$(CH_3)_3-C-O-O-H \rightarrow (CH_3)_3-C-\dot{O} + \dot{O}H$; $(CH_3)_3-C-\dot{O} \rightarrow (CH_3)(CH_3)C{=}O + \dot{C}H_3$
(e) Hydrogen peroxide (H_2O_2)	"	$H-O-O-H \rightarrow 2\dot{O}H$
(f) Persulphate ion ($S_2O_8^{=}$)	"	${}^-O_3S-O-O-SO_3^- \rightarrow 2O_3S\dot{O}^-\ (2\,S\bar{O}_4^{\bullet})$
(g) Peroxy dicarbonate	"	$R-O-\overset{O}{\overset{\|}{C}}-O-O-\overset{O}{\overset{\|}{C}}-O-R \rightarrow 2RO\overset{O}{\overset{\|}{C}}-\dot{O} \rightarrow 2R\dot{O} + 2CO_2$

Contd.

Table 3.2 (Contd.)

Initiator System	*General Mechanism*	*Primary Radical Generation Reaction*
(h) Azobisisobutyronitrile (AIBN)	Cleavage of C—N bond by thermal/photo activation	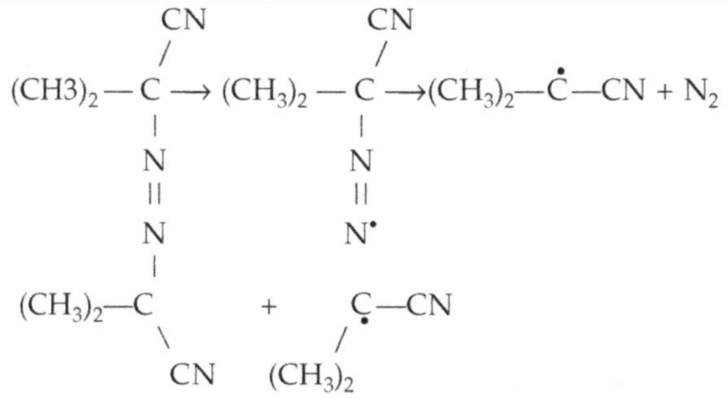
Bimolecular initiator system (redox initiator)	Reduction activation (thermal/photochemical)	
(a) Reducing metal ion and peroxide or hydroperoxide	Electron transfer	$M^{(n-1)+} + R—O—O—R \rightarrow M^{n+} + \dot{O}R + \bar{O}R$ (R = H atom, alkyl/aryl group or $^{-}O_3S$—group, etc.)
(b) Oxidizing metal ion and	"	$S_xO_y^{n-} + M^{m+} \longrightarrow \dot{S}_xO_y^{(n-1)-} + M^{(m-1)+}$
reducing (Sulphoxy) compounds ($SO_3^{=}$, $S_2O_3^{=}$, $S_2O_4^{=}$, etc.)	"	
(c) Amines and peroxides	Complexation followed by proton transfer	$R_2NH + R'—O—O—R' \rightleftharpoons [R_2NH(OR')]^+ (OR')^- \rightarrow R_2\dot{N} + \dot{O}R' + R'OH$
Metal, metal salts, complexes or chelates, etc.		
(a) Ion-pair complex of Fe^{3+} ion	Electron transfer	$(Fe^{3+} X^-) \rightarrow Fe^{2+} + \dot{X}$ ($\dot{X}$ = Cl, Br, OH, etc.)
(b) Some metal ions in higher oxidation states (Mn^{4+}, Ce^{4+} Co^{3+}, Ag^{3+}, etc.) as salts or complexes, in aqueous medium	"	$M^{(n+1)+} + H_2O \rightarrow M^{n+} + H^+ + \dot{O}H$

Contd.

Table 3.2 *(Contd.)*

Initiator System	*General Mechanism*	*Primary Radical Generation Reaction*
(c) Metallic copper or cuprous ion in acid aqueous medium	Electron transfer	(i) $Cu^{0+} + H^+ \rightarrow Cu^+ + \dot{H}$ (ii) $Cu^+ + H^+ \rightarrow Cu^{2+} + \dot{H}$ (iii) $Cu^+ + O_2 + H_2O \rightarrow Cu^{2+} + \bar{OH} + H\dot{O}_2$ $Cu^+ + H\dot{O}_2 + H_2O \rightarrow Cu^{2+} + \bar{OH} + H_2O_2$ $Cu^+ + H_2O_2 \rightarrow Cu^{2+} + \bar{OH} + \dot{O}H$
(d) Oxalato complex of transition metals	"	$[M(C_2O_4)_x]^{n-} \rightarrow [M(C_2O_4)_{x-1}]^{(n-1)-} + \dot{C}_2O_4^-$
(e) Metal carbonyl (say, manganese carbonyl) in combination with CCl_4	Faster decomposition of metal carbonyl via complexation with CCl_4	$Mn_2(CO)_{10} \underset{k_2}{\overset{k_1}{\rightleftharpoons}} Mn_2(CO)_9 + CO$ $\xrightarrow[CCl_4]{k_3} [complex]_1 + CO$ $\xrightarrow{k_4} {}^{\bullet}CCl_3 + [complex]_2$
Lone Sulphur Compounds		
(a) Dibenzoyl disulphide (Bz_2S_2)	Cleavage of S—S bond	$C_6H_5-\overset{O}{\overset{\parallel}{C}}-S-S-\overset{O}{\overset{\parallel}{C}}-C_6H_5 \longrightarrow 2C_6H_5-\overset{O}{\overset{\parallel}{C}}-S^{\bullet}$
(b) Sulphuric acid (aq) (H_2SO_4)	A case of photodissociation of water	$SO^{=}_4 \xrightarrow{h\nu} S\dot{O}_4^- + e$ $\dot{S}O_4^- + H_3O^+ \longrightarrow H_2SO_4 + \dot{O}H$ $H_3\overset{+}{O} + e \longrightarrow \dot{H} + H_2O$
(c) Sulphur dioxide (SO_2)	Complexation with monomer (M) and biradical formation	$SO_2 + M \longrightarrow [SO_2 \ldots M] \longrightarrow \dot{S}O_2-\dot{M}$
(d) Hydrogen sulphide (H_2S)	Cleavage of H—S bond (photoactivation)	$H_2S \xrightarrow{h\nu} \dot{H} + H\dot{S}$

Contd.

Table 3.2 (Contd.)

Initiator System	General Mechanism	Primary Radical Generation Reaction
(e) Benzene sulphonyl azide	Cleavage of N—N bond	$C_6H_5SO_2N_3 \longrightarrow C_6H_5SO_2\dot{N} + N_2$
(f) Sulphuryl chloride (SO_2Cl_2)	Cleavage of S—Cl bond	$SO_2Cl_2 \longrightarrow \dot{S}O_2Cl + \dot{C}l$
(g) Dithionite ion ($S_2O_4^=$)	Cleavage of S—S bond	$S_2O_4^= \longrightarrow 2\dot{S}O_2^-$ ($\bar{O}_2S$—SO_2^-)
Halogen and halogen compounds		
(a) Halogen (X_2) (Cl_2, Br_2 and I_2)	Complexation with monomer M, and subsequent photolysis	$X_2 + nM \longrightarrow$ [halogen–monomer complex] $\xrightarrow{h\nu}$ radicals
(b) Interhalogen compounds (ICl, IBr, ICl_3, etc.)	"	Similar as for halogens
(c) Hydrogen halide (HX) (aqueous)	Dissociative electron transfer (photoactivation)	$HX \rightleftharpoons \overset{+}{H} + \bar{X} \xrightarrow{h\nu} \dot{H} + \dot{X}$
(d) Alkyl pyridinium/ammonium bromide	Electron transfer and cleavage of N—R bond (photoactivation)	$\left[\begin{array}{c} \backslash \;| \\ —N^+ \\ / \end{array} \right] \bar{Br} \xrightarrow{h\nu} \begin{array}{c}\backslash \\ —N \\ / \end{array} + R^\bullet + Br^\bullet$ (R on N^+)
Complex initiator systems		
(a) Halogen (X_2) complexes (X_2 + amine/ether)	Charge transfer (CT) interaction and ready decomposition of complex formed, specially on photoactivation	(i) $X_2 + :N\langle \rightarrow [X_2 \leftarrow N\langle]$ $X_2 + :O\langle \rightarrow [X_2 \leftarrow O\langle]$ (CT complex)

Contd.

Table 3.2 (Contd.)

Initiator System	*General Mechanism*	*Primary Radical Generation Reaction*
		(ii) CT complex + monomer (M) $\longrightarrow \left[>N\ldots \rightarrow X_2\ldots M \right] \rightarrow\ >N: + X\dot{M} + \dot{X}$ or $\left[>O\ldots \rightarrow X_2\ldots M \right] \rightarrow\ >O: + X\dot{M} + \dot{X}$
(b) Sulphur dioxide (SO_2) complexes (SO_2 + weak amines such as pyridine, quinoline, etc., and ethers)	Charge transfer (CT) interaction and ready decomposition of complex formed, specially on photoactivation	(i) formation of $\left(SO_2 \leftarrow N< \right)$ or $\left(SO_2 \leftarrow O< \right)$ CT complex as in (a)—(i) above (ii) CT complex + monomer (M) $\longrightarrow \left[>N\ldots \rightarrow SO_2\ldots M \right] \rightarrow\ >N: + \dot{S}O_2—\dot{M}$ or, $\left[>O\ldots \rightarrow SO_2\ldots M \right] \rightarrow\ >O: + \dot{S}O_2-\dot{M}$
(c) Nitrobenzene (ϕNO_2) complexes (ϕNO_2 + amines)	″	$\left[C_6H_5 \cdot NO_2 \leftarrow \ldots N(RH)< \right] \rightarrow \phi NO_2 + R^{\bullet}N< + \dot{H}$ (CT complex formed between ϕNO_2 and amine)

chain initiating radicals as typified by the general reaction (3.3) are exemplified by such compounds as peroxides, hydroperoxides and azo compounds. Decomposition of some of these classes of compounds and the transient radicals and other products that they generate are listed in Table 3.2. In many instances, such as with benzoyl peroxide (Bz_2O_2), acetyl peroxide (Ac_2O_2) or azobisisobutyronitrile (AIBN), the radicals from the unimolecular decomposition step further undergo decomposition leading to loss of a small molecule. It is necessary to consider[5] a somewhat complex kinetic scheme as detailed in the sequence of reactions (3.23) for even a first approach to a detailed decomposition mechanism.

$$\begin{array}{ccccc} & & \mathrm{I} \underset{}{\overset{k_d}{\rightleftharpoons}} (\mathrm{R}^\bullet + \mathrm{R}^\bullet)_{cage} & \xrightarrow{k_D} & 2\mathrm{R}^\bullet \\ & & k_x \downarrow & & \\ \mathrm{R} - \mathrm{R}' + \mathrm{X} & \xleftarrow{k_C} & (\mathrm{R}^\bullet + \mathrm{R}'^\bullet + \mathrm{X})_{cage} & \xrightarrow{k_D} & \mathrm{R}^\bullet + \mathrm{R}'^\bullet + \mathrm{X} \end{array} \tag{3.23}$$

The initiator molecule's net rate of decomposition, characterized by the rate constant k_d produces a pair of free radicals ($R^\bullet$) which are entrapped in a cage whose walls are the solvent molecules that surround the original initiator molecule. Each of the radicals may undergo a net translational diffusion out of the cage (characterized by the rate constant k_D), or a small molecule such as N_2 or CO_2 may be released within the cage through further decomposition of one or both the radicals present (characterized by the rate constant k_x) before they diffuse out of the solvent cage. Should the release of a small molecule occur as described, combination of the resultant radical $R'^\bullet$ with $R^\bullet$ might produce a totally inactive or stable species for radical initiation. This is one of the major causes of initiator efficiency and at the same time a source of additional information giving the details of the initiation process.

3.7.2 The Initiator Efficiency

The initiator efficiency is not an exclusive property of the initiator molecules alone, but it depends on the prevailing condition of polymerization, including the solvent used. It is possible that the decomposition of a compound such as AIBN or Bz_2O_2 to produce two radicals and a small molecule of N_2 or CO_2 is not exactly a two-step reaction, but proceeds via a concerted mechanism.

A relationship between the ratio k_x/k_D with the initiator efficiency f is due to O'Driscoll *et al.*[6] which is expressed as:

$$\frac{k_x}{k_D} = \frac{1-f}{f} \tag{3.24}$$

In many experimental conditions, the value of f ranges between 0.3–0.8. Several methods are employed for the evaluation of f, the initiator efficiency. One method, probably the most elegant, depends on the direct analysis of initiator fragments as end groups in the polymer formed compared to the amount of initiator consumed.

For the success of this method, very sensitive and accurate methods of end-group analysis must be employed and this requirement attaches limitations to widespread adoption of this approach. The use of isotopically labelled initiators such as [^{14}C]-benzoyl peroxide and other related peroxides[7,8] and C^{14} labelled AIBN[3,9] and S^{35} labelled potassium persulphate[10] or sulphur dioxide[11] provides appropriate sensitive methods for the determination of number of initiator fragments trapped as end groups in the resulting polymers.

A second method involves measurement and comparison of both polymer production and initiator decomposition. Reaction environment has a significant effect on initiator decomposition reaction. It is, therefore, necessary that measurement of the initiator decomposition is done during actual polymerization, as an independent measurement in the absence of monomer may lead to significant errors. Study of the decomposition of AIBN is relatively simple and it is readily done by following the evolution of N_2. Determination of the number average molecular weight, and hence, the number of polymer molecules formed in a given time and comparing it with the number of radicals produced over the same period allows a determination of f, provided a knowledge about termination by coupling or by disproportionation is available, since the former gives rise to two initiator fragments trapped as end-groups per polymer molecule, while the latter results in only one such end-group per polymer molecule. It is important to take induced decomposition of initiator into consideration in determining the number of radicals produced. Induced decomposition is practically nonexistent in the case of azonitriles, but it takes place appreciably in many peroxides. Brominated benzoyl peroxides incorporate more initiator fragments in the polymer[12] such as polystyrene than expected as per true initiation reaction. The anomaly is explained and also supported by experiments by considering significant induced decomposition of the brominated peroxy initiators in the presence of chain radicals as shown below:

$$— \sim\sim — M^{\bullet} + I \xrightarrow{k_{tr,I}} — \sim\sim — M - R + R^{\bullet} \tag{3.25}$$

A third approach to the counting of radicals in a system makes use of radical scavengers or inhibitors of polymerization which act effectively by stopping chain growth. The stable free radical diphenylpicrylhydrazyl (DPPH) has been widely used for this purpose.[13,14] The DPPH radical (purple or deep violet) reacts with other radicals to form a non-radical adduct (light brown yellow).

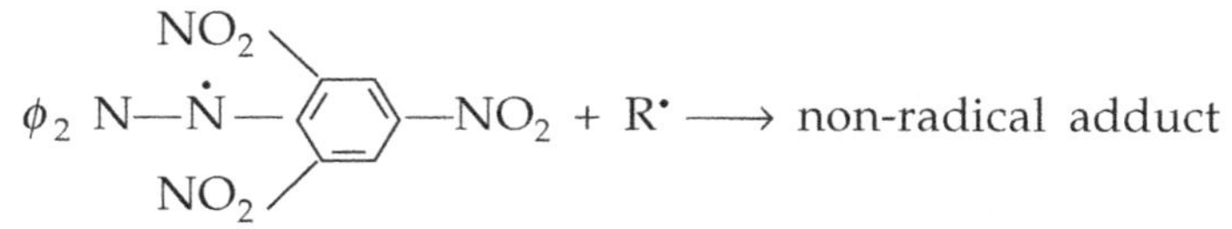

(3.26)

(DPPH, purple or deep violet) (light brown yellow)

The colour change associated with the reaction permits one to follow the reaction spectrophotometrically. Other radical scavengers that have been used in similar counting of radicals and determination of initiator efficiency are ferric chloride[15] and benzoquinone[16]. This approach of determining f using the radical scavengers mentioned here is unfortunately not very useful and leaves much to be desired in view of the reaction between the scavengers and radicals being often not quantitative.[17,18]

A fourth and probably the most useful method of determination of initiator efficiency is based on the dead-end effect in polymerization technique which will be treated separately later in this chapter. This technique allows treatment of kinetic data obtained under dead-end conditions to evaluate both k_d, the rate constant of initiator decomposition reaction and f, the initiator efficiency.

3.7.3 Bimolecular Initiating Systems

A good many number of free radical polymerizations, particularly those accomplished at low temperatures are induced by two-component initiator systems. They can be activated both thermally and photochemically. Many of such bimolecular initiator systems generate radicals by redox or electron transfer reaction mechanism involving the two initiator components or via formation of intermolecular complexes of different degrees of heat or light stability, often through donor-acceptor or charge transfer mechanism. Examples of such different types of initiator systems and the respective radical generation reactions are listed in Table 3.2. In aqueous media, initiation of polymerization by $\dot{O}H$ radicals, at least in part, is a good possibility, since ions and radicals in the system have ample chance to react with H_2O molecules to produce $\dot{O}H$ radicals which may then survive long in water by a regenerative transfer mechanism.

For a bimolecular initiator system, such as the common redox initiator systems, aqueous or non-aqueous, ordinary second order kinetics are generally obeyed if monomer concentration is high[19] and in such a case, R_p may be expressed as in Eq. (3.27), analogous to Eq. (3.21), showing the initiator concentration [I] substituted by the product of the reducing agent concentration, a, and the oxidizing agent concentration, b:

$$R_p = \frac{k_p}{k_t^{1/2}}(fk_d)^{1/2}\,(a \cdot b)^{1/2}\,[M] \tag{3.27}$$

3.7.4 Metals, Metal Ions and Metal Complexes or Chelates as Initiators of Free Radical Polymerization

Metals in their zero oxidation state are not generally suitable as radical initiators, with some rare exceptions. In intermediate or high oxidation states such as in the form of per(oxy) compounds, salts, complexes or chelates, they have been widely studied and used for radical vinyl polymerization both in aqueous and non-aqueous systems. In non-aqueous media, free radicals are usually formed by the decomposition of initiator species either on surfaces or homogeneously by heat, light or other radiation. The situation is much the same in aqueous media, where, however, the radicals may also be formed from or by ions via electron transfer reactions in which the solvent often plays an important role. Some of these initiator systems and the radicals that they generate to induce vinyl polymerization are listed in Table 3.2.

3.8 Autoacceleration or Gel Effect in Radical Polymerization

Considering normal behaviour for the first order transformation of monomer to polymer including general independence of initiator efficiency, f on monomer concentration and gradual depletion of initiator during polymerization, one would normally expect the reaction rate, R_p to fall with time, particularly after an initial short steady zone or period. But in many instances, just the opposite pattern of behaviour is observed, i.e. R_p follows an increasing trend, often very sharp, with increasing conversion or time of polymerization either from the very onset of polymerization or from an intermediate stage of conversion depending on the nature of the monomer–polymer system. The reaction medium transforms fast into a highly viscous mass not only as a consequence of increase in reaction rate but also due to sharp rise in molecular weight of the product polymer accompanying this phenomenon, commonly termed as the autoacceleration effect or gel effect,[20,21] which is also referred to as Trommsdorff effect or Norrish–Smith effect.

The gel effect is more effectively manifested in undiluted monomers than in diluted systems where the non-ideal or unusual effect may be minimized or even eliminated. In some monomer systems the autoacceleration effect may be so high as to be accompanied with large rise in temperature[22] and using undiluted methyl acrylate, acrylic acid and acrylonitrile, the autoacceleration effect associated with the polymerization process may even lead to explosion. The deviation from normal kinetics would not be eliminated even if isothermal conditions are maintained; the effect of autoacceleration and its lowering or elimination with progressive dilution with benzene as the solvent for polymerization of methyl methacrylate, shown in Fig. 3.3, is a typical example.[4]

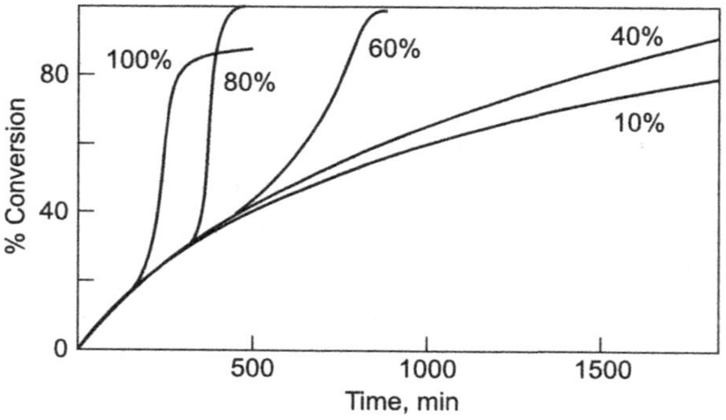

Fig. 3.3 *Influence of dilution in controlling auto-acceleration: per cent conversion vs time plots for benzoyl peroxide-initiated polymerization of methyl methacrylate[4] at 50°C (monomer concentration shown along each curve was varied using benzene as the solvent) (Schulz, 1947;* **Courtesy,** *Huthig & Wepf Verlag, Basel)*

In explaining the autoacceleration effect, keeping the rate Eq. (3.21) in view, alternative causes may be examined. The effect being independent of initiator, an increase in the value of $(f \cdot k_d)$ to explain the effect seems untenable. The other idea that appears more reasonable and acceptable is a sharp increase in the value of the kinetic parameter $(k_p/k_t^{1/2})$, which may arise as a consequence of increase in k_p or a decrease in k_t or both. However, a decrease in the termination rate constant with increase in conversion appears to be the most convincing reason. The decreasing trend in the k_t value with conversion, usually beyond a critical conversion range is believed to be due to progressive increase in viscosity of the monomer–polymer mixture. It is known that the k_t values are higher than k_p values by a factor of 10^7 -10^5; so perturbation of the kinetics from what is expected from Eq. (3.21) due to enhanced propagation would be of much lesser significance, if it occurs at all, than that due to hindered termination, the result being a significant gain in $(k_p/k_t^{1/2})$ value. Termination, considered to be bimolecular in nature, involves reaction between two very big chain radicals and the propagation reaction involves the reaction between a macroradial and a small monomer molecule. Naturally, the higher medium viscosity at higher conversion would hinder the termination reaction much more than affecting the propagation reaction. It is likely that the intrinsic reactivity of the chain radicals practically remains unchanged, but with the increasing medium viscosity acting as a much greater physical barrier for the diffusion of macroradicals than for the diffusion of small monomer molecules, there is sharp reduction in the likelihood of reaction between two macro (chain) radicals, the net result being a reduction in the value of k_t and a rise in the concentration of chain radicals as a consequence. R_p, therefore, sharply rises and an increase in the degree of polymerization with conversion follows. All these factors combine into an autoacceleration effect.[23–26]

3.9 Thermal Initiation in Absence of an Initiator or Catalyst

Some monomers are known to undergo spontaneous polymerization when carefully purified and heated in the absence of a catalyst, while many others, on similar purification, fail to undergo self-initiated polymerization at elevated temperatures. Styrene and methyl methacrylate belong to the first type of monomers. In most cases of thermal spontaneous polymerization, the initiation mechanism and the kinetics involved appear obscure and sometimes irreproducible because of the presence of traces or uncertain amount of impurities and their direct or indirect role in radical forming reactions with or without involvement of monomer molecules.

The radical generation or chain initiation in self-initiated polymerization of pure styrene is thought to be due to thermal activation of the monomer on heating leading to its decomposition into radicals, the exact mechanism of which is yet uncertain. The rate of polymerization is of second order in the monomer (styrene) and the rate is not much influenced by the use of solvent of different nature to provide a homogeneous system.[27–31] Second order dependence of R_p on styrene concentration also indicates the rate of initiation, R_i to be of second order in styrene in view of the relationship given by Eq. (3.22). Initiation rates of spontaneous thermal polymerization of styrene are far smaller than those for polymerization in the presence of an initiator. Bimolecular initiation mechanism resulting in the formation of a biradical shown in reaction scheme (3.28) explains the experimental observations reasonably well

$$2CH_2 = CHX \xrightarrow{k_i} {}^{\bullet}CH_2{-}CH_2{-}\underset{|}{\overset{X}{C}}H{-}\ldots \quad \text{(with X on both carbons)}$$

$$2CH_2 = CHX \xrightarrow{k_i} {}^{\bullet}\overset{X}{\overset{|}{C}}H_2{-}CH_2{-}\overset{X}{\overset{|}{C}}H^{\bullet} \tag{3.28}$$

The rate of initiation R_i can then be expressed as:

$$R_i = 2k_i\,[M]^2 \tag{3.28a}$$

Considering bimolecular termination giving rate of termination, $R_t = 2k_t\,[M^{\bullet}]^2$ shown in Eq. (3.16), and under steady-state assumption, giving $R_i = R_t$, the chain radical concentration is expressed as:

$$[M^{\bullet}] = (k_i/k_t)^{1/2}\quad [M] \tag{3.29}$$

The rate of polymerization, R_p is thus expressed as:

$$R_p = k_p\,[M^{\bullet}]\,[M] = k_p(k_i/k_t)^{1/2}\,[M]^2 \tag{3.30}$$

which is in agreement with the general experimental observations.

The bimolecular biradical mechanism is subject to severe criticism and objections in view of the biradicals being highly vulnerable to cyclization and in view of unsuccessful attempts to induce polymerization by biradicals formed by decomposition of cyclic peroxy, disulphide and azo compounds.[32] Alternatively, the bimolecular mode of chain initiation forming a pair of monoradicals shown below appears to be more sound and acceptable.[33]

$$2CH_2 = CHX \xrightarrow{k_i} CH_3{-}\dot{C}HX + CH_2 = \dot{C}X \tag{3.31}$$

3.10 Photo-Initiation of Polymerization

Photochemical reactions can also lead to initiation of free radical polymerization[34,35]. Photo-initiation may be considered under three broad categories: (i) Uncatalyzed, (ii) Catalyzed, and (iii) Sensitized photo-initiation.

(a) *Uncatalyzed Photopolymerization* On photo-activation, the monomer M may generate excited species M*, by absorbing light quanta of specific wavelength and subsequently decompose into radicals by homolysis or related mechanism which then contribute to chain initiation:

$$M \xrightarrow{h\nu} M^* \tag{3.32}$$

$$M^* \longrightarrow R^{\bullet} + R'^{\bullet} \tag{3.33}$$

Just as in purely thermal initiation, so also in uncatalyzed photo-initiation the exact identity of the primary radicals as in Eq. (3.33) is not yet established with any degree of certainty.

The rate of photochemical initiation may be expressed as:

$$R_i = 2\phi I_a \tag{3.34}$$

simply by replacing k_d [I] of Eq. (3.15) with the intensity of active radiation absorbed, I_a, and replacing f with ϕ for the photochemical polymerization where ϕ is interpreted as the quantum yield for chain initiation or radical generation which actually stands for the number of pairs of chain radicals generated per quantum of light absorbed.

Assuming, for simplicity, that the incident light intensity does not measurably vary with thickness of the reaction mass, I_a will be proportional to the product of the intensity of incident light, I_0 and the monomer concentration, i.e.

$$I_a = \varepsilon I_0 \text{ [M]} \tag{3.35}$$

combining Eqs (3.34), (3.35) and (3.22), one may write for uncatalyzed photo-polymerization:

$$R_p = (k_p/k_t^{1/2})\ (\phi\varepsilon I_0)^{1/2}\ [\text{M}]^{3/2} \tag{3.36}$$

Here, ε is the molar absorption coefficient for the active radiation.

(b) ***Catalyzed Photopolymerization*** In the case of photopolymerization using a photo-active additive as a catalyst or initiator in the monomer, the photo-initiator readily undergoes photolysis to generate radicals which then initiate the chain polymerization. Using a simple carbonyl compound such as a ketone of the general formula **R—(C : O)—R**, the radical generation reaction showing photolysis of the ketone may be written as:

$$\text{R—}\overset{\overset{\displaystyle \text{O}}{\|}}{\text{C}}\text{—R} \rightarrow \text{R—}\overset{\overset{\displaystyle \text{O}}{\|}}{\text{C}}{}^{\bullet} + \text{R}^{\bullet} \tag{3.37}$$

Many thermal initiators which produce specific radicals on thermal homolytic cleavage of specific bonds also undergo photolytic decompositions to produce the same radicals. Photoactivation allows the use of a wide range of chemicals as polymerization initiators in comparison with the thermal catalyzed process. This is consequent to the higher selectivity of photolytic homolysis. For many compounds other than the more common initiators mentioned earlier in this chapter, thermal homolysis occurs at too high a temperature and generally leads to the generation of a wide spectrum of radicals as various chemical bonds break up randomly. A few useful photo-initiator systems and the radicals that they generate on photolysis are described in Table 3.2.

The rate of initiation using a photo-initiator, I, is given by:

$$R_i = 2\phi \varepsilon I_0[I] \tag{3.38}$$

Hence, under steady-state and assuming bimolecular termination, $[M^\bullet] = (\phi \varepsilon I_0/k_t)^{1/2}$ $[I]^{1/2}$, the rate of related photopolymerization is given by

$$R_p = \frac{k_p}{k_t^{1/2}} (\phi \varepsilon I_0)^{1/2}[I]^{1/2}[M] \tag{3.39}$$

(c) *Photopolymerization using a Photosensitizer* Photosensitizers are generally employed in order to bring about effective homolysis of monomers or initiators which otherwise do not undergo sufficient photoexcitation at the frequency of light made available to the system. The photosensitizer (Z), because of its being highly photosensitive, readily gets excited on exposure to light to Z*

$$Z \xrightarrow{h\nu} Z^* \tag{3.40}$$

which then carries the energy absorbed to the initiator or monomer molecule and transfers energy to form excited initiator or monomer species:

$$Z^* + I \text{ (or M)} \Rightarrow Z + I^* \text{ (or M}^*) \tag{3.41}$$

The excited species, thus formed, undergo homolysis to produce chain-initiating radicals. Aromatic ketones such as benzophenone is one of the most widely used photosensitizers in organic photochemistry and it has also been used by many workers as a sensitizer in many uncatalyzed or catalyzed photo (vinyl) polymerizations.[36–42]

3.11 Polymerization Initiated by High Energy Radiations

High energy radiations including particular radiations (α-particles, neutrons, electrons or β-rays) and electromagnetic radiations (X-rays and γ- rays) readily interact with materials of different chemical structures in a manner much more complicated than that produced by exposure of these materials to visible or UV light[43].

Photolysis usually generates radical reactions through homolysis of bonds or through electron transfer reactions involving opposite ions or ions and neutral species. On radiolysis or exposing the material systems with radioactive emanations or high energy radiations, molecular excitations and subsequent formation of radicals may take place in much the same manner as in the case of photolysis, though probably at a much faster rate. As a consequence of the higher energies of these radiations, the substrate species S may suffer ionization by ejection of electrons.

$$S \xrightarrow[\text{radiation}]{\text{high energy}} S^+ + e^- \tag{3.42}$$

A series of reactions involving splitting of the cations, absorption of electrons by cationic or neutral species to form radicals or anions and anion splitting into radical and electron may then follow:

$$S^+ \longrightarrow R_1^{\bullet} + B^+ \tag{3.43}$$

$$B^+ + e^- \longrightarrow B^{\bullet} \tag{3.44}$$

$$S + e^- \longrightarrow S^- \tag{3.45}$$

$$S^- \longrightarrow R^{\bullet}_2 + A^- \tag{3.46}$$

$$A^- \longrightarrow A^{\bullet} + e^- \tag{3.47}$$

Thus, on radiolysis of a vinylic or olefinic monomer, a host of reactive and transient species including free radicals, anions, cations and electrons are likely to be formed into the system and the polymerization may ensue by radical or ionic mechanism or combinations thereof, depending on prevailing conditions, including temperature of polymerization. However, radiation-induced polymerization takes place by and large by radical mechanisms and ionic propagation assumes significance only at low temperatures. Kinetically, the radiation-induced polymerization with or without added catalysts has characteristics similar to those applicable to photopolymerization when done at ambient or higher temperatures.

3.12 Evaluation of the Kinetic Parameter, k_p^2/k_t

Equation (3.22) may be rearranged to give:

$$\frac{k_p^2}{k_t} = \frac{2R_p^2}{(R_i)[M]^2} \tag{3.48}$$

R_i, which is equal to $2fk_d[I]$, may be evaluated from the rate of decomposition of the initiator or from a knowledge of k_d supplemented by a measure of the initiator efficiency, f. Experimental determination of R_p and a knowledge of the monomer concentration used, allow calculation of the kinetic parameter k_p^2/k_t in litre mol^{-1} s^{-1}. The parameter may also be conveniently evaluated from graphical treatment of degree of polymerization ($\overline{X}_n$) and R_p data as discussed later in Sec. 3.15.

Using a given initiator I at a specified concentration and a fixed monomer concentration, [M] at a given temperature for different monomers, different R_p values are observed because of different k_p^2/k_t values; higher k_p^2/k_t value produces a higher rate of polymerization. Comparison of k_p^2/k_t values for different monomers under comparable conditions of polymerization gives a relative idea of the ease of polymerizability of the monomers by free radical mechanism. Measurements of k_p^2/k_t as a function of temperature of polymerization furnish useful data about the thermochemistry of polymerization.

3.13 Energetic Characteristics of Free Radical Polymerization

The examination of Eq. (3.21) clearly indicates that R_p depends on a combination of rate constants, k_d, k_p and k_t and the quantitative effect of temperature on R_p is rather complex. On treatment of the rate Eq. (3.21) in accordance with the Arrhenius concept, one obtains:

$$\ln R_p = \ln\left[A_p\left(\frac{A_d}{A_t}\right)^{1/2}\right] + \ln\left[(f[\text{I}])^{1/2}[\text{M}]\right] - \frac{[E_p + (E_d - E_t)/2]}{RT} \quad (3.49)$$

The apparent energy of activation, E_a for the overall rate of polymerization, R_p is given by

$$E_a = E_p + (E_d - E_t)/2 \quad (3.50)$$

A plot of ln R_p *vs* $1/T$, where T is the absolute temperature (K) allows evaluation of the overall frequency factor $A_p(A_d/A_t)^{1/2}$ and the apparent activation energy (E_a) from the intercept and slope of the plot. For photochemical polymerization, considering rate of photoinitiation to be independent of temperature, the increase in R_p with gain in temperature must be entirely due to the overall change in ($k_p/k_t^{1/2}$). The slope of the plot of ln R_p *vs* $1/T$ for photopolymerization would yield the apparent activation energy,

$$E_a = E_p - E_t/2$$

For most commonly used free radical initiators such as the peroxides, hydroperoxides or azo compounds, the activation energy for initiator decomposition, E_d is about 30–40 kcal/mol while the E_p and E_t values for most of the monomers are in the range of 4–10 kcal/mol and 1–5 kcal/mol respectively[44]. The apparent activation energy, E_a for overall polymerization is therefore about 20 kcal /mol in most cases. For photochemical polymerization, the value of E_a, ($E_a = E_p - E_t/2$) ranges between 3–6 kcal/mol.[34,41,45]

The activation energy for thermal (uncatalyzed) polymerization is of the same order as for thermal catalyzed polymerization using lone initiators such as Bz_2O_2 or AIBN. Very slow rates of purely thermal polymerization arise as a consequence of extremely low frequency factors (10^4–10^6) associated with the process. For redox polymerization, the radical generation reaction is characterized by an E_d value of only about 10–15 kcal/mol, so that the E_a value for the overall polymerization is close to 10 kcal/mol.

3.14 Chain Length and Degree or Polymerization

It is relevant at this point to examine if the chain length or degree of polymerization, $\overline{X}_n$ of a polymer product could be predicted or calculated from measurement of rate of polymerization with a knowledge of kinetic constants. The number average degree of polymerization is given by the average number of monomer molecules consumed per polymer molecule. For evaluation of the $\overline{X}_n$ from the kinetics of polymerization, it is necessary to define and find an expression for the kinetic chain length, ν.

3.14.1 Kinetic Chain Length

The kinetic chain length, ν represents the average number of monomer molecules consumed by a given active centre from its initiation to its termination. The kinetic chain length is viewed as the ratio of the rate of propagation, R_p to the rate of initiation, R_i or to the rate of termination, R_t, since $R_i = R_t$ under the steady state condition. Thus

$$\nu = R_p/R_i = R_p/R_t \tag{3.51}$$

combining Eqs (3.17) and (3.16), we get

$$\nu = \frac{k_p}{2k_t} \cdot \frac{[M]}{[M^\bullet]} \tag{3.52}$$

Eliminating the radical concentration term $[M^\bullet]$ with the help of Eq. (3.17)

$$\nu = \frac{k_p^2}{2k_t} \cdot \frac{[M]^2}{R_p} \tag{3.53}$$

Equation (3.53) is an important expression for free radical polymerization. It is applicable for all cases of bimolecular termination irrespective of the exact mechanism (combination or disproportionation) and also irrespective of the nature of the initiation process. The kinetic chain length is inversely proportional to chain radical concentration, Eq. (3.52), and hence it is also inversely proportional to the rate of polymerization, Eq. (3.53).

For polymerization initiated through radicals generated by thermal decomposition of an initiator, I, Eqs (3.22) and (3.15) may be combined to give an alternative expression for ν:

$$\nu = \left[\frac{k_p}{2(fk_d k_t)^{1/2}}\right] \cdot \frac{[M]}{[I]^{1/2}} \tag{3.54}$$

and for purely thermal polymerization, an alternative expression for ν may be obtained by combining Eqs (3.30) and (3.28a):

$$\nu = k_p/2(k_i k_t)^{1/2} \tag{3.55}$$

3.14.2 Kinetic Chain Length and Degree of Polymerization

In the absence of side reactions or reactions other than those considered so far, the number average degree of polymerization, $\overline{X}_n$ will have a direct relation with the kinetic chain length, ν. For termination by combination written in Eq. (3.56),

$$\begin{aligned} &R{-}(CH_2{-}CHX)_{n-1}{-}CH_2{-}\dot{C}HX + \dot{C}HX{-}CH_2{-}(CHX{-}CH_2)_{m-1}{-}R \\ &\longrightarrow R{-}(CH_2{-}CHX)_n{-}(CHX{-}CH_2)_m{-}R \end{aligned} \tag{3.56}$$

which was also briefly expressed by Eq. (3.6), the product polymer molecule having a degree of polymerization $\overline{X}_n = (n + m)$ is made up of two kinetic chains such that the kinetic chain length $\nu = (m + n)/2$, thus giving

$$\overline{X}_n = 2\nu \tag{3.57}$$

For termination by disproportionation written in Eq. (3.58),

$$\begin{aligned} &R{-}(CH_2{-}CHX)_{n-1}{-}CH_2 - \dot{C}HX + R{-}(CH_2 - CHX)_{m-1}{-}CH_2 \\ &{-}\dot{C}HX \rightarrow R{-}(CH_2 - CHX)_{n-1} - CH_2 - CH_2X + R{-}(CH_2 - CHX)_{m-1} \\ &{-}CH = CHX \end{aligned} \tag{3.58}$$

which was also briefly expressed by Eq. (3.7), the two product polymer molecules giving an average degree of polymerization $\overline{X}_n = (n + m)/2$, are also made up of two kinetic chains such that the value of $\nu = (m + n)/2$, and hence

$$\overline{X}_n = \nu \tag{3.59}$$

Thus, if the termination mechanism (combination or disproportionation) is known with certainty, the measurements of R_p and $\overline{X}_n$, and a knowledge of the parameter (k_p^2/k_t) readily allow calculation of the kinetic chain length, ν.

In case termination occurs simultaneously by combination and disproportionation the kinetic chain length ν can be calculated from the degree of polymerization $\overline{X}_n$ according to the following considerations.

If the average number of initiator fragments appearing as end groups per polymer molecule is n, then the fraction of polymer molecules having two such end groups per chain is $(n - 1)$ and the fraction of those with only one end group per chain is $(2 - n)$. Considering that only one molecule is formed on combination against two molecules formed on disproportionation of a pair of radicals, the relative number of radicals terminated by combination and disproportionation are $2(n - 1)$ and $(2 - n)$ respectively. Therefore, the fraction of radicals terminated by combination and by disproportionation will be $2(n - 1)/n$ and $(2 - n)/n$ respectively. If, further, the ratio of incidences of termination reaction by combination and by disproportionation is given by y,

then,

$$y = \frac{2(n-1)}{(2-n)} \tag{3.60}$$

which may be rearranged as

$$n = \frac{2(y+1)}{(y+2)} \tag{3.60a}$$

The number average degree of polymerization $\overline{X}_n$ will then be related to the kinetic chain length ν by the relation:

$$\overline{X}_n = \frac{2(y+1)}{(y+2)}\nu \tag{3.61}$$

Regardless of the proportionate significance of disproportionation and combination, in view of direct proportionality between $\overline{X}_n$ and ν and in accordance with Eq. (3.53), $\overline{X}_n$ should increase with $[M]^2$ and vary inversely with R_p. This is true in a general manner, but in many systems, the degree of polymerization is measurably lower[46] than expected on the basis of the general relationship given by Eq. (3.61).

Similar lowering of $\overline{X}_n$ or molecular weight occurs more commonly in presence of many solvents.

Combining Eqs (3.53) and (3.61) one obtains:

$$\overline{X}_n = \frac{(y+1)}{(y+2)} \cdot \frac{k_p^2}{k_t} \cdot \frac{[M]^2}{R_p} \tag{3.62}$$

Figure 3.4 shows plots of $1/\overline{X}_n$ *vs* R_p for polymerization of undiluted methyl methacrylate at 60°C using *t*-butyl hydroperoxide (TBHP), cumene hydroperoxide (CHP), benzoyl peroxide (Bz_2O_2) and azobisisobutyronitrile (AIBN) as initiators, the data having been obtained by carrying the polymerization[46] to low conversions. For Bz_2O_2 and AIBN systems, the expected direct proportionality according to Eq. (3.62) was observed. But with the hydroperoxides as initiators, a sharp deviation from the linear relationship is clearly apparent from the plot. At comparable rates of polymerization, $1/\overline{X}_n$ (giving a measure of the number of polymer molecules formed per monomer unit polymerized) is much higher indicating that for the hydroperoxide initiated systems, $\overline{X}_n$ is not proportional to v, and that additional processes, not taken into consideration so far, contribute to measurable increase in the number of polymer molecules formed.

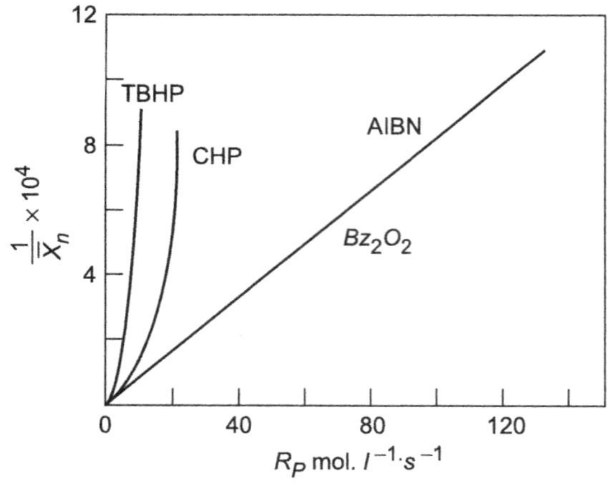

Fig. 3.4 ***Plot of reciprocal of the average degree of polymerization (1/$\overline{X}_n$) vs rate of polymerization, R_p for polymerization of undiluted methyl methacrylate at 60°C using azobisisobutyronitrile (AIBN), benzoyl peroxide (Bz_2O_2), cumene hydroperoxide (CHP) and t-butyl hydroperoxide (t-BHP) as initiators[46] (Baysal & Tobolsky, 1952; Courtesy, Wiley-Interscience, New York)***

3.15 Chain Transfer

The lack of correlation between $\overline{X}_n$ and R_p and hence between $\overline{X}_n$ and v cannot be regarded as a mere exception. Wide studies and experience indicate that this feature relevant to polymerization in general arises due to occurrence of certain reactions limiting the growth of polymer molecules without reducing the number of active centres of comparable reactivity and hence leaving the rate of polymerization practically unaffected. The monomer (M), initiator (I), solvent (S) or other additives present in the polymerization system or even the product polymer being formed may terminate a growing radical rather prematurely through transfer of a labile atom (hydrogen or other atom) to it, and consequently, the reacting molecule acquires a radical site which normally starts a new chain and then continues to grow at the same rate. The reactions are commonly known as chain transfer reactions which can be generally depicted as:

$$M^{\bullet} + XA \xrightarrow{k_{tr}} MX + A^{\bullet} \tag{3.63}$$

where, $M^{\bullet}$ is the growing radical and XA may be the monomer, initiator, solvent or any other substance and X is the atom or species transferred.

The rate of chain transfer reaction R_{tr} is given by

$$R_{tr} = k_{tr}[M^{\bullet}] \cdot [XA] \tag{3.64}$$

where k_{tr} is the rate constant for the chain transfer reaction. The new substrate radical $A^{\bullet}$ produced in the transfer reaction then adds to a monomer molecule to start a new chain

$$A^{\bullet} + M \xrightarrow{k_i'} M^{\bullet} \tag{3.65}$$

The unmistakable consequence of chain transfer reaction is a lowering of the $\overline{X}_n$ value as long as the chain transfer agent is a monomer, solvent, initiator or some other small compound. Chain transfer with the polymer formed may often lead to branching and to an overall increase in $\overline{X}_n$. The effect of chain transfer on R_p depends on the value of the reinitiation rate constant k_i' relative to the propagation rate constant k_p. For $k_i' \ll k_p$, XA would have the role of an inhibitor or retarder rather than that of a chain transfer agent, and R_p would be adversely affected as a consequence.

Considering Eqs (3.63) and (3.65), it becomes obvious that each incidence of a chain transfer reaction produces an extra pair of chain ends in the system. Assuming that the termination occurs exclusively by combination, the only other process from which chain ends originate is the initiation process. The average degree of polymerization can now be redefined from the kinetic viewpoint as the ratio of the overall rate of polymerization, R_p to the total rate of production of pairs of chain ends, i.e.

$$\overline{X}_n = \frac{R_p}{fk_d[I] + k_{tr,M}[M^\bullet][M] + k_{tr,s}[M^\bullet][S] + k_{tr,I}[M^\bullet][I]} \tag{3.66}$$

The first term in the denominator is also equal to $R_t/2$, i.e., $k_t[M^\bullet]^2$; the terms $k_{tr,M}$, $k_{tr,S}$ and $k_{tr,I}$ are the rate constants of chain transfer reaction with monomer, solvent and initiator respectively. The chain transfer constant C for a transfer agent is defined as the ratio of the rate constant k_{tr} for the chain transfer reaction to the rate constant k_p for the propagation reaction. The chain transfer constants for the monomer (C_M), solvent (C_S) and initiator (C_I) are expressed as:

$$C_M = \frac{k_{tr,M}}{k_p}; \quad C_S = \frac{k_{tr,S}}{k_p}; \quad C_I = \frac{k_{tr,I}}{k_p} \tag{3.67}$$

Thus, combining Eqs (3.17), (3.66) and (3.67)

$$\frac{1}{\overline{X}_n} = \frac{fk_d[I]}{R_p} + C_M + C_S \cdot \frac{[S]}{[M]} + C_I \cdot \frac{[I]}{[M]} \tag{3.68}$$

which may also be expressed alternatively as:

$$\frac{1}{\overline{X}_n} = \frac{k_t}{k_p^2} \cdot \frac{R_p}{[M]^2} + C_M + C_I \frac{[I]}{[M]} + C_S \frac{[S]}{[M]} \tag{3.69}$$

Equation (3.69), applicable to a case of termination exclusively by combination will assume the following expression of general applicability, taking Eq. (3.62) into consideration:

$$\frac{1}{\overline{X}_n} = \frac{(y+2)}{(y+1)} \cdot \frac{k_t}{k_p^2} \cdot \frac{R_p}{[M]^2} + C_M + C_I \frac{[I]}{[M]} + C_S \frac{[S]}{[M]} \tag{3.70}$$

where, y is the ratio of termination by combination to that by disproportionation; in the absence of all chain transfer reactions, Eq. (3.70) simply reduces to Eq. (3.62).

3.15.1 Chain Transfer to Monomer and Initiator

In the absence of a solvent and any other additive in a catalyzed thermal polymerization, Eq. (3.70) changes to

$$\frac{1}{\overline{X}_n} = \frac{(y+2)}{(y+1)} \cdot \frac{k_t}{k_p^2} \cdot \frac{R_p}{[M]^2} + C_M + C_I \frac{[I]}{[M]} \tag{3.71}$$

which may be conveniently rearranged to

$$\left[\frac{1}{\overline{X}_n} - \frac{(y+2)}{(y+1)} \cdot \frac{k_t}{k_p^2} \cdot \frac{R_p}{[M]^2}\right] = C_M + C_I \frac{[I]}{[M]} \tag{3.72}$$

For polymerization at a given temperature to a low conversion and in the absence of initiator transfer or at low R_p, Eq. (3.71) permits a plot of $(1/\overline{X}_n)$ *vs* R_p or $R_p/[M]^2$ which would give a straight line, from the slope of which the value of the parameter k_p^2/k_t may be calculated. In the absence of any monomer transfer, the linear plot would pass through the origin leaving no intercept on the $(1/\overline{X}_n)$ axis. For significant occurrence of monomer transfer the intercept on the $(1/\overline{X}_n)$ axis gives a direct measure of the monomer transfer constant, C_M. In case initiator transfer is significant in the undiluted polymerization, Eq. (3.72) permits a plot of left hand side of the equation *vs* [I]/[M] which would give a straight line, the slope of which gives a direct measure of the initiator transfer constant, C_I and the intercept gives a direct measure of C_M.

The monomer transfer constants are usually small for most monomers (Table 3.3). The initiator transfer constant for several initiator systems are shown in Table 3.4. Even if the C_I value for a given initiator is high, its effect on the polymer molecular weight may not be very pronounced in view of the fact that it is the product of C_I and ([I]/[M]) and not C_I only that affects $\overline{X}_n$. In most polymerization, the initiator concentrations used are usually very low (10^{-4} – 10^{-2} molar) giving still lower values of [I]/[M], as a consequence.

Table 3.3 *Some values of chain (monomer) transfer constants (C_M)*[44]

Monomer	*Temperature, °C*	$C_M \times 10^4$
Acrylamide	60	0.6
Acrylonitrile	60	0.26–0.3
	40	0.17
Ethyl methacrylate	60	0.259
Methacrylonitrile	60	5.81
	80	10.05
Methyl acrylate	60	0.036–0.325
	75	0.22–0.25

Contd.

Table 3.3 *(Contd.)*

Monomer	*Temperature, °C*	$C_M \times 10^4$
Methyl methacrylate	60	0.07–18
	75	0.27–0.70
Styrene	60	0.6–1.1
Vinyl acetate	60	1.75–2.8
Vinyl chloride	60	10.8–12.8

Table 3.4 *Some values of chain (initiator) transfer constants (C_I)*[44]

Initiator	*Temperature, °C*	$C_I \times 10^4$ *for the polymerization of*	
		Styrene	*Methylmethacrylate*
t-Butyl hydroperoxide (TBHP)	60	350	12,700
Cumyl hydroperoxide (CHP)	60	630	3,300
Benzoyl peroxide (Bz_2O_2)	60	480–550	200
t-Butyl peroxide (TBP)	60	3–13	—
Hydrogen peroxide (H_2O_2)	40[94]	—	8
Hydrogen peroxide (H_2O_2)	60	—	460
Azo-bisisobutyronitrile (AIBN)	60	0	0
Sulphur dioxide	40[65]	—	16,000
Quinoline–bromine complex	40[95]	—	340

3.15.2 Solvent Transfer

In the presence of solvents or similar additives (S), the last term in Eq. (3.70) may assume significance, provided both C_S and [S]/[M] values are significant. By proper choice of polymerization conditions and carrying the polymerization at a given temperature to low conversions, the value of C_S, the solvent transfer constant for different solvents or chain transfer agents can be determined from a knowledge of $\overline{X}_n$ under different conditions of [S]/[M] in each case. Using an initiator of a very low or negligible C_I value and if necessary, by adjusting the initiator concentration with variation of [S] so as to keep $(R_p/[M]^2)$ effectively constant, Eq. (3.70) may be written in the form:

$$\frac{1}{\overline{X}_n} = \left(\frac{1}{\overline{X}_n}\right)_0 + C_S \frac{[S]}{[M]} \tag{3.73}$$

Here, $(1/\overline{X}_n)_0$ is the reciprocal degree of polymerization in the absence of the solvent or the chain transfer agent, and it combines the values of the first three terms on the right hand side of Eq. (3.70). The slope of the linear plot of $(1/\bar{X}_n)$ *vs* [S]/[M] gives the measure of C_S. Figure 3.5 shows such plots for several aromatic solvents as chain transfer agents in the polymerization of styrene. The plots shown in Fig. 3.5 clearly indicate dependence of chain transfer constants on the chemical structure of the chain transfer agents.

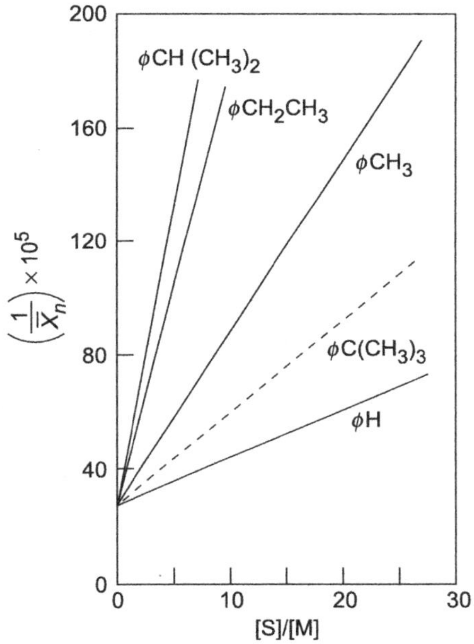

Fig. 3.5 *Chain transfer effect: Plot of* $(1/\bar{X}_n)$ vs *([S]/[M]) for several solvents (S) in the polymerization of styrene* **(M)**[31] *at 100°C (Gregg & Mayo, 1947:* **Courtesy,** *Royal Society of Chemistry, London)*

For uncertainty about constancy of $(R_p/[M]^2)$, a plot of

$$\{(1/\bar{X}_n) - [(y + 2)k_t R_p/(y + 1)k_p^2\ [M]^2\} \quad vs \quad [S]/[M]$$

would give the value of C_S from the slope of the linear plot.

3.15.3 Regulation or Control of Molecular Weight

Chain transfer reactions, initially viewed as additional reactions disturbing the normal features of radical polymerization, gradually came to be recognized as useful and often advantageous in limiting the growth of polymer chains by design and hence in controlling their molecular weight to desirable ranges, and also in widespread studies of structure–reactivity relationships in radical displacement reactions. The transfer constants for a number of solvents/additives for polymerization of styrene, methyl methacrylate and vinyl acetate are listed in Table 3.5. Strong C—H bonds in hydrocarbons show low transfer constant for them. In comparison with benzene the presence of progressively weak benzylic hydrogens in toluene, ethyl benzene and isopropyl benzene (the reactivity increases with increasing substitution at the α - carbon) leads to C_S values increasing in the order benzene < toluene < ethylbenzene < isopropyl benzene; the effect of a phenyl substituent is much more prominent than that of a methyl substituent. The C_S value for *t*-butyl benzene, however, drops to a low value because of its having no benzylic or α-hydrogens, thus rendering it relatively non-reactive despite nine available hydrogens (β-hydrogens) in the three methyl groups.

Table 3.5 *Some values of chain (solvent) transfer constants (C_S)*[44]

Transfer Agent (solvent/additive)	*$C_S \times 10^4$ at 60°C or as mentioned*		
	Styrene	*Methylmethacrylate*	*Vinyl acetate*
Benzene	0.023	0.04	1.2
Toluene	0.125	0.20	21.6
Ethylbenzene	0.67	1.38 (80°C)	55.2
Isopropyl benzene	0.82	1.9	89.9
t-Butyl benzene	0.06	0.074	3.61
Cyclohexane	0.024–0.063	12.0	7.0–100.0
Water	—	0	—
Chloroform	0.5	1.77	150
Carbon tetrachloride	84–148	0.93–5.0	7,300–10,000
Carbon tetrabromide	17,800–22,000	2,700	28,740(70°C)
n-Butyl chloride	0.04	1.20 (80°C)	10
n-Butyl bromide	0.06	—	50
n- Butyl iodide	1.85	—	800
Chlorine[47]	—	38	—
Bromine[47]	—	70	—
Iodine[47]	—	6,000	—
Pyridine[48]	4.0	0	12.5
n-Butyl amine	0.5	—	—
n-Butyl mercaptan	220,000	6700	480,000
t-Butyl mercaptan	37,000	—	—
Hydrogen sulphide[49]	19,900	810	—

Some compounds such as carbon tetrabromide, pentaphenyl ethane, and various mercaptans exhibit high chain transfer activity; in fact, they surpass the monomers, such as styrene, in reactivity toward the chain radical, thus giving C_S values greater than unity. Such compounds are particularly useful in controlling the degree of polymerization. Only a very small quantity of such a chain transfer agent is required to bring down the polymer molecular weight to a desired level. They are extensively used in commercial polymerizations particularly in emulsion polymerization systems to meet this objective and they are commonly known as (chain length) modifiers or regulators. C_S values for such regulators or modifiers are better obtained from evaluation of the rate of consumption of the transfer agent ($-d[S]/dt$) and the rate of monomer consumption ($—d[M]/dt$) so that one may write:

$$\frac{d[S]/dt}{d[M]/dt} = \frac{k_{tr,S}[S]}{k_p[M]} = C_S \frac{[S]}{[M]} \tag{3.74}$$

or, $$d \log [S]/d \log [M] = C_S \tag{3.75}$$

Hence, for such a system, the slope of the linear plot of log [S] *vs* log [M] passing through the origin would give the value of C_S.

3.15.4 Polymer Transfer

In the beginning of Sec. 3.15, mention was made about possible chain transfer reactions involving the polymer being formed but this specific feature of chain transfer was subsequently ignored in the expressions relating the effect of chain transfer reactions on $\overline{X}_n$, particularly for polymerizations done to low conversions in view of the fact that the effect of polymer transfer under this condition is negligible. A polymer transfer reaction and subsequent reinitiation reaction may be generally expressed as

$$\sim\!\!\sim M^{\bullet} + \sim\!\!\sim CH_2 - CHX \sim\!\!\sim \xrightarrow{k_{tr,p}} \sim\!\!\sim MH + \sim\!\!\sim CH_2 - \underset{\bullet}{C}X \sim\!\!\sim \tag{3.76}$$

$$\sim\!\!\sim CH_2 - \underset{\bullet}{C}X \sim\!\!\sim + n\,M \rightarrow \ldots \rightarrow \sim\!\!\sim CH_2 - \underset{\substack{| \\ (M)_{n-1} \\ | \\ M^{\bullet}}}{CX} \sim\!\!\sim \tag{3.77}$$

Transfer to polymer is, however, likely to be significant for practical polymerizations done to high or almost complete conversions. As mentioned earlier and as can be seen from reactions (3.76) and (3.77), polymer transfer leads to chain branching which normally makes the polymer less crystallizable, weaker in mechanical properties and less resistant to heat, solvents and chemicals.

The simple inclusion of an additional term, $C_P([P]/[M])$ in the right hand side of Eq. (3.70) [where C_P is the polymer transfer constant ($k_{tr,P}/k_P$) and $[P]/[M]$ is the molar ratio of the repeat units in the polymer to the monomer] cannot generally explain the overall effect of the polymer transfer process. Polymer transfer may also lead to limited cross-linking through occasional coupling of two branch radical units. If both the monomer transfer constant and the respective polymer transfer constant are relatively high, such as in the case of vinyl acetate, polymerization results in extensive branching, particularly in the high range of conversion.

As second kind of polymer transfer involves intramolecular chain transfer leading to many short branches in a given polymer molecule (as many as 30–50 branches per 1000 carbon atoms in the main chain), most vividly exemplified by the high pressure radical polymerization of ethylene yielding what is known as the low density polyethylene. Short branches such as butyl and ethyl branches arise as a consequence of the so-called "back biting" mechanism:

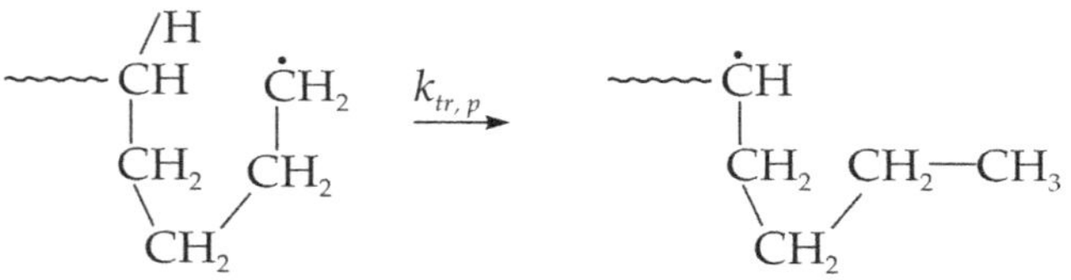

(3.78)

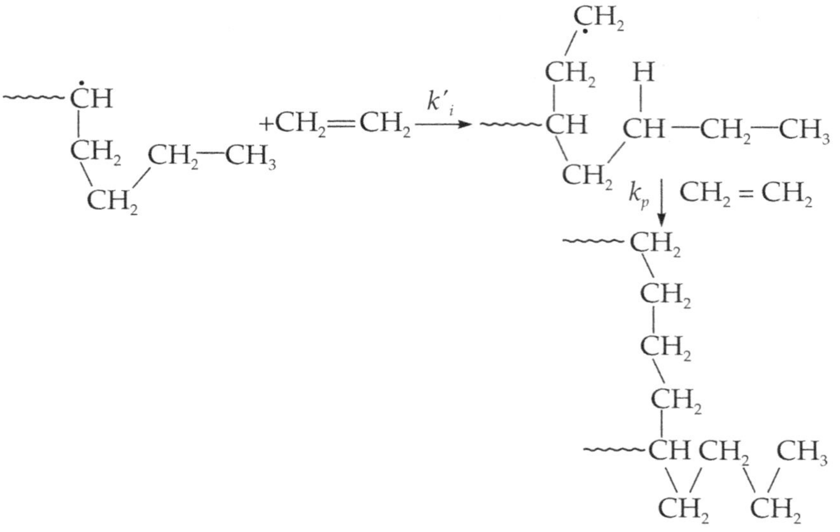

(3.78a)

(butyl branch) (3.78b)

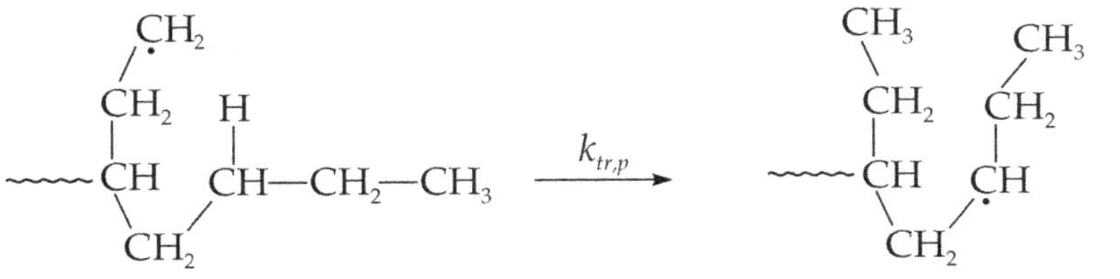

(ethyl branch)

(3.78(c))

3.16 Inhibition and Retardation

There are certain substances which when present in the monomer or the polymerization system, suppress the chain growth process to different degrees, usually by reacting with the initiating radicals or with the chain radicals (before they could grow to polymeric size of any consequence) and converting them to non-radical species or inactive radicals or radicals too low in reactivity to undergo meaningful propagation. Such substances or radical scavengers are classified as inhibitors if they effectively consume every radical generated in the system and completely suppress the polymerization and reduce the rate of polymerization substantially to zero. If, on the other hand, they act not as effectively or much less efficiently so that both the rate of polymerization as also the degree of polymerization are reduced quite measurably, without completely halting the propagation process, they are classified as retarders.

The inhibitors or retarders get consumed by reacting with radicals during the inhibition or retardation process. The inhibition or total suppression of polymerization continues till all the inhibitor molecules present are consumed. At the end of what is known as the inhibition or induction period, polymerization proceeds at a rate comparable to what is observed in the absence of the inhibitor. Examples of typical inhibitors are benzoquinone, *t*-butyl catechol, oxygen and the stable free radical diphenylpicrylhydrazyl (DPPH). One can observe the inhibition or retardation processes using very small concentration (0.001–01%) of the inhibitor or retarder in the polymerization systems.[49–59] Figure 3.6 shows[57] inhibition effect with 0.1% benzoquinone (curve 2), and retardation effect in the presence of 0.5% nitrobenzene as a typical retarder (curve 3). Nitrosobenzene (0.2%) produces a complex behaviour (curve 4), showing an inhibition period followed by polymerization at a rate lower than that for the pure monomer (curve 1). It is evident, therefore, that the product arising from reaction between chain radicals and nitrosobenzene primarily behave as a retarder.

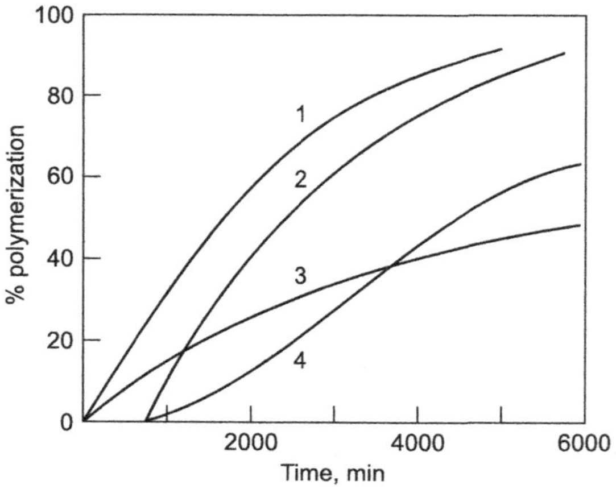

Fig. 3.6 *Thermal polymerization of styrene at 100°C under different conditions*[57]*: Curve 1, pure monomer; Curve 2, in presence of 0.1% benzoquinone, Curve 3, in presence of 0.5% nitrobenzene, and Curve 4, in presence of 0.2% nitrosobenzene*[57] *(Schulz, 1947;* **Courtesy,** *VCH Verlagsgesellschaft, Weinheim)*

Among the various stable free radicals, DPPH is the only one which is of practical utility as an inhibitor of polymerization. It is too stable to act as a chain initiator but still has the capacity to react with radicals efficiently to act as a radical terminator or inhibitor [*See* Sec. 3.7.2 and Eq. (3.26)].

The most widely used inhibitors are the quinones such as benzoquinone, duroquinone (tetramethyl benzoquinone) and chloranil (tetrachloro benzoquinone). For an ideal inhibition process, the rate of consumption of inhibitor is independent of inhibitor concentration and it is dependent on the rate of generation of radicals in the system on initiator decomposition or simply by a thermal process if the initiator is absent. The radical generation in either case is, however, independent of the inhibitor and thus, the inhibitor consumption is of the zero order. Under normal situations, the inhibition period is proportional to the initial concentration of the inhibitor.[49]

The mechanism of inhibition by the quinones is rather complex. The number of kinetic chains terminated per molecule of quinone consumed is not the same for all quinones. With benzoquinone as the inhibitor the stoichiometry is 1 : 1 for many monomers. The termination of a chain radical by a quinone molecule may involve addition to an oxygen atom or to a ring site of the quinone:

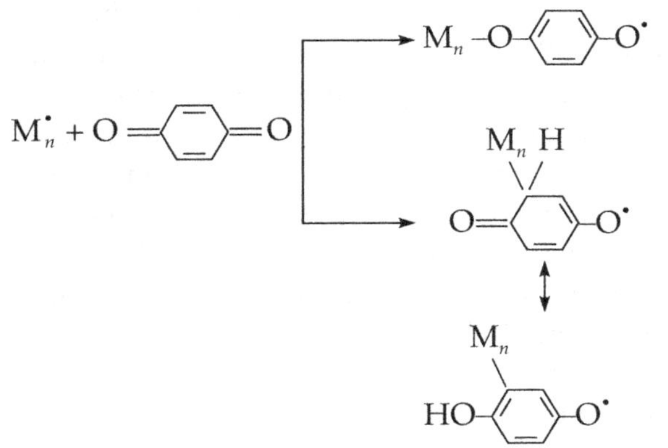

(3.79a)

(3.79a)

(Inactive radical from inhibitor)

The interaction between a chain radical and a quinone molecule may also involve disproportionation:

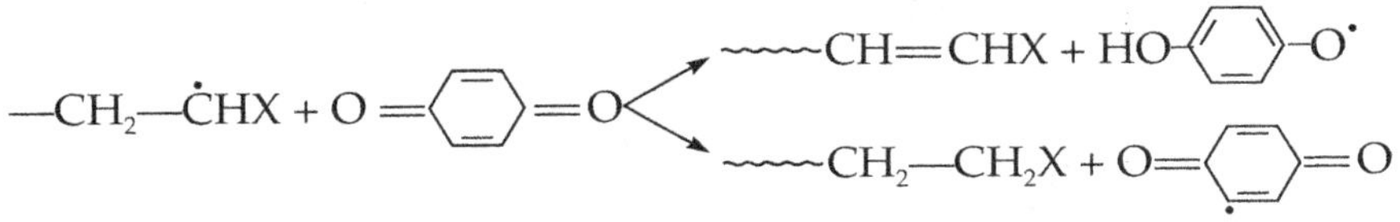

(3.80a)

(3.80b)

A pair of inhibitor radicals may either dimerize or disproportionate to form non-radical products or they may further react with other chain radicals. Only dimerization leads to 1 : 1 inhibitor stoichiometry while other two alternatives would produce a higher ratio (2 : 1). Both 1 : 1 and 2 : 1 stoichiometric ratios have been observed.

Hydroquinone and *t*-butyl catechol are also used as inhibitors, but their inhibitory action is prominent only in the presence of oxygen, as they become effective due to their oxidation to quinones.[53] In fact, hydroquinone is one of the identified products of quinone inhibition.[54]

The mechanism of inhibtory action, nature of the non-radical products formed through reactions between pairs of inhibitor radicals or between inhibitor radicals and chain radicals in the subsequent steps, sometimes even leading to copolymerization and the stoichiometric ratio referred to above, do not follow a uniform pattern for the same inhibitor and they largely depend on the nature of the monomer and the overall polymerization environment including the initiator, solvent or other additives.[16, 55–56]

Aromatic nitro-compounds also intercept chain radicals when present in a polymerization system, depressing the rate of polymerization, without causing its total suppression and often without exhibiting an induction period.[57–59] Their action in different monomer systems is also not uniform in pattern and observed effects. Their varied roles are exemplified by retardation for styrene polymerizaion[57], inhibition for vinyl acetate polymerization[33] and practically ineffective in either of these roles for methacrylate or acrylate polymerization.[60] Nitrobenzene, on the other hand, has also been used as an initiator component along with an amine in the studies of photopolymerization of methyl methacrylate,[61] where radical generation takes place by photolysis of a charge transfer complex intermediate formed between nitrobenzene (acceptor) and the amine (donor) used.

For retarder action, relatively high quantities of the nitroaromatic compounds are required to produce prominent reductions in the rate of polymerization, and the retardation effect continues throughout the polymerization. The radical displacement reaction involving nitrobenzene may take place by radical attack on either the aromatic ring on one instance or the nitro group on the other.

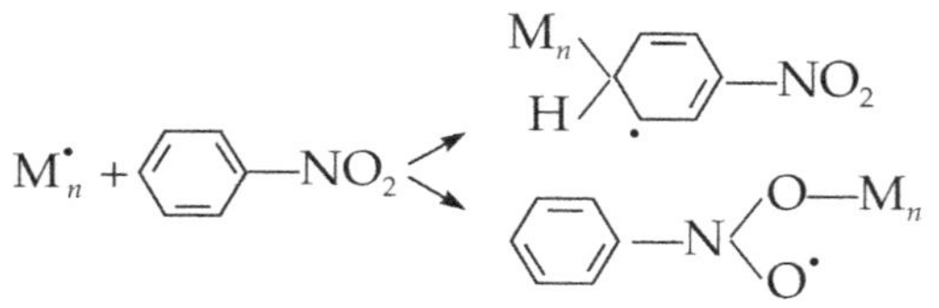

(3.81a)

(3.81b)

Oxygen is also a well-known inhibitor of polymerization.[62,63] Oxygen reacts with a chain radical to produce a peroxy radical of low reactivity in the first step. However, the peroxy radical slowly reacts with a monomer molecule to effect regeneration of the usual chain radical and the two processes may recur with rigorous alternation to produce a copolymer of a significant length of 1 : 1 sequence $(\text{—M—O—O—})_n$ but usually of low degrees of polymerization.[64]

$$M_x^{\bullet} + O_2 \xrightarrow{\text{very fast}} M_x\text{—O—O}^{\bullet} \qquad (3.82a)$$

$$M_x^{\bullet}\text{—O—O}^{\bullet} + M \xrightarrow{\text{very slow}} M_x\text{—O—O—M}^{\bullet} \qquad (3.82b)$$

It is interesting to note that the rate of polymerization usually observed at the end of the inhibition period due to oxygen often exceeds that for the uninhibited monomer under otherwise comparable conditions. The polyperoxides or related peroxy species formed appear to act as additional sources of free radicals. Thus, oxygen plays the role of an inhibitor, a comonomer and an initiator intermediate, i.e., an indirect initiator. Commercial high pressure polymerization of ethylene often utilizes oxygen as the initiator in less than a critical concentration. In this case, initiation results from generation of radicals by thermal decomposition of the peroxy radical or other peroxy species formed in the system under the reaction condition.

3.16.1 A Brief and Critical Review of Chain Transfer, Inhibition and Retardation

From analysis of the principles, mechanisms and observed effects, it becomes quite apparent that the differences among chain transfer agents, inhibitors, retarders and the like which may be present in the polymerization system either by incidence or by design are not very intrinsic; the differences are rather more of degree than of kind. In each case, a radical displacement reaction involving commonly a chain radical and the substrate (S) concerned and characterized by the rate constant, say k_{tr} takes place. The reacting chain radical species is replaced by a new radical species in the system as a consequence, while itself undergoing termination resulting in a dead species. The newly formed radical species may have reactivities of different orders toward reinitiation involving the monomer molecules (rate constant, k_i') and toward radical annihilation reaction through mutual interactions or through capture of other (chain) radicals present. By judging the relative values of k_p, k_{tr} and k_i' a rough classification of the substrates may be made. For $k_p \gg k_{tr}$, one may have k_i' either equal to or less than k_p. The former is commonly a case of normal chain transfer characteristically featuring little perturbation of R_p but measurable lowering in $\overline{X}_n$, the effect being more pronounced for higher (k_{tr}/k_p) or transfer constant (C) value. The latter, however, is a case of typical retardation causing lowering of both R_p and $\overline{X}_n$. For $k_p \ll k_{tr}$ also, one may have k_i' equal to or less than k_p and in either case, there is significant to sharp lowering in $\overline{X}_n$, and R_p may or may not suffer a change depending on the C, i.e. (k_{tr}/k_p) value. Very high C value ($C > 1$) would measurably lower R_p unless the [S]/[M] value is drastically low. For k'_i close or equal to zero, and a very high value of C, it is always a case of inhibition. Under other relative orders of the rate constants, a substrate behaves as a normal chain transfer agent for $k_i' \simeq k_p$, and it is only necessary to select an appropriate [S]/[M] range to see R_p remain practically unaffected. It is also not a very uncommon experience to see some substrates act as meaningful initiators of polymerization when used at relatively low concentrations and then also assume the role of effective retarders if not of inhibitors at progressively higher ranges of concentration or in different sets of condition, much as a consequence of effective degradative chain (initiator) transfer along with probably drastic fall in initiator efficiency.[1,65–71]

3.16.2 Kinetics of Inhibition

The kinetics of polymerization for a system containing an inhibitor can be analyzed considering the reaction scheme described by Eqs (3.3) to (3.7) and the additional reaction given in Eq. (3.83), where Z represents an inhibitor molecule:

$$M_n^{\bullet} + Z \xrightarrow{k_z} M_n + Z^{\bullet} \tag{3.83}$$

Considering, for the sake of simplicity, that reinitiation by the inhibitor radicals $Z^{\bullet}$ is negligible and that their termination does not lead to any regeneration of the original inhibitor molecule, the usual steady-state assumption for chain radicals gives

$$2k_t[M^{\bullet}]^2 + k_z[Z][M^{\bullet}] = R_i \tag{3.84}$$

Substituting the term $(R_p/k_p[M])$ for $[M^{\bullet}]$, we get

$$\frac{2k_t}{k_p^2} \cdot \frac{R_p^2}{[M]^2} + \frac{k_z}{k_p} \cdot \frac{[Z]}{[M]} \cdot R_p = R_i \tag{3.85}$$

or

$$\frac{2k_t}{k_p^2} \cdot \frac{R_p^2}{[M]^2} + C_Z \cdot \frac{[Z]}{[M]} \cdot R_p = R_i \tag{3.85a}$$

Equation (3.85) is useful[72] for kinetic analysis of the inhibition process and evaluation of the inhibitor constant C_z given by (k_z/k_p).

For uninhibited polymerization given by $[Z] \simeq 0$ or $k_z \simeq 0$, Eq. (3.85) reduces to Eq. (3.22) derived in Sec. (3.4.3) for such a system. But when $[Z] \neq 0$, and $(k_z/k_p) \gg 1$, severe retardation or a typical case of inhibition results and normal bimolecular termination of chain radicals recedes to insignificance such that

$$\frac{k_z}{k_p} \cdot \frac{[Z]}{[M]} \cdot R_p = R_i \tag{3.86}$$

In contrast to square root dependence given by Eq. (3.22), R_p for strongly retarded or inhibited polymerization is dependent on first power of R_i as shown by Eq (3.86).

The rate constant ratio or the inhibitor constant (k_z/k_p) can also be deduced from relative rates of reaction of the inhibitor and monomer with chain radicals. Thus,

$$-\frac{d[M]}{dt} \simeq k_p\,[M][M^{\bullet}]$$

and $$-\frac{d[Z]}{dt} \simeq k_z[Z][M^\bullet]$$

Equation (3.86a) then readily follows:

$$\frac{d[Z]}{d[M]} = \frac{k_z}{k_p} \cdot \frac{[Z]}{[M]} = C_Z \frac{[Z]}{[M]} \tag{3.86a}$$

For $C_Z \gg 1$, the polymerization or monomer consumption rate will be negligible for a significant concentration of the inhibitor or strong retarder. With progressive consumption of Z during the inhibition period, the chain propagation assumes a competitive position only when [Z] reduces to a very low or insignificant value. Integration of Eq. (3.86a) yields

$$\log ([Z]/[Z]_0) = C_Z \log ([M]/[M]_0) \tag{3.87}$$

where $[Z]_0$ and $[M]_0$ are the initial concentrations of Z and M.

For an effective inhibitor, C_Z being very large, it is evident from Eq. (3.87) that the inhibitor must almost be totally consumed before the monomer concentration drops measurably. A plot of log [Z] *vs* log [M] permits determination of the inhibitor constant, C_Z with data collected within the range showing measurable polymerization, i.e. in the range of transition between total inhibition and pick up of normal rate of polymerization. Value of C_Z for a number of inhibitors is listed in Table 3.6. It can be clearly seen that the inhibitor constants for a given substrate differ widely depending on the reactivity of the chain radical. Thus, nitrobenzene behaves as a strong chain transfer agent for the polymerization of methyl acrylate and as a retarder for styrene and vinyl acetate polymerization, while 1,3,5-trinitrobenzene acts as an inhibitor for vinyl acetate polymerization, as a strong retarder for styrene polymerizations and as a much less strong retarder for methyl acrylate polymerization. The reactivity is in the order: vinyl acetate radical $\gg$ styrene radical $\gg$ methyl acrylate radical.

3.17 Equilibrium in Chain Polymerization

In practice, the vast majority of polymerization reactions proceed under normal conditions to what may be termed as 'completion'. But polymerization, like many other chemical reactions must be considered as a reversible process. There are a number of monomeric compounds under normal conditions and many other monomers under special reaction conditions for which the polymerization process can only be understood in terms of equilibria. Most notable among them are highly

Table 3.6 *Some inhibitors and inhibitor constants (C_Z)*[44]

Inhibitor	*Temperature, °C*	*Monomer*	$C_Z = (k_z/(k_p)$
Nitrobenzene	50	Methyl acrylate	0.00464
	50	Styrene	0.326
	50	Vinyl acetate	11.2
m-Dinitrobenzene	50	Methyl acrylate	0.0309 (0.204)
(1,3,5-trinitrobenzene)*	50	Styrene	5.17 (64.2)
	50	Vinyl acetate	66.0 (404)
p-Benzoquinone	50	Acrylonitrile	0.91
	50	Styrene	518
	60	Methyl methacrylate	4.5
Tetrachlorobenzoquinone	44	Methyl methacrylate	0.26
(chloranil)			
	50	Styrene	2040
Oxygen	50	Methyl methacrylate	33000
	50	Styrene	14600
$FeCl_3$ in DMF solution	60	Acrylonitrile	3.33
	60	Methacrylonitrile	3.08
	60	Styrene	536
DPPH	44	Methyl methacrylate	2000

*C_Z values for 1,3,5-trinitrobenzene are given in parentheses.

substituted vinyl monomers, heterocyclic ring compounds such as trioxane, tetrahydrofuran and caprolactam, and elemental sulphur which exist as rings. Many ordinary monomers exhibit characteristic features of polymerization under extreme conditions of temperature and/or pressure which are understood and interpreted from an equilibrium viewpoint.

The impact of equilibrium condition is felt on rate and degree of polymerization and also on chain length distribution. The limiting conversion of monomer to polymer is also dependent on it. In the case of copolymerization, the equilibrium controlled chain propagation can affect the composition of the copolymer significantly.

3.17.1 Thermodynamic Considerations

Polymerizations, like all other reactions, are observed under conditions when they are thermodynamically possible and kinetically probable. For most polymerizations, the reversible character becomes apparent and detectable at and beyond some temperature. The propagation (forward) reaction in equilibrium with the depropagation (reverse) reaction may be written as

$$M_n^{\bullet} + M \underset{k_{dp}}{\overset{k_p}{\rightleftharpoons}} M_{n+1}^{\bullet} \quad (3.38)$$

The equilibrium point for the monomer polymer equilibrium is dependent on the temperature. The propagation reaction being exothermic, increase in temperature will result in the shifting of the equilibrium in favour of the reverse reaction. When the polymerization temperature is initially increased from a low level, R_p increases due to gain in the value of k_p. The effect of k_{dp} is not felt initially for many monomers because the value of k_{dp} is initially zero, but it becomes progressively significant with higher temperatures, Fig. 3.7. For such systems a critical temperature, called the "ceiling temperature", T_c is finally attained at which the propagation rate (R_p) and the depropagation rate (R_{dp}) are equal. Thus, at the ceiling temperature, the overall polymerization rate reduces to zero.

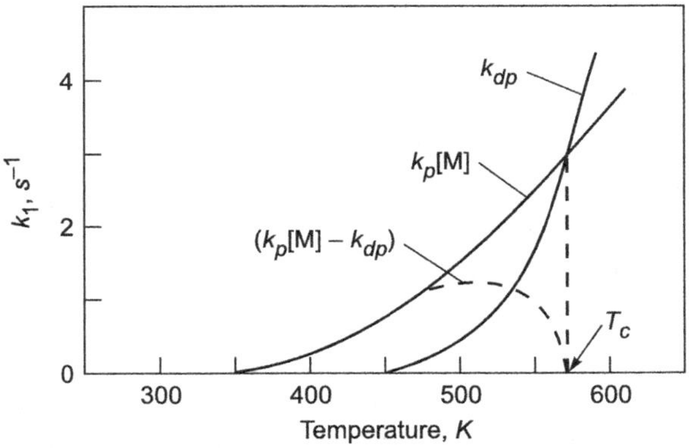

Fig. 3.7 ***Equilibrium in polymerization of styrene showing ceiling temperature phenomenon: Plot shows variation of k_p [M] and k_{dp} with change of temperature***[73] ***(Dainton & Ivin, 1958;*** **Courtesy, *Royal Society of Chemistry, London)***

At equilibrium, the free energy change for the polymerization, $\Delta G = 0$, such that one may write $\Delta G = = \Delta G° + RT \ln K = 0$, where $\Delta G°$ is the ΔG for the transformation of the monomer to polymer for both in respective standard states, and $K = (k_p/k_{dp})$ is the equilibrium constant. Hence, at equilibrium

$$\Delta G° = \Delta H° - T_c \Delta S° = -RT_c \ln K \quad (3.89)$$

and, $k_p[M^\bullet][M] = k_{dp}[M^\bullet]$, so that $k_p[M] = k_{dp}$, and

$$K = \frac{k_p}{k_{dp}} = \frac{1}{[M]_c} \tag{3.90}$$

Thus, combining Eqs (3.89) and (3.90), one obtains

$$T_c = \frac{\Delta H^\circ}{\Delta S^\circ - R \ln K} = \frac{\Delta H^\circ}{\Delta S^\circ + R \ln [M]_c} \tag{3.91}$$

The equilibrium monomer concentration, $[M]_c$ may also be expressed as

$$\ln [M]_c = \frac{\Delta H^\circ}{RT_c} - \frac{\Delta S^\circ}{R} \tag{3.91a}$$

The critical temperature T_c for polymerization may be of two kinds. If both ΔH° and ΔS° values are negative, T_c is called the "ceiling temperature" or upper limiting temperature (for a given [M] value, i.e. $[M]_c$) above which it is thermodynamically impossible to convert monomer to polymer; if both are positive, it is a "floor temperature" below which it is thermodynamically impossible to convert a monomer to a polymer.

Equations (3.91) and (3.91a) clearly show the dependence of T_c on ΔH° and $[M]_c$. If one makes a graphical plot of $k_p[M]$ and k_{dp} (each expressed in the unit of s^{-1}) against temperature, T in K (Fig. 3.7) one can readily find out the value of ceiling temperature (T_c) from the temperature corresponding to the point of intersection of the k_{dp} and $k_p[M]$ curves.[73] The net rate of polymer formation is zero at the ceiling temperature where $k_p[M] = k_{dp}$. Equations (3.91) and (3.91a) reveal several points about T_c. It is affected by conditions which affect monomer concentration or the heat of polymerization. T_c is obviously affected by the reaction medium. The polymerization of a given monomer solution at a given temperature proceeds until equilibrium is reached at which point the monomer concentration decreases to the critical (equilibrium) value of $[M]_c$ corresponding to that temperature (T_c). For every temperature of polymerization there will be a corresponding equilibrium monomer concentration. The $[M]_c$ values at 25°C for polymerization of most common vinyl monomers are so low (10^{-9} to 10^{-3} mol 1^{-1}) that they mostly defy precise measurements and hence, the polymerizations apparently proceed to completion. Polymerizations of some pure vinyl monomers show ceiling temperatures in the range of 60–300°C (ΔH value ranging between –14 and –20 kcal/mol and ΔS ranging between –20 and –30 cal/degree/mol). Among ring compounds, tetrahydrofuran (THF) shows a ceiling temperature of nearly 70°C, while elemental sulphur (S_8 ring molecules) exhibits a floor temperature of nearly 160°C.[74] Viscosity of molten sulphur

(S_8 ring) [Fig. 3.8], passes through a minimum at nearly 160°C which then rises very sharply. The physical property changes are explained in terms of homolytic cleavage of S_8 ring to yield a diradical which then gives rise to free radical polymerization on further heating. The existence of a floor temperature near 160°C is obvious from Fig. 3.8.

$$\underset{\text{S—S}}{\overset{\text{S}_6}{/\backslash}} \rightleftharpoons {}^{\bullet}\text{S} - \text{S}_6 - \text{S}^{\bullet} \xrightleftharpoons{\text{S}_8} {}^{\bullet}\text{S} - \text{S}_{14} - \text{S}^{\bullet} \ldots \rightarrow \text{S}^{\bullet} - \text{S}_n - \text{S}^{\bullet} \tag{3.92}$$

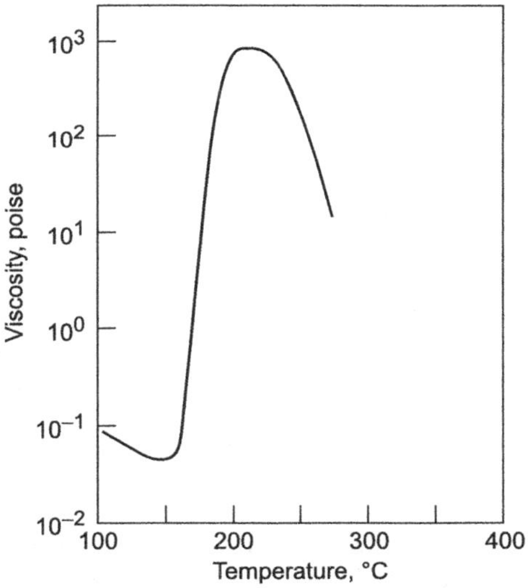

Fig. 3.8 *Plot of viscosity* vs *temperature for molten sulphur showing floor temperature near 160°C*

3.18 Dead-end Effect in Radical Polymerization

In a radical-induced vinyl polymerization, maximum possible conversion on the basis of the concept of $[M]_c$ corresponding to the temperature of polymerization may be attained if only the initiator molecules do not get depleted during the process to below a minimum required to sustain the polymerization. Under normal situation, one starts with a finite quantity of an initiator which slowly gets consumed in the formation of chain-initiating radicals. In the event of a limiting condition leading to a large depletion or complete consumption of the initiator before maximum conversion of monomer to polymer is accomplished, it is quite likely to observe what is known as the dead-end phenomenon.

The dead-end phenomenon results in a limiting conversion of monomer to polymer consequent to depletion of the initiator to such a low concentration as to render the half life of the kinetic chains approximate that of the initiator, and the effect is shown schematically in Fig. 3.9. However, under conditions of retardation of termination due to autoacceleration effect or gel effect leading to sharp rise in rate of polymerization, medium viscosity and degree of polymerization, pure dead-end effect cannot be observed.

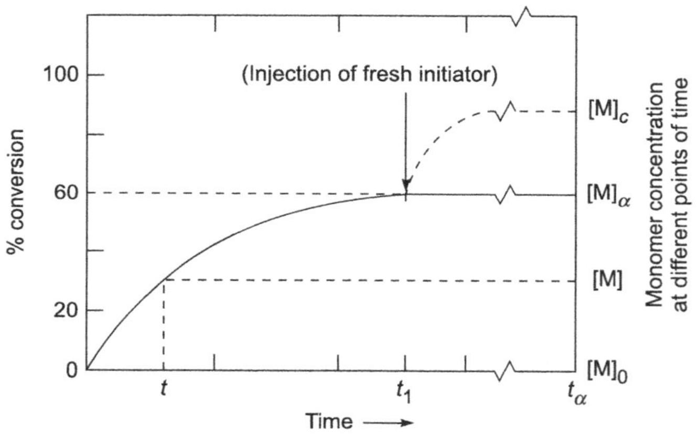

Fig. 3.9 ***Schematic representation of dead-end effect in radical polymerization showing a limiting conversion of monomer to polymer (due to initiator concentration used being low and insufficient). $[M]_0$ = initial monomer concentration; [M] = monomer concentration at time t; $[M]_\alpha$ = monomer concentration after attainment of limiting conversion beyond time t_1; and $[M]_c$ = equilibrium monomer concentration ($[M]_0 > [M] > [M]_\alpha > [M]_c$). Injection of the initiator in adequate amounts in the system at time t_1 immediately causes formation of more polymers up to a maximum conversion corresponding to $[M]_c$***

In the case of a dead-end polymerization initiated by the thermal unimolecular decomposition of an initiator I, the rate of initiator depletion is given by Eq. (3.12), i.e. $-(d[I]/dt) = k_d[I]$ and on integration one gets

$$[I] = [I]_0 \exp(-k_d t) \tag{3.93}$$

where, $[I]_0$ is the initial concentration of the initiator. The rate of polymerization, R_p given by Eq. (3.21) can then be written as

$$R_p \simeq -\frac{d[M]}{dt} = k_p\left(\frac{fk_d[I]_0}{k_t}\right)^{1/2}[M]\exp(-k_dt/2) \tag{3.94}$$

so that on rearranging one obtains

$$-\frac{d[M]}{[M]} = k_p\left(\frac{fk_d[I]_0}{k_t}\right)^{1/2}\exp(-k_dt/2)dt \tag{3.95}$$

On integration, Eq. (3.95) reduces to

$$-\ln\frac{[M]}{[M]_0} = 2k_p\left(\frac{f[I]_0}{k_tk_d}\right)^{1/2}[1-\exp(-k_dt/2)] \tag{3.96}$$

Let $[M]_0$ and [M] be the concentrations of the monomer at the start and at time t corresponding to an extent of conversion, p from monomer to polymer, such that $p = \{([M]_0 - \{M])/[M\}_0\}$ or $(1-p) = ([M]/[M]_0)$. Equation (3.96) may then be written in the form:

$$-\ln\frac{[M]}{[M]_0} = -\ln(1-p) = 2k_p\left(\frac{f[I]_0}{k_tk_d}\right)^{1/2}[1-\exp(-k_dt/2)]$$

$$= -\ln(1-p_\alpha)[1-\exp(-k_dt/2)] \tag{3.97}$$

At long times of reaction ($t = \alpha$), [M] and p attain the limiting values of $[M]_\alpha$ and p_α respectively such that

$$-\ln\frac{[M]_\alpha}{[M]_0} = -\ln(1-p_\alpha) = 2k_p\left(\frac{f[I]_0}{k_tk_d}\right)^{1/2} \tag{3.98}$$

Combining (3.97) and (3.98), one obtains

$$\frac{\ln(1-p)}{\ln(1-p_\alpha)} = [1-\exp(-k_dt/2)] \tag{3.99}$$

Equation (3.99) may be rearranged and expressed in a useful form taking logarithm:

$$\ln\left[1-\frac{\ln(1-p)}{\ln(1-p_\alpha)}\right] = -k_dt/2 \tag{3.100}$$

A plot of the left hand side of Eq. (3.100) which is equivalent to the expression $\ln[(\ln[M]_\alpha - \ln[M])/(\ln[M]_\alpha - \ln[M]_0)]$ *vs* time as shown[75] in Fig. 3.10 permits ready evaluation of k_d, the rate constant of initiator decomposition reaction from the slope. With a knowledge of $k_p/k_t^{1/2}$ determined from a study of degree of polymerization or molecular weight as a function of R_p, the rate of polymerization (*See* Sec. 3.15.1), the parameter f/k_d may be experimentally determined from limiting conversion of monomer to polymer making use of Eq. (3.98). Thus, study of polymerization by the dead-end technique permits evaluation of k_d and the efficiency of initiation, f.

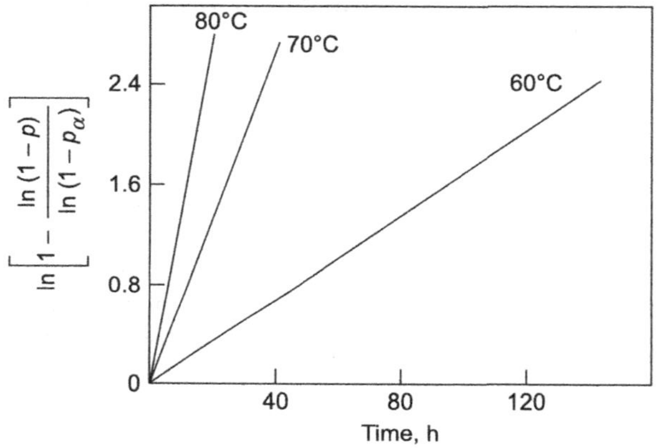

Fig. 3.10 ***Dead-end polymerization of isoprene initiated by azobisisobutyronitrile at several temperatures showing plot of*** $\ln\left[1 - \frac{\ln(1-p)}{\ln(1-p_\alpha)}\right]$ ***vs time***[75] ***(Gobran et al., 1960;* Courtesy, *Wiley-Interscience, New York)***

3.19 Allylic Polymerization and Autoinhibition

Allylic monomers of the general formula ($CH_2 = CH - CH_2X$), such as allyl acetate, allyl alcohol or allyl chloride ($-X = -OCOCH_3$, $-OH$ or $-Cl$ respectively) undergo polymerization exhibiting the characteristics of high monomer transfer with little reinitiation effect leading to what may be termed as autoinhibition. The result is very low rates of polymerization producing polymers of very low degrees of polymerization. In allylic polymerization the rate is roughly proportional to the first power of the initiator concentration[76] and the degree of polymerization is quite independent of the rate or initiator concentration. The chain (monomer) transfer reaction assumes degradative character because of high resonance stability of the allylic radical that is formed as a consequence:

$$\sim\!\!\sim CH_2-\dot{C}H-CH_2X+CH_2=CH-CH_2X$$

$$\rightarrow \sim\!\!\sim CH_2-CH_2-CH_2X+CH_2=CH-\dot{C}HX$$

$$\updownarrow \qquad (3.101)$$

$$\dot{C}H_2-CH=CHX$$

The fate of the inactive (resonance stabilized) allylic radicals formed is not clearly known, but they are stable enough to cause termination of the kinetic chains. The chain length equals the ratio of propagation and termination (degradative monomer transfer) rate constants. The low reactivity of different α-olefins such as propylene, octene-1 and 1,1-dialkyl olefins such as isobutylene is because they undergo radical polymerization with difficulty to yield low molecular weight products much as a consequence of autoinhibition.[77]

3.20 Non-ideal Kinetics in Radical Polymerization

A given polymerization conforming strictly to the reaction scheme (3.3) to (3.7) in Sec. 3.2.1 is commonly considered as an ideal polymerization. Any system deviating from this pattern of reaction is to be considered as a case of non-ideal polymerization. The ideal behaviour prescribes, according to Eq. (3.21), constancy of the term $(R_p^2/[I]\,[M]^2)$ expressed as:

$$\frac{R_p^2}{[I][M]^2}=\frac{k_p^2}{k_t}\cdot fk_d \qquad (3.102)$$

But studies of the kinetics of polymerization of different monomers, each under different conditions and chemical environments, indicate that ideal behaviour is probably more an exception than the rule. There are abundant reports on deviations from the ideal relationship given by Eq. (3.102). A number of side effects or additional reactions, not taken into consideration in the normal polymerization scheme, have been put forward in order to explain the observed kinetic deviations, and they are:

- Complex formation between monomer and initiator, and monomer dependent initiation
- Cage effect

- Initiator dependent termination: Primary radical termination and degradative initiator transfer
- Retardation by solvents
- Enhancement of the rate of initiation by solvents or additives
- Solvent/additive influencing the propagation rate constant
- Retarded termination and dependence of termination rate on medium viscosity

3.20.1 Non-ideal Chain Initiation

(a) ***Initiator—Monomer Complexation*** Initiator-monomer complexation leads to non-ideal kinetic behaviour in many cases.[78–80]

A simplified scheme of chain initiation based on this concept is given below:

$$\mathrm{I + M \underset{}{\overset{K}{\rightleftharpoons}} C \text{ (initiator-monomer complex)}} \tag{3.103}$$

$$\mathrm{C \xrightarrow{k_d} 2\,R^{\bullet}} \tag{3.103a}$$

$$\mathrm{R^{\bullet} + M \xrightarrow{k_i} M^{\bullet}} \tag{3.103b}$$

where K is the equilibrium constant of the initiator-monomer complexation reaction (3.103) and C is the true source of initiating radicals. For usual conditions of $[M] \gg [C]$,

$$[C] = K\,([I]_0 - [C])\,[M] \tag{3.104}$$

which on rearrangement gives:

$$[C] = \frac{K[M][I]_o}{1 + K[M]} \tag{1.104a}$$

where $[I]_o$ is the initial concentration of the catalyst. The rate of polymerization R_p may then be expressed as

$$R_p = -\frac{d[M]}{dt} = k_p \left(\frac{k_d}{k_t}\right)^{1/2} \left(\frac{K}{1 + K[M]}\right)^{1/2} [M]^{3/2}\,[I]_0^{\,1/2} \tag{3.105}$$

considering the usual bimolecular termination as the only prevailing mode of termination. Equation (3.105) describes a change in the reaction order with respect to monomer from 1.5 to 1.0 with increasing [M]. Equation (3.105) may be transformed as follows:

$$\frac{[M]^3}{R_p^2} = \frac{k_t}{k_p^2 k_d K [I]_o} + \frac{k_t [M]}{k_p^2 k_d [I]_o} \tag{3.106}$$

Equation (3.106) permits a plot of $[M]^3/R_p^2$ *vs* [M] to give a straight line such that the quotient of the slope and the intercept is equal to *K*. There are many reports on polymerization of different monomers showing different degrees of kinetic complexity under undiluted or different solvent environments where the observed effects are well explained and understood on the basis of equilibrium complex formation between the initiator used and the monomer.[22,41,67,81–83] Some of the reports also give spectroscopic or other evidences for the complex formation.

(b) ***Cage Effect*** Matheson[84] suggested an alternative mechanism to reach a kinetic equation much the same as Eq. (3.105). According to this concept, on decomposition of an initiator molecule, the radicals formed stay close in the surrounding monomer (monomer cage) where they may suffer many collisions leading to their mutual deactivation before being finally separated or reacting individually with monomer. Jenkins[85] considered the following scheme of reactions to develop the kinetic relationship, neglecting chain transfer:

$$\text{I} \xrightarrow{k_d} (2\ \text{R}^\bullet) \tag{3.107}$$

$$(2\ \text{R}^\bullet) \xrightarrow{k_r} \text{Q} \tag{3.107a}$$

$$(\text{R}^\bullet) + \text{M} \xrightarrow{k_x} \text{M}^\bullet \tag{3.107b}$$

$$(\text{R}^\bullet) \xrightarrow{k_a} \text{R}^\bullet \tag{3.107c}$$

$$\text{R}^\bullet + \text{R}^\bullet \xrightarrow{k'_r} \text{Q}' \tag{3.107d}$$

$$\text{R}^\bullet + \text{M} \xrightarrow{k_x} \text{M}^\bullet \tag{3.107e}$$

$$\text{M}^\bullet + \text{M} \xrightarrow{k_p} \text{M}^\bullet \tag{3.107f}$$

$$\text{M}^\bullet + \text{M}^\bullet \xrightarrow{k_t} \text{P} \tag{3.107g}$$

Here $M^\bullet$ stands for chain radicals, $R^\bullet$ for primary radicals derived from initiator decomposition, Q and Q′ for products of reaction between a pair of primary radicals within and outside the cage of their formation and P for dead polymer molecules. Brackets indicate the species in the cage provided by the medium. For all practical values of [M], reaction (3.107d) is negligible and considering that reactions generating $M^\bullet$ are analogous, the rate of polymerization may be expressed as:

$$R_p = k_p \left(\frac{k_d}{k_t}\right)^{1/2} \left(\frac{k_a + k_x [M]}{k_r + k_a + k_x [M]}\right)^{1/2} [I]^{1/2}[M] \tag{3.108}$$

For $k_a \gg [M]\, k_x$, monomer exponent remains unity:

$$R_p = k_p \left(\frac{k_d}{k_t}\right)^{1/2} \left(\frac{k_a}{k_r + k_a}\right)^{1/2} [I]^{1/2}\, [M] \qquad (3.109)$$

For $k_a \ll [M]\, k_x$, a rate expression, Eq. (3.110), similar to Eq. (3.105), is obtained:

$$R_p = k_p \left(\frac{k_d}{k_t}\right)^{1/2} \left(\frac{k_x}{k_r + k_x\,[M]}\right)^{1/2} [I]^{1/2}\, [M]^{3/2} \qquad (3.110)$$

The only difference is that here the ratio of (k_x/k_r) is used in place of K in Eq. (3.105).

3.20.2 Non-ideal Chain Termination

(a) *Unimolecular Chain Termination* In heterogeneous polymerization, particularly where the polymer being formed is insoluble in the monomer or the monomer-solvent media, bimolecular termination gets highly restricted and unimolecular termination by some obscure mechanism may assume prominence. For exclusive unimolecular termination, the termination rate R_t will be expressed as $R_t = k_{t_1}\, [M^{\bullet}]$ where k_{t_1} is the rate constant of unimolecular termination. The rate of polymerization may then be expressed as:

$$R_p \doteq \frac{k_p}{k_{t_1}}\,(2fk_d)\,[I]\,[M] \qquad (3.111)$$

Unimolecular termination is thus manifested through an initiator order >0.5 and in the limiting situation of exclusive unimolecular termination, the initiator order would be equal to one as shown in Eq. (3.111).

(b) *Initiator Dependent Chain Termination*

(i) *Primary Radical Termination* Primary radicals under ideal conditions would contribute to chain initiation only. But it is not altogether unlikely in certain systems and under special conditions in certain others that they also contribute to chain termination, partly or exclusively, giving rise to significant deviations from the ideal kinetics. In the extreme case of termination exclusively by primary radicals ($R^{\bullet}$) characterized by the rate constant k_{prt} and

$$M^{\bullet} + R^{\bullet} \xrightarrow{k_{prt}} P \qquad (3.112)$$

giving rate of termination as $k_{prt}\,[M^{\bullet}]\,[R^{\bullet}]$, the rate of polymerization, R_p, on the basis of the steady-state concept, may be expressed as:

$$R_p = k_p \cdot \frac{k_i}{k_{prt}} \cdot [M]^2 \tag{3.113}$$

Thus, with significant primary radical termination, and considering other side reactions as inconsequential, the initiator order would be less than 0.5 and the monomer order greater than unity. In the limiting case of exclusive primary radical termination, R_p is independent of initiator concentration and it depends on the square of the monomer concentration as shown in Eq. (3.113).[86–89]

If bimolecular and primary radical terminations occur simultaneously, then considering steady concentrations of $[R^\cdot]$ and $[M^\cdot]$ separately, one may obtain:

$$\frac{2k_t [M^\bullet]^2}{2fk_d [I]} = \frac{k_i [R^\bullet][M] - k_{prt}[R^\bullet][M^\bullet]}{k_i [R^\bullet][M] + k_{prt}[R^\bullet][M^\bullet]} \tag{3.114}$$

Substituting $(R_p/k_p [M])$ for $[M^\bullet]$ and rearranging gives:

$$\frac{R_p^2}{[I][M]^2} = \left(\frac{k_p^2}{k_t} \cdot fk_d\right) \left(\frac{1 - \dfrac{k_{prt}}{k_i k_p} \cdot \dfrac{R_p}{[M]^2}}{1 + \dfrac{k_{prt}}{k_i k_p} \cdot \dfrac{R_p}{[M]^2}}\right) \tag{3.115}$$

The second factor in the right hand side of Eq. (3.115) is obviously <1 and it is evidently clear that due to introduction of non-ideality as a result of significant primary radical termination, the constancy of the parameter $(R_p^2/[I]\ [M]^2)$ is disturbed[90] and it becomes dependent on R_p or $R_p/[M]^2$. Considering that for vinyl polymerization, $(k_{prt}/k_i k_p) . (R_p/[M]^2) \ll 1$, and taking logarithm, Eq. (3.115) reduces to the following simplified and useful form for the analysis of primary radical termination effect:

$$\ln \frac{R_p^2}{[I][M]^2} = \ln \left(\frac{k_p^2}{k_t} \cdot fk_d\right) - 2\, \frac{k_{prt}}{k_i k_p} \cdot \frac{R_p}{[M]^2} \tag{3.116}$$

Equation (3.116) permits a plot of the left hand side against $(R_p/[M]^2)$ and a negative slope for the linear plot, as shown in Fig. 3.11 is indicative of primary radical termination.[38,83,91] From the magnitude of the slope, a measure of the parameter $(k_{prt}/k_i k_p)$ is readily obtained.

(ii) *Degradative Initiator Transfer* The degradative process may be of two kinds: (1) with little reinitiation effect, and (2) with reinitiation effect.

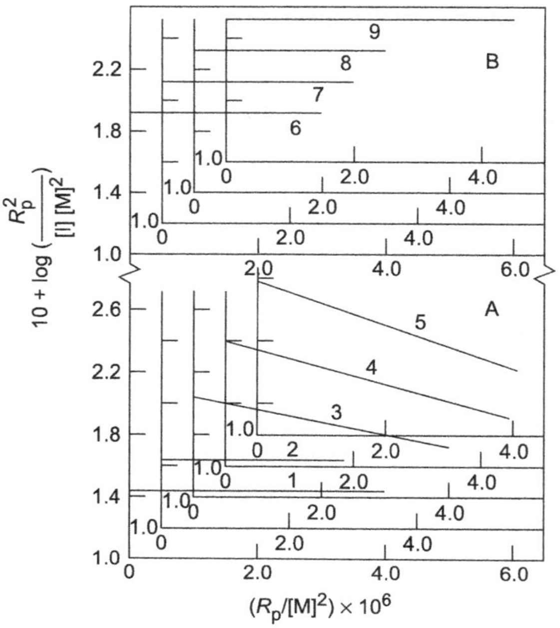

Fig. 3.11 *Plot of log $R^2_p/[M]^2$ [I] vs $R_p/[M]^2$ for solution polymerization of styrene at 60°C using* **A**—*benzoyl peroxide and* **B**—*azobisisobutyronitrile as the initiators and pyridine as the solvent*[91]*: [Pyridine] in mol l^{-1} are 1.2, 2.4, 3.6, 4.8 and 6.0 for Curves 1,2,3,4 and 5, respectively under* **A** *and 1.2, 2.4, 3.6 and 6.0 for curves 6,7,8 and 9 respectively under* **B.** *A negative solpe is indicative of primary radical termination (Ghosh and Mukhopadhyay, 1979;* **Courtesy;** *Wiley-Interscience, New York)*

For case (1), the initiator-dependent termination may be written as:

$$\left.\begin{aligned} M^{\bullet} + I \xrightarrow{k'_t} \text{Polymer} + \text{Inactive radical} \\ \text{or non - radical} \\ \text{product from I} \end{aligned}\right\} \quad (3.117)$$

The rate of termination by the degradative process given in Eq. (3.117) is equal to k'_t [M$^{\bullet}$] [I] where k'_t is the corresponding rate constant. Using the steady state concept for the chain radicals, and termination of the kinetic chains exclusively by the degradative initiator transfer process, the polymerization rate, R_p will be given by:

$$R_p = \frac{k_p}{k'_t} (2\, f k_d)\ [M] \tag{3.118}$$

Thus, with significant degradative initiator transfer and in the absence of primary radical termination, the non-ideality is reflected in an initiator order < 0.5, and in the limiting case of termination of the kinetic chains exclusively by the degradative initiator transfer process, R_p is independent of initiator concentration but it still retains first order dependence on monomer concentration.

Again, under steady state, the following relationship holds good for simultaneous occurrence of termination by the degradative process and by bimolecular mechanism:

$$2k_t\ [M^\bullet]^2 + k'_t[I]\ [M^\bullet] = 2f k_d\ [I] \tag{3.119}$$

On elimination of the radical concentration term ($[M^\bullet] = R_p/k_p\ [M]$) and on rearrangement, one obtains

$$2\ \frac{k_t}{k_p^2} \cdot \frac{R_p^2}{[I]\,[M]^2} = 2f k_d - \frac{k'_t}{k_p} \cdot \frac{R_p}{[M]} \tag{3.120}$$

Equation (3.120) allows a plot of left hand side against $R_p/[M]$, and a negative slope of the linear plot is indicative of termination by the degradative initiator transfer reaction, Fig. 3.12. The slope of the plot gives a measure of the parameter k'_t/k_p and the value of $f k_d$ can be calculated from the intercept.[61, 92–93] When R_p becomes independent of [I] for a given [M], Eq. (3.120) is unsuitable for the analysis of the degradative effect.

For case (2), i.e. for degradative initiator transfer with reinitiation effect, the transfer reaction and other reactions that may follow as a consequence are shown in the following reaction scheme:

$$\left.\begin{array}{lll} \text{(a)} & M^\bullet + I \xrightarrow{k_{trI}} P + I^\bullet & \text{(transfer)} \\ \text{(b)} & I^\bullet + M^\bullet \xrightarrow{k_{rtI}} P & \text{(termination by } I^\bullet) \\ \text{(c)} & I^\bullet + M \xrightarrow{k_{iI}} M^\bullet & \text{(reinitiation following transfer)} \end{array}\right\} \tag{3.121}$$

Degradative effect will be prominent due to the occurrence of reaction (b); in the absence of reaction (b), consideration of reaction (c) becomes more relevant and important; degradative effect would still be measurable and prominent for $k_{iI} < k_p$. For significant termination via degradative initiator transfer of the above kind along with the usual bimolecular termination and assuming absence of other chain transfer

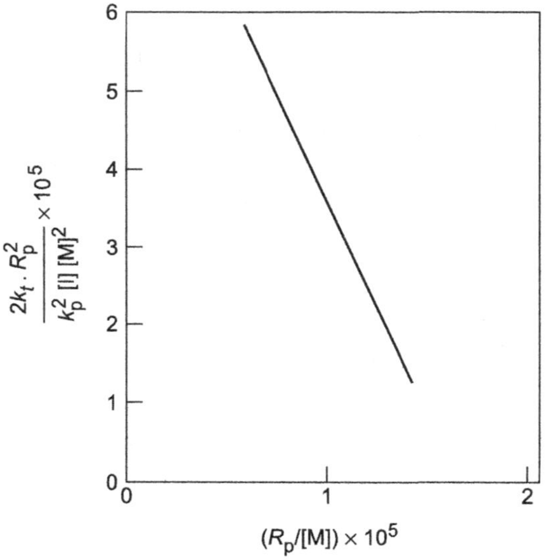

Fig. 3.12 ***Analysis of degradative initiator transfer with little reinitiation effect[92] in the photopolymerization of methyl methacrylate using poly (N-vinyl carbazole)—bromine complex as the photoinitiator, I : Plot of $2k_tR^2_p/k^2_p$ [I] [M]^2 vs R_p/[M] (Ghosh & Ghosh, 1984;* Courtesy, *Wiley-Interscience, New York)***

reactions and primary radical termination, the following equation may be deduced for analysis of the degradative effect[94] where C_I stands for k_{trI}/k_p:

$$\frac{R_p^2}{[I]\,[M]^2}\left\{1 + C_I \cdot \frac{k_p^2}{k_t}\cdot\frac{[I]}{[M]}\cdot\left[\frac{k_{rtI}/k_{iI}\cdot k_p}{1+(k_{rtI}\cdot R_p/k_{iI}k_p[M]^2)}\right]\right\} = \frac{k_p^2}{k_t}\cdot f k_d \qquad (3.122)$$

For $(R_p/[M]^2) <<< 1$, Eq. (3.122) reduces to:

$$\frac{R_p^2}{[I]\,[M]^2}\cdot\left\{1+\frac{k_p^2}{k_t}\cdot\frac{k_{rtI}}{k_{iI}\cdot k_p}\cdot C_I\cdot\frac{[I]}{[M]}\right\} = \frac{k_p^2}{k_t}\cdot f k_d \qquad (3.123)$$

On rearrangement of Eq. (3.123) and taking logarithms, a more useful expression, Eq. (3.124) is obtained for kinetic analysis of the degradative effect:[1, 95–96]

$$\ln\frac{R_p^2}{[I]\,[M]^2} = \ln\frac{k_p^2}{k_t}\cdot f k_d - \frac{k_p^2}{k_t}\cdot\frac{k_{rtI}}{k_{iI}\cdot k_p}\cdot C_I\cdot\frac{[I]}{[M]} \qquad (3.124)$$

The convenient form of Eq. (3.124) permits a plot of the left hand side against [I]/[M] and a negative slope for the linear plot is indicative of the degradative initiator transfer under consideration (Fig. 3.13). With a knowledge of the initiator transfer constant, C_I, i.e. (k_{trI}/k_p) and k_p^2/k_t, the plot gives a measure of the kinetic parameters $(k_{rtI}/k_{iI}{\cdot}k_p)$ and f_{kd} from the slope and the intercept respectively.

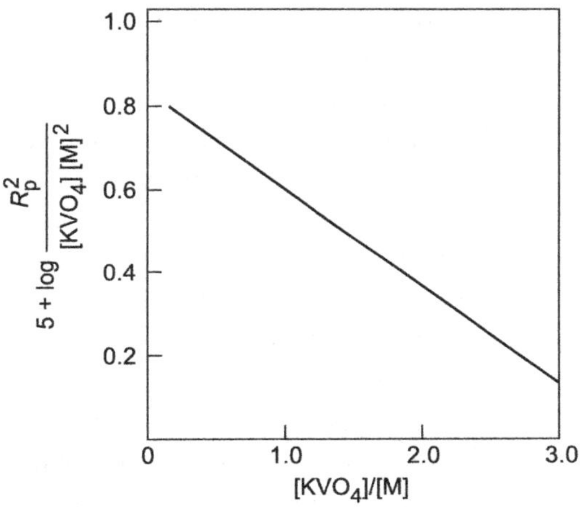

Fig. 3.13 *Analysis of degradative initiator transfer with reinitiation effect in the polymerization of methyl methacrylate at 50°C using KVO_4 as the initiator in acidic (H_2SO_4) aqueous media*[1]*: Plot of log R_p^2 / $[KVO_4]$ $[M]^2$ vs $[KVO_4]/[M]$. (Ghosh & Paul, 1983;* **Courtesy,** *Marcel Dekker, New York)*

3.20.3 Kinetic Non-ideality due to Abnormal Solvent Effect

Solvents used in polymerization may act normally as mere diluents and sometimes also as chain transfer agents without affecting rate and thus featuring as inert additives kinetically. But it is otherwise very likely that different solvents bring about different degrees of deviation from normal kinetics for a given initiator–monomer combination, depending on chemical nature of the solvents and level of dilution, and in fact varied degrees of non-ideality caused or magnified by the presence of solvents have been widely observed and reported.

(a) ***Solvent Influence on Chain Initiation*** Enhancement of rate of chain initiation is manifested through an enhanced rate of polymerization and this may arise as a consequence of solvent-induced increase in initiator efficiency,[97] or due to the increase in the value of the initiator decomposition rate constant, k_d, i.e., for solvents causing an increase in rate of radical generation.[80,91,98–100]

(b) *Solvent Modification of Chain Propagation and Chain Termination* Non-ideal kinetics in polymerization more widely observed in diluted systems, have variously been explained and understood on the basis of modification of the reactivity of chain radicals to different degrees under different environments, i.e. monomer–solvent combinations, both in kinds and in volume proportions for a given initiator system. This solvent-modification of radical reactivity may arise due to complex interactions between the macroradicals and solvent molecules or simply due to solvation of the radical end and not necessarily of the whole macroradical. The degree of interaction or solvation would certainly depend on monomer–solvent combination and their volume proportions in a given combination. Solvent-modification of monomer reactivity may also be a factor, but general experience and observed effects do not firmly confirm this concept.[101–102]

It is also likely that the solvent modification of chain propagation reaction would be differently blended with modification of the termination reaction causing k_t to have different values in view of the viscosity of the polymerization medium having a major influence on the termination process, and termination being generally recognized as a diffusion controlled process.[91, 103–104]

Perturbation of k_p value under the influence of solvents, however, is generally of more limited relevance or significance than of the k_t value, k_t having an inverse relationship with medium viscosity.[105–106] Solvents may also effect retardation of polymerization by directly causing degradative chain transfer and even limited copolymerization.[107–108] Still more interesting cases may arise where solvents play effective roles in modifying the initiator species and thereby bring into effect prominent degradative chain transfer, more so in the higher range of solvent concentrations,[95] even though the kinetic parameter k_p^2/k_t remains practically unchanged over the whole range of dilution (from undiluted to highly diluted systems).

3.21 Determination of Absolute Rate Constant of Propagation Reaction

The value of the kinetic parameter k_p^2/k_t can be conveniently determined from determination of rate of polymerization R_p with a knowledge of R_i, the rate of initiation or $\overline{X}_n$, the average degree of polymerization (*See* Secs. 3.12 and 3.15.1). But the measurement of the absolute value of k_p, the rate constant of propagation is one of the essential problems. A knowledge of the absolute rate constant of propagation is required for a better appreciation and correlation of the reactivities of a monomer and the derived radical with their molecular structures. Due to the lack of convenient techniques, evaluation of k_p remained an unsolved problem till about 1945.

In radical polymerization, the average lifetime of a growing radical, τ_s under steady state is given by the ratio of the steady state concentration of the radical and the rate of its termination:

$$\tau_s = \frac{[M^\bullet]_s}{2k_t[M^\bullet]_s^2} = \frac{1}{2k_t[M^\bullet]_s} \tag{3.125}$$

Eliminating the radical concentration term using Eq. (3.17), one obtains

$$\tau_s = \frac{k_p[M]}{2k_t(R_p)_s} \tag{3.126}$$

Thus, measurements of τ_s and $(R_p)_s$ at a given [M] value enable calculation of k_p/k_t. A knowledge of k_p^2/k_t calculated from other approaches then permits evaluation of the individual rate constants k_p and k_t.

Measurement of average lifetime of a radical is conveniently done by photo-initiated polymerization using a light source which blinks on and off at fixed intervals in what is known as the rotating sector technique. A rotating opaque disc from which one or more sectors of equal size and spacings are sliced out is placed between the light source and the reaction vessel. The system passes alternately through periods of light and darkness as the notched parts and solid parts of the disc coming in the way of the light path permit the light to pass and block its passage respectively. Variation of the notch area and velocity of rotation of the disc admit variation and control of relative lengths of dark and light periods and the frequency of blinking.

The relation between τ_s and frequency of blinking is rather complex. The polymerization rate depends on this frequency and also on the incident light intensity, I_o. A short but finite period of time of illumination, t is required for the radical concentration to reach the steady state concentration $[M^\bullet]_s$, ($[M^\bullet]_s = \{\phi I_a/k_t\}^{1/2}$, *See* Sec. 3.10) and a short but finite period of time, t' is required for it to drop to zero after the light goes out.

During illumination period, the rate of radical generation is given by

$$d[M^\bullet]/dt = 2k_t([M^\bullet]_s^{2+} - [M^\bullet]^2) \tag{3.127}$$

which, on integration, gives:

$$\ln\left[\frac{(1+[M^{\bullet}]/[M^{\bullet}]_s)}{(1-[M^{\bullet}]/[M^{\bullet}]_s)}\right] = 4k_t[M^{\bullet}]_s(t - t_o) \tag{3.128}$$

Here, t_o is the integration constant such that $[M^{\bullet}] = 0$ for $t = t_o$. Combining Eqs (3.128) and (3.125) and rearranging, one obtains,

$$\tanh^{-1}([M^{\bullet}]/[M^{\bullet}]_s) = (t - t_o)\tau_s \tag{3.129}$$

The ratio of rate of polymerization over a period short of attainment of steady state, (R_p) to the steady rate $(R_p)_s$ would then be given by the expression:

$$\frac{(R_p)}{(R_p)_s} = \frac{[M^{\bullet}]}{[M^{\bullet}]_s} = \tanh[(t - t_o)/\tau_s] \tag{3.130}$$

since the rate is proportional to the radical concentration. In the above expressions t refers to the time of illumination.

During the time period (t) of continued illumination and after radical concentration has reached $[M^{\bullet}]_s$, radicals are formed and terminated at equal rate. If the light is turned off at this stage, the radical formation reaction stops in the dark, but the radicals continue to undergo termination according to Eq. (3.16) and the radical concentration $[M^{\bullet}]$ falls from $[M^{\bullet}]_s$ value to zero over a finite time period of darkness, t'. A qualitative sketch of how the actual radical concentration reaches the stationary state value $[M^{\bullet}]_s$ during a long illumination period (t) and how the radical concentration decays from the steady value to zero over a subsequent long dark period (t') is shown in Fig. 3.14.

If the light source is switched on and off and held for long but equal time periods of light and darkness, the radical concentration will alternately build up from zero to $[M^{\bullet}]_s$ and decay from $[M^{\bullet}]_s$ to zero as can be conceived from repetition of the sequence (t and t') of illumination and darkness, (Fig. 3.14, Curve A), over which the intensity of illumination is either I_o or zero and the radical concentration is either $[M^{\bullet}]_s$ or zero with short zones of transition in between. The radical concentration is consistent with I_o, but radicals are present only half of the time and hence the rate of polymerization observed corresponds to one half of that to be observed for the same intensity on the basis of continuous illumination.

When the frequency of blinking is high giving very short but equal light and dark periods then, even if one starts with $[M^{\bullet}] = 0$ at the beginning of the first light period, the radical build up is interrupted by the interception of light before it reaches $[M^{\bullet}]_s$. $[M^{\bullet}]$ reaches a maximum value, say, $[M^{\bullet}]_1 < [M^{\bullet}]_s$ at the end of the illumination period (t). The decay of radical concentration from the $[M^{\bullet}]_1$ value

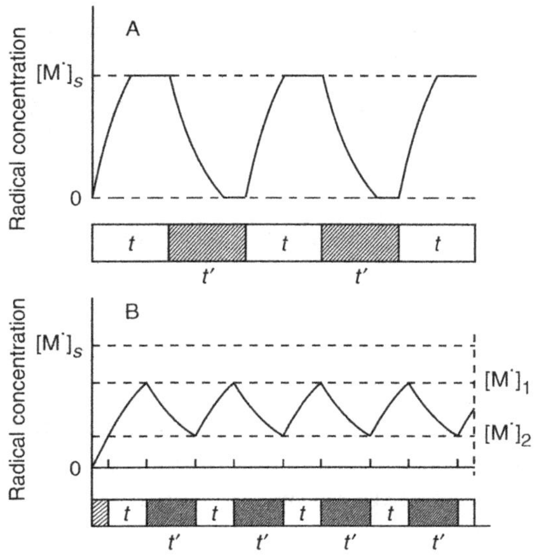

Fig. 3.14 ***Schematic representation of variation of (chain) radical concentration [M˙] over: A—cycles of long illumination period (t) and dark period (t′), and B—cycles of short (intermittent) illumination period (t) and dark period (t′)***

begins immediately thereafter as the dark period commences. At the end of the short dark period (t'), radical concentration drops to a minimum value, say $[M^{\bullet}]_2 > 0$, and before it could drop to zero, the illumination is on again. With frequent blinking, the concentration of radicals alternates between $[M^{\bullet}]_1$ and $[M^{\bullet}]_2$ as the system passes through the end of a light period to the end of a dark period and so on in successive cycles of illumination and darkness (Fig. 3.14, Curve B).

With progressive increase in the frequency of blinking the difference between $[M^{\bullet}]_1$ and $[M^{\bullet}]_2$ will be progressively narrow and in the limiting case of a very fast flashing the radical concentration would reach a constant or plateau value below $[M^{\bullet}]_s$. Under this limiting condition, the flashing is fast enough to cause little fluctuation in radical concentration which in that event, virtually corresponds to a continuous illumination of intensity $I_o/2$, as only 50% of the irradiation is received by the system.

Now, the rate of polymerization, R_p is proportional to the concentration of radicals, $[M^{\bullet}]$ present in a system. According to analysis and consideration detailed above for photo-initiated polymerization, R_p or $[M^{\bullet}]$ can be varied by varying the frequency of blinking. Under otherwise comparable conditions, the average rate (R_p) at different flashing conditions for an intensity of illumination I_o will show proportionality as follows: (a) For steady conditions under continuous illumination (i.e., no blinking), $(R_p)_s \propto (I_o)^{1/2}$; (b) for slow blinking with equal light and dark periods, $(\overline{R}_p) \propto \frac{1}{2}(I_o)^{1/2}$, and (c) for very fast blinking with equal light and dark periods, $(\overline{R}_p) \propto (I_o/2)^{1/2}$ such that

$$\frac{(\overline{R}_p)}{(R_p)_s} = \frac{1}{2} \text{ (for slow blinking)} \tag{3.131}$$

and

$$\frac{(\overline{R}_p)}{(R_p)_s} = \frac{1}{(2)^{1/2}} \text{ (for very fast blinking)} \tag{3.132}$$

For unequal light and dark periods such that $t' = rt$, i.e. the dark period is r times longer than the period of illumination in all successive cycles ($r = 1$, for the systems considered above), the general expressions for the relative rates may be written as: *Case I*:

$$\frac{(\overline{R}_p)}{(R_p)_s} = \frac{1}{r+1} \text{ (for slow blinking)} \tag{3.133}$$

Case II:

$$\frac{(\overline{R}_p)}{(R_p)_s} = \frac{1}{(r+1)^{1/2}} \text{ (for very fast blinking)} \tag{3.134}$$

The relative rate equals 1/4 and 1/2 for Case I and II respectively for a value of $r = 3$. Thus, the average rate, $(\overline{R}_p)$ increases from one-fourth to one-half of the steady-state rate $(R_p)_s$ as the frequency of flashing $\{1/(t + rt)\}$ increases from a much lower value to much higher value in comparison with $(1/\tau_s)$.

When the light is turned off, the radical concentration decays according to the relationship $d[M^\bullet]/dt' = -2k_t[M^\bullet]^2$ which, on integration gives:

$$(1/[M^\bullet]) - (1/[M^\bullet]_1) = 2k_t t' \tag{3.135}$$

Here, $[M^\bullet]_1$ is the radical concentration at the beginning of the dark period. Combining with Eq. (3.125) and multiplying by $[M^\bullet]_s$, one obtains:

$$([M^\bullet]_s/[M^\bullet] - [M^\bullet]_s/[M^\bullet]_1) = t'/\tau_s = rt/\tau_s \tag{3.136}$$

Again, considering Eq. (3.129), if $[M^\bullet] = 0$ at $t = 0$, then $t_o = 0$; but for $[M^\bullet] = [M^\bullet]_2$, where $[M^\bullet]_2 > 0$ at the beginning of the illumination period, then

$$t_o/\tau_s = -\tanh^{-1}([M^\bullet]_2/[M^\bullet]_s)$$

and hence,

$$\tanh^{-1}([M^{\bullet}]/[M^{\bullet}]_s) - \tanh^{-1}([M^{\bullet}]_2/[M^{\bullet}]_s) = t/\tau_s \quad (3.137)$$

The mathematical treatment of Briers *et al.*, for chain reactions in gaseous systems induced photochemically under intermittent illumination and involving bimolecular termination[109] is also applicable in the present case. The maximum and minimum radical ratios, $[M^{\bullet}]_1/[M^{\bullet}]_s$ and $[M^{\bullet}]_2/[M^{\bullet}]_s$, are related to τ_s by the following two expressions similar in form as shown by Eqs (3.137) and (3.136) respectively:

$$\tanh^{-1}([M^{\bullet}]_1/[M^{\bullet}]_s) - \tanh^{-1}([M^{\bullet}]_2/[M^{\bullet}]_s) = t/\tau_s \quad (3.138)$$

and

$$([M^{\bullet}]_s/[M^{\bullet}]_2) - ([M^{\bullet}]_s/[M^{\bullet}]_1) = t'/\tau_s = rt/\tau_s \quad (3.139)$$

Equations (3.138) and (3.139) allow evaluation of maximum and minimum radical concentration ratios for a given value of r, ($r = t'/t$) and of t/τ_s. The average radical concentration, $[\overline{M^{\bullet}}]$, over a complete light-dark cycle will be expressed as

$$[\overline{M^{\bullet}}] = \left(\int_0^t [M^{\bullet}]\,dt + \int_0^{rt} [M^{\bullet}]\,dt'\right)\Big/(t + rt) \quad (3.140)$$

Here, $[M^{\bullet}]$ in the first integral covering the period of illumination is given by Eq. (3.137) and that in the second integral covering the dark period is given by Eq. (3.136). Evaluation of the integrals yields the following expression[110]:

$$\frac{(\overline{R}_p)}{(R_p)_s} = \frac{[\overline{M^{\bullet}}]}{[M^{\bullet}]_s} = \frac{1}{r+1}\left\{1 + \left(\frac{\tau_s}{t}\right)\ln\left[\frac{[M^{\bullet}]_1/[M^{\bullet}]_2 + [M^{\bullet}]_1/[M^{\bullet}]_s}{1 + [M^{\bullet}]_1/[M^{\bullet}]_s}\right]\right\} \quad (3.141)$$

Thus, for a given ratio of dark period to illumination period ($r = t'/t$), values of $[\overline{M^{\bullet}}]/[M^{\bullet}]_s$ or $(\overline{R}_p)/(R_p)_s$ may be calculated for different values of t/τ_s. A semilog plot of relevant data for $r = 3$ is shown[111] in Fig. 3.15. The average radical concentration falls from one-half of the steady-state value for fast flashing (low t/τ_s) to one-fourth of the same for slow flashing, as can be seen from Fig. 3.15, in full conformity with Eqs (3.133) and (3.134).

The steady-state polymerization rate $(R_p)_s$ is measured without the use of the rotating sector. Then the average rate $(\overline{R}_p)$ is measured using the rotating sector at increasing speed of sector rotation. The blinking frequency as well as t and t' are controlled by controlling the speed of sector rotation. The polymerization is conveniently followed by dilatometric technique. Thus, a number of rate ratios $(\overline{R}_p)/(R_p)_s$ are obtained for a given sector ratio, r, the only variable being the sector speed which in its turn varies the flash time. The rate ratios are then plotted against

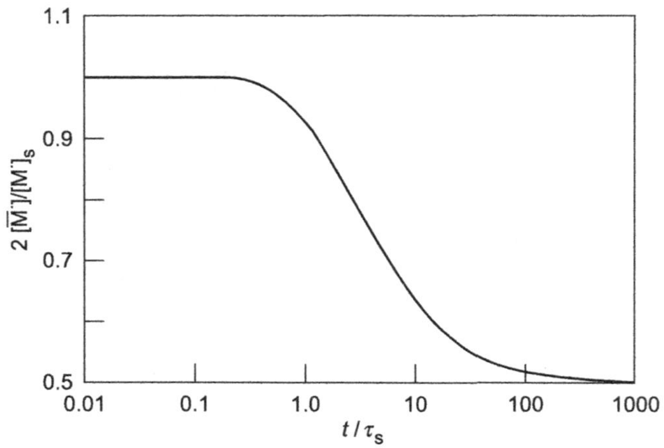

Fig. 3.15 *Semilog plot[111] of 2 [M˙] / [M˙]$_s$ vs. t/τ_s for the ratio of dark to light intervals, r = 3 (Matheson et al., 1949;* **Courtesy,** *American Chemical Society, Washington)*

log, t, (Fig. 3.16). The theoretical curve corresponding to the same value of r (r = 3, as in Fig. 3.15) is then superimposed on the experimental points by shifting it along the time scale (log t) to obtain the best fit. The displacement of one curve relative to the other along the abscissa yields log τ_s, since the abscissa scale for the theoretical curve represents (log t – log τ_s).

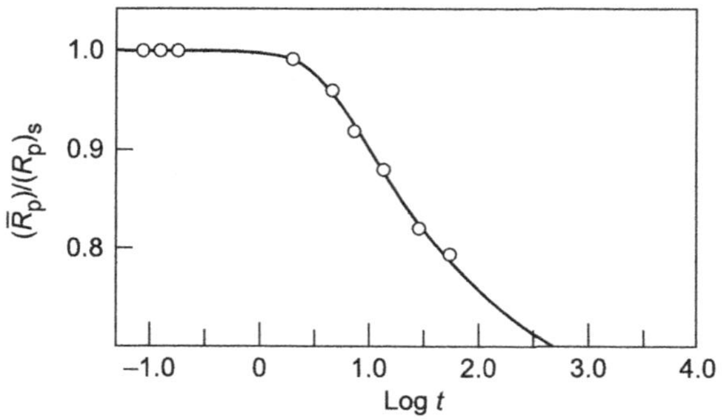

Fig. 3.16 *The rate ratio for the polymerization of methyl methacrylate for intermittent illumination with light periods of t seconds under conditions of very high sector speed plotted against log t:[35] (o), experimental points: (—), theoretical curve (Dainton and Sisley, 1963;* **Courtesy,** *Royal Society of Chemistry, London)*

Evaluation of τ_s then easily allows calculation of absolute rate constants k_p and k_t and hence, k_{tr} and k_z values, making use of Eq. (3.126) relating τ_s with k_p/k_t and of equations and approaches discussed in Secs 3.12 and 3.15 leading to evaluation of the parameter k^2_p/k_t. In radical polymerization, k_p and k_t for different systems generally fall[35,44] in the range 10^2–10^4 $(1\ mol^{-1} \cdot s^{-1})$ and 10^6–10^8 $(1\ mol^{-1} \cdot s^{-1})$, respectively, and τ_s values generally range between 0.1–10 s.

3.22 Olefin Metathesis and Metathesis Polymerization

3.22.1 Olefin Metathesis

Discovery of olefin metathesis (OM) reaction is recognized as a significant turning point in organic reactions in recent decades.[112–116] Olefin metathesis has many applications in the low molecular weight range as well as in the field of oligomers and polymers. The key point in this context is the use of catalysts with novel selectivity and high potential for effective chemical transformations for olefin metathesis and metathesis polymerization systems.

In relation to olefin chemistry, the term metathesis commonly means interchange of carbon atoms between a pair of double bonds as illustrated below:

A.
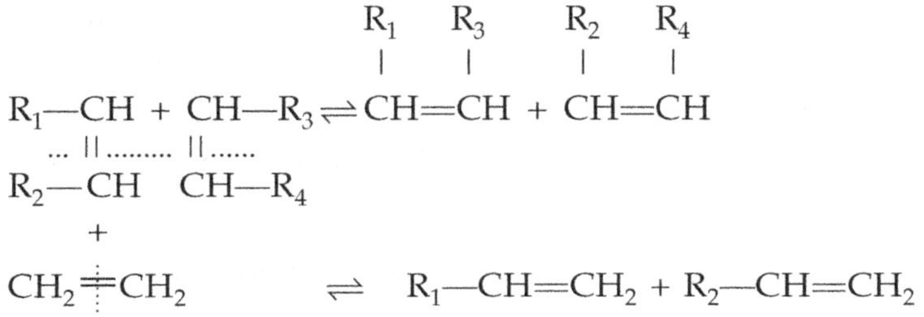
(3.142)

(Olefin Metathesis, OM) (3.143)

B.
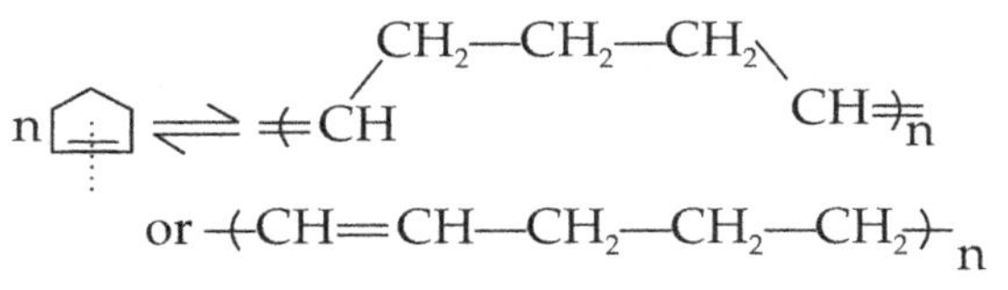

(Ring opening metathesis Polymerization, ROMP) (3.144)

C. (i)
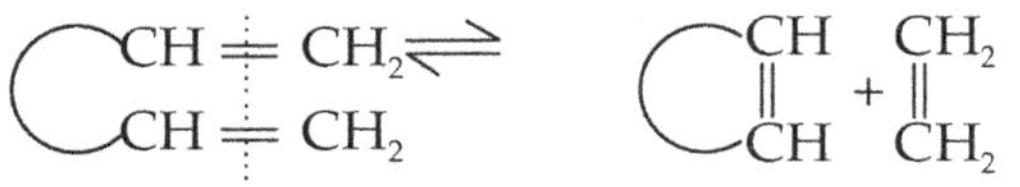

(Cyclization)

(Degradative metathesis, DM) (3.145)

(ii)

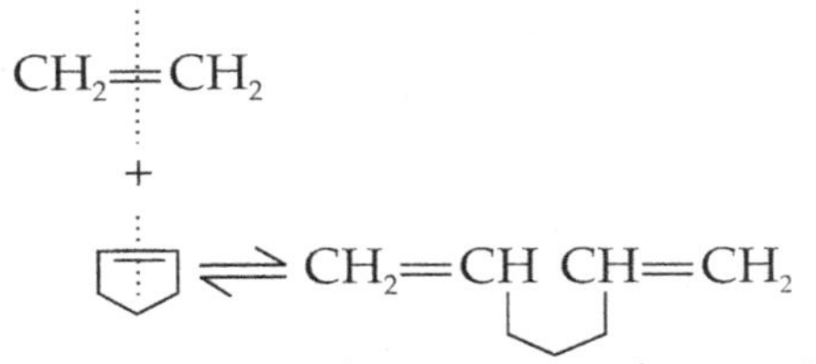

(Cross metathesis, CM, involving ring opening) (3.146)

Reactions (A) and (C) are viewed as olefin exchange or olefin disproportionation while reaction (B) is viewed as ring opening metathesis polymerization (ROMP) of a cycloalkene.[117–120]

3.22.2 Mechanism

Olefin metathesis would seldom take place spontaneously. Catalysts in common use are based on transition-metal compounds frequently in combination with a cocatalyst and sometimes a third compound as a promoter. Catalysts based on W and Mo are particularly effective and those based on Ti, Nb, Ta, Re, Os and Ir are also useful. It is widely evidenced that a metal carbene complex of the type $M = CR_1\ R_2$ acts as the active species for OM and ROMP reactions. There may be other ligands present round the central metal M forming the coordinatively unsaturated metal carbene complex (C). The complex (C) gets further complexed with the olefins (O) used to form an addition complex of the metallacyclobutane type (C_1). C_1 subsequently rearranges to yield a new metal carbene complex (C'_1) which in turn propagates the reaction by adding a further molecule of the olefin (O) in a manner as shown for the ROMP of norbornene:

(A) Metal carbene/olefin adduct (C_1) formation (initiation) and its rearrangement (propagation) to a new metal-carbene complex:

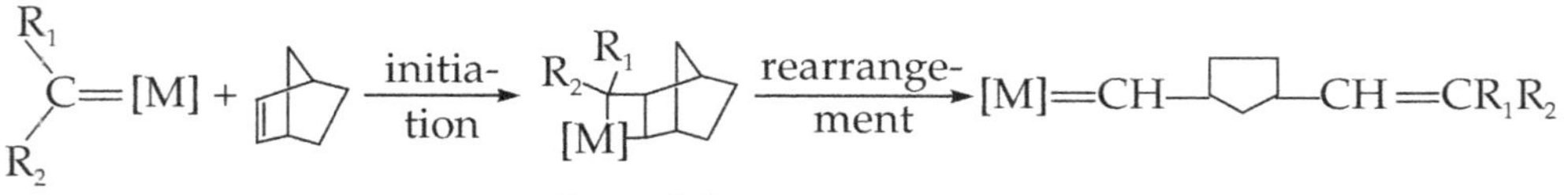

metallacyclobutane complex (3.147)

(C) (O) (C_1) (C'_1)

Scheme-I

(B) Further olefin addition (initiation) and rearrangement (propagation) via ring opening:

$$C'_1 + O \xrightarrow[\text{initiation}]{\text{fresh}} C_2 \xrightarrow[\text{ment}]{\text{rearrange-}} C'_2 \text{ (repetition of Scheme-I)} \quad (3.148)$$

The numerical subscripts denote the number of olefin (O) units incorporated in the respective complex/propagating species.

The successive initiation and propagation reactions may be generally written as follows according to Scheme I

$$\text{(C)} \quad \underset{\text{(Metal carbene)}}{C'_n} + \underset{\text{(Olefin)}}{O} \xrightarrow{\text{Initiation (fresh)}} \underset{\text{(Metallacyclo-butane complex)}}{C_{n+1}} \xrightarrow{\text{Propagation}} \underset{\text{(Metal carbene)}}{C'_{n+1}}$$

3.22.3 Intramolecular Reaction and Formation of Cycloolefins from Diolefins

Metal carbene-catalysed degradative metathesis (DM). (Equation 3.145) is viewed as an intramolecular reaction (Eq. 3.149) that successively occurs via a back-biting reaction mechanism involving the metal–carbene complex; the metal carbene complex is formed *in-situ* as an intermediate by cross metathesis (CM) as shown in Eq. (3.150).

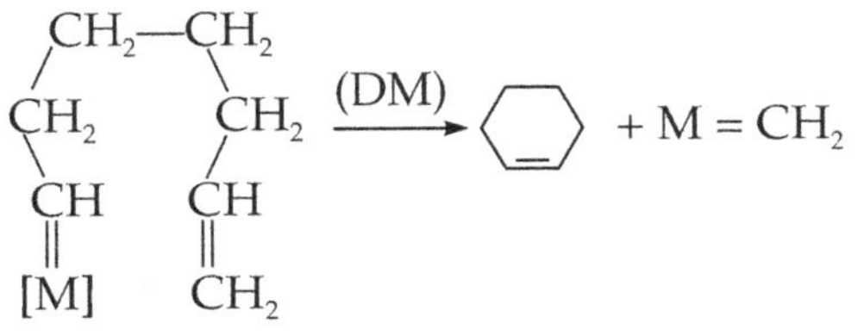
(3.149)

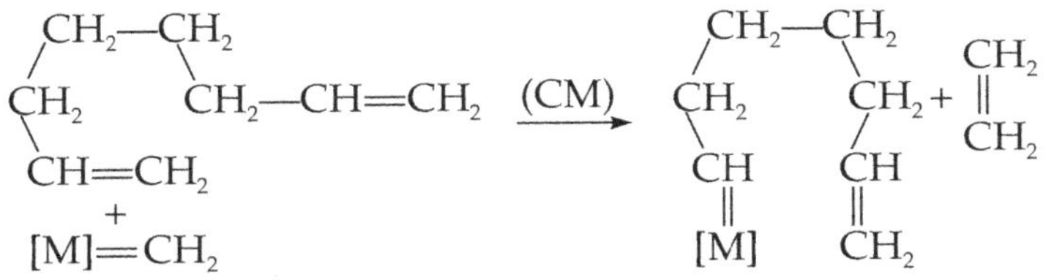
(3.150)

By proper choice of ligands, the catalytic activity of the metal carbene catalyst can be so controlled as to reduce the scope for the intramolecular reaction (DM) to insignificance, thus rendering the macromolecular species C'_n truly "living" with a capacity for unending linear propagation till complete monomer (olefin) depletion to yield polymers of low polydispersity. Living metathesis polymerization is a late edition of living polymerization earlier demonstrated for anionic polymerization,[121] cationic polymerization,[122] group-transfer polymerization[123] and radical polymerization.[124]

3.22.4 Formation of Block and Graft Copolymers

Olefin metathesis polymerization achieved by admitting first a cycloolefin (O_1), then a second cycloolefin (O_2) and finally again the cycloolefin (O_1) may generate in turns the respective propagating species from the admitted monomer in the given sequence to finally form a tri block copolymer bearing units of O_1, O_2 and O_1 again in the structure of the type:

$$-\!\!\left(O_1-O_1-O_1-O_1 \ldots O_1-O_1\right)\!\!\left(O_2-O_2-O_2-O_2 \ldots O_2-O_2\right)\!\!\left(O_1-O_1-O_1 \ldots O_1-O_1\right)\!\!-$$

For use of first norbornene, then cyclopentene and again norbornene, synthesis of such block copolymers with no lowering in the overall metal carbene concentration was reported.[125]

Graft copolymers having a macromolecular backbone constituted of norbornene units with grafted-on chains of polystyrene may be synthesized by the ring opening metathesis copolymerization of norbornene and polystyrene having a norbornene moiety as end group prepared from a sequence of reactions[126] shown below:

$$-\!\!\left(CH_2-\underset{C_6H_5}{CH}\right)_{n-1}\!\!-CH_2-\underset{C_6H_5}{\overset{(-)}{CH}}\;\overset{(+)}{Li} \xrightarrow{\text{ethylene oxide}} \left(CH_2-\underset{C_6H_5}{CH}\right)_n\!-CH_2-CH_2\overset{(-)}{O}\;\overset{(+)}{Li} \quad \text{I} \tag{3.151}$$

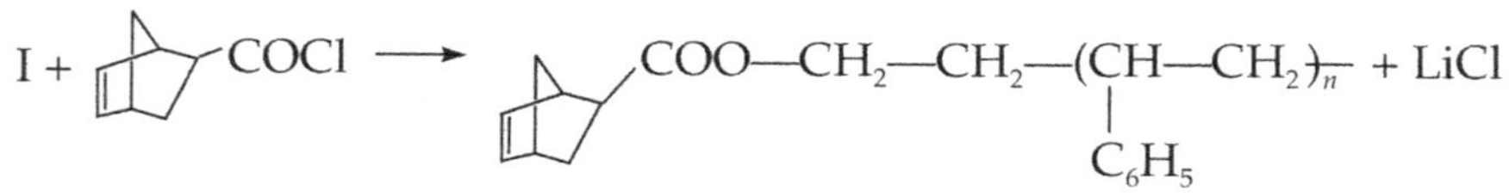

(3.152)

3.22.5 Some Other Olefin Metathesis Reactions

Metathesis polymerization of acetylenes may likewise be propagated by metal-carbene complexes as for olefin metathesis:

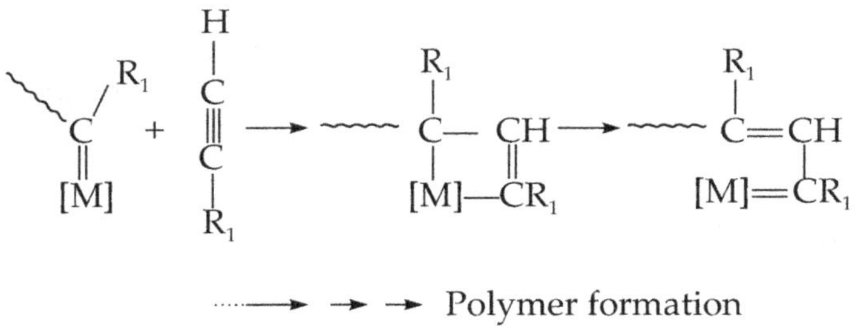

(3.153)

However, the actual alkyne metathesis propagated by selected metal carbyne complexes may be written as:

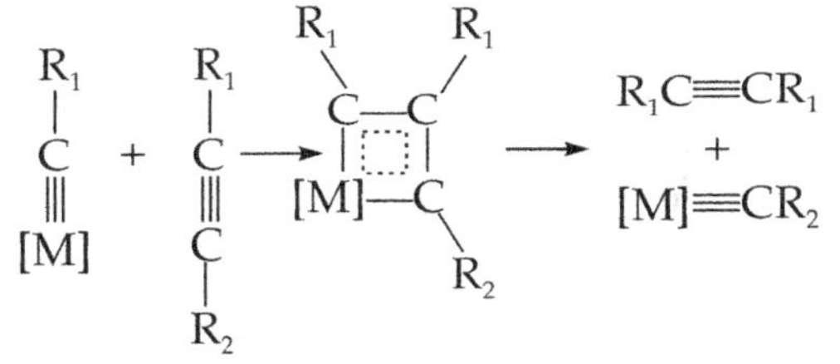

(3.154)

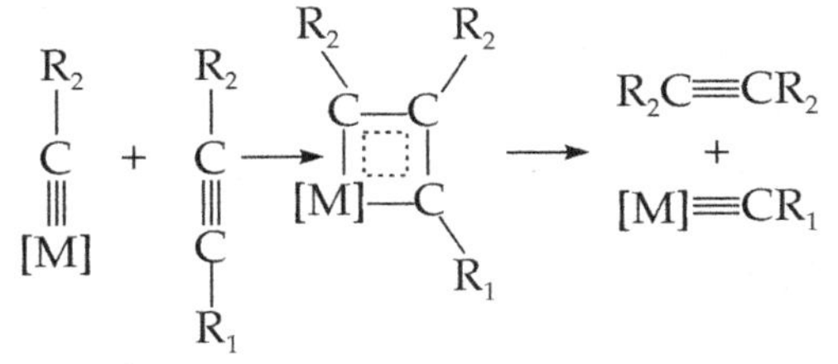

(3.155)

Metathesis of unsaturated fatty acid esters has brightened the prospect of synthesizing symmetrical unsaturated diesters and olefins[127]:

$$2CH_3(CH_2)_7\text{—}CH{=}CH\text{—}(CH_2)_7\text{—}COOCH_3 \xrightleftharpoons{WCl_6/Sn(CH_3)_4}$$

$$\begin{array}{c} CH_3\text{—}(CH_2)_7\text{—}CH \\ \| \\ CH_3\text{—}(CH_2)_7\text{—}CH \end{array} + \begin{array}{c} CH\text{—}(CH_2)_7COOCH_3 \\ \| \\ CH\text{—}(CH_2)_7COOCH_3 \end{array} \qquad (3.156)$$

Metathesis of polyunsaturated fatty acid esters gives a variety of reaction products, due to the presence of more than one double bond in each molecule:

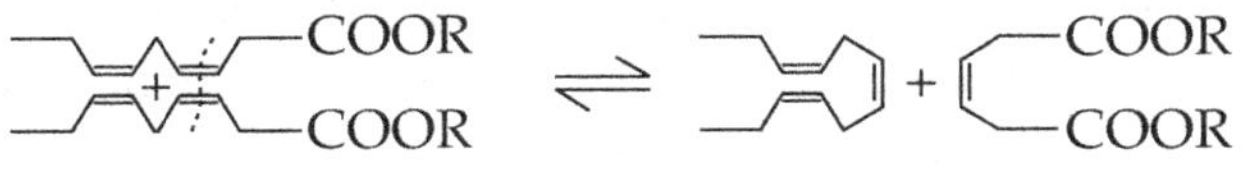

(3.157)

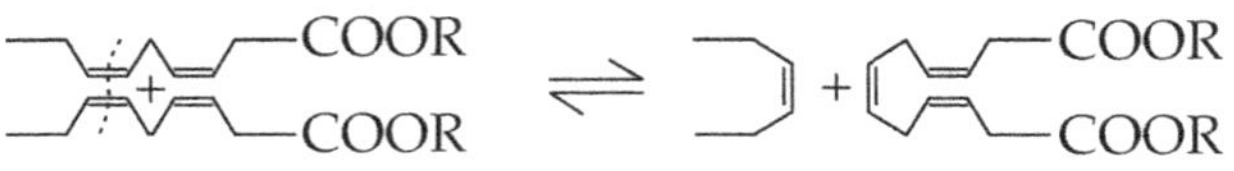

(3.158)

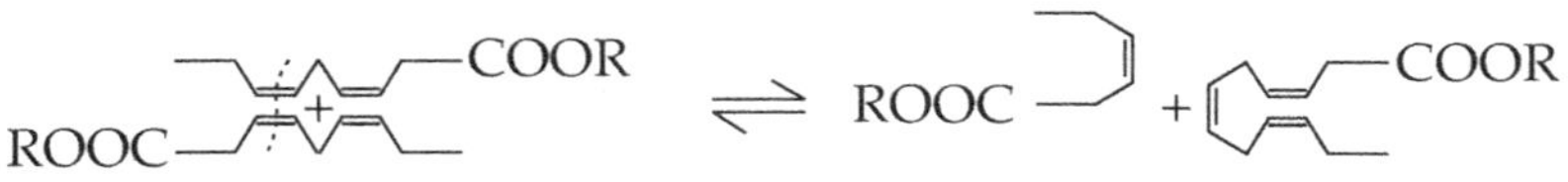

(3.159)

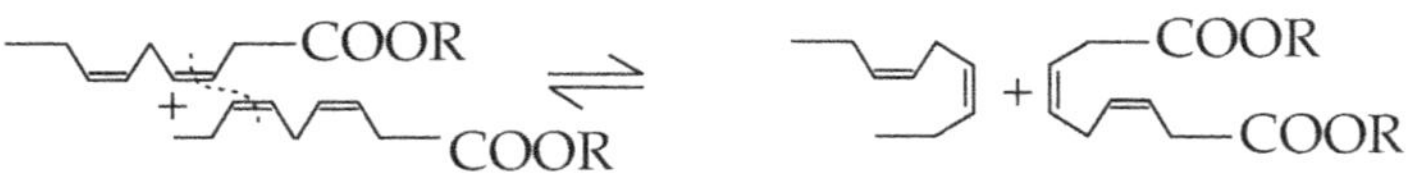

(3.160)

Likewise, all types of polyacetylene can be polymerized by olefin metathesis catalysts by ROMP of cyclo octatetraenes.[128]

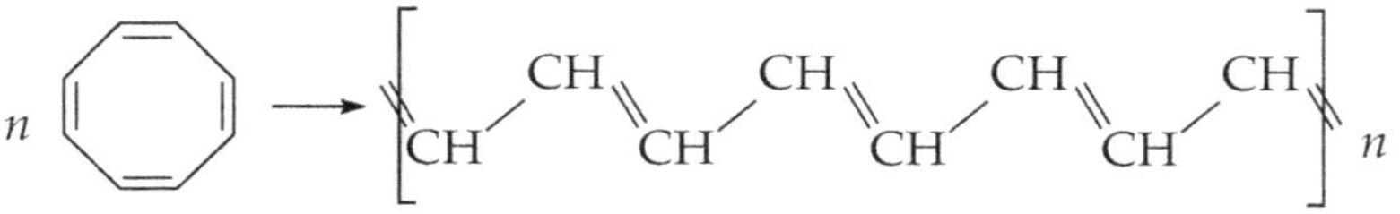

(3.161)

An important question may arise as to the nature of the initiating metal-carbene complex formed *in-situ* from a non-carbene catalyst—cocatalyst—olefin system. Often, an alkyl group attached to the cocatalyst becomes the source of the alkylidene group attached to the metal and this is very much evidenced from the initial products of OM or from the endgroups in the product of ROMP. In the absence of a catalyst bearing an alkyl group, the olefin itself must be involved in generating the initiating metal-carbene complex

$$\underset{\downarrow \atop M}{H_2C{=}CH.CH_3} \longrightarrow \underset{| \atop M{-}H}{HC{=}CH.CH_3} \longrightarrow \underset{\| \atop M}{HC.CH_2.CH_3} \qquad (3.162)$$

Me_4Sn-activated MoO_3 catalysts on Al_2O_3, TiO_2, SnO_2 or SiO_2 supports are usually prepared by impregnation of the support with ammonium molybdate, molybdenum tetraalkyl or $Mo(CO)_6$ followed by heat treatment and alternate exposure to reducing (CO/H_2) and oxidising (O_2/N_2O) environments. [Mo]-carbene complex is believed to be formed *in-situ* during propene metathesis by a 1,2 hydrogen shift mechanism as above.[129]

Polymer degradation by olefin metathesis may yield significant information about structure of polymers, copolymers and technical products containing polymers. In most cases, the test polymers are subjected to degradative olefin metathesis to yield degraded products of lower molecular weight which are identified and quantitatively determined employing gas chromatography and mass spectrometry. Such investigations are also useful in finding evidence for cross-links, determination of fillers and characterization of blends. In some cases, metathesis degradation studies are superior to investigations employing spectroscopic methods. However, the destructive nature of the metathesis degradative process, its time consuming and cost-intensive nature (catalysts and symmetrical olefins needed and the special analytical equipments used are all expensive) obscure much of its advantages.

References

1. Ghosh, P. and S.K. Paul, *J. Macromol. Sci. Chem.* **A20** (1983), 247.
2. Ghosh P. and M.S. Banerjee, *J. Appl. Poylm. Sci.*, **23** (1979), 1923.
3. Arnett, L.M. and J.H. Peterson, *J. Am. Chem. Soc.*, **74** (1952), 2031.
4. Schulz, G.V. and G. Harborth, *Makromol. Chem.*, *1* (1947), 106.
5. O'Driscoll, K.F. and P. Ghosh, *Structure and Mechanism in Vinyl Polymerization*, T. Tsuruta and K.F. O'Driscoll, Eds., Marcel Dekker, New York, 1969, Ch. 3, pp. 59–98.
6. O'Driscoll, K.F. and P.J. White, *J. Polym. Sci.*, **A3** (1965), 283.
7. Bevington J.C., *Proc. Roy, Soc.* (London), **A239** (1957), 420; *Makromol. Chem.*, **34** (1959), 152.
8. Allen, J.K. and J.C. Bevington, *Polymer*, **2** (1961), 265.
9. Deschrijver, F. and G. Smets, J. *Polym. Sci.*, **A1** (4) (1966), 2201.
10. Smith, W.V., *J. Am. Chem. Soc.*, **71** (1949), 4077.
11. Ghosh, P. and K.F. O'Driscoll, *J. Macromol. Sci. Chem.*, **A1** (1967), 1393.
12. Bevington, J.C., J. Toole and L. Trossarelli, *Makromol. Chem.* **32** (1959), 57.
13. Bartlett, P.D. and H. Kwart, *J. Am. Chem. Soc.*, **72** (1950), 1051.
14. Matheson, M.S., E.E. Auer, E.B. Bevilacqua and E.J. Hart, *J. Am. Chem. Soc.*, **73** (1951), 1700.
15. Bamford, C.H., A.D. Jenkins and R. Johnston, *Nature*, **177** (1956), 992.
16. Cohen, S.G., *J. Am. Chem. Soc.*, **67** (1945), 17; **69** (1947), 1057.
17. Mackey, M.H. and H.W. Melville, *Trans. Faraday Soc.*, **46** (1950), 63.
18. Mayo, F.R. and R.A. Gregg, *J. Am. Chem Soc.*, **70** (1948), 1284.
19. Meltzer, T.H. and A.V. Tobolsky, *J. Am. Chem. Soc.*, **76** (1954) 5178.
20. Trommsdorff, E., H. Kohle and P. Lagally, *Makromol Chem.* **1** (1948), 169.
21. Norrish, R.G.W. and R.R. Smith, *Nature*, **150** (1942), 336.
22. Schulz, G.V. and F. Blaschke, *Z. Physik. Chem.*, **B51** (1942), 75.
23. Hayden, P. and H.W. Melville, *J. Polym. Sci.*, **43** (1960), 201.
24. Naylor, M.A. and F. W. Billmeyer, Jr., *J. Am. Chem. Soc.*, **75** (1953), 2181.
25. North, A.M., *Progress in High Polymers*, J.C. Robb and F.W. Peaker, Eds., CRC Press, Cleveland, Ohio, 1968, Vol. 2, 95–133.
26. Ghosh, P. and P.S. Mitra, *J, Polym. Sci. Polym. Chem. Ed.*, **13** (1975) 921.
27. Schulz, G.V., A. Dinglinger and E. Husemann, *Z. Physik. Chem.*, **B43** (1939), 385.

28. Suess, H., K. Pitch and H. Rudorfer, *Z. Physik. Chem.* **A179** (1937), 361.
29. Suess, H. and A. Springer, *Z. Physik Chem.*, **A181** (1937), 81.
30. Schulz, G.V. and E. Husemann, *Z. Physik Chem.* **B36** (1937), 184.
31. Gregg, R.A. and F.R. Mayo, *Disc. Faraday Soc.*, **2** (1947), 328.
32. George, M.H. *Vinyl Polymerization*, G.E. Ham Ed., Mercel Dekker, New York (1967), Vol. 1, Pt. I, Ch. 3.
33. Walling, C., *Free Radical Solution*, John Wiley, New York, (1957), Chs. 3–5.
34. Burnett, G.M. and H.W. Melville, *Proc. Roy. Soc* (London) **A189** (1947), 456, 481 & 494.
35. Dainton, F.S. and W.D. Sisley, *Trans. Faraday Soc.*, **59** (1963), 1369.
36. Block, H., A. Ledwith and A. Taylor, *Polymer*, **12** (1971), 271.
37. Braun, D. and K.H. Beker, *Angew Makromol. Chem.*, **6** (1969), 186; *Makromol. Chem.*, **91** (1971), 147.
38. Ghosh, P. and R. Ghosh, *Eur. Polym. J.*, **17** (1981) 545, & 817.
39. Ghosh, P., R. Ghosh and G. Mukhopadhyay, *Eur. Polym J.*, **16** (1980), 457.
40. Ghosh, P., D.K. Chowdhury and A.R. Bandyopadhyay, *J. Macromol. Sci., Chem.*, **A20** (1963), 549.
41. Ghosh, P., S. Jana and S. Biswas, *J. Polym. Sci., Polym. Chem. Ed.*, **21** (1983) 3347.
42. Ledwith, A., J.A. Boseley and M.D. Purbrick, *J. Oil Colour Chem. Assocn.*, **61** (1978), 95.
43. Chapiro, A., *Radiation Chemistry of Polymer Systems*, Wiley-Interscience, New York 1962, Ch. IV.
44. Brandrup, J. and E.H. immergut, Eds., *Polymer Handbook*, Wiley-Interscience, New York 1966, pp. II (1–133).
45. Burnett, G.M., *Quart Revs.* **4** (1950) 292.
46. Johnston, D.H., and A.V. Tobolsky, *J. Am. Chem. Soc.*, **74** (1952), 983 and Baysal, B., and A.V. Tobolsky, J. Polym. Sci., **8** (1952), 529.
47. Ghosh, P., A.N. Banerjee, P.S. Mitra and S. Chakraborty, *J. Polym. Sci., Polym. Letters Ed.*, **13** (1975), 35.
48. Ghosh, P. and G. Mukhopadhyay, *Makromol. Chem.*, **180** (1979), 2253.
49. Ghosh, P., *J. Polym. Sci.*, **B7** (1969), 293.
50. Goldfinger, G., I. Skeist and H. Mark, *J. Phys. Chem.*, **47** (1943), 578.
51. Bartlett, P.D., G.S. Hammond and H. Kwart, *Disc. Faraday Soc.*, **2** (1947), 342.
52. Schulz, G.V. and H. Kammerer, *Chem. Ber.*, **80** (1947), 327.
53. Georgieff, K.K., *J. Apply. Polym. Sci.*, **9** (1965), 2009.
54. Breitenbach, J.W., A. Springer and K. Horeischy, *Ber*, **71** (1938), 1438; **74** (1941), 1386.
55. Hauser, C.F. and N.L. Zutty, *Polymer Preprints*, **8**(1) (1967), 369.
56. Mayo, F.R. and R.A. Gregg, *J. Am. Chem. Soc.*, **70** (1948), 1284.
57. Schulz, G.V., *Chem. Ber.*, **80** (1947), 232.
58. Price, C.C. and D.A. Durham, *J. Am. Chem. Soc.*, **65** (1943), 757.

59. Price, C.C., *J. Am. Chem. Soc.*, **65** (1943), 2380.
60. Tabata, Y., K. Ishigura, K. Oshima and H. Sobue, *J. Polym. Sci.*, **A2** (1964), 2445.
61. Ghosh, P. and N. Mukherji, *Eur. Polym. J.*, **17** (1981), 541; **19** (1985) 493.
62. Barnes, C.E., *J. Am. Chem. Soc.*, **67** (1945), 217.
63. Barnes, C.E., R.M. Elofson and G.D. Jones, *J. Am Chem. Soc.*, **72** (1950), 210.
64. Bovey, F.A. and I.M. Kolthoff, *J. Am. Chem. Soc.*, **69** (1947), 2143.
65. Ghosh, P. and S. Chakraborty, *Eur. Polym. J.*, **15** (1979), 137.
66. Ghosh, P. and S.N. Maity, *Eur. Polym. J.*, **15** (1979), 787.
67. Ghosh, P. and S. Biswas, *J. Macromol. Sci. Chem.*, **A16** (1981), 1033.
68. Mulier, M.E., *J. Polym. Sci.*, **58** (1962), 1009.
69. Ghosh, P. and D.K. Chowdhury, *J. Polym. Sci., Polym. Chem. Ed.*, **23** (1985), 1407.
70. Ferrington, T.E. and A.V. Tobolsky, *J. Am. Chem. Soc.*, **77** (1955), 4510; **80** (1958), 3215.
71. Otsu, T. and K. Nayatani, *Makromol. Chem.*, **27** (1958), 149.
72. Melville, H.W. and W.F. Watson, *Trans. Faraday Soc.*, **44** (1948), 886.
73. Dainton, F.S. and K.J. Ivin, *Quart. Revs.* **12** (1958), 61.
74. Brandrup, J. and E.H. Immergut Eds. *Polymer Handbook*, Wiley-Interscience, New York, 1966, pp. II. 390–394.
75. Gobran, R.H., M.E. Berenbaum and A.V. Tobolsky, *J. Polym. Sci.*, **46** (1960), 431.
76. Bartlett, P.D. and R. Altschul, *J. Am. Chem. Soc.*, **67** (1945), 812 & 816.
77. Nozaki, K., *Disc. Faraday Soc.*, **2** (1947), 337.
78. Josefowitz, D. and H. Mark, *Polymer Bull.*, **1** (1945), 140.
79. Ghosh, P. and F.W. Billmeyer, Jr., *Advances in Chemistry Series*, **91** (1969), 75.
80. Ghosh, P. and A.N. Banerjee, *J. Polym. Sci., Polym. Chem. Ed.*, **12** (1974), 375; P. Ghosh et al., *ibid*, **11** (1973), 2021.
81. Kamenskaya, S. and S.S. Medvedev, *Acta Physicochim.* USSR, **13** (1940), 565.
82. Walling, C. and L. Heaton, *J. Am. Chem. Soc.*, **87** (1965), 48.
83. Ghosh, P. and S. Biswas, *Makromol. Chem.*, **182** (1981), 1985.
84. Matheson, M.S., *J. Chem. Phys.*, **13** (1945), 584.
85. Jenkins, A.D., *J. Polym. Sci.*, **29** (1958), 245.
86. Bamford, C.H., A.D. Jenkins and R. Johnston, *Trans. Faraday Soc.*, **55** (1959), 1454.
87. Allen, P.E.M. and C.R. Patrik, *Makromol. Chem.*, **47** (1961), 154.
88. Baldwin, M.G., *J. Polym. Sci.*, **A1** (1963), 3209.
89. Ito, K. and T. Matsuda, *J, Appl. Polym. Sci.*, **14** (1970), 311.
90. Deb, P.C. and G. Meyerhoff, *Eur. Polym. J.*, **10** (1974), 709.
91. Ghosh, P. and G. Mukhopadhyay, *J. Polym. Sci. Polym. Chem. Ed.*, **17** (1979), 583; **18** (1980), 283.
92. Ghosh, P. and T.K. Ghosh, *J. Polym. Sci. Polym. Chem. Ed.*, **22** (1984), 2295.
93. Ghosh, P., S. Biswas and S. Jana, *J. Polym. Materials*, **1** (1984), 167.

94. Deb, P.C., *Eur. Polym. J.*, **11** (1975), 31.
95. Ghosh, P. and P.S. Mitra, *J. Polym. Sci., Polym. Chem. Ed.*, **15** (1977), 1743.
96. Ghosh, P., P.K. Sengupta and N. Mukherji, *J. Polym. Sci. Polym. Chem. Ed.*, **17** (1979), 2119.
97. Burnett, G.M., W.S. Daily and J.M. Pearson, *Trans. Faraday Soc.*, **61** (1965), 1216.
98. Anderson, D.B., G.M. Burnett and A.C. Gowan, *J. Polym. Sci.*, **A1** (1963), 1465.
99. Tobolsky, A.V. and L.R. Matlak, *J. Polym. Sci.*, **55** (1961), 49.
100. Nozaki, K. and P.D. Bartlett, *J. Am. Chem. Soc.*, **68** (1946), 1686.
101. Bamford, C.H. and S. Brumby, *Makromol. Chem.*, **96** (1966), 221; **105** (1967), 122.
102. Henrici-Olive, G. and S. Olive, *Makromol. Chem.*, **68** (1963), 219; *Z. Physik. Chem.*, **47** (1965), 285; **48** (1966), 35, 51.
103. Zafar, M.M., *Makromol. Chem.*, **157** (1972), 219.
104. Burnett, G.M., G.G. Cameron and M.M. Zafar, *Eur. Polym. J.*, **6** (1970), **823**.
105. North, A.M. and G.A. Reed, *Trans. Faraday Soc.*, **57** (1961), 859: *J. Polym. Sci.*, **A1** (1963), 1311.
106. North, A.M. and D. Postlethwaite, *Polymer*, **5** (1964), 237; *Trans. Faraday Soc.*, **62** (1966), 2843.
107. Stockmayer,, W.H. and L.H. Peebles, *J. Am. Chem. Soc.*, **75** (1953), 2278.
108. Ghosh, P. and G. Mukhopadhyay, *Eur. Polym. J.*, **15** (1979), 141.
109. Briers, F., D.L. Chapman and E. Walters, *J. Chem. Soc.*, (1926), 562.
110. Noyes, W.A. and P.A. Leighton, *The Photochemistry of Gases*, Reinhold, New York, 1941, p. 202.
111. Matheson, M.S., E.E. Auer, E.B. Bevillaqua and E.J. Hart, *J. Am. Chem. Soc.*, **71** (1949), 497.
112. Ivin, K.J., *Olefin Metathesis*, Academic Press, London 1983.
113. Banks, R.L., *Discovery and Development of Olefin Disproportionation (Metathesis)*, ACS Symposium Series, **222** (1983) 403.
114. Dolgoplosk, B.A. and Y.V. Korshak, *Russ. Chem. Rev.* **53** (1984), 36.
115. Imamoglu, Y., B. Zumreoglu-Karan and A. J. Amass, Eds., *Olefin Metathesis and Polymerization Catalysts*, NATO ASI Series C 326, Kluwer Academic, Dordrecht, 1990.
116. Schubert, U. Ed., *Advances in Metal Carbene Chemistry*, NATO ASI Series **C 269**, Kluwer Academic, Dordrecht 1989.
117. Dall'Asta, G. Mazzanti, G. Natta and L. Porri, *Makromol. Chem.*, **56** (1962), 224.
118. Natta, G, G. Dall'Asta and G. Mazzanti, *Amgew. Chem. Int. Ed.*, **3** (1964) 723.
119. Natta, G., G. Dall'Asta, and L. Porri, *Makromol. Chem.*, **81** (1965) 253.
120. Rinchart, R.E., and H.P. Smith, *J. Polym. Sci., Polym. lett. Ed.*, **B3** (1965), 1049.
121. Szwarc, M., *Adv. Polym. Sci.*, **49** (1983) 1.
122. Higashimura, R., M. Miyamoto and M. Sawamoto, *Macromolecules*, **18** (1985), 611.
123. Sogah, D. K., W.R. Hertler, O.W. Webster and G. M. Cohen, *Macromolecules*, **20** (1987) 1973.
124. Otsu, T., K. Yamashita and K. Tsuda, *Macromlecules*, **19** (1986) 287.
125. Kress, J., J. A. Osborn, R.M.E. Greene, K.J. Ivin, and J.J. Rooney, *J. Chem. Soc., Commun.* (1985) 874.

126. Norton, R.L. and T. J. McCarthy, *Macromolecules*, **22** (1989) 1022.
127. Van Dam, P.B., M.C. Mittelmeijir and C. Boelhower, *J. Chem. Soc., Chem. Commun.*, (1972) 1221.
128. Korshak, Y. V., V.V. Korshak, G. Kanischka and H. Hockner, *Makromol. Chem., Rapid Commun.*, **6** (1985) 685.
129. Iwasawa, Y. and H. Hamamura, *J. Chem. Soc., Chem. Commun*, (1983) 130.

CHAPTER 4

Copolymerization and Techniques of Polymerization

4.1 Concept of Copolymerization

A polymer whose chain molecules are composed of more than one kind of repeating chemical units is commonly called a copolymer. A simple case of a step-growth copolymer is the linear unsaturated polyester formed by intermolecular condensation of a mixture of phthalic acid (anhydride) (PA) and maleic acid (anhydride) (MA) with a given diol, say ethylene glycol (EG) resulting in what may be termed as a copolyester with at least two types of distinctly identifiable repeating units in the chain molecules:

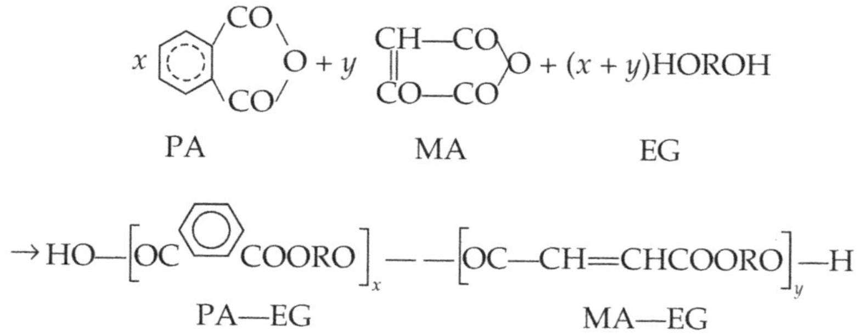

(4.1)

Equation (4.1), showing formation of a copolyester by a polycondensation process clearly identifies two types of repeating chemical units, (PA–EG) and (MA–EG), though in a very oversimplified manner. In the actual copolymer the two types of distinguishable chemical units would normally appear at random along the chain, and the overall composition would depend on the relative values of x and y.

Similarly, one can prepare step-growth copolymers with different interunit chemical linkages by proper selection of combinations of different bifunctional monomeric species bearing, —OH, —COOH, —NH_2, —NCO or other functional groups, and using two, three or all the functional species simultaneously or in a phased manner in the presence or absence of a related polyfunctional and monofunctional species during the synthesis. However, polycondensation leading to formation of copolymers from this kind of complicated mixtures of monomeric species is of little or rare practical importance or relevance. More relevant step-growth copolymerization systems involve formation of copolymer structures with a specific interunit linkage such as amide (—CONH—), ester (—COO—), urethane (—NHCOO—), etc., following the approach as in Eq. (4.1). It may also involve chain extension and/or cross-linking reaction introducing similar or different kinds of interunit linkages, using low molecular weight linear or branched polymers and selected bifunctional monomeric species. Examples of this kind of step-growth copolymers are found in the making of polyester urethanes or polyether urethanes, using low molecular weight preformed linear polyesters or polyethers with hydroxyl end groups and allowing them to further react with each other through diisocyanates leading to chain extension and producing useful elastomeric products commonly known as

$$n\ \text{HO-polyester—OH} + n\ \text{OCNRNCO} \rightarrow \sim\sim[\text{polyester—OOCNHRNHCOO}]_n\sim\sim \quad \text{(polyesterurethane)} \tag{4.2}$$

$$n\ \text{HO-polyether—OH} + n\ \text{OCNRNCO} \rightarrow \sim\sim[\text{polyether—OOCNHRNHCOO}]_n\sim\sim \quad \text{(polyetherurethane)}$$

polyurethane rubbers, which may be vulcanized or cross-linked when heated with excess diisocyanates in a mould, e.g.,

```
OCNRNCO+ ~~~RNHCOO—polyester ~~~
      →  ~~~RNCOO—polyester ~~~                    (4.3)
            \
             CO
               \
                NH
                  \
                   R
                    \
                     NH
                     |
                     CO
                       \
  ~~~polyester—OOCNR ~~~
```

An altogether different kind of copolymer is obtained by polymerization of two or more vinyl and related monomers, all taken together in a reaction vessel. The copolymer molecules contain segments of all the monomers used in their structure, usually in a calculable proportion depending on many factors including the feed monomer composition, the chain-growth mechanism and the inherent reactivities of the monomers with a given chain carrying species.

The number of reactions to be necessarily considered to describe the copolymerization of more than one monomer increases geometrically with increase in the number of participating monomers, giving increasingly varied and complicated structures of the copolymer molecules at the same time. A case of a binary copolymerization involving the two monomers M_1 and M_2 may be simply represented as:

$$\underset{\text{monomer mixture}}{nM_1 + mM_2} \rightarrow \underset{\text{copolymer}}{\sim\!\sim(M_1)_n\text{—}(M_2)_m\sim\!\sim} \qquad (4.4)$$

The structure of the vinyl copolymer shown in Eq. (4.4) is far too simplified, however; the two monomer units appear in the copolymer structure in a random sequence in general.

4.2 Binary Copolymerization of Vinyl Monomers by Free Radical Mechanism

4.2.1 Analysis of the System and the Reactions Involved

In a free radical copolymerization, the number of chain radicals to be considered is given by the number of monomers present and the reactivity of a chain radical is practically determined by the free radical bearing terminal monomer unit and the size and chemical nature of those preceding the radical bearing terminal unit in the chain are considered to be of little consequence in this context.

In a binary copolymerization, therefore, two different chain radicals can be identified and each radical being capable of adding both the monomers, though not usually or necessarily with equal or comparable ease, the system is characterized by four simultaneously occurring propagation reactions. With two different kinds of chain radicals, and considering termination taking place only bimolecularly, the process is associated with three different chain termination reactions along with the possibility of several chain transfer reactions. Chain initiation takes place by reaction of primary radicals (derived from decomposition of the radical initiator used) with either of the monomers present with almost equal ease and efficiency in general or else, in the absence of an initiator, by thermal or photo activation of either monomer with different relative ease depending on the chemical nature of the monomers used.

4.2.2 Kinetics of Chain Propagation in Binary Copolymerization and Copolymer Composition

The four chain-propagation reactions in a binary copolymerization may be described by the reaction scheme (4.5) shown below:

$$
\begin{aligned}
M_1^{\bullet} + M_1 &\xrightarrow{k_{1,1}} M_1^{\bullet} && \text{(i)} \\
M_1^{\bullet} + M_2 &\xrightarrow{k_{1,2}} M_2^{\bullet} && \text{(ii)} \\
M_2^{\bullet} + M_2 &\xrightarrow{k_{2,2}} M_2^{\bullet} && \text{(iii)} \\
M_2^{\bullet} + M_1 &\xrightarrow{k_{2,1}} M_1^{\bullet} && \text{(iv)}
\end{aligned}
\qquad (4.5)
$$

Here, M_1 and M_2 are the two monomers, and $M_1^{\bullet}$ and $M_2^{\bullet}$ represent the chain radicals of all sizes with free radical bearing terminal units M_1 and M_2, respectively. $k_{1,1}$ and $k_{1,2}$ are rate constants for homopropagation and cross-propagation reactions respectively involving $M_1^{\bullet}$; likewise, $k_{2,2}$ and $k_{2,1}$ refer to the homopropagation and cross-propagation rate constants respectively involving $M_2^{\bullet}$. For a given rate constant, the first numeral is indicative of the reacting chain radical and the second numeral indicates the reacting monomer. Homopropagation relates to reaction between like radical and monomer and cross propagation relates to reaction between unlike radical and monomer.

Neglecting monomer consumption in the initiation step, the rates of consumption of the two monomers are written as

$$-(d[M_1]/dt) = k_{1,1}[M_1^{\bullet}][M_1] + k_{2,1}[M_2^{\bullet}][M_1] \qquad (4.6)$$

$$-(d[M_2]/dt) = k_{1,2}[M_1^{\bullet}][M_2] + k_{2,2}[M_2^{\bullet}][M_2] \qquad (4.7)$$

An expression describing relative incorporation of the two monomeric units in the copolymer may be conveniently obtained by elimination of the radical concentration terms using the steady-state concept for each chain radical and then combining Eqs (4.6) and (4.7). Under steady state, the concentration of $M_1^{\bullet}$ and $M_2^{\bullet}$ remaining constant, the rates of interconversion of chain radicals must be equal, which prescribes that rates of reactions (ii) and (iv) of scheme (4.5) are equal, i.e.

$$k_{1,2}[M_1^{\bullet}][M_2] = k_{2,1}[M_2^{\bullet}][M_1] \qquad (4.8)$$

Now, combining Eqs (4.6) and (4.7) and expressing $[M_1^{\bullet}]$ in terms of $[M_2^{\bullet}]$ (or vice versa) with the help of Eq. (4.8), one obtains:

$$\frac{d[M_1]}{d[M_2]} = \frac{k_{1,1}[M_1^{\bullet}][M_1] + k_{2,1}[M_2^{\bullet}][M_1]}{k_{2,2}[M_2^{\bullet}][M_2] + k_{1,2}[M_1^{\bullet}][M_2]} \tag{4.9}$$

or,

$$\frac{d[M_1]}{d[M_2]} = \frac{\dfrac{k_{1,1} \cdot k_{2,1}[M_2^{\bullet}][M_1]^2}{k_{1,2}[M_2]} + k_{2,1}[M_2^{\bullet}][M_1]}{k_{2,2}[M_2^{\bullet}][M_2] + \dfrac{k_{1,2} \cdot k_{2,1}[M_2^{\bullet}][M_1][M_2]}{k_{1,2}[M_2]}}$$

or,

$$\frac{d[M_1]}{d[M_2]} = \frac{[M_1]}{[M_2]} \cdot \frac{k_{2,1}\dfrac{k_{1,1}}{k_{1,2}} \cdot \dfrac{[M_1]}{[M_2]} + k_{2,1}}{k_{2,2} + k_{2,1}\dfrac{[M_1]}{[M_2]}} \tag{4.10}$$

Dividing the numerator and the denominator of the right hand side of Eq. (4.10) by $k_{2,1}$, one obtains

$$\frac{d[M_1]}{d[M_2]} = \frac{[M_1]}{[M_2]} \cdot \frac{\dfrac{k_{1,1}}{k_{1,2}} \cdot \dfrac{[M_1]}{[M_2]} + 1}{\dfrac{k_{2,2}}{k_{2,1}} + \dfrac{[M_1]}{[M_2]}} \tag{4.11}$$

In Eq. (4.11), the left hand side represents relative molar increment of the two monomers in the copolymer. For copolymerization to low conversions, this may be taken as the molar ratio of the two monomer units incorporated in the copolymer, i.e., $([M_1]/[M_2])_{copolymer}$, while the ratio $[M_1]/[M_2]$ on the right hand side of the equation relates to the feed monomer ratio, i.e. molar ratio of the two monomers in the feed monomer mixture, $([M_1]/[M_2])_{feed}$. Expressing the ratio of the rate constants by the parameters r_1 and r_2 as shown below:

$$\frac{k_{1,1}}{k_{1,2}} = r_1, \quad \text{and} \quad \frac{k_{2,2}}{k_{2,1}} = r_2 \tag{4.12}$$

one may readily express Eq. (4.11) as

$$\left(\frac{[M_1]}{[M_2]}\right)_{copolymer} = \left(\frac{[M_1]}{[M_2]}\right)_{feed} \cdot \frac{r_1\left(\dfrac{[M_1]}{[M_2]}\right)_{feed} + 1}{r_2 + \left(\dfrac{[M_1]}{[M_2]}\right)_{feed}} \tag{4.13}$$

Equation (4.13) is known as the copolymer composition equation or simply the copolymer equation relating the copolymer composition and feed monomer composition with the help of the kinetic parameters r_1 and r_2, commonly called the monomer reactivity ratios. Equation (4.13) clearly expresses that the copolymer composition is dependent on the molar ratio of the two monomers in the feed (but not actually on the absolute concentrations of the two monomers) and on the kinetic parameters r_1 and r_2, i.e. the monomer reactivity ratios. The copolymer equation may also be expressed as:

$$\left(\frac{[M_1]}{[M_2]}\right)_{\text{copolymer}} = \frac{[M_1]}{[M_2]} \cdot \frac{r_1[M_1]+[M_2]}{r_2[M_2]+[M_1]} \qquad (4.14)$$

where $[M_1]$ and $[M_2]$ on the right hand side stand for molar concentrations of monomer M_1 and M_2 in the reactant or feed monomer mixture.

The copolymer equation may also be alternatively expressed using copolymer and monomer compositions as mole fractions instead of mole ratios. If F_1 and F_2 are the mole fractions of the monomer segments M_1 and M_2 respectively in the copolymer formed ($F_1 = 1 - F_2$) and if f_1 and f_2 are the mole fractions of the monomers M_1 and M_2 in the reactant monomer mixture ($f_1 = 1 - f_2$) then one obtains from Eq. (4.11) or (4.13) the following useful expression:

$$F_1 = \frac{(r_1 f_1^2 + f_1 f_2)}{(r_1 f_1^2 + 2f_1 f_2 + r_2 f_2^2)} \qquad (4.15)$$

4.2.3 Significance of Monomer Reactivity Ratios

Examination of one or the other forms of copolymer equation clearly shows that with a knowledge of the feed monomer composition, the copolymer composition may be easily calculated if the monomer reactivity ratios r_1 and r_2 are known. The value of $r_1 > 1$ means that the chain radical $M_1^{\bullet}$ has a preference to add the like monomer, i.e. M_1 over the unlike monomer, M_2, and r_1 value < 1 means just the opposite. The same concept applies to r_2, which gives relative preference of chain radical $M_2^{\bullet}$ to add the like monomer (M_2) over the unlike monomer (M_1).

The absence of the initiation and termination rate constants in the copolymer composition equation, makes it readily apparent that the (instantaneous) copolymer composition is independent of rate of initiation and hence of the overall rate of polymerization and of initiator concentration. The composition is also not measurably affected in most cases due to the presence of solvents or other additives, indicating that the kinetic factors controlling the copolymer composition, i.e. the monomer reactivity ratios r_1 and r_2, remain unaffected by variations in initiator concentration, the overall rate of reaction and by the presence of various additives and solvents. They also usually remain unaffected even in heterogeneous systems as long as the reacting monomers are evenly distributed between the phases. However, the composition of copolymer would vary for differential distribution of monomers between phases.[1,2] The values of r_1 and r_2 also change significantly for a drastic change in the mechanism of chain propagation from free radical to ionic, and even for ionic mechanism, from cationic to anionic mode of chain propagation.[3] The copolymer equation has been found applicable to almost all comonomer combinations and equally applicable and relevant to radical, cationic and anionic systems, despite wide variations in r_1 and r_2 values depending on the mode of chain propagation. In radical copolymerization, the effect of temperature on monomer reactivity ratios is small or marginal in view of relatively low activation energies of radical propagation reactions and small differences in activation energies of different propagation reactions.

4.3 Types of Copolymerization

4.3.1 Ideal Copolymerization

It is the widespread observation and experience that the mole fraction F_1 is seldom equal to f_1 and as a consequence of this feature, both f_1 and F_1 keep on changing with progress of copolymerization. The copolymer produced over a specified time period or range of conversion will consist of a mixture of copolymer molecules differing progressively in their compositions (F_1).

A copolymerization is said to be ideal when each of the two radicals exhibits the same preference to add one of the monomers over the other, i.e.

$$k_{1,1}/k_{1,2} = k_{2,1}/k_{2,2}$$

or

$$r_1 r_2 = 1 \tag{4.16}$$

Equation (4.14) then reduces to:

$$\left(\frac{[\mathrm{M}_1]}{[\mathrm{M}_2]}\right)_{\text{copolymer}} = r_1 \left(\frac{[\mathrm{M}_1]}{[\mathrm{M}_2]}\right)_{\text{feed}} \tag{4.17}$$

and Eq. (4.15) takes the form:

$$F_1 = r_1 f_1/(r_1 f_1 + f_2) \tag{4.18}$$

In Fig. 4.1, plots of F_1 vs f_1 for several cases of ideal copolymerization are shown and the values of $r_1(= 1/r_2)$ for each curve are given. For $r_1 = 1$, the composition curve reduces to a straight line given by the diagonal and representing a special case in which $r_1 = r_2 = 1$ and hence $F_1 = f_1$. For $r_1 = r_2 = 1$, $k_{1,1} = k_{1,2}$ and $k_{2,2} = k_{2,1}$ or in other words, the two monomers are equally reactive with each radical and the copolymer composition, arising out of a random placement of the two monomers along the chain molecule, is equal to the feed monomer composition over the full range of monomer composition. In case of r_1 or $r_2 \neq 1$, i.e. for $r_1 > 1$ and $r_2 < 1$, or $r_1 < 1$ and $r_2 > 1$, one of the reacting monomers is more reactive than the other toward the two propagating species, and consequently, the copolymer formed will incorporate randomly a higher proportion of the more reactive monomer in its chain structure. The copolymer is proportionately richer in M_1 in comparison with the feed monomer mixture for $r_1 > 1$, and just the opposite composition characteristics would result for $r_2 > 1$, i.e., for $r_1 < 1$. With large differences in the reactivities of the two monomers, (i.e., in the r_1 and r_2 values) the formation of copolymer sequences becomes progressively difficult as a consequence of very fast depletion of the more reactive monomer from the monomer mixture. Thus, for $r_1 = 0.1$ and $r_2 = 10$, giving $r_1r_2 = 1.0$ copolymers incorporating good amounts of M_1 in its structure is difficult to obtain.

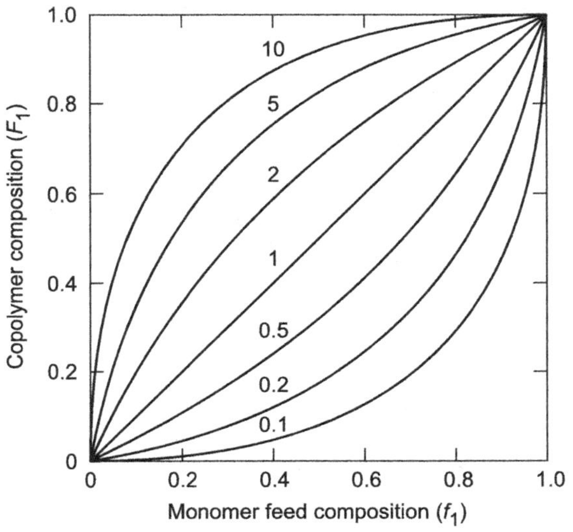

Fig. 4.1 *Plot of instantaneous copolymer composition (mole fraction F_1) as a function of monomer feed composition (mole fraction f_1) for indicated values of r_1 for ideal copolymerization giving $r_1r_2 = 1$; values of r_1 shown*

4.3.2 Alternating Copolymerization

The other typical case of copolymerization of widespread practical relevance is given by $r_1r_2 < 1$. In radical copolymerization, one would seldom experience an instance for which $r_1r_2 > 1$. Although, one finds limited instances of copolymerization approaching ideal behaviour ($r_1r_2 = 1$), the product of the reactivity ratios, i.e. r_1r_2 is less than unity in most cases.

At one extreme, a real case of radical copolymerization may be such as to give very small or even negligible values for r_1 and r_2, which, in other words, means that the system is characterized by very low or negligible values of $k_{1,1}$ and $k_{2,2}$. The shift in the value of the r_1r_2 product from unity to very close to zero leads to higher trends toward alternation, i.e. alternate sequence of the two monomer units in the copolymer structure. The alternation is perfect for both r_1 and r_2 being equal to zero, where homopropagation is strictly prohibited and chain growth takes place only by cross propagation. In this limiting case, the copolymerization equation reduces to:

$$\left(\frac{[M_1]}{[M_2]}\right)_{\text{copolymer}} = 1; \quad \text{or} \quad F_1 = 0.5 \tag{4.19}$$

The copolymer then assumes a 1 : 1 alternating structure as

$$—M_1—M_2—M_1—M_2—M_1—M_2—M_1—M_2— \tag{4.20}$$

shown in chain sequence (4.20) irrespective of the feed monomer composition. Copolymerization of stilbene ($C_6H_5—CH = CH—C_6H_5$) or of allyl acetate with maleic anhydride may be cited as examples closely resembling the case of alternation in copolymerization.[4,5] Almost a perfect brand of alternation is shown by copolymers of sulphur dioxide and certain olefins, cycloolefins or vinyl compounds; they are commonly known as polysulphones and are obtained by radical polymerization of the respective monomer in liquid sulphur dioxide.[6-8] The alternation is thought to arise as a consequence of self-polymerization of the 1 : 1 complex formed *in situ* between the olefin or related monomer (donor) and sulphur dioxide (acceptor), (*See* also Eq. (4.31) in Sec. 4.7.4).

4.3.3 Azeotropic Copolymerization

Some theoretical curves showing copolymer compositions (F_1) for various feed monomer compositions (f_1) corresponding to different r_1 and r_2 values, where $r_1r_2 <$ 1 (and also for some hypothetical cases given by $r_1r_2 > 1$), are shown in Fig. 4.2. It may be seen that for both r_1 and $r_2 < 1$ (or > 1), the curves intersect the diagonal.

i.e. the line representing $F_1 = f_1$. The point of intersection for a given system corresponds to a condition in which polymerization proceeds without change in composition and the composition of the polymer being formed is just the same as that of the monomer mixture at that point. Copolymerization of these critical mixtures of monomers is commonly known as azeotropic copolymerization. The critical monomer composition is then given by:

$$\left(\frac{[M_1]}{[M_2]}\right)_{\text{feed}} = \frac{1-r_2}{1-r_1} \tag{4.21}$$

or,

$$(f_1)_c = \frac{1-r_2}{2-r_1-r_2} \tag{4.22}$$

The calculated value of $(f_1)_c$ is obtained within the premissible range of $0 < (f_1)_c < 1$ only when both r_1 and r_2 are less than unity (representing the real azeotropic systems) or greater than unity (representing hypothetical systems). No critical composition exists for $r_1 > 1$ and $r_2 < 1$ or vice versa (Fig. 4.2).

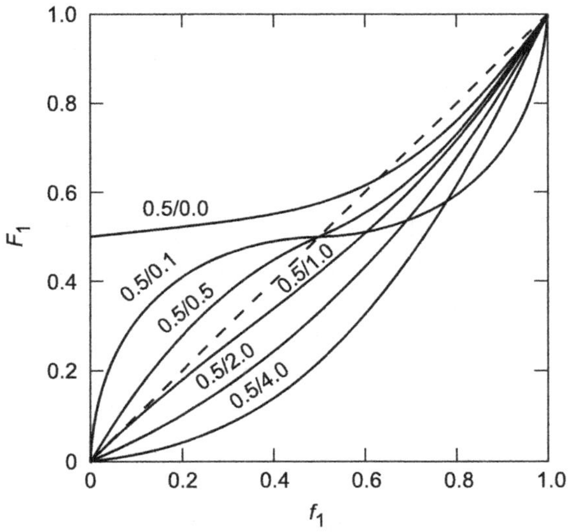

Fig. 4.2 *Plot of instantaneous copolymer composition (mole fraction F_1) as a function of monomer feed composition (mole fraction f_1) for indicated values of r_1/r_2. The broken line or the diagonal is for $F_1 = f_1$, corresponding to $r_1r_2 = 1$*

4.4 Average Copolymer Composition

The average composition of a binary copolymer obtained by polymerization of the two monomers M_1 and M_2 together over a finite range of conversion is conveniently calculated following the method and approach of Skeist.[9] If [M] is the total or overall moles of the two monomers in the initial mixture, then $[M] = [M_1] + [M_2]$, and initially $[M_1]$ is $f_1[M]$. For $d[M]$ moles of monomers converted to polymer, the number of moles of M_1 polymerized is $F_1 d[M]$. As during the process f_1 changes by df_1, and considering $F_1 > f_1$, i.e., the copolymer formed is richer in M_1 than the feed monomer mixture, the unreacted monomer mixture will contain $(f_1 - df_1)([M] - d[M])$ moles of monomer M_1. The moles of M_1 going into the copolymer structure being equal to the difference in the moles of M_1 in the monomer mixture before and after the reaction, one readily obtains

$$f_1[M] - (f_1 - df_1)([M] - d[M]) = F_1 d[M] \tag{4.23}$$

On simplification (neglecting the very small term $df_1 d[M]$), Eq. (4.23) reduces to:

$$d[M]/[M] = df_1/(F_1 - f_1) \tag{4.24}$$

The integral form of Eq. (4.24) from initial feed composition $(f_1)_0$ to the value f_1 is given by:

$$\int_{[M]_0}^{[M]} d[M]/[M] = \ln([M]/[M]_0) = \int_{(f_1)_0}^{f_1} df_1/(F_1 - f_1) \tag{4.25}$$

calculation of F_1 as a function of f_1 may be readily done with the help of Eq. (4.15) for a given set of values of r_1 and r_2. A knowledge of $(F_1 - f_1)$ then permits graphical integration of Eq. (4.25) to give the required degree of conversion $\{([M]_0 - [M])/[M]_0\}$ where $[M]_0$ is the initial value of [M], for a change in the feed composition from $(f_1)_0$ to f_1. By successive application of this method for different values of f_1 a relationship between f_1 and degree of conversion may be obtained. The integral composition of the cumulative mass of copolymer formed during a given conversion stage is readily calculated from the shift in the value of f_1 over the corresponding time span. This allows graphical expression of the instantaneous composition of the copolymer as a function of conversion.

Figure 4.3 shows how the average composition of the copolymer changes with degree of conversion[10] for the styrene (M_1)-methyl methacrylate (M_2) system for a case specified by $(f_1)_0 = 0.80$ and $(f_2)_0 = 0.20$ and $r_1 = 0.53$ and $r_2 = 0.46$. The instantaneous value of F is seen to follow f to unity (for M_1) or to zero (for M_2), but the average copolymer composition is seen to change marginally in comparison, becoming slightly but progressively richer in styrene with conversion.

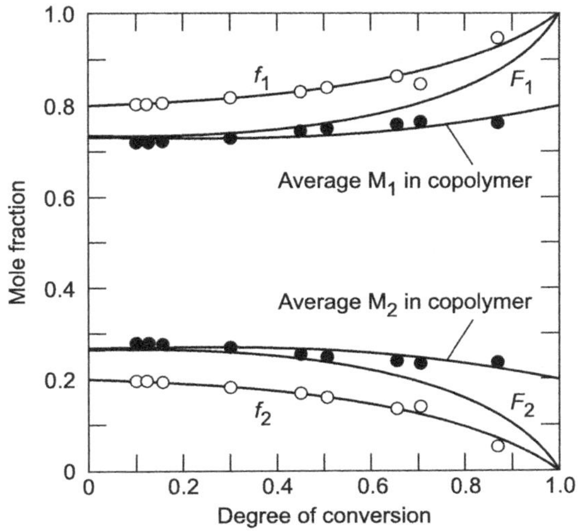

Fig. 4.3 *Plot of copolymer and monomer feed composition as a function of conversion*[10] *for styrene* (M_1)*–methyl methacrylene* $(M)_2$ *system with* $(f_1)_0 = 0.80$ *and* $(f_2)_0 = 0.20$ *and* $r_1 = 0.53$ *and* $r_2 = 0.46$ *(Meyer and Chan, 1967)*

For laboratory experiments, copolymerizations done to low conversions (within 5%) are better understood with the help of Eq. (4.15), in view of f and F values as well as the average copolymer composition remaining practically unchanged over the limiting conversion zone. But this is impractical for an industrial process where replacement or replenishment of the depleting monomer at regular intervals is usually practised to minimize or eliminate variations in the monomer composition. For styrene–methyl methacrylate system, the reactivity ratios are quite close; so, it gives rise to rather limited variations in the average copolymer composition. But even such systems are also subject to the question as to just how much variation in the average composition can be tolerated without detrimental effects on the useful or desirable properties of the copolymer product. The replenishment of the depleting monomer (i.e. the more reactive monomer) often through continuous addition is usually worth the added cost that it entails.

4.5 Determination of Monomer Reactivity Ratios

For experimental evaluation of the monomer reactivity ratios, r_1 and r_2, copolymerization of the two monomers M_1 and M_2 at a number of feed compositions are done to low conversions, usually within 5%; the copolymers formed are then isolated, purified usually by repeated precipitation and finally analyzed for their compositions employing elemental analysis, determination of reactive groups, tracer (radioactive) technique, spectroscopic methods, etc. The feed and copolymer composition data may then be treated according to the following approaches for evaluation of r_1 and r_2.

(i) *Trial and error or curve-fitting method* The method utilizes the experimental data on theoretical curves (F_1 *vs* f_1) to find which theoretical curve gives the best fit on trial selection of r_1 and r_2 values. This turns out to be a poor and tedious approach and precise determination of r_1 and r_2 is often difficult in view of insensitiveness of the composition curve to small changes of r_1 and r_2.

(ii) *Method of intersection* For this method Eq. (4.13) is rearranged to express r_2 in terms of r_1 by the linear relationship given by Eq. (4.26):

$$r_2 = \left(\frac{[M_1]}{[M_2]}\right)_{\text{feed}} \cdot \left[\left(\frac{[M_1]}{[M_2]}\right)_{\text{copolymer}} \cdot \left\{1 + r_1\left(\frac{[M_1]}{[M_2]}\right)_{\text{feed}}\right\} - 1\right] \tag{4.26}$$

Plot of calculated values of r_2 for assumed values of r_1 or vice versa for each experiment with a given feed composition gives a straight line. Theoretically, different linear plots for different experiments would intersect at a common point to allow evaluation of the r_1 and r_2 values for the given monomer pair from the intersection point, Fig. 4.4. In case there are experimental errors, the lines may not

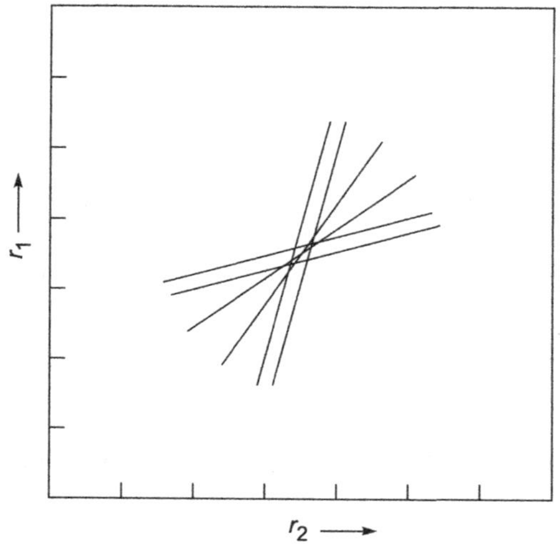

Fig. 4.4 *Nature of plots, r_1 vs r_2 for evaluation of monomer reactivity ratios (method of intersection)*

intersect at a single point and in that event the actual r_1 and r_2 values are obtained by averaging or from the central point of the contour given by the different points of intersection.

(iii) *The Method of Fineman and Ross*[11] This method allows a straightforward analysis of the copolymerization data. Equation (4.15) is rearranged to the linear form, Eq. (4.27) where the mole fraction f_2 is replaced by $(1 - f_1)$:

$$\frac{f_1(1-2F_1)}{F_1(1-f_1)} = r_2 + \left\{\frac{f_1^2(F_1-1)}{F_1(1-f_1)^2}\right\} r_1 \tag{4.27}$$

With known values of f_1 and corresponding F_1 values determined by experimental analysis, left hand side of Eq. (4.27) when plotted against the coefficient of r_1 would yield a linear plot and the r_1 and r_2 values for the monomer pair would be given by the slope and intercept of the plot, respectively.

(iv) *The method of limiting feed composition*[1,12] Success of this method depends on finding very sensitive method of analysis of composition of copolymer prepared under conditions of very low values of $([M_1]/[M_2])_{feed}$ such that the feed ratio progressively approaches zero. Equation (4.13) or its alternative form, Eq. (4.26), would then reduce to a more simple form given by Eq. (4.28). Likewise, the ratio given by feed composition $([M_2]/[M_1])_{feed}$

$$\text{Lt}\left(\frac{[M_1]}{[M_2]}\right)_{feed} \rightarrow 0, \left(\frac{[M_1]}{[M_2]}\right)_{copolymer} = \frac{1}{r_2} \cdot \left(\frac{[M_1]}{[M_2]}\right)_{feed} \tag{4.28}$$

to copolymer composition $([M_2]/[M_1])_{copolymer}$ may be equated to r_1 for $[M_2] \rightarrow 0$. Plot of the left hand side of Eq. (4.28) against the feed ratio, $([M_1]/[M_2])_{feed}$ yields a straight line passing through the origin and r_2 is calculated from the slope of the linear plot. Calculation of r_1 then readily follows from a knowledge of r_2.

Monomer reactivity ratio of some selected binary copolymerization systems are listed in Table 4.1.

4.6 Rate of Copolymerization

For treatment of copolymer composition, it was sufficient to consider the four propagation reactions or the two relative propagation rate constants, i.e. the monomer reactivity ratios, r_1 and r_2 for a binary system. Evaluation of overall rate of copolymerization requires consideration of the rates of initiation (R_i) and termination (R_t) at the same time. Primary radicals released into the monomer

Table 4.1 *Monomer reactivity ratios[13] for some binary copolymerization induced by free radical mechanism*

M_1	M_2	$T°, C$	r_1	r_2
Acrylic acid	Acrylonitrile	50	1.15	0.35
	Styrene	60	0.25	0.15
	Vinyl acetate	70	2	0.1
Acrylonitrile	1,3-Butadiene	40	0.02	0.3
	Methyl acrylate	50	1.5	0.84
	Methyl methacrylate	80	0.15	1.22
	Styrene	60	0.04	0.40
	Vinyl acetate	50	4.2	0.05
Methacrylic acid	Styrene	60	0.7	0.15
	Vinyl acetate	70	20	0.01
	2-Vinyl pyridine	70	0.58	1.55
Methyl methacrylate	Styrene	60	0.46	0.52
	Vinyl acetate	60	20	0.015
	Vinyl chloride	68	10	0.1
	Vinylidene chloride	60	2.53	0.24
Styrene	Isoprene	50	0.8	1.68
	Maleic anhydride	50	0.04	0.015
	Vinyl acetate	60	55	0.01
	2-Vinyl pyridine	60	0.55	1.14

mixture through initiator decomposition react with each monomer with nearly the same ease and efficiency so that consideration of two separate types of initiation is not necessary. Three types of termination must be considered assuming bimolecular mechanism involving all possible pairs of radical types, as has already been mentioned in Sec. 4.2.1. Under steady state condition of all the chain radicals, R_i being equal to R_t, one may write:

$$R_i = 2k_{t_{1,1}}[M_1^{\bullet}]^2 + 2k_{t_{1,2}}[M_1^{\bullet}][M_2^{\bullet}] + 2k_{t_{2,2}}[M_2^{\bullet}]^2 \quad (4.29)$$

where $k_{t_{1,1}}$ and $k_{t_{2,2}}$ are termination rate constants involving pairs of like radicals and $k_{t_{1,2}}$ is the rate constant for cross-termination. The overall rate of propagation, R_p is expressed as:

$$R_p = -\left(\frac{d[M_1]}{dt} + \frac{d[M_2]}{dt}\right)$$

$$= k_{1,1}[M_1^{\bullet}][M_1] + k_{2,1}[M_2^{\bullet}][M_1] + k_{1,2}[M_1^{\bullet}][M_2] + k_{2,2}[M_2^{\bullet}][M_2] \quad (4.30)$$

Eliminating radical concentrations from Eq. (4.30) by taking help of Eqs (4.8) and (4.29) and using the concept of r_1 and r_2 one may express R_p as

$$R_p = \frac{(r_1[M_1]^2 + 2[M_1][M_2] + r_2[M_2]^2)\, R_i^{1/2}}{\{r_i^2\,\delta_1^2[M_1]^2 + 2\phi r_1 r_2 \delta_2 \delta_1 [M_1][M_2] + r_2^2 \delta_2^2 [M_2]^2\}^{1/2}} \qquad (4.30a)$$

where $\delta_1 = \left(\dfrac{2k_{t_{1,1}}}{k_{1,1}^2}\right)^{1/2}$, $\delta_2 = \left(\dfrac{2k_{t_{2,2}}}{k_{2,2}^2}\right)^{1/2}$

and

$$\phi = \frac{k_{t_{1,2}}}{2(k_{t_{1,1}}\, k_{t_{2,2}})^{1/2}}$$

The δ values simply represent the reciprocals of the well-known kinetic parameter $k_p/(2k_t)^{1/2}$ for homopolymerization of the individual monomers and ϕ represents the ratio of half the cross-termination rate constant to the geometric mean of the termination rate constants for like radicals; $\phi > 1$ indicates that cross-termination is favoured and $\phi < 1$ means just the opposite. The factor 1/2 appearing in the ratio indicates that termination involving pairs of unlike radicals is statistically favoured by a factor of two compared with termination involving pairs of like radicals.

The values of δ_1 and δ_2 can be conveniently obtained from the studies of rates of homopolymerization of M_1 and M_2 using the general relation ($-d \ln[M]/dt = R_i^{1/2}/\delta$). The monomer reactivity ratios r_1 and r_2 are readily determined from studies of copolymer composition. A measurement of rate of copolymerization, then, allows calculation of ϕ using Eq. (4.30a). A falling trend in the rate of copolymerization as compared to the mean of the rates of homopolymerization of the two monomers used follows as a consequence of the usual preference for cross-termination in copolymerization. Figure 4.5 shows how rate of copolymerization varied with feed monomer composition for a binary monomer system involving styrene and methyl methacrylate for (a) $\phi = 1$, showing the theoretical curve and (b) $\phi = 13$, showing the experimental points and the theoretical curve.[14] The widespread experience is that the experimental rates of copolymerization fall below the $\phi = 1$ line,[15–18] indicating that ϕ is usually > 1 and that the cross-termination rate exceeds the geometric mean for rate of termination involving like radicals.

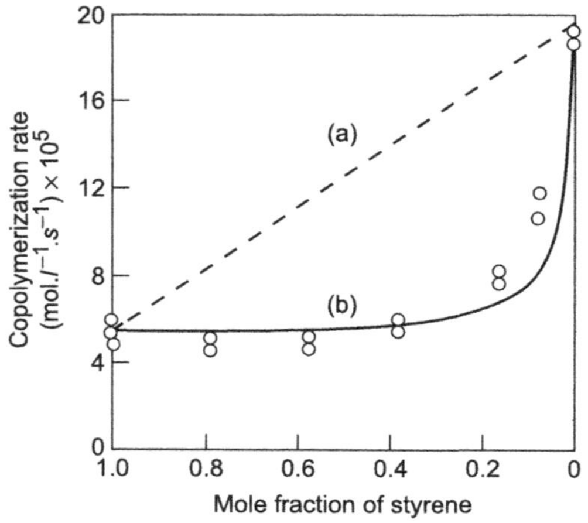

Fig. 4.5 *The rate of copolymerization of styrene and methyl methacrylate at 60°C in presence of azobisisobutyronitrile as a function of mole fraction of styrene.[14] (Walling, 1949;* Courtesy, *American Chemical Society, Washington)*

4.7 Structure and Reactivity of Monomers and Radicals

Nature and number of substituents and the site of the substituents in case the unsaturated (olefinic) monomer bears more than one substituent on the unsaturation have decisive roles to play in determining the reactivity or stability of a specific monomer and the corresponding radical. The substituents influence or modify reactivity or stability by three basic approaches: (i) by activating or deactivating the π bond and thus rendering the monomer less stable and hence more reactive or vice versa, (ii) by imparting enhanced or reduced stability to the derived radical depending on better or limited prospects for resonance, and (iii) by providing different degrees of steric hindrance at the reaction site.

4.7.1 Structure and Reactivity of Monomers

A relative order of reactivities of a series of monomers (M_2) is conveniently obtained by comparing the reciprocal monomer reactivity ratio ($1/r_1 = k_{1,2}/k_{1,1}$) for each of the monomers in the given series with a specific chain radical, $M_1^{\bullet}$. If the rate constant for the reaction of the reference radical with its own monomer is taken as unity, relative monomer reactivities can be readily examined and compared. Related data for a few monomer–radical combinations are listed in Table 4.2. Thus, the first column under polymer radical of Table 4.2 compares reactivities of different monomers toward a chain radical having styrene unit as the radical bearing terminal unit, the second column gives a comparative idea of the reactivities of different monomers toward butadiene chain radical and so on. In view of different basis taken for each radical, values for different radicals given in different columns cannot be compared. The order of monomer reactivities is roughly parallel for all radicals, with some exceptions, here and there. The general order of enhancement of reactivity of a monomer by a substituent toward radical attack is as follows:

$$—C_6H_5,\ —CH = CH_2 > —COR,\ —CN > —COOH,\ —COOR$$
$$> —Cl > —OCOR > —OR$$

Table 4.2 *Relative reactivities of monomers with different polymer radicals[13] at 60°C*

Monomer	*Polymer Radical*					
	Styrene	*Butadiene*	*Methyl methacrylate*	*Acrylonitrile*	*Vinyl chloride*	*Vinyl acetate*
Styrene	(1.0)	0.4	2.2	25	50	100
Butadiene	1.7	(1.0)	4.0	50	29	—
Acrylonitrile	2.5	3.3	0.82	(1.0)	25	20
Methyl methacrylate	1.9	1.3	(1.0)	6.7	10	67
Vinyl chloride	0.059	0.11	0.10	0.37	(1.0)	4.4
Vinyl acetate	0.019	–	0.05	0.24	0.59	(1.0)

4.7.2 Resonance Stabilization

The relative order of monomer reactivities toward a radical generally corresponds to the relative resonance stabilization of the radical formed on addition of the monomer. The more reactive monomer generally yields a radical of higher stability through resonance. A second substituent on the same carbon atom usually leads to an additive effect on the monomer reactivity or on the resonance stabilization of the corresponding radical.

Comparison of reactivities of different radicals toward the same monomer is conveniently done by making use of the absolute propagation rate constant, $k_{1,1}$ and the reactivity ratio r_1. Division of $k_{1,1}$ by r_1 gives an indirect measure of the rate constant $k_{1,2}$ for reaction between radical $M_1^{\bullet}$ and monomer M_2. Absolute (propagation) rate constants for several radicals with each of the respective monomers are listed in Table 4.3. It is clearly seen that the reactivity of the radicals is in the reverse order of that for the monomers. Among the vinyl monomers shown, styrene is the most and vinyl acetate is the least reactive, while styrene radical is the least and vinyl acetate radical the most reactive. The styrene radical owes a high order of stability due to the possible existence of three quinonoid resonance structures, and as

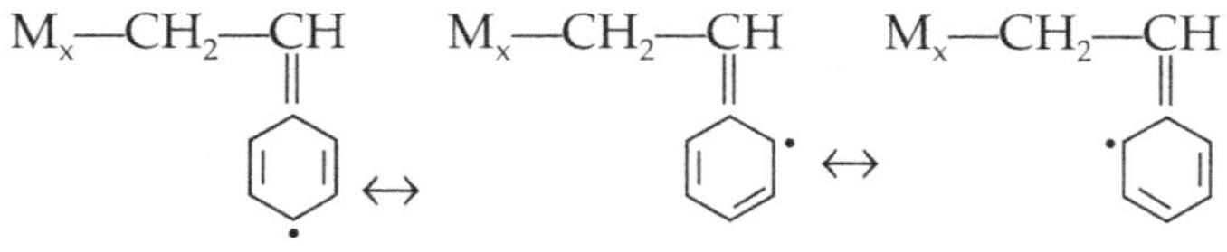

Table 4.3 *Rate constants[13] for radical–monomer reactions at 60°C (1 mol^{-1} s^{-1})*

Monomer	*Radical*					
	Butadiene	*Styrene*	*Methyl methacrylate*	*Acrylonitrile*	*Methyl acrylate*	*Vinyl acetate*
Butadiene	100	250	2,820	98,000	42,000	—
Styrene	74	145	1,550	49,000	14,000	2,30,000
Methyl methacrylate	134	276	705	13,000	4,100	1,50,000
Acrylonitrile	330	435	578	2,000	2,500	46,000
Methyl acrylate	132	205	370	1,300	2,100	23,000
Vinyl acetate	—	2.6	35	230	520	2,300

a consequence, the radical stability corresponds to a resonance energy of nearly 20 kcal/mol. In comparison, substituents having no unsaturation in conjugation with the ethylenic or vinylic double bond, such as —Cl or —OR, produce radicals of resonance energy as low as 1–4 kcal/mol, as because only polar or nonbonded forms contribute to resonance.[16]

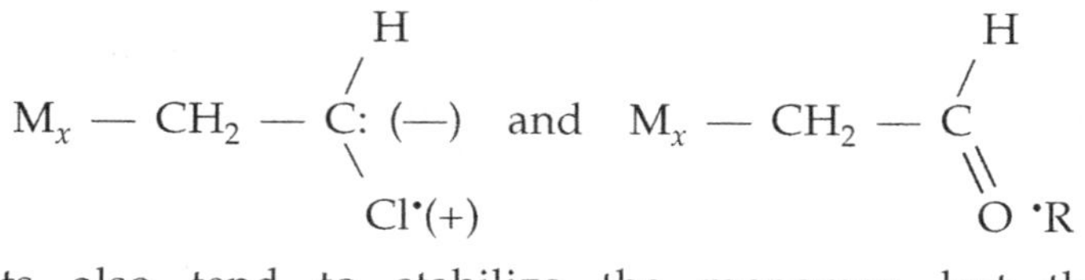

The substituents also tend to stabilize the monomer, but the degree of stabilization is much smaller for the monomer than for the radical. The resonance stabilization due to conjugation amounts to only about 3 kcal/mol for styrene and butadiene as against a resonance stabilization to the tune of about 20 kcal/mol for the corresponding radicals.

4.7.3 Radical Reactivity and Steric Effects

The reactivities of different radicals toward a given monomer and vice versa can be judged and appreciated from a comparison of the respective reaction rate constants listed in Table 4.3. Quite expectedly, a substituent rendering the monomer more stable confers on the corresponding radical higher reactivity and vice versa. A substituent's role in enhancing the reactivity of a radical is much greater than its effective role in depressing the reactivity of the monomer. Thus, the vinyl acetate radical is apparently 500–1500 times more reactive than the styrene radical toward a given monomer, but the styrene monomer is only about 50–100 times more reactive than the vinyl acetate monomer. If $R^{\bullet}$ and $R_s^{\bullet}$ represent an unstabilized and a stabilized radical and M and M_s represent an unstabilized and a resonance-stabilized monomer, then the radical monomer reactivity follows the order:

$$(R^{\bullet} + M_s) > (R^{\bullet} + M) > (R_s^{\bullet} + M_s) > (R_s^{\bullet} + M)$$

Substituents are also known to influence the radical monomer reaction rate constants from the viewpoint of steric hindrance. This can be well appreciated by considering some of the related rate constant data for monomers such as vinyl chloride (monochloroethylene), vinylidene chloride (1,1-dichloroethylene), 1,2-dichloroethylene and polychloroethylenes, given in Table 4.4. The effect of a second substituent on the monomer reactivity produces roughly an additive or reinforcing effect for the 1,1-disubstituted monomer, but in the case of a 1,2-disubstituted monomer, there is a significant decrease in reactivity principally due to steric hindrance.

Table 4.4 *Rate constants $k_{1,2}$ for some radical-monomer reaction at 60°C showing steric effect of substituents in copolymerization (1 mol^{-1} s^{-1})*

Monomer (M_2)	*Radical* ($M_1^{\bullet}$)	
	Vinyl acetate	*Styrene*
Vinyl chloride	43,000	10.35
Vinylidene chloride	1,00,000 (68°C)	95.14
cis-1,2-Dichloroethylene	1,580	0.83
trans-1,2-Dichloroethylene	9,900	4.75
Trichloroethylene	15,200	11.0
Tetrachloroethylene	1,470	0.95

$k_{1,2}$ ($= k_{1,1}/r_1$) values were calculated using $k_{1,1}$ i.e. k_p value for vinly acetate[19] as about 10,000 and for styrene[20] as 176 and the respective r_1 values from the literature.[13]

Among the two 1,2-dichloroethylenes, the *cis*-isomer is much less reactive than the *trans*-isomer toward reaction with a given radical in copolymerization because the *cis*-isomer is less stable and also because the *cis*-isomer is unable to achieve a completely coplanar conformation in the transition state during reaction with a radical, thus impairing resonance stabilization of the derived radical by the substituent.[21]

The lower reactivity of tetrachloroethylene in comparison with that of trichloroethylene is analogous to the difference in reactivities between vinyl chloride and 1,2-dichloroethylene. Trichloroethylene is more reactive than 1,2-dichloroethylene but is less reactive than the 1,1-dichloro derivative, i.e. vinylidene chloride.

4.7.4 Polar Effects and Alternation

Examination of Tables 4.2 and 4.3 reveals that the reactivities of a given monomer to different radicals or of a given radical to different monomers do not exactly follow the same order for different reference monomers or for different reference radicals respectively. Monomer and radical reactivities are interdependent. Deviation of the product $r_1 \cdot r_2$ from unity and its closeness to zero is taken as a measure of alternating tendency and this enables one to tabulate monomers in order of their $r_1 r_2$ values with other monomers, as shown in Table 4.5 in a diagonal

Table 4.5 *Product of reactivity ratios, $r_1 r_2$ in radical chain copolymerization[13,14] for monomer pairs arranged in order of increasing alternating tendency by placing them farther apart*

Vinyl acetate								
—	Butadiene							
0.55	0.98	Styrene						
0.39	0.31	0.34	Vinyl chloride					
0.30	0.19	0.24	1.0	Methyl methacrylate				
0.6	0.1	0.16	0.96	0.61	Vinylidene chloride			
0.35	—	0.10	0.83	—	0.99	Methyl vinyl ketone		
0.21	0.0006	0.016	0.11	0.18	0.34	1.1	Acrylo nitrile	
0.0049	—	0.021	0.056	—	0.55	—	—	Diethyl fumarate

manner such that further apart the two monomers are the greater in their tendency to alternate. Monomers with electron releasing type of substituents are placed at the top and those with electron withdrawing substituents at the bottom part of the table. Thus, a specific nature of radical–monomer reaction manifested through alternation is favourably achieved for monomer pairs with dissimilar polarization properties of the substituents.

Thus, acrylonitrile copolymerizes randomly (ideal copolymerization) with methyl vinyl ketone (r_1r_2 = 1.1) while it undergoes alternating copolymerization with alkyl vinyl ethers (r_1r_2 = 0.0004) or butadiene (r_1r_2 = 0.0006). Monomers which do not homopolymerize will readily copolymerize with strict regularity of alternation, as exemplified by the copolymerization of stilbene (electron donor) and maleic anhydride (electron acceptor). Such copolymerization stands as a fine illustration of polar effect and alternation.

Alternating or 1 : 1 copolymer may also result more as a consequence of 1 : 1 donor–acceptor complex formation between electron-donor and electron-acceptor monomers and homopolymerization of the complex by a head to tail mechanism than due to the alternate addition of the donor and acceptor monomers to the propagating radical. Catalyzed or spontaneous polymerization of suitable olefins or cycloolefins (donor) in liquid sulphur dioxide (acceptor) to produce 1 : 1 olefin/ cycloolefin–sulphur dioxide copolymers is believed to proceed via initial 1 : 1 donor–acceptor complexation and homopolymerization of the complex in the subsequent step.[6,22,23]

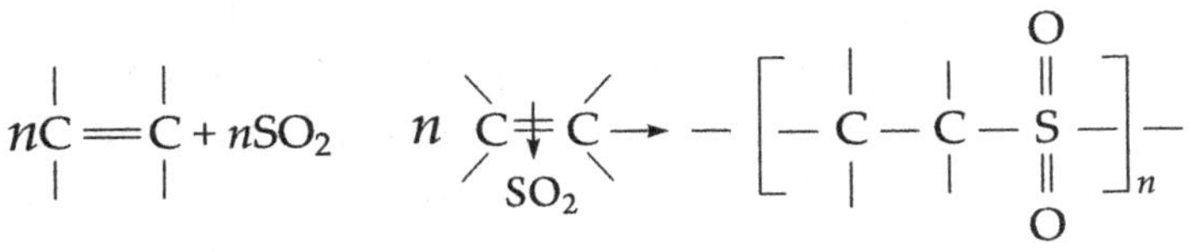

(4.31)

4.8 The *Q–e* Scheme

In an attempt to correlate structure and reactivity of radical–monomer systems, various approaches have been developed and tried, including the Hammett σ–ρ approach, with fair degree of success. The *Q—e* scheme of Alfrey and Price developed for this correlation is of more general applicability. According to this scheme, the rate constant $k_{1,2}$ for copolymerization involving reaction between radical $M_1^{\bullet}$ and monomer M_2 is written as

$$k_{1,2} = P_1Q_2 \exp(-e_1e_2) \tag{4.32}$$

where P_1 and Q_2 relate to the general reactivities and e_1 and e_2 give measures of the polarities of the radical $M_1^{\bullet}$ and monomer M_2 respectively. Resonance effects in the radical and the monomer are defined and expressed by the P and Q terms and assuming that the same e value applies to both monomers, M and its radical, $M^{\bullet}$ one can write[24]:

$$k_{1,1} = P_1Q_1 \exp(-e_1^2) \tag{4.33}$$

$$k_{2,2} = P_2Q_2 \exp(-e_2^2) \tag{4.34}$$

$$k_{2,1} = P_2Q_1 \exp(-e_2e_1) \tag{4.35}$$

$$r_1 = (Q_1/Q_2) \exp[-e_1(e_1 - e_2)] \tag{4.36}$$

$$r_2 = (Q_2/Q_1) \exp[-e_2(e_2 - e_1)] \tag{4.37}$$

$$r_1r_2 = \exp[-(e_1 - e_2)^2)] \tag{4.38}$$

Thus the Q—e scheme correlates monomer–radical reactivities using the parameters Q_1, Q_2, e_1 and e_2. Reactivity ratios for copolymerization of any pair of monomers can be computed by assigning appropriate Q and e values to each set of monomers. However, because of its semiempirical approach and the associated uncertainties in the assignments of Q and e values, the Q—e scheme is subject to much criticism on theoretical grounds. The assumption that the same e value applies to both the monomer and the radical derived from it is made without much justification. Moreover, the concept that alternation in copolymerization is electrostatic in origin is not largely supported by experiments, as the alternating effect does not generally depend on the reaction medium. Thus, the electrostatic basis or polarity concept behind the Q—e scheme and the equations based on the scheme should be judged and used with all these limitations in view. Despite all criticisms, the Q—e scheme has proved quite useful as a basis for correlating a wide expanse of otherwise incoherent kinetic data (r_1, r_2, $k_{1,2}$ and r_1r_2) relating to radical monomer reactions in a satisfactory and even splendid manner. The Q—e scheme may be considered to be an effective transcription to equation form of the reactivity and polarity series of Tables 4.2 and 4.5. The scheme has also been conveniently employed for correlation of data relating to chain transfer involving a transfer agent and a radical.[25,26]

In conclusion it may be said that irrespective of the approach or basis of analysis and interpretation adopted, the reactivity of a monomer in copolymerization depends primarily on two factors: (i) The first factor relates to the intrinsic behaviour of the monomer (resonance factor). The order of reactivity of the monomer is determined most significantly by its capacity for resonance stabilization in the transition state. (ii) The second factor is, however, connected with the specific interaction between a monomer–radical pair (polar effect). The second factor modifies the first in a manner that depends roughly on the degree of dissimilarity in the polarization of the two reacting species—the higher the dissimilarity, the more favourable the reaction.

4.9 Technical Significance of Copolymerization

The modification of a polymer by introducing a second monomer during the polymerization process has long been used to alter its physical properties and sometimes to achieve improved processing characteristics. Incorporation of a second monomer structure as repeating units in the polymer, i.e. the copolymer molecules, reduces overall molecular symmetry leading to lowering (or loss) of degree of crystallinity and of softening temperature. Copolymerization is therefore, considered as a case of internal plasticization. The modification of poly(vinyl chloride) (PVC) by taking small proportions of vinyl acetate (VA) monomer with vinyl chloride (VC) during polymerization to make more flexible and soluble resins of easier processability is well known. Similarly, straight polybutadiene has many technical disadvantages arising from its poor tack and difficult processability. But copolymerization of butadiene with styrene, acrylonitrile and other monomers has led to many technically important resinous and rubbery products. There is practically unlimited scope for variation in polymer structure and polymer properties through copolymerization. In fact, the number of technically important copolymers far surpasses that of important homopolymers. Besides the synthesis of copolymers having random or alternate sequences of monomeric units which have so far been considered and discussed, it is also possible to recognize and synthesize copolymers having two other different kinds of sequences, and they are commonly known as block and graft copolymers.

4.10 Block and Graft Copolymers

The probability of finding long sequences of one monomer in an ordinary random copolymer is very remote except in the trivial case where one monomer is present in a large excess. Methods of synthesis of polymer containing such long sequences of specific monomer units are of great interest, since they often lead to polymers with properties widely different from those of the two homopolymers or the corresponding random copolymers or their blends. Polymers with long sequences of two monomers can have two distinctive arrangements of chains:

(i) *"Block copolymers"*—where the sequences follow one another along the polymer chain axis:

~~~ AAABBB ~~~ BBBAAA ~~~ AAABBB ~~~

(ii) *"Graft copolymer"*—where long or short sequences of one monomer appear as grafted or pendent chains linked to a "backbone" of long sequences of another monomer:

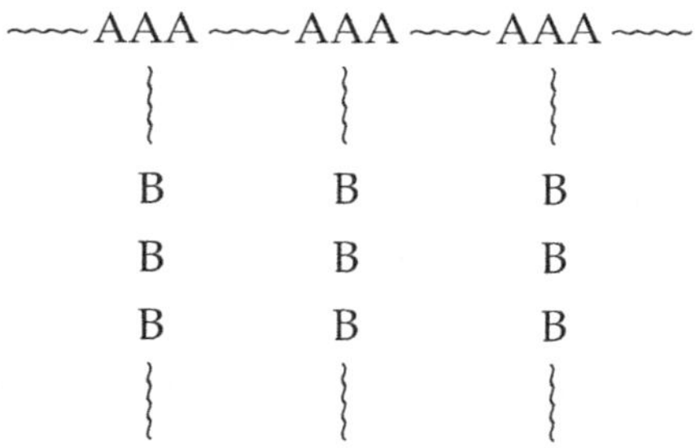

The ultimate aim in preparing a graft or block copolymer is to obtain the desired copolymer free from homopolymers as far as practicable. Although the copolymer structures resulting from the graft and block copolymerization processes are quite different, the former being a chain-branching process and the latter a chain extension process, both may be considered as a chain modification process differing only in direction. They are usually prepared by polymerizing a second monomer in presence of the preformed polymer of a different kind of monomer, and the process generally depends on the presence of labile sites particularly sensitive to radical attacks at different points along the chain axis or at chain ends of the reactant macromolecule. Reactive groups, according to their presence as end groups or as side groups along the chain may also lead to formation of block or graft copolymers by appropriate design of the overall process. Polymer–polymer reaction through selected functional groups present or incorporated on them may also lead to block or graft copolymer structure.

Studies of block copolymers are very much limited in comparison with those of graft copolymers because formation of block copolymers solely depends on the presence of reactive groups or labile atoms at chain ends of linear polymers. The physical properties of high polymers containing blocks of two or more repeat units with two, three or many blocks in a chain have not been studied and examined extensively, even though modification of polymers to block copolymer structures is being used commercially. A commercial product commonly known as "thermoplastic elastomer" is a triblock copolymer containing a middle block of polybutadiene or polyisoprene linked in a linear fashion to two end blocks of polystyrene and is
~~~

prepared by anionic polymerization. Formation of polyurethane rubbers by chain extensions of different polyesters or polyethers as depicted by Eq. (4.2) also involves an approach to block copolymerization.

Graft copolymerization usually involves polymerization of a monomer onto a different polymer backbone; the various reactions associated with polymerization processes can be utilized for the synthesis of graft copolymers. Most graft copolymers are formed by free radical polymerization, and often the major activation involves chain transfer. The reaction shown involves generation of a radical site on the preformed polymer by a chain-transfer mechanism and then growth of a new polymer structure from that new radical site:

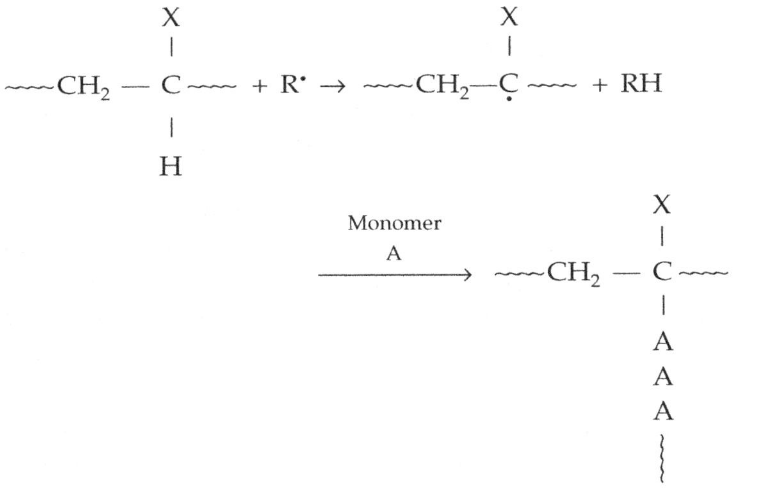

(4.39)

(Graft Copolymer)

4.11 Techniques of Polymerization

For laboratory and industrial purposes, polymerization is practised by a variety of approaches and techniques that require different design considerations with respect to recipe of polymerization and with respect to physical conditions for the process and process equipment. Generally, polymerizations are studied and practised by four different techniques:

1. Bulk or Mass Polymerization
2. Solution Polymerization
3. Suspension Polymerization and
4. Emulsion Polymerization

Of the four techniques, the last two are essentially heterogeneous polymerization systems containing a large proportion of a non-solvent, usually water, acting as a dispersion medium for the immiscible liquid monomer. Bulk and solution polymerization may be either homogeneous all the way or homogeneous to begin with, but becoming heterogeneous with progress of polymerization due to the polymer formed being insoluble in its monomer (for bulk polymerization) or in the solvent used to dilute the monomer (for solution polymerization). Polymerization of monomers used in the gaseous phase, to begin with, also usually ends up in a heterogeneous system. Most condensation polymerizations are conveniently done by employing bulk or the solution techniques.

4.11.1 Bulk Polymerization

Polymerization in bulk, i.e. of undiluted monomer is apparently the most obvious if not the most simple method or technique of polymer synthesis. In terms of polymerization recipe, this obviously provides the most simple system, requiring the use of a very low concentration of an initiator as the only additive in the monomer and that again decomposes and reaches almost a vanishing concentration at the end of the polymerization process. For uncatalyzed vinyl polymerization (using no catalyst or initiator) done thermally, photochemically or otherwise, the polymer produced would be the most pure, free from any (additive) contaminants except probably traces of unreacted monomer.

Vinyl polymerizations are highly exothermic. Again, to obtain polymer at convenient rates, temperature in the range 80–200°C is needed. For dissipation of the heat liberated with progress of polymerization, continuous stirring of the monomer–polymer mixture is essential. But the stirring process and heat dissipation become progressively difficult with conversion due to gain in the medium viscosity due to solubilization or precipitation of polymer in the monomer, more so in the high conversion zones. The autoacceleration or gel effect often makes the problem more acute. If left uncontrolled, excessive rise of temperature at local points known as "hot spots" may lead to discolouration, thermal degradation, branching, development of chain unsaturation or even cross-linking, thus giving rise to irreproducible and often inferior product quality. Because of these problems and disadvantages, bulk polymerization technique finds rather limited use in commercial production of vinyl polymers, e.g. in the production of poly-(methyl methacrylate) as acrylic castings (sheets), and in the making of certain grades of polystyrene, poly (vinyl chloride) and (low density) polyethylenes. The heat dissipation problem in such cases is kept to a minimum and within tolerable limits by two approaches: (i) by carrying out the polymerization to low conversions as in high pressure polymerization of ethylene using narrow but long tubular reactor at temperatures far higher than the melting temperature of the polymer; and (ii) by accomplishing the polymerization in stages—first up to about 20–30% conversion in a stirred reactor at 80–100°C and subsequently allowing the monomer–polymer syrup (viscous mass) poured in an appropriate mould assembly to polymerize to (near) completion at progressively higher temperatures in stages (e.g. preparations of acrylic castings from methyl methacrylate monomer) or allowing the monomer–polymer syrup, called the prepolymer to flow under gravity aided by some positive pressure down a vertical tower reactor with appropriate design for heat transfer or heat dissipation and having a temperature gradient with increasing temperature zones downward for the reaction mixture till it reaches a stage of near-complete conversion (manufacture of polystyrene from styrene monomer).

4.11.2 Solution Polymerization

Solution polymerization, i.e. polymerization of a monomer in the presence of a diluent miscible in all proportions with the monomer and usually with the ability to dissolve the polymer is conceived as a means to overcome much of the problems and disadvantages of the bulk method. The solvent reduces the viscosity gain with conversion, allows more efficient agitation or stirring of the medium, thus effecting better heat transfer and heat dissipation and minimization or avoidance of local overheating or heat accumulation. Although the solution technique allows much better control of the process, it has its own demerits. The solution method often requires handling of flammable or hazardous solvents and removal or recovery of the solvent to isolate the polymer after the polymerization is over. Chain transfer to solvent may also pose a problem and purity of the polymer may suffer due to retention of last traces of solvent in the isolated product. Solution polymerization is, however, advantageous if the polymer formed is to be applied in solution by brushing or spraying as in the case of making of coating (lacquer) grade poly (methyl methacrylate) resins from methyl methacrylate and related monomers. In certain cases, the solvents generally used may not be able to dissolve the polymer being formed and so, the system becomes heterogeneous in character with polymer formation (polymerization of acrylonitrile in water or in some common organic solvents).

4.11.3 Suspension Polymerization

This is a case of heterogeneous polymerization right from the beginning. This technique is designed to combine the advantages of both the bulk and solution techniques, and it is one of the extensively employed techniques in the mass production of vinyl and related polymers. This technique primarily involves breaking down or dispersion of the liquid monomer mass into separate tiny droplets in a large continuous mass of a nonsolvent commonly referred to as the dispersion or suspension medium, by efficient agitation. Water is invariably used as the suspension medium for all water insoluble monomers because of many advantages that go with it, viz. ready availability at low cost, non-toxic nature, ease of storage and handling without the necessity of recovery, etc. Moreover, the boiling temperature of water is a convenient upper limit for radical vinyl polymerization and in aqueous suspension system the rise in temperature in any location due to exothermic nature of polymerization cannot go beyond the boiling temperature of the medium.

The size of the monomer droplets usually range between 0.1–5 mm in diameter. Continuous agitation is allowed to prevent coalescence of the droplets. Low concentrations of suitable water-soluble polymers such as carboxymethyl cellulose (CMC) or methyl cellulose, poly(vinyl alcohol) (PVA), gelatin, etc., are used to raise the medium viscosity and they play the role of suspension stabilizers particularly in the intermediate stages of conversion when the tendency of the suspended droplets to coalesce together becomes high as the polymer beads become sticky. The water-soluble polymeric stabilizers form a thin layer on the monomer–polymer droplets and stabilization is effected due to repulsion between like charges thus gathered on different droplets. Finely divided suspended clay particles such as kaolin or china clay in small amounts is sometimes used to interfere mechanically with the agglomeration of beads. In suspension polymerization, oil-soluble initiators such as organic peroxides, hydroperoxides or azocompounds are used and thus, each tiny droplet behaves as a miniature bulk polymerization system. At the end of the process, polymers appear in the form of tiny beads or pearls and hence, the process is also known as bead or pearl polymerization. The polymers are filtered, washed profusely with water to remove the water soluble stabilizer as far as practicable, and dried. They, however, usually retain traces of stabilizers besides the residual initiator as contaminants. A typical recipe for suspension polymerization is given in Table 4.6.

4.11.4 Emulsion polymerization

(a) ***Description of the System*** The use of water as a medium of polymerization by the solution and suspension techniques has been described and discussed above. A special feature about water is that it provides us with a medium characterized by

Table 4.6 *A typical recipe for suspension polymerization of a vinyl monomer*

Ingredients	*Parts by Weight*
Monomer	100
Water	200–400
Initiator	
(peroxides, azo compounds)	0.2–0.5
Stabilizers	
(CMC, PVA, gelatin, etc.)	0.02–1.5

zero chain transfer constant.[27] Another polymerization technique, in which water is used as the medium, is the "emulsion polymerization". Polymers from water-insoluble monomers are usually obtained at faster rates in the aqueous medium by the emulsion technique using water soluble initiators. For efficient making of an emulsion, it is necessary to agitate the monomer–water mixture in presence of a measured dose of an emulsifying agent, i.e. a soap or a detergent. To minimize fluctuations in surface tension of the emulsion as polymerization progresses, small proportion of a surface tension regulator (usually a long chain fatty alcohol such as cetyl alcohol) is used. Small volume of a selected buffer solution is added to minimize or eliminate variations of *pH* of the system due to hydrolysis or other reactions. A chain length regulator or chain transfer agent of high transfer constant, such as dodecyl mercaptan is often used in low concentrations to permit optimum control of polymer chain length. A selected inhibitor, known as a short-stop may be added in small amounts late into the system in order to stop the reaction at a desired extent of polymerization.

A low concentration of a water-soluble polymer such as carboxymethyl cellulose (CMC), poly(vinyl alcohol) (PVA), gelatin, dextrin, etc. is normally used in the emulsion polymerization system not to control the polymerization process with respect to rate or degree of polymerization but with the purpose of obtaining stable emulsion and to prevent emulsion break-down with progress of polymerization; this ingredient is commonly called a protective colloid or (emulsion) stabilizer.

In principle and practice, and as a physical system, emulsion polymerization is distinctly different from suspension polymerization. It is associated with almost all the advantages of suspension polymerization and moreover, using selected efficient redox initiator systems, emulsion polymerization is conveniently accomplished at ambient or slightly elevated temperatures giving very high rates and degrees of polymerization at the same time. The rate and degree of polymerization in the emulsion system can be varied in a large measure independently of each other. The emulsion technique is extensively employed for the polymerization of dienes such as isoprene and butadiene, and vinyl compounds such as vinyl chloride, vinyl acetate, styrene, various acrylates and methacrylates and for making various copolymers of these and other related monomers. A typical recipe for emulsion polymerization of a vinyl monomer is shown in Table 4.7.

Table 4.7 *A typical recipe for emulsion polymerization of a vinyl monomer*

Component	*Parts by Weight*
Monomer	100
Water	180–200
Surfactant/Emulsifier (soap/detergent)	4–8
Initiator: (i) $K_2S_2O_8$	0.1–0.25
(ii) $NaHSO_3$	0.1–0.2
Surface tension regulator (octanol/lauryl alcohol)	0.5
Emulsion stabilizer (CMC/PVA, etc.)	0.5–1.5
Buffer solution	5–10
Chain length regulator (a mercaptan)	0.3–0.8

(b) *Principle and Theory* The surfactant or the emulsifier plays an important role. When soap or detergent is taken in an aqueous system in excess of a low critical concentration, micellar or layered aggregates consisting of 50, 100 or even more soap/detergent molecules are formed. They may be lamellar micelles or spherical micelles with their polar ends (heads) facing water remaining in the outer surface and the long hydrophobic hydrocarbon 'tails' meeting at the central or interior part of the micelle[28,29]. The diameter or thickness of the initial micelles approximates twice the length of the soap/detergent molecules (40–80 Å). When water insoluble monomer is present, part of it gets "solubilized" and finds room within the micelles, presumably among the hydrophobic tails of the soap/detergent molecules forming the micellar aggregates. The solubilization leads to some swelling of the initial micelles. A major part of the monomer taken, however, remains initially in macroscopic droplets. The micelles offer a far greater total surface area than the droplets, even though the total volume of the micelles is significantly lower than that of the monomer droplets. A schematic representation of micelles with or without solubilized monomer, surfactant molecules or ions in aqueous solution and monomer droplets is given in Fig. 4.6.

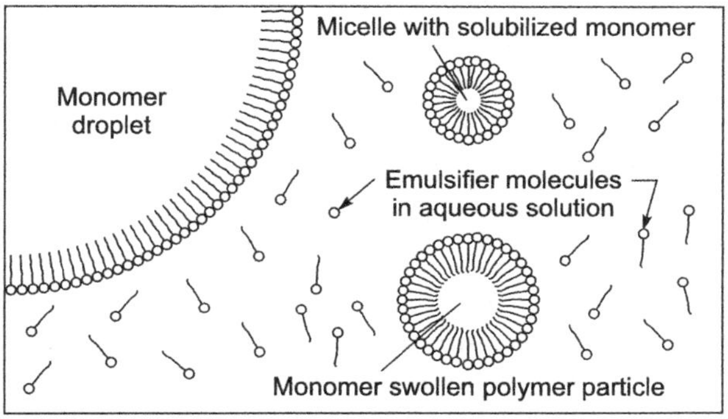

Fig. 4.6 *Schematic representation of an emulsion system in an emulsion polymerization*

Experiments clearly demonstrate that polymer is not formed in the monomer droplets. This leaves the aqueous phase and the micelles as the locus of polymerization in the emulsion system. Aqueous polymerization can be readily effected without the use of an emulsifier and without having excess monomer as droplets (i.e. using the monomer within its solubility limit in water), though at much lower rates. But this alone is far insufficient to account for the bulk of the polymer formed in a typical emulsion system. In presence of a surfactant, the polymerization rate increases with surfactant concentration.

Polymerization begins in the aqueous phase as that is where the initiators occur and react or decompose into radicals. The primary radicals immediately pick up monomers dissolved in water to bring about chain initiation. With very low concentration of monomer in the aqueous phase, chain growth at relatively low rates proceeds till the radicals enter micelles which are present abundantly in the system. It is also not altogether unlikely that the primary radicals would find straight entry into the micelles being attracted by their high monomer reserve. Within the micelles, polymerization or chain growth takes place much rapidly in view of high monomer concentration due to accumulation of solubilized monomer there. As polymer is formed due to monomer depletion in the micelles, more monomer is transferred to the micelles from the reservoirs of monomer droplets. The former expands at the expense of the latter.

Soon after a low conversion stage (a few per cent) in the emulsion system, the low surface tension of the initial emulsion rises quite sharply evidently as a consequence of a fall in surfactant concentration in the aqueous phase of the emulsion, leading to breakdown or disruption of the micelles and consequent instability of the monomer droplets. Micelles give way to numerous polymer particles which at this stage average to not more than 200–400 Å in diameter. Nearly all of the surfactant used get adsorbed on the surfaces of the polymer particles. A redistribution of surfactant in favour of polymer particles makes the monomer droplets unstable at this stage. If agitation is discontinued, the monomer droplets cluster together into a continuous phase with zero polymer content. Monomer is now fed to the growing polymer particles by diffusion through the aqueous medium. As polymerization proceeds further, the polymer particles swell and grow in size even though they remain sensibly constant in their number. The rate of polymerization is practically constant over most part till about 60–80% conversion, excepting the initial zone up to several per cent. From the stage the monomer droplets completely disappear, the rate follows a fast decreasing trend with further conversion. In the final analysis, the emulsified polymer particles grow to a size in the range 500–1500 Å in diameter which is much larger than the size range of micelles initially formed but much smaller than the size of initially formed monomer droplets (> 10,000 Å). After polymerization to optimum conversion, appropriate dose of a short-stop or inhibitor is added to stop further conversion. The polymer is isolated by breaking the emulsion using salt or acid, coagulating, filtering, washing and drying.

(c) ***Rate of Emulsion Polymerization*** Under general conditions employed in emulsion polymerization, the radicals are generated from the aqueous initiator at rates of the order of 10^{13} per cubic centimetre per second. The number of polymer particles amount to about 10^{14} per cubic centimetre. If all the radicals generated eventually enter polymer particles, the rate of radical entry in a particle will average out to about one every 10 seconds. The equilibrium radical concentration in the aqueous medium $[R^\bullet]_{aq}$ has been shown to be about 10^8 radicals per cubic centimetre. Judging against the radicals being generated at a rate of 10^{13} per cubic centimetre per second, the average lifetime[30] of a radical $[R^\bullet]_{aq}$ till its entry into a polymer particle comes out to only about 10^{-5}s. Rate of radical termination in the aqueous phase $(R_t)_{aq}$, expressed as $(R_t)_{aq} = 2\,k_t\,[R^\bullet]^2_{aq}$, calculated using literature value of the termination rate constant k_t comes out to be of the order of 10^3 radicals per cubic centimetre per second which is less than the rate of radical generation by many orders of magnitude. Thus, radical termination in the aqueous phase must be negligibly small, indicating with good degree of certainty that almost every radical generated finds entry into a micelle or a polymer particle. Diffusion currents are quite adequate to permit fast diffusion of radicals into the polymer particles.

Calculation using known k_t values also predicts that two radicals cannot coexist long in the same polymer particle and they would terminate mutually within a few thousandths of a second. Over a given time period, a particle must bear either one free radical or none, and in any case, the very short intervals in which a particle may have two radicals are negligible. Subsequent to termination by capture of a second radical, a specific particle will remain dormant for an average interval of about 10 s until another radical finds entry into it and the particle becomes active again. After another interval of equal average time period a second radical will enter the particle and mutual annihilation of the radical pair will follow almost instantly. At any time, therefore, one-half of the polymer particles remains active for having only one radical in each of them while the other half remains dormant for having no radical in them.[31] The rate of polymerization per cubic centimetre of the aqueous phase is then given by

$$R_p = k_p\,(N/2)\,[M] \tag{4.40}$$

where N is the number of polymer particles per cubic centimetre of the aqueous phase, and [M] denotes the monomer concentration in the polymer particles. Thus, the rate, R_p, primarily depends on the number of particles, N, as [M] is practically invariable as long as excess monomer droplets remain, and R_p is independent of the rate of radical generation, R_i. An increase or decrease in R_i simply causes an increase or decrease in the frequency of alternation of active and inactive periods for each polymer particle. R_p increases with the surfactant concentration since N increases as the concentration of the surfactant increases.

Once inside the polymer particle, a radical propagates at a rate v_p given by

$$v_p = k_p\,[M] \tag{4.41}$$

and the rate v_i at which a primary radical enters a polymer particle is given by

$$v_i = \frac{R_i}{N} \tag{4.42}$$

v_i is also equal to the rate of termination of a polymer chain since termination takes place as soon as a radical enters an active polymer particle, i.e. in which a polymer chain is propagating. Hence, the degree of polymerization $\bar{x}_n$ will be given by

$$\bar{x}_n = \frac{v_p}{v_i} = \frac{k_p N\,[M]}{R_i} \tag{4.43}$$

assuming, however, that chain transfer of any kind is negligible.

Thus, the degree of polymerization, $\overline{x}_n$, like the rate of polymerization R_p is directly dependent on the number of polymer particles, but unlike R_p which is independent of R_i, $\overline{x}_n$ has an inverse relationship with R_i. Examination of Eqs (4.40) and (4.43) shows that unlike polymerization by the (homogeneous) bulk and solution techniques and by the suspension technique, that by the emulsion technique permits increase in both rate and degree of polymerization simultaneously by increasing the number of polymer particles, N at a fixed rate of initiation.[32] The number of micelles initially formed is determined by the emulsifier concentration and the number of polymer particles formed depends on the number of micelles initially present and on the rate of radical generation. Smith and Ewart[31] deduced the relationship given by Eq. (4.44):

$$N = k\left(\frac{R_i}{\mu}\right)^{2/5} (a_s c_s)^{3/5} \tag{4.44}$$

where k is a constant having a value between 0.37 and 0.53, μ is the rate of increase in the volume of a polymer particle, c_s is the concentration of the surfactant in g/cm^3 and a_s represents the area occupied by a gram of surfactant at the time of exhaustion of micellar surfactant and their distribution as a continuous monolayer over polymer particles, $a_s c_s$ is the total surface area of the particles when this happens. Most observed effects in emulsion polymerization are explained and understood on the above quantitative relationships [Eqs (4.40), (4.43) and (4.44)] with some deviations, here and there, however. Styrene and vinyl acetate are the two widely used monomers in the studies of emulsion polymerization.

References

1. Sen Gupta, P.K., A.R. Mukherjee and P. Ghosh, *J. Macromol. Chem.*, **1** (1966), 481.
2. Fordyce R.G. and G.E. Ham, *J. Am. Chem. Soc.*, **69** (1967) 695.
3. Pepper, D.C., *Quart. Revs.*, **8** (1954), 88.
4. Lewis F.M. and F.R. Mayo, *J. Am. Chem. Soc.*, **70** (1948) 1533.
5. Bartlett P.D. and K. Nozaki, *J. Am. Chem. Soc.*, **68** (1946), 1495.
6. Ghosh, P. *Macromol. Rev.*, **5** (1971), 195.
7. Overberger, C.G. and J.A. Moore, *Chem. Ind.* (1968), 24.
8. Iwatsuki, S., T. Kokubo and Y. Yamashita, *J. Polym. Sci.*, **A1** (6) (1968), 2481.
9. Skeist, I., *J. Am. Chem. Soc.*, **68** (1946), 1781.
10. Meyer, V.E. and R.K.S. Chan., *Polym. Preprint*, **8** (1967), 209.
11. Fineman, M. and S.D. Ross, *J. Polym. Sci.*, **5** (1950), 250.

12. Palit, S.R. and P. Ghosh, *J. Polym. Sci.*, **58** (1962), 1225.
13. Brandrup, J. and E.H. Immergut, Eds., *Polymer Handbook*, Wiley-Interscience, New York, (1966), II, pp. (141–291.
14. Walling, C., *J. Am. Chem. Soc.*, **71** (1949), 1930.
15. Melville, H.W. and L. Valentine, *Proc. Roy. Soc.* **A200** (1950), 337 & 358.
16. Mayo, F.R. and C. Walling, *Chem. Rev.*, **46** (1950), 191.
17. Arlman, E.J. and H.W. Melville, *Proc. Roy. Soc.* **A203** (1950), 301.
18. Bonsall, E.P., L. Valentine and H.W. Melville, *J. Polym. Sci.*, **7** (1951), 39; *Trans. Faraday Soc.*, **48** (1952), 763.
19. Schulz, G.V. and D.J. Stein, *Makromol. Chem.* **52** (1962), 1.
20. Matheson, M.S., E.E. Auer, E.B. Bevilaqua and E.J. Hart, *J. Am. Chem. Soc.*, **73** (1951), 1700.
21. Walling, C., *Free Radicals in Solution*, Wiley-Interscience, New York 1957, Ch. 4.
22. Zutty, N.L. and C.W. Wilson, *Tetrahedron Letters* (1963), 2181.
23. Zutty, N.L., C.W. Wilson, G.H. Potter, D.C. Priest and C.J. Whiteworth, *J. Polym. Sci.*, **A3** (1965), 2781.
24. Alfrey, T. Jr. and C.C. Price, *J. Polym. Sci.*, **2** (1947), 101.
25. Fuhrman, N. and R.B. Mesrobian, *J. Am. Chem. Soc.*, **76** (1954) 3281.
26. Ghosh, P. *J. Polym. Sci.*, **B7** (1969), 293.
27. Nandi, U.S., P. Ghosh and S.R. Palit, *Nature*, **195** (1962), 1197.
28. McBain, J.W., *Advances in Colloid Science*-Wiley-Interscience, New York, 1942, Vol. 1.
29. Hartley, G.S., *Aqueous Solutions of Paraffin—Chain Salts*, Herman et. Cie, Paris, 1936.
30. Smith, W.V., *J. Am. Chem. Soc.*, **70** (1948), 3695.
31. Smith, W.V. and R.H. Ewart, *J. Chem. Phys.*, **16** (1948), 592.
32. Bovey, F.A., I.M. Kolthoff, A.I. Medalia and E.J. Meehan, *Emulsion Polymerization*, Wiley-Interscience, New York, 1955.

CHAPTER 5

Ionic and Stereo-regular Polymerization

5.1 Ionic Polymerization: Introduction

Different features of ionic polymerization will be taken up for discussion and analysis in this chapter. Unlike free radical polymerization, ionic polymerization is associated with what may be termed as monomer selectivity. Generally, monomers with electron-releasing substituents such as alkoxy, 1,1 dialkyl or aryl-alkyl, phenyl and vinyl groups exemplified by isobutyl vinyl ether, isobutylene or α-methyl styrene, styrene and butadiene are readily susceptible to cationic polymerization and not really or not as readily to anionic polymerization. The comparatively basic character of their double bonded carbon atoms induces them to share a pair of electrons with an electrophilic reagent, and the propagation takes place by a carbocationic or carbonium ion mechanism. Monomers with electron-withdrawing substituents, such as nitrile, carboxyl, ester groups, etc. (exemplified by acrylonitrile, acrylic acid, methyl acrylate, etc.), are readily susceptible to anionic polymerization. Phenyl and vinyl substituents also make the monomer (styrene, butadiene, etc.) capable of polymerizing anionically. Mayo and Walling[1] gave the following order of reactivity: Vinyl ethers > isobutylene > α-methyl styrene > isoprene > styrene > butadiene for cationic polymerization, and acrylonitrile > methacrylonitrile > methyl methacrylate > styrene > butadiene for anionic polymerization. Interestingly, styrene and the dienes, capable of polymerizing by radical means are also polymerized by both cationic and anionic routes.[1] However, they are placed low in each series in the order of reactivity.

Polarity differences of monomers play more significant roles in their reaction with an ionic chain-propagating centre than with a free radical centre. The relative reactivities for different monomers are nearly independent of terminal unit carrying the charge, as the charge on the propagating centre is much the same for different terminal units.

Thus, monomer reactivities can be understood without reference to the specific unit bearing the carbonium ion or carbanion and ideal copolymerization is approached more closely by the ionic propagation than by free radical propagation. Bimolecular termination as prevalent in the free radical polymerization is totally prohibitive in the ionic polymerization, as like charges repel each other and cannot come close enough for mutual interaction and annihilation. Unimolecular termination or termination due to transfer to monomer or solvent are the consequential mechanisms of termination of ionic polymerization. Ionic polymerizations are generally not favoured or rather inhibited or strongly retarded in aqueous medium, although traces of moisture have sometimes been reported to produce significant cocatalytic effects in some cationic polymerization. Basic solvents or additives prevent or inhibit the cationic propagation while acidic solvents or additives inhibit anionic growth. Solvents of such polarity as to favour formation of ion-pairs and as not to permit large scale ion dissociation offer different degrees of stability and reactivity for the propagating ion centres; often very low temperatures, usually in the range of –40 to –100°C, are needed for ionic polymerization at desirable or optimum rates and to obtain optimum or high degrees of polymerization. Methyl chloride, ethylene dichloride, pentane, hexane, nitrobenzene offer good media for ionic polymerization. Usually, a counter ion stays close to the propagating ion throughout its lifetime. Initially, limited separation between the two ions of the ion pair helped by increased polarity or solvating power of the medium becomes more favourable for ionic polymerization, but a highly polar medium or additive such as water, alcohol, etc., reacts with and destroys most ionic catalysts or reduces their catalytic activity to insignificance.

5.2 Cationic Polymerization

Protonic acids such as H_2SO_4 and $HClO_4$ or Lewis acids such as BF_3, $SnCl_4$, $AlCl_3$, $AlBr_3$, $TiCl_4$, etc., which are strong electron acceptors, readily induce polymerization of monomers having electron releasing substituents by cationic (carbonium ion) mechanism. Almost all Lewis acids or Friedel-Crafts catalysts become effective in presence of a cocatalyst which complexes with the catalyst and acts as a proton donor. The cocatalyst may be water, alcohol, organic acid, alkyl halide or even a hydrocarbon. Early studies on cationic polymerization were done using boron trifluoride, BF_3 as the catalyst for the polymerization of isobutylene $CH_2 = C(CH_3)_2$.

Polymerization occurs almost instantaneously at –100°C to give polymers of high degrees of polymerization (~ 10^5). Cationic polymerizations of this kind are characterized by a low or negative temperature coefficient—both rate and degree of polymerization dropping down sharply with rise in temperature in some cases.[2]

BF_3 was found to be ineffective to induce polymerization of isobutylene when the reactants were taken in a high degree of purity and free from moisture. The catalyst is however very effective in the presence of traces of moisture or similar polar additives as a cocatalyst.[3,4]

Traces of moisture (H_2O) react with BF_3 to form the catalyst-cocatalyst complex, i.e., boron trifluoride monohydrate, $BF_3 \cdot H_2O$. The mechanism of polymerization of isobutylene in the presence of $BF_3 \cdot H_2O$ is as follows:[5,6]

(i) Chain initiation:

$$BF_3 \cdot H_2O \rightleftharpoons H^+(BF_3 \cdot OH)^- \quad \text{(ion pair formation)} \tag{5.1}$$

$$H^+(BF_3 \cdot OH)^- + CH_2 = C(CH_3)_2 \xrightarrow{k_i} CH_3-\overset{\overset{\large CH_3}{|}}{\underset{\underset{\large CH_3}{|}}{C^+}} + (BF_3 \cdot OH)^- \tag{5.2}$$

(protonation of monomer)

(ii) Chain propagation:

$$CH_3-\overset{\overset{\large CH_3}{|}}{\underset{\underset{\large CH_3}{|}}{C^+}} + (BF_3 \cdot OH)^- \cdots \xrightarrow[k_p]{nCH_2=C(CH_3)_2}$$

$$H-\left(CH_2-\overset{\overset{\large CH_3}{|}}{\underset{\underset{\large CH_3}{|}}{C}}-\right)_n CH_2-\overset{\overset{\large CH_3}{|}}{\underset{\underset{\large CH_3}{|}}{C^+}} + (BF_3 \cdot OH)^- \tag{5.3}$$

counter ion remaining close to the growing carbonium ion

(iii) Chain termination:

$$\text{(a)} \quad \sim\sim CH_2-\overset{\overset{\large CH_3}{|}}{\underset{\underset{\large CH_3}{|}}{C^+}} + (BF_3 \cdot OH)^- \xrightarrow{k_t} \sim\sim CH_2-\overset{\overset{\large CH_2}{||}}{\underset{\underset{\large CH_3}{|}}{C}} + BF_3 \cdot H_2O \tag{5.4}$$

proton abstraction by the gegen ion

(b)

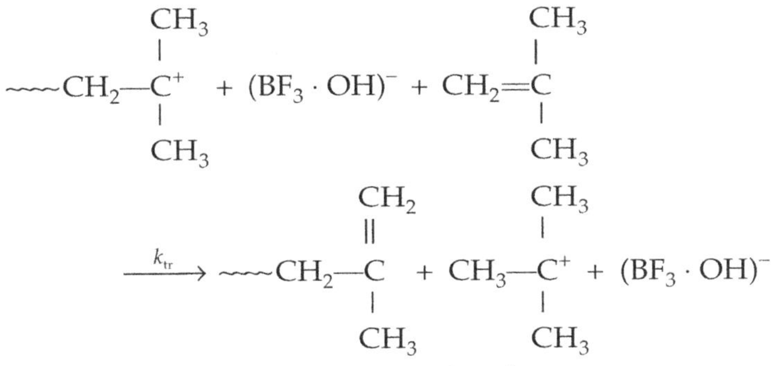

(5.5)

monomer transfer

The catalyst–cocatalyst complex releases a proton (or some other cation as the case may be) that adds to the monomer to generate a carbonium ion and leaves the counter ion or gegen ion in the vicinity of the carbocationic centre [Eq. (5.2)]. As the solvent media generally used are of low dielectric constant, the gegen ion stays close to the carbonium ion centre throughout the growth process [Eq. (5.3)]. Cessation of the growth of a given chain usually takes place by abstraction of a proton by the gegen ion [Eq. (5.4)] or by monomer transfer [Eq. (5.5)], and in either case, unsaturation arises at the terminating end. The mechanism is also substantiated by evidences for the presence of $(CH_3)_3$—C— and —CH_2—$C(CH_3)$ = CH_2 end groups in the polymers as obtained from analysis employing infrared spectroscopy [Dainton and Sutherland, (1949)][6]. Low proportions of end groups, such as —CH = $C(CH_3)_2$, formed by proton abstraction from the β-methylene group, and ~~~CH_2—$C(CH_3)_2$—OH, resulting from transfer of a hydroxyl ion from the closeby counter ion, may also be present in the polymers formed. The cocatalyst fragments only get incorporated into the polymer as end groups, but the catalyst does not combine with the polymer and usually, a large number of polymer molecules are formed for each catalyst–cocatalyst complex species taken. A more active catalyst is usually less active as a chain terminator.

5.3 Kinetics of Cationic Polymerization

Cationic initiator systems (catalyst–cocatalyst complex C) may be either soluble or insoluble in the monomer–solvent media used. The kinetics of polymerization is better studied and understood using soluble catalyst complexes and homogeneous systems. A generalized overall reaction scheme may be written as:

$$
\begin{aligned}
&\mathrm{A + SH \overset{K}{\rightleftharpoons} A \cdot SH \xrightarrow{k_i} H^+AS^-} && \text{(a)} \\
&\mathrm{(H^+\ AS^-)\ A \cdot SH + M \xrightarrow{k_i} HM^+ + AS^-} && \text{(b)} \\
&\mathrm{M_n^+ + AS^- + M \xrightarrow{k_p} M_{n+1}^+ + AS^-} && \text{(c)} \qquad (5.6)\\
&\mathrm{M_n^+ + AS^- \xrightarrow{k_t} P_n + A \cdot SH} && \text{(d)} \\
&\mathrm{M_n^+ + AS^- + M \xrightarrow{k_{tr}} P_n + HM^+ + AS^-} && \text{(e)}
\end{aligned}
$$

Here, A represents the catalyst, SH the cocatalyst and P_n, the polymer molecule of degree of polymerization n, and all other terms have their usual significances. Initiation of polymerization involves a proton transfer from the complex $A \cdot SH$ (or H^+AS^-) to the monomer M. Termination of a growing chain may take place by abstraction of a proton from the terminal unit by the counter ion AS^- or by monomer transfer. The exact nature and concentration of the complex are often uncertain, and for kinetic purposes, the concentration of the catalyst component or the cocatalyst whichever is present in stoichiometric deficiency is taken to represent the concentration of the complex formed. The rates of initiation, propagation and termination are given by

$$R_i = \frac{d[M^+]}{dt} = k_i[C][M] \tag{5.7}$$

$$R_p = -\frac{d[M]}{dt} = k_p[M^+][M] \tag{5.8}$$

$$R_t = -\frac{d[M^+]}{dt} = k_t[M^+] \tag{5.9}$$

In view of the presence of the counter ion AS^- in the vicinity of the cationic centre throughout its growth and at the point of termination as conceived in the reaction scheme (5.6), the concentration of AS^- is not considered in the expressions for R_p and R_t, and [C] in Eq. (5.7) would be given by $[C] = K[A] \cdot [SH]$. In case the initiating complex $A \cdot SH$ is largely split into ions as in Eq. (5.6)(a), R_i would be independent of [M] and the overall kinetics should be modified accordingly.

Despite some uncertainty about attainment of steady state, one may find a simple and workable expression for R_p assuming a steady state concentration of M^+ such that

$$[M^+] = \frac{k_i}{k_t} \cdot [C][M] \tag{5.10}$$

and hence,

$$R_p = \frac{k_i k_p}{k_t} \cdot [C][M]^2 \tag{5.11}$$

The average degree of polymerization $\overline{X}_n$, assuming little chain transfer, may be expressed as

$$\overline{X}_n = \frac{R_p}{R_t} = \frac{k_p [M^+][M]}{k_t [M^+]} = \frac{k_p}{k_t} \cdot [M] \tag{5.12}$$

On the other hand, for termination exclusively by monomer transfer [Eq. (5.6)(e)], one may write:

$$\overline{X}_n = \frac{k_p [M^+][M]}{k_{tr} [M^+][M]} = \frac{k_p}{k_{tr}} \tag{5.13}$$

Therefore, for cationic polymerization, the overall rate of polymerization is proportional to the first power of catalyst concentration and to the square of monomer concentration. Thus, in respect of rate-dependence on catalyst or initiator concentration, cationic polymerization differs from the free radical polymerization for which R_p normally follows a square root dependence. The difference arises as a consequence of the difference in their termination mechanism which is second order in the growing species in free radical polymerization and first order in ionic polymerization.

The degree of polymerization in cationic polymerization is independent of the catalyst concentration and for true termination, $\overline{X}_n$ is directly proportional to [M], while for termination exclusively by monomer transfer, $\overline{X}_n$ is a constant given by the ratio of propagation rate constant and the rate constant for the transfer reaction. The rate of cationic polymerization is determined by the ratio k_p/k_t (or k_p/k_{tr}, as the case may be), while that of radical polymerization is determined by the parameter $k_p/k_t^{1/2}$. The k_p and k_t values in cationic polymerization of a monomer are far less than those in free radical polymerization. However, the rates of polymerization in cationic systems, are usually much faster than those in radical polymerization. The former represents a case favouring polymerization at a high rate and the concentration of the propagating species (carbonium ions) in cationic polymerization (10^{-3} molar) is generally much higher than the same (concentration of radicals) in the case of a radical polymerization (10^{-8} molar). In ionic polymerization, large increases in the rate and degree of polymerization are usually effected through increases in the solvating power of the polymerization medium.

In cationic polymerization, the effective catalyst or initiator species and the actual propagation centres are mostly ion pairs (closely bound or solvent separated) and existence and role of free ions must be very limited. Mean separation or the binding energy between the propagating ion and the gegen ion is largely dependent on the nature of the reaction media and the gegen ion and on the temperature of reaction. Propagation through monomer entry between the carbonium ion and the gegen ion is facilitated by a lowering in the binding energy between the two associated ions.

The activation energy for the overall rate and degree of polymerization is given by $[(E_i + E_p) - E_t]$ and $(E_p - E_t)$ respectively where the subscripts i, p and t relate to initiation, propagation and termination respectively. For termination by transfer reaction, E_{tr} would be used in place of E_t. Propagation involving addition of an ion to a neutral monomer in a medium of low polarity is characterized by a low or very little activation energy, and in most cases E_i and E_t are higher than E_p. The activation energy of overall polymerization[7,8] ranges between ±10 kcal/mol. The activation energy for the degree of polymerization is, however, always negative irrespective of mode of termination since $E_t > E_p$. Since E_{tr} is usually greater than E_t, the magnitude of the negative activation energy for the degree of polymerization is higher for termination by transfer reactions than when it is a case of true termination, either partly or exclusively. Limitations or restrictions to chain growth becomes more severe with rise in temperature due to shifts in mode of terminations—from true termination to transfer or from one transfer mode to another.[9]

5.4 Anionic Polymerization

Basic compounds such as amides, alkoxides, alkyls and aryls of alkali metals and different Grignard reagents have been used as catalysts for anionic polymerization of monomers having electron-withdrawing substituents. The catalysts (BA) break up into cations (B^+) and anions (A^-) under the reaction conditions. In anionic polymerization, the anion adds to the double bond of the monomer to generate a carbanion as the active centre for chain-growth.

$$CH_2 = CHX + BA \rightarrow A—CH_2—\overset{(-)}{C}HX + B^{(+)} \tag{5.14}$$

Different catalysts exhibit different reactivities depending on their basicities producing different rates of chain initiation in the presence of an appropriate monomer. The ease of polymerization is dependent not only on the catalyst system used for a given monomer, but also on the structure of the monomer for a given catalyst.

The propagation occurs by addition of monomer molecules, M, to the carbanion by the entry of successive monomers between the ions of the active species:

$$A—CH_2—\overset{(-)}{C}HX + B^{(+)} + (n - 1)M \rightarrow A—(CH_2—CHX)_{n-1}—CH_2—\overset{(-)}{C}HX + B^{(+)} \quad (5.15)$$

The nature of the medium is highly influential at this stage as it determines the distance of separation between the ions at the active site. Termination often takes place by chain transfer involving solvent or some other additive (SH) used such as water, alcohol, etc.

$$A \sim CH_2—\overset{(-)}{C}HX + B^{(+)} + SH \rightarrow A \sim CH_2—CH_2X + BS \quad (5.16)$$

The kinetics of anionic polymerization may be considered and appropriate expressions for rate and degree of polymerization may be deduced using the following scheme of polymerization for an appropriate monomer, say styrene, using potassium amide catalyst in liquid ammonia.[10]

$$KNH_2 \overset{K}{\rightleftharpoons} K^+ + NH_2^-$$

$$NH_2^- + M \xrightarrow{k_i} H_2N—M^-$$

$$H_2N—M^- + (n - 1)M \xrightarrow{k_p} H_2N—(M)_{n-1}—M^-$$

$$H_2N—M_n^- + NH_3 \xrightarrow{k_{tr}} H_2N—M_n—H + NH_2^- \quad (5.17)$$

The rates of initiation (R_i), propagation (R_p) and termination by solvent transfer (R_{tr}) may then be expressed as

$$R_i = k_i[NH_2^-][M] = \frac{k_i K[M][KNH_2]}{[K^+]} \quad (5.18)$$

$$R_p = k_p[M^-][M] \quad (5.19)$$

$$R_{tr} = k_{tr}[M^-][NH_3] \quad (5.20)$$

Using the steady state assumption for the active chain ends (M^-), and combining the Eqs (5.18–20):

$$R_p = \frac{Kk_i k_p [M]^2 [KNH_2]}{k_{tr} [K^+][NH_3]} = \frac{k_i k_p [M]^2 [NH_2^-]}{k_{tr} [NH_3]} \quad (5.21)$$

The number average degree of polymerization $\overline{X}_n$ is given by (R_p/R_{tr}), i.e.,

$$\overline{X}_n = \frac{k_p [M^-][M]}{k_{tr} [M^-][NH_3]} = \frac{k_p}{k_{tr}} \cdot \frac{[M]}{[NH_3]} \quad (5.22)$$

The gegen ion, K^+, has not been included in the reaction scheme (5.17) in view of high dielectric constant of the liquid ammonia medium. Termination of a given chain essentially involves chain transfer with solvents, thereby releasing a new anionic centre at the incidence of termination of each growing chain.

5.5 Living Polymers

There are many cases of anionic polymerization which are not associated with termination reaction and for which propagation continues till complete consumption of the monomers. The carbanions would remain active without termination in such solvents as tetrahydrofuran and dioxane, because of non-occurrence of termination due to proton transfer from the solvent. The polymers thus obtained are termed as 'living polymers'. The propagating carbanions are sometimes coloured and the colour and its intensity remain undiminished throughout polymerization and even beyond complete conversion. Addition of more monomer to the living polymer system leads to its polymerization also quantitatively causing further increase in degree of polymerization, the number of growing species remaining unchanged. Polymerization of methyl methacrylate with metal alkyls produces such enhancement of molecular weight with conversion through the second stage polymerization on fresh monomer addition after complete conversion in the first stage.[11]

5.5.1 Initiation by Sodium Naphthalene

An interesting case of living polymerization is provided by such organometallic initiators as sodium naphthalene.[12,13] Sodium naphthalene in tetrahydrofuran (THF) solution readily forms naphthalene radical anions by an electron transfer mechanism. Studies of living polymer system have been made in detail in the context of polymerization of styrene by sodium naphthalene in THF at low temperatures. Formation of the active catalyst, i.e. the naphthalene radical-ion may be written as:

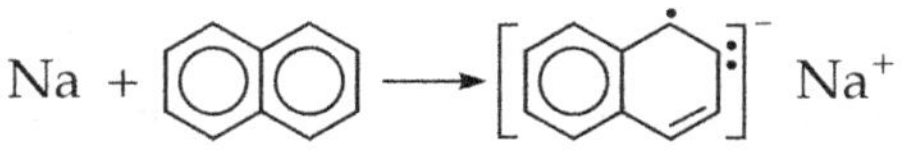 (5.23)

(naphthalene anion -radical, green)

The solution looks green due to the green colour of the naphthalene anion-radical. On addition of styrene, the colour of the medium turns orange due to formation of orange coloured styryl radical anions by transfer of an electron to the styrene monomer from the naphthalene anion-radical.

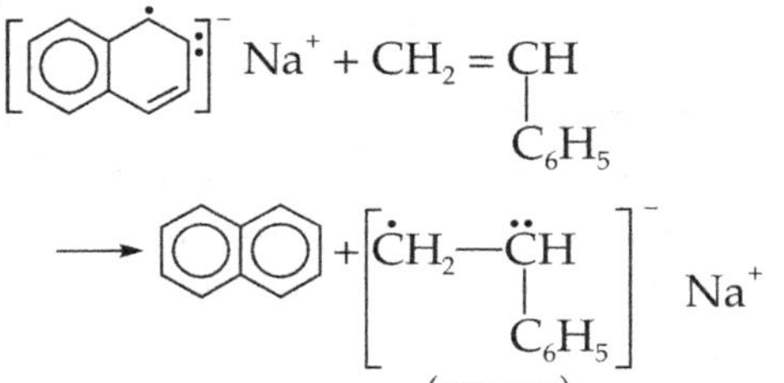 (5.24)

(orange)

The styryl (radical) anion formed in the initiation step may carry on chain propagation by anionic mechanism at one end and by radical mechanism at the other. It is, however, usual and more likely that the radical anions initially formed dimerize by coupling of the radical ends to form dianions which then allow anionic propagation at both ends.

$$2\left[\dot{C}H_2\text{—}\ddot{C}H(C_6H_5)\right]^- Na^+ \longrightarrow Na^+\left[{}^-\ddot{C}H(C_6H_5)\text{—}CH_2\text{—}CH_2\text{—}\ddot{C}H(C_6H_5)^-\right]Na^+ \qquad (5.25)$$

(orange) (orange)

$$\longrightarrow Na^+\left[{}^-\ddot{C}H(C_6H_5)\text{—}CH_2\text{—}CH_2\text{—}\ddot{C}H(C_6H_5)^-\right]Na^+ + (m+n)\,CH_2{=}CH(C_6H_5)$$

$$\longrightarrow Na^+\left[{}^-\ddot{C}H(C_6H_5)\text{—}CH_2\text{—}\left(CH(C_6H_5)\text{—}CH_2\right)_m\text{—}\left(CH_2\text{—}CH(C_6H_5)\text{—}\right)_n\text{—}CH_2\text{—}\ddot{C}H(C_6H_5)^-\right]Na^+ \qquad (5.26)$$

(orange)

Absence of radical centres, evidenced by results of electron spin resonance studies of the living polymer system largely substantiates the dimerization concept.

5.5.2 Initiation by Solvated Electron

Polymerization of methacrylonitrile or styrene initiated by sodium or potassium in liquid ammonia proceeds in much the same manner as for initiation by alkali metal amides in liquid ammonia, as described earlier (initiation by amide ion). However, polymerization of methacrylonitrile using lithium metal in liquid ammonia has been reported to take place at a much faster rate than that induced by the corresponding amide ($LiNH_2$) in liquid ammonia.[14] The polymerization by metallic lithium is initiated through the agency of solvated electrons formed in the ammonia solution which assumes a typical deep blue colour due to the presence of the solvated electrons. The solvated electron is then captured by the monomer to form the corresponding radical anion:

$$Li + NH_3 \rightarrow Li^+ (NH_3) + e^-(NH_3) \quad (5.27)$$

$$CH_2 = CHX + e^-(NH_3) \rightarrow \dot{C}H_2\text{—}\overset{(-)}{\ddot{C}}HX \ (NH_3) \quad (5.28)$$

Propagation follows in much the same manner as in the case of polymerization initiated by sodium naphthalene discussed earlier in this section. Dispersion of alkali metals in hydrocarbon solvents also brings about polymerization of appropriate monomers by a similar mode of initiation involving electron transfer. In the hydrocarbon media of low polarity, however, it is required to consider the presence of the gegen ion (i.e., the metal ion) in the vicinity of the propagating carbanion.

In cases where anionic propagation at two ends follows, the average degree of polymerization $\overline{X}_n$ is equal to $2\overline{v}$, when $\overline{v}$ is the average kinetic chain length, in view of the dimerization by radical combination following the chain initiation step.

5.5.3 Initiation by Metal Alkyls and Metal Aryls

Of the various organometallic compounds used for studies of anionic polymerization, metal alkyls and metal aryls occupy the most prominent position.[15] Among the metal alkyl initiators, butyl lithium has received wider attention and has been used most extensively.[16–18].

Polymerization by a metal alkyl such as butyl lithium proceeds as follows:

(a) Chain initiation

$$BuLi + CH_2{=}CHX \rightarrow Bu{-}CH_2{-}\ddot{C}H\overline{X}\ (Li^+) \tag{5.29}$$

(b) Chain propagation

$$Bu{-}CH_2{-}\ddot{C}H\overline{X}\ (Li^+) + n\ CH_2{=}CHX$$
$$\rightarrow Bu{-}(CH_2{-}CHX)_n{-}CH_2{-}\ddot{C}H\overline{X}\ (Li^+) \tag{5.30}$$

Butyl lithium-initiated vinyl polymerization takes place at unusually slow rates in nonpolar media and studies by well designed experiments indicate initiator association of the following kind, where the value n may be as high as 6.

$$n\ BuLi \rightleftharpoons (BuLi)_n \tag{5.31}$$

Similarly, in non-polar media there are indications for association of the propagating centres in pairs. The immediate consequence of the association of the butyl lithium catalyst and of the corresponding propagating species in non-polar media such as in benzene or toluene is a much slower polymerization rate and a much longer time to complete conversion when compared with similar polymerization using some other anionic initiators, such as sodium naphthalene. However, in presence of a polar solvent, such as tetrahydrofuran, partly or fully replacing the non-polar solvent, the association of butyl lithium catalyst or of the propagating centres becomes less consequential or they are totally eliminated, allowing the polymerization to take place much more rapidly.

5.5.4 Living Ends and Their Termination

The living polymerizations considered above do not involve spontaneous terminations of any kind. A direct consequence of the absence of terminations is the production of a near monodisperse polymer of a remarkably narrow molecular weight distribution. The sharpness results from almost simultaneous initiation of all chains and continuance of simultaneous growth of all chains as long as the monomer is not fully consumed. Even after a long period of storage, injection of a fresh dose of pure monomer leads to further growth by addition of the monomer molecules at the living ends.

It is, however, essential that such polymerizations are done in virtual absence of all reactive impurities. The presence of such impurities as water, oxygen or carbon dioxide, even in traces, will transform the situation to obscurity by causing termination of the kinetic chains and thus limiting the growth in size of the molecular chains by turning the propagating anions unreactive or annihilating the carbanions by proton transfer, as the case may be. On the other hand, living polymers, formed in pure and uncontaminated systems, can be terminated after complete conversion of a monomer as and when desired by the addition of water or other proton transferring agents.

5.6 Ionic Copolymerization

The copolymerization equation derived and discussed in the context of radical polymerization of mixture of two monomers is equally applicable to cationic and anionic copolymerization. The copolymerization equation is not dependent on the initiation and termination rate constants and consequently, the copolymer composition is independent of rates of initiation, termination or chain transfer and related reactions and in most cases, of the degree of polymerization. For a given type of chain initiation and chain propagation (radical, cationic and anionic), the copolymer composition is practically independent of the nature of the initiator. The monomer reactivity ratios (r_1 and r_2) and hence the copolymer composition for a specific composition of a monomer pair suffer changes, significantly or even drastically depending on the propagation mode. The difference in copolymer composition curves for styrene (M_1) and methyl methacrylate (M_2) to be obtained by radical, cationic and anionic copolymerizations[19,20] are shown in Fig. 5.1. The r_1 and r_2 values for the monomer pair are 0.52 and 0.46 for *free* radical copolymerization, 10 and 0.1 for cationic copolymerization and 0.1 and 6 for anionic copolymerization. Even polymerization by a *complex* radical mechanism may produce measurable differences in copolymer composition in comparison with that produced by the usual *free* radical mechanism.[21,22] The selective nature of ionic polymerization should be viewed against the universal character of free radical polymerization to appreciate the difference. Methyl methacrylate monomer expectedly shows much

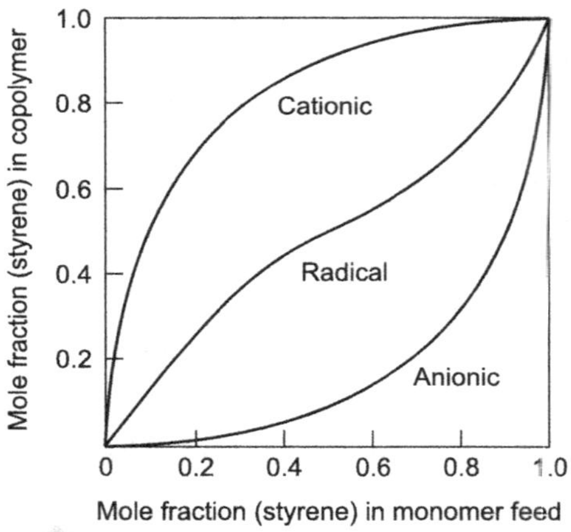

Fig. 5.1 *Copolymer composition (mole fraction F_1) as a function of monomer feed (mole fraction f_1) composition using styrene (M_1) and methyl methacrylate (M_2) for polymerization by radical, cationic and anionic mechanisms (Schematic)*

higher order of reactivity than styrene in anionic copolymerization and the order of reactivity is just the opposite in cationic copolymerization. The selective nature of ionic copolymerization is a major consideration that goes against their widespread practical exploitation.

5.7 Stereo-Regular Polymerization

Stereo-regularity in olefin or vinyl polymerization that proceeds anionically is largely dependent on the physical conditions of the reaction and the chemical environment including the solvent and the positive counterion.

The statistical nature of polymers and the polymerization reaction is important in relation to stereo-regularity, and for quantitative purposes it is required to describe a polymer in terms of relative orders or percentages of isotactic, syndiotactic and atactic sequences.

Stereo-regular polymerization is mostly characterized by (a) anionic propagation, (b) hair-like growth, and (c) exclusive head-to-tail arrangement of repeating monomer units. The features have been widely studied and better understood using the so-called Ziegler catalysts or Ziegler–Natta catalysts. Such catalysts are prepared by interaction between an aluminium alkyl, such as aluminium triethyl, $Al(C_2H_5)_3$ and a transition metal halide, such as titanium tetrachloride, $TiCl_4$. The exact nature of the catalyst activity is contingent upon the tendency of the transition metal and aluminium to operate with coordination valencies. Usually, the Ziegler–Natta catalysts are of high activity allowing low temperature initiation and imparting stability to the growing chain end. This is a prerequisite for the production of polymers of very high molecular weight.

In polymerization with Ziegler-Natta catalysts, a radical mechanism is not tenable as it cannot explain the stereo–regularity and formation of isotactic polymers. An ionic mechanism is, of necessity, widely favoured and accepted.

In a truly anionic initiation, the anion from the initiator system would have to add an additional electron to the already existing π electron pair of an olefin, say ethylene, to generate a carbanion that may lead to chain propagation. In a system of poor resonance, such as with ethylene, this is energetically unfavourable under normal or moderate conditions. Thus, metal alkyl catalysts do not favourably initiate polymerization of ethylene and related monomers by their carbanions but do so by their metal cations.

Alkali metals would polymerize monomers only when they are able to form polar or ionic adducts through establishment of polarized or ionized metal–

carbon (M–C) bonds which would permit addition of large number of monomer units in quick succession to yield polymers. Li and Na fail to polymerize ethylene in view of their reactivity (i.e. electropositivity) being not strong enough to produce M–C bonds of adequate reactivity and the resulting carbanion is rather unstable due to poor resonance. Other alkali metals also behave in much the same manner.

In contrast, styrene and conjugated dienes can be polymerized with the alkali metals as the adducts formed are associated with enhanced mesomeric possibilities, thereby conferring stability to the resulting carbanion to a level allowing less frequent termination and carrying the propagation to yield high polymers.

In presence of a cation centre as the initiator, there would be attractions between the cation centres and the π electrons of the olefinic monomers and under a favourable situation an anchorage of the monomer on the cation would create a carbonium ion at the remote end of the monomer segment due to acceptance of a π electron by the cation.

$$\begin{array}{l} \quad\;\; CH_2 \\ (—)\ \| \\ \quad\;\; CH_2 \\ \\ \quad\;\; M^+ \end{array} \longrightarrow \begin{array}{c} \overset{+}{C}H_2 \\ | \\ CH_2 \\ | \\ M \end{array} \qquad (5.32)$$

The strength of the carbonium ion would depend on how strongly the cations (M^+) can bind the π electrons. Transition elements offer high resonance to accepted electrons because not only the outer electron shell but also the inner orbitals can participate in resonance stabilization. A cation of a transition metal would, therefore, have an exceptionally high ability to bind the electrons of the monomer and to induce a strong carbonium ion.

The M—C bond in alkali metal alkyls such as lithium alkyl ($Li^+...R^-$) has a salt-like character. Its strong reactivity permits ethylene to polymerize to limited extents, involving cationic initiation and anionic propagation. Only higher olefins or low polymers are obtained, as the Li cation frequently intercepts propagation by hydride (H^-) extraction causing termination.

$$\begin{array}{l} \quad\;\; CH_2 \\ (—)\ \| \\ \quad\;\; CH_2 \\ \\ \quad\;\; Li^+...R^- \end{array} \longrightarrow \begin{array}{l} \overset{+}{C}H_2 \\ | \\ CH_2 \\ | \\ Li...R^- \end{array}$$

$$\longrightarrow \overset{+}{Li}\ldots\overline{C}H_2—CH_2—R \qquad (5.33)$$

$$\big\downarrow x\,(CH_2{=}CH_2)$$

$$\overset{+}{Li}\ldots\overline{C}H_2—CH_2—(CH_2—CH_2)_x—R$$

$$\downarrow$$

$$LiH + CH_2{=}CH—(CH_2—CH_2)_x—R$$

The lifetime of the ions during migration is very short due to poor resonance. A similar mechanism is involved in the Ziegler aluminium alkyl process. Aluminium has an additional driving force to react with the π electrons of ethylene leading to the formation of an electron octet.

The individual components of the Ziegler–Natta catalyst system for stereo-regular polymerization can separately induce polymerization of some form but cannot account for the stereo-regularity. The role of metal alkyl has already been discussed above. $TiCl_4$ can readily initiate cationic polymerization. The stereo-regulating influence of the catalyst combination is derived from a coordination mechanism, whereby both the growing chain and the monomer coordinate with the catalyst.

The catalyst combination forms ionic complexes that precipitate out when the two components are mixed together in a hydrocarbon solvent. The initial complex formation using aluminium triethyl ($AlEt_3$) and titanium tetrachloride ($TiCl_4$) or the trichloride ($TiCl_3$) may be written as:

$$TiCl_4 + AlEt_3 \rightarrow (TiCl_3)^+(AlEt_3Cl)^-$$

$$TiCl_3 + AlEt_3 \rightarrow (TiCl_2)^+(AlEt_3Cl)^- \qquad (5.34)$$

When a non-polar monomer such as an α-olefin is used, initiation of polymerization is performed by the transition metal cation; propagation then follows in the negative complex ion of aluminium and only heterogeneous Ziegler–Natta catalysts yield isotactic polymers. Titanium based complexes are the most common of the heterogeneous catalysts. Among the other transition metal, vanadium has been more widely used, and vanadium-based systems commonly provide homogeneous catalysts.

For stereo-regular polymerization, the catalysts must not only initiate polymerization, but they must also effectively regulate: (a) the entry and orientation of the successive monomer units during initiation (anchorage) on the cation site and

(b) the migration and propagation of the successive incoming monomer units on the anionic site. The prerequisite for a constant specific orientation of the monomer is a fixed location of both cation and anion; if either is soluble or dispersible, poor steric regularity, or atactic poly (α-olefin) structures would result. Syndiotacticity is obtained in a few instances also by soluble catalysts. The chain propagation via the anionic coordination mechanism involves insertion of an oriented monomer molecule at a carbanion centre permitting head-to-tail addition and hair-like growth on the anionic site, the rate determining step being the electrophilic attack by the catalyst fragment.

Mechanisms proposed for different systems are based on the overall principle detailed above but often they are so specific in detail that they can seldom be generalized beyond the systems for which they are suggested. The course of isotactic placement of successive monomer units during polymerization of propylene using a bimetallic catalyst complex may be depicted in a simplified manner by reaction steps in Fig. 5.2, highlighting bimetallic mechanism.

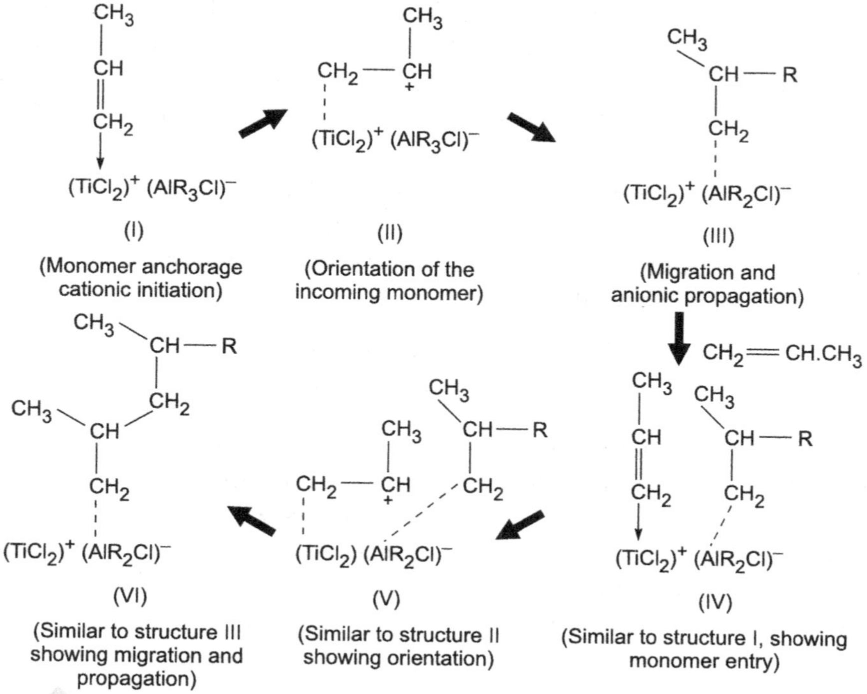

Fig. 5.2 *Formation of isotactic polypropylene using the catalyst system $AlEt_3$–α–$TiCl_3$ (bimetallic mechanism)*

The catalyst complex of $TiCl_3$ and $AlEt_3$ essentially acts as a template for the successive orientation and isotactic placements of the incoming monomer units. The crystal structure of the catalyst surface and the chemical structure of the catalyst complex determine the orientation(s) permitted to a given monomer as it adds to a propagating chain. One concept ascribes isotactic placement of monomers during chain propagation to steric and electrostatic interactions between the substituent of the incoming monomer and the ligands of the transition metal at the reactive site of the crystal surface, while another concept attributes the isotactic placement to the interactions between the incoming monomer and the Group I-III metallic part of the bimetallic catalyst. There can be little reservation about either concept if both bimetallic and monometallic mechanisms are involved in the Ziegler–Natta polymerization. The monometallic mechanism describes the transition metal (in the present case, titanium) part of the catalyst complex as the only active site for overall polymerization. The monomer is coordinated to the vacant orbital of the transition metal complex and then inserted into the polymer chain at the transition metal—carbon bond giving rise to regeneration of the vacant orbital with a different orientation. Propagation with the species would result in a syndiotactic sequence. Isotactic polymerization is allowed by migration of the polymer chain to its original location causing regeneration of the original vacant orbital, as shown in reaction scheme in Fig. 5.3, using $TiCl_4$ complex for the polymerization of propylene, where the (vacant) square indicates a vacant orbital of the transition metal.

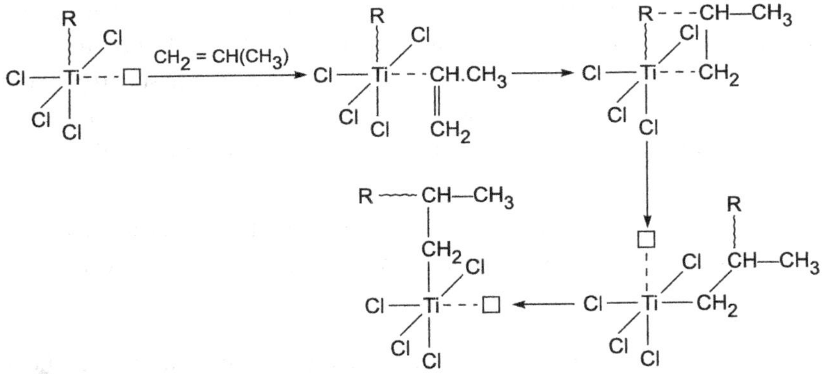

Fig. 5.3 ***Monometallic mechanism for stereo regular (isotactic) polymerization of propylene (□ indicates vacant orbital)***

Non-polar monomers, such as the α-olefins, having poor coordinating powers with the catalyst, require such catalysts as would specifically hold them in place with strong stereo-regulating influence to give isotactic sequences for the polymer, thus necessitating the use of heterogeneous catalysts. For polar monomers (acrylates, methacrylates, vinyl ethers, etc.) with strong coordinating powers with the catalyst, heterogeneity of the catalyst is not a prerequisite for isotacticity. When non-polar solvents are employed, the polymerization for polar monomers proceeds by anionic coordination mechanism, the gegen ion directing the orientation of each approaching monomer unit into the polymer chain and the ability of the gegen ion to coordinate determining the steric regularity. In polar solvents, the gegen ion is removed from the vicinity of the propagating centre and so the stereochemical influence on the approaching monomer unit is lost resulting in random structures. For such systems, rise of temperature favours atacticity, while lower temperature favours syndiotacticity.[23,24]

5.8 Stereo-Structures of Polymers of Disubstituted Ethylenes

Stereo-regularity or lack of it for polymers of disubstituted ethylenes is dependent on the position and identity of the substituents. For polymers of 1, 1-disubstituted ethylene, stereoisomerism is non-existent when the two substituents are identical such as for vinylidene chloride, $CH_2{=}CCl_2$ or isobutylene, $CH_2{=}C(CH_3)_2$. In case the substituents are different, such as in methyl methacrylate, $CH_2{=}C(CH_3)(COOCH_3)$, the possibility of stereo-isomerism in the polymer is just the same as for monosubstituted olefins, and the introduction of the second substituent makes no difference as steric disposition of one substituent instantly fixes that of the other. If a structure is isotactic for the first substituent, it is so for the second; if it is syndiotactic or atactic for the first substituent, it is automatically syndiotactic or atactic for the second.

Polymers of 1,2-disubstituted ethylenes, CHX=CHY provide two sites of steric isomerism in view of the presence of two asymmetric carbon atoms in each of the repeat units $\sim\!\!\begin{array}{c}H\\|\\C\\|\\X\end{array}\!\!-\!\!\begin{array}{c}H\\|\\C\\|\\Y\end{array}\!\!\sim$ with the capacity to display ditacticity. Four different stereo-regular structures indicating the placements of the substituents can be written and identified as: (a) erythro-di-isotactic, (b) threo-di-isotactic, (c) erythro-di-syndiotactic and (d) threo-di-syndiotactic structures as shown in Fig. 5.4.

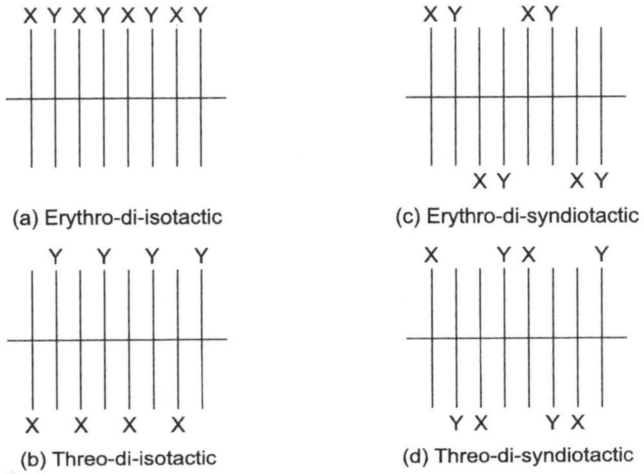

Fig. 5.4 *Four major stereo-regular structures for polymers of 1, 2-disubstituted ethylene, —(CHX–CHY)$_n$—*

5.9 Geometrical and Optical Isomerism of Polymers of 1, 3-Diene Monomers

The common 1,3-dienes used to produce industrial polymers are butadiene, isoprene and chloroprene and the related polymers may

$$\underset{1}{CH_2}{=}\underset{2}{CH}{-}\underset{3}{CH}{=}\underset{4}{CH_2} \qquad \underset{1}{CH_2}{=}\underset{2}{\overset{\overset{CH_3}{|}}{C}}{-}\underset{3}{CH}{=}\underset{4}{CH_2} \qquad \underset{1}{CH_2}{=}\underset{2}{\overset{\overset{Cl}{|}}{C}}{-}\underset{3}{CH}{=}\underset{4}{CH_2}$$

1, 3-butadiene — isoprene — chloroprene

feature both optical and geometrical isomerism. When one of the double bonds is involved in polymerization (1, 2 or 3, 4 polymerization) leaving the other intact, the polymer shows only optical isomerism. For butadiene, 1, 2 and 3, 4 polymerization produces identical structures —[CH_2—CH(CH=CH_2)]$_n$—, much alike the structure of an α-olefin, thus allowing three stereo-structural possibilities for the polymer (isotactic, syndiotactic and atactic). For the 2-substituted polybutadienes (e.g., beginning with isoprene or chloroprene as the monomer), there are larger possibilities of isomerism since polymerizations through (a) 1,2 and (b) 3,4 unsaturations yield two different structures, Fig. 5.5, with three stereo sequences in either case corresponding to isotactic, syndiotactic and atactic placements giving a total of six possible structures. 1,4-polymerization of a conjugated diene such as isoprene is, however, associated with a rearrangement of bonds leading to a polymer structure with an unsaturation along the chain backbone in the 2,3 position in each repeat unit, which for 1,4-polyisoprene may be represented as $—[CH_2—C(CH_3)=CH—CH_2]_n—$. The reformed unsaturation in each repeat unit of the chain molecule provides a centre of steric isomerism giving either cis or trans configuration as shown below. Natural rubber and *gutta percha* are mainly (a) 1,4-cis-polyisoprene and (b) 1,4-trans-polyisoprene respectively.

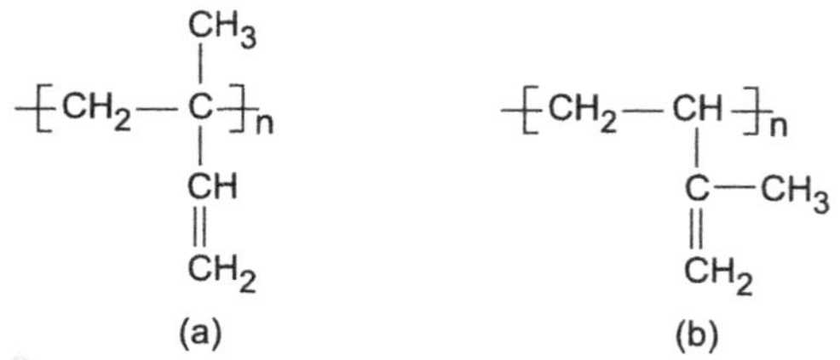

Fig. 5.5 *Structure of (a) 1, 2 polyisoprene and (b) 3, 4-polyisoprene*

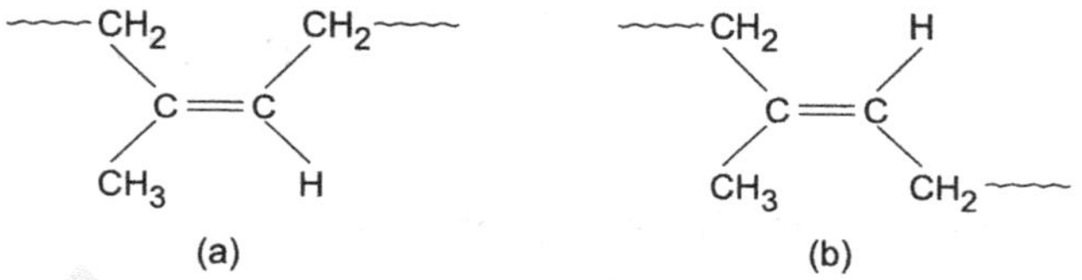

Fig. 5.6 *Structure of a segment of (a) 1,4-cis-polyisoprene, and (b) 1,4-trans-polyisoprene*

5.10 Polymerization of 1, 3-Dienes

Polymerization of 1,3-dienes can be carried out by radical or ionic initiators. Radical initiators favour formation of 1,4-trans structure mostly, at low temperatures.[25–27] Delocalization of the radical over the 2–4 carbon atoms of the terminal (radical bearing) monomer unit ($\sim\underset{1}{C}H_2–\underset{2}{C}H–\underset{3}{C}H–\underset{4}{C}H_2$) and a higher degree of steric hindrance at carbon 2 than at carbon 4, more so for 2-substituted 1,3-dienes, leads to predominance of 1,4-propagation over 1,2-propagation and also over 3,4-propagation (for the 2-substituted monomers). Furthermore, in view of predominance of (a) *s*-trans conformation of the monomer against (b) *s*-cis conformation as shown in Fig. 5.7 for 1,3 butadiene, and in view of retention of the monomer conformation as the diene monomer adds to the propagating radical, the radical polymerization results in predominance of the 1,4-trans configuration.[28]

Fig. 5.7 *Conformations of 1, 3-diene (e.g. butadiene)*

Different anionic catalysts may allow polymerization of 1,3-dienes with different degrees of coordination between the propagating centre and the gegen ion, yielding different polymer structures. For isoprene, polymerization in a nonpolar solvent using metallic lithium or butyl lithium as the catalyst, gives overwhelming predominance of 1,4 cis structure, producing what may be termed as "synthetic natural rubber". It is believed that the lithium ion is co-ordinated simultaneously with the approaching monomer and the propagating carbanion chain end.[29] The monomer—Li^+—propagating end coordinating species schematically shown in Fig. 5.8, then leads to chain extension by addition of the monomer to the carbanion centre by 1,4-mechanism. The monomer coordination locks the diene into a cis-structure before and after its addition to the propagating chain.[30] For strong coordination, such as in non-polar solvents, rotation about the bond linking carbon atoms 2 and 3 of the propagating chain end is not permissible, thus largely favouring, cis-1,4-polymerization. For relatively weak coordination, trans-1,4-polymerization also assumes significance, while for very weak coordination linkage between carbon 4 of the monomer and carbon 3 of the allylic carbanion is favoured resulting in primarily 3,4-polymerization. Most Ziegler–Natta catalysts produce very sharp or exclusive stereo specificity for diene polymers.

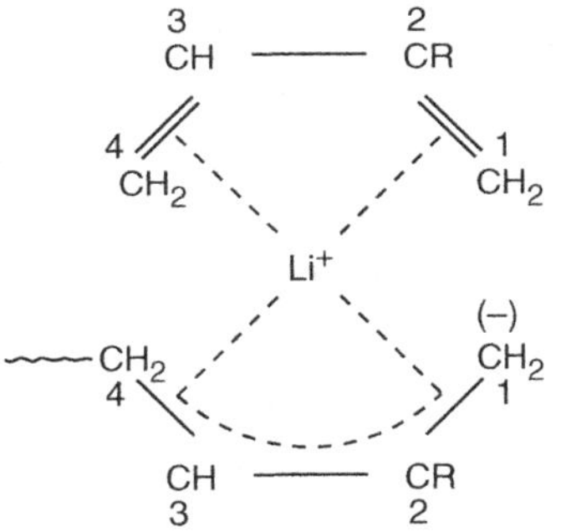

Fig. 5.8 *Schematic representation of the monomer and the propagating carbanion coordinating simultaneously with the gegen ion in the polymerization of 2-substituted-1, 3-diene*

1,3-dienes can also be polymerized by cationic catalysts which usually yield polymers of lower molecular weight, often with condensed cyclic structures. Such cyclopolymerization of conjugated dienes proceeds by a Diels–Alder type reaction via charge transfer mechanism. The formation and structure of the cyclopolymer viewed essentially as a ladder-type polymer[31] and showing high thermal stability may be represented by the reaction shown in Fig. 5.9.

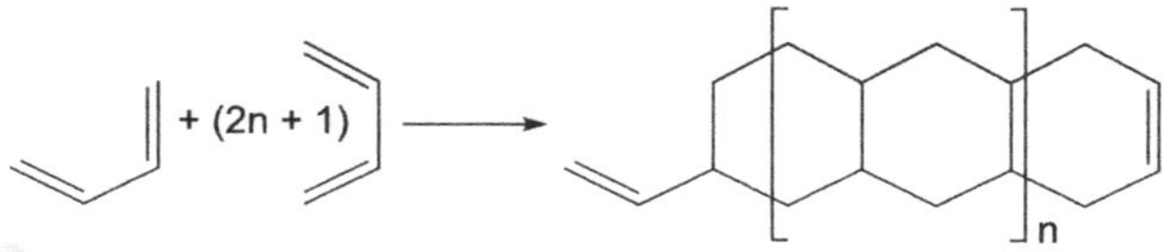

Fig. 5.9 *Structure of a ladder polymer from cationic polymerization of a 1,3-diene monomer*

5.11 Electroinitiated Polymerization

Electrochemical or electrolytic initiation of polymerization is conveniently carried out in a conducting medium in an electric field in presence of different additives. It was a natural outcome of electro-organic chemistry and the first monomer to be studied being the one bearing a carbonyl group (i.e., methyl methacrylate). The electro-initiated polymerization is rather complicated in the sense that simple hydrogenation of olefinic unsaturation, reduction of other functional groups, decomposition of electrolytes present, etc., usually take place in addition to the desired polymerization of the monomer employed.[32]

The electrode material, current density, electrolyte additives should therefore be properly selected for a given monomer system so as to enable control of the electrolytic process to give high yield of the desired product.

There can be at least two types of electrochemical initiation of polymerization—direct and indirect. By the direct process the active initiating species from the monomer or additive is formed in the system by an electrode reaction, cathodic or anodic, such as shown in reactions below:

Cathodic

$$H^+ + e^- \rightarrow H^\bullet \tag{5.35}$$

$$RCOR' + H_2O + e^- \rightarrow RC^\bullet(OH)R' + OH^- \tag{5.36}$$

Anodic

$$RCOO^- \rightarrow RCOO^\bullet \text{ (or } R^\bullet + CO_2) + e^- \tag{5.37}$$

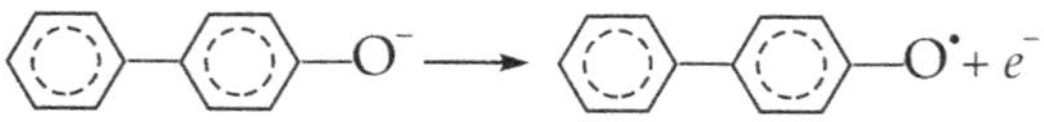

(5.38)

The radicals generated at the electrodes initiate polymerization. It is also possible that an electrode behaves as a Lewis acid or base of different strengths and initiates polymerization by ionic mechanism. Even if ionic polymerization in aqueous medium is not altogether unlikely, such polymerization by electro-initiation would favourably occur where the monomer or a nonaqueous compound is the solvent; the cathode may initiate anionic polymerization and the anode cationic polymerization if appropriate monomer is used and right conditions are maintained. A chemical initiating species formed *in situ* by the electrochemical process gives rise to what is known as the indirect initiation. A good example of indirect electrochemical initiation is the electrolytic reduction of O_2 to H_2O_2 which then polymerizes the monomer used.[33]

A large number of useful polymers are insoluble in water, while water is the most suitable solvent for the ionic additives which provide the much desired electroconductivity. Monomer emulsions or protonic solvents such as alcohols have been used by many to improve miscibility. Emulsions of monomers lead to deposition of polymer coatings on the electrode, thereby greatly reducing conductivity and the protonic solvents often interfere with the reactions at the electrode. Related problems are solved more satisfactorily using nonprotonic, water-miscible solvents such as dimethyl sulphoxide, propylene carbonate, dimethyl formamide, etc. It is generally experienced by quantitative kinetic analysis and other studies that in many electrochemical systems, polymerization is initiated by both radical and ionic means, sometimes by both mechanisms, at the same time.[34,35] Mechanism of polymerization in a given electrolyte-solvent-monomer system is adjudged[36–38] by examining the effect of radical scavengers, zone of polymer formation (cathodic or anodic zone), variation of temperature, copolymer composition using binary monomer combination, and doing polymerization in special divided cells and noting the formation of colour due to generation of carbanions from specific monomers.

References

1. Mayo, F.R. and C. Walling, *Chem. Rev.*, **46** (1950), 277.
2. Plesch, P.H., *Research*, **2**, (1949), 267; *J. Chem. Soc.*, (1950), 543.
3. Norrish, R.G.W. and K.E. Russel, *Trans. Faraday Soc.*, **48** (1952), 91.
4. Evans, A.G. and G.W. Meadows., *Trans. Faraday Soc.*, **46** (1950), 327.
5. Evans, A.G. and M. Polanyi, *J. Chem. Soc.*, (1947), 252.
6. Dainton, F.S, and G.B.B.M. Sutherland, *J. Polym. Sci.*, **4** (1949) 37.

7. Plesch, P.H., *J. Chem. Soc.* (1953), 1653.
8. Brown, C.P. and A.R. Matheson, *J. Chem. Soc.*, 1957, pp. 3608, 3612, 3620, 3631; 1958, p. 3445.
9. Kennedy, J.P. and R.G. Squires, *Polymer*, **6** (1965), 579.
10. Higginson, W.C.E. and N.S. Wooding, *J. Chem. Soc.* (1952), 760.
11. L'Abbe, G. and G. Smets, *J. Polym. Sci.*, **45** (1960), 127.
12. Szwarc, M., M. Levy and R. Milkorich, *J. Am. Chem. Soc.*, **78** (1956), 2656.
13. Szwarc, M., R. Waak, A. Rembaum and J.D. Coombar, *J. Am. Chem. Soc.*, **79** (1957), 2026.
14. Overberger, C.G., E.M. Pearce and N. Mayes, *J. Am. Chem. Soc.*, **34** (1959), 109.
15. Bywater, S., *Fortschr. Hochpolymer Forsch* (*Adv. Polymer Sci.*), **4** (1965), 66.
16. Bywater, S. and D.J. Worsfold, *J. Phys. Chem.*, **70** (1966), 162.
17. Worsfold, D.J. and S. Bywater, *Can. J. Chem.*, **38** (1960), 1891.
18. O'Driscoll, K.F. and A.V. Tobolsky, *J. Polym. Sci.*, **35** (1959), 259.
19. Landler, Y., *Compt. Rend.*, **230** (1950), 539.
20. Pepper, D.C., *Quart. Revs.* **8** (1954), 88.
21. Ghosh, P. and K.F. O'Driscoll, *J. Polym. Sci.*, **B4** (1966), 519.
22. Ghosh, P., *J. Polym. Sci.*, **B7** (1969), 293.
23. Braun, D., M. Herner, U. Johnson and W. Kern, *Makromol. Chem.*, **52** (1962), 15.
24. Hatada, K., K. Ota and H. Yuki, *J. Polym. Sci.*, **B5** (1967), 225.
25. Condon, F.E., *J. Polym. Sci.*, **11** (1953), 139.
26. Richardson, W.S., *J. Polym. Sci.*, **13** (1954), 229.
27. Pollock, D.J., L.J. Elyash and T.W. Dewitt, *J. Polym. Sci.*, **15** (1955), 86.
28. Craig, D., J.J. Shipman and R.B. Fowler, *J. Am. Chem. Soc.*, **83** (1961), 2885.
29. O'Driscoll, K.F., T. Yonezawa and T. Higashimura, *J. Macromol. Chem.*, **1** (1966), 1.
30. Stein, R.S. and L.E. Forman, *J. Polym. Sci.*, **41** (1959), 381.
31. Gaylord, N.G. and M. Svestha, *J. Polym. Sci.*, **B7** (1969), 55.
32. Dineen, I.E., T.C. Schwan and C.L. Wilson, *Trans. Electrochem. Soc.*, **96** (1949), 226.
33. Fedorova, A.I., I.V. Shelepin and N.B. Moiseeva, *Dokl Akad. Nauk* (*SSSR*), **138** (1961), 165.
34. Yang, J.Y., W.E. McEwen and J. Kleinberg, *J. Am. Chem. Soc.*, **79** (1957), 5833.
35. Funt, B.C. and F.D. Williams, *J. Polym. Sci.*, **A2** (1964), 865.
36. Breitenbach, J.W. and Ch. Srna, *Pure and Appl. Chem.* **4** (1962), 245.
37. Funt, B.L. and K.C. Yu, *J. Polym. Sci.*, **62** (1962), 359.
38. Yamazaki, N., S. Nakahama and S. Kambara, *Polym. Letters*, **3** (1965), 57.

CHAPTER 6

Polymer Characteristics and Polymer Characterization

6.1 The Structure of Vinyl and Related Polymers

The selective nature of ionic polymerization in respect of the nature of monomer, solvents and additives, the complexities and variations in the mechanistics and kinetics depending on the nature and identity of the ionic catalyst including the different structural (steric) sequences imparted under the influence of different gegen ions are generally established and widely recognized. For all these and associated reasons, ionic polymerization is not favoured and widely adopted for industrial polymerization. It is employed only in a few instances where the objective is the synthesis of a polymer of high structural (steric) regularity. Most industrial polymerizations of vinyl and related monomers are done by free radical means because of many practical and economic advantages and versatility associated with the free radical process.

6.2 Prevalence of Head-to-Tail Structure in Vinyl Polymers

Addition of a free radical $R^{\cdot}$ (primary radical or chain radical) to a vinyl monomer may take place according to either of the two ways shown below:

$$R^{\cdot} + CH_2 = CHX \longrightarrow R - CH_2 - \overset{X}{\overset{|}{\dot{C}H}} \quad \text{(I)} \qquad (6.1)$$

$$R^{\cdot} + CH_2 = CHX \longrightarrow R - \underset{X}{\underset{|}{CH}} - \dot{C}H_2 \quad \text{(II)} \qquad (6.2)$$

The relative stabilities of the product radicals I and II determine the relative rates of reactions (6.1) and (6.2). In structure I, the presence of the substituent X on the radical-bearing carbon atom (α-carbon atom of the monomer) offers a much higher scope for stability of the radical and the degree of stabilization would depend to a large extent on the capacity of the substituent for resonance. The scope for such stabilization is poor in structure II in view of the non-availability of the substituent for taking part in resonating structures involving the lone electron as the substituent is attached to the carbon atom in the β-position with respect to the radical site. Moreover, on consideration of steric hindrance to the approaching radical offered by the substituent X of the vinyl monomer, reaction (6.1) appears to be more consequential than reaction (6.2). The steric factor would, however, be dependent on the size of the substituent. Successive addition of monomers exclusively by the favoured reaction route (6.1) in the propagation step would result in a head-to-tail structure, i.e., the polymer formed would exhibit 1,3 structure with respect to the placement of the substituent, X, along the chain axis. For monomer addition exclusively by reaction route (6.2), the resultant polymer would also exhibit head-to-tail or 1,3 structure (III):

head-to-tail structure (III)

On the other hand in an extreme case where monomer addition is permitted alternately by routes (6.1) and (6.2), the resultant polymer would appear as having head-to-head, tail-to-tail or 1,2–1,4 structure (IV).

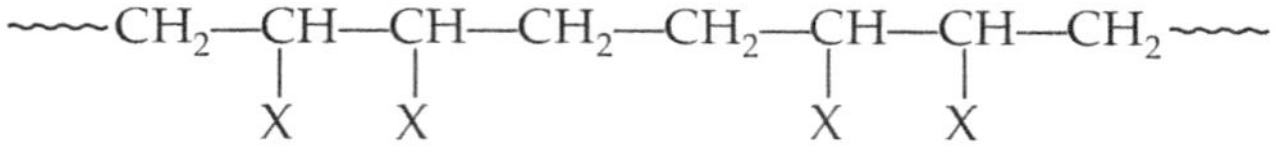

head-to-head, tail-to-tail structure (IV)

But from different theoretical considerations, the head-to-tail structure is expected to be the prevalent structure for all cases where the monomer structure is such as to permit addition of the chain radical to the monomers by only one route to the virtual exclusion of the other. For monomers with a substituent X producing little directive influence, random addition giving all kinds of possible arrangements of the repeat units in the polymer chain (1,3 and 1,2–1,4 arrangements) may take place.

However, the structural pattern of vinyl polymers, as confirmed by different experiments using polymers of different monomeric origin, appears to be mostly head-to-tail in nature. Pyrolysis or destructive distillation of polystyrene at about

300°C produced 1,3-diphenylpropane (V), 1,3,5-triphenylpentane (VI) and 1,3,5-triphenyl-benzene (VII).

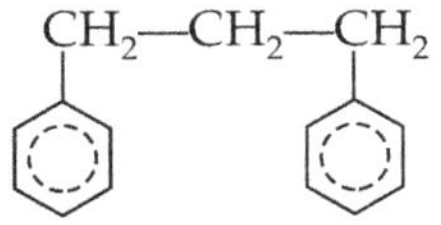

(V)

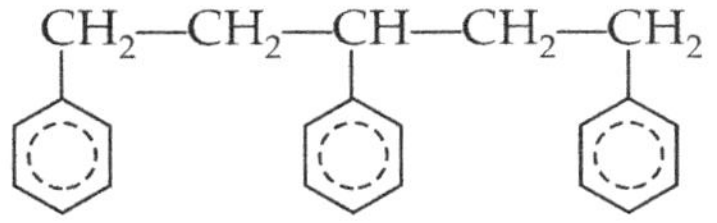

(VI)

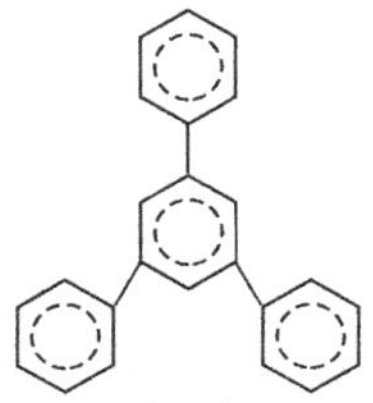

(VII)

In these products the phenyl groups are attached to alternate carbon atoms as in polystyrene having head-to-tail structure and no products having phenyl groups on adjacent carbon atoms are obtained.[1]

Treatment of a dilute solution of poly(vinyl chloride) in dioxane with zinc dust leads to removal of chlorine producing cyclopropane rings along the chain. However, a maximum of about 84–86% of chlorine can be normally removed in this manner. The retention of nearly 14% chlorine on extended reaction is attributed to the occasional isolation of a lone chlorine substituent between neighbours that have paired at random and reacted:

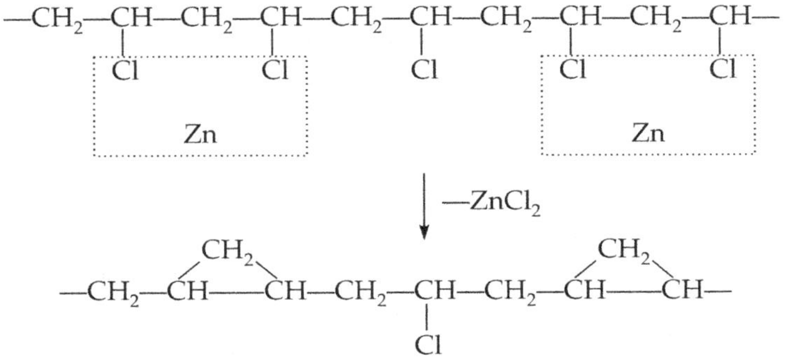

(6.3)

The observed extent of halogen removal is in good agreement with the value (86.5%) obtained by statistical calculations assuming head-to-tail structure[2–4]. Successive head-to-head, tail-to-tail structure (IV) without a break in the sequence, on the other hand, would have produced a polyene structure (—CH_2—CH = CH—

$CH_2—CH_2—CH = CH—CH_2—$); for a random structure, however, theoretical calculation predicts that 18.4% of chlorine should escape elimination and would be retained in the polymer, considering that chlorine atoms in 1,2 and 1,3 positions only are removable in pairs by zinc.

Likewise, poly(methyl vinyl ketone) transforms into a product consisting of long sequence of condensed cyclohexene rings interrupted by statistically isolated groups at random intervals through internal aldol condensation, much in conformity with a head-to-tail structure for the polymer. Extended reaction results in progressive removal of oxygen for which the experimental upper limit[5] is about 79–85% against a calculated value of 81.6% for intramolecular reaction of the kind in a head-to-tail polymer. For 1,2 placements of all successive pairs of the substituents leading to formation of 1,2 and 1,4 diketone structures (IV), the intramolecular reaction should, expectedly, yield furan ring structures with only 50% loss of the total oxygen.

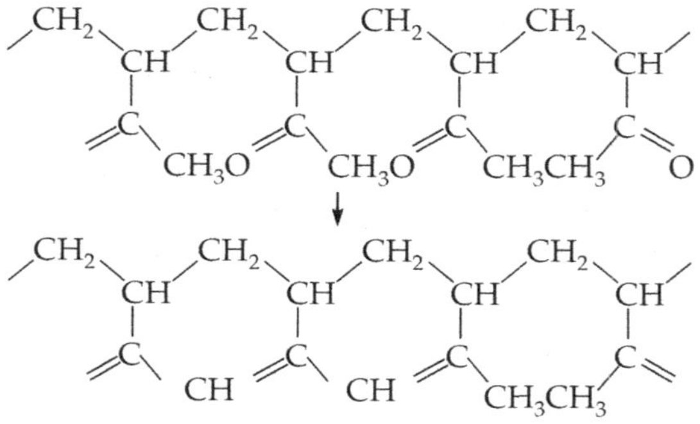

(6.4)

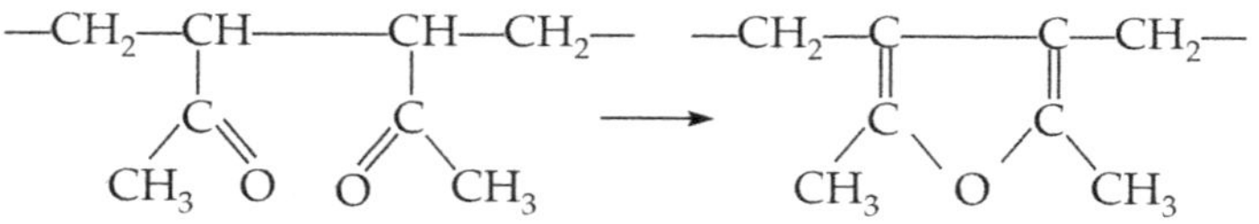

(6.5)

In a similar manner, poly (vinyl acetate) also appears to be largely characterized by a head-to-tail arrangement of the repeating units in the chain molecules as revealed by treatment of the hydrolyzed polymer, i.e., poly(vinyl alcohol) with periodic acid reagent which attacks only the 1,2-glycols and gets consumed in the process. Experimental reports, however, indicate little consumption of the reagent substantiating successive monomer addition by head-to-tail mechanism despite the fact that the substituent $—OCOCH_3$ offers very limited stability to the free radical through resonance.[6] At the same time, periodic acid or periodate ion treatment in aqueous solution causes rather notable lowering in molecular weight of the poly(vinyl alcohol) within a few minutes at room temperature, giving evidence for a minor proportion of head-to-head/tail-to-tail structure. It is important to bear in mind that even for head-to-tail propagation sequences, termination by combination results in a break in the monotony of the 1,3 structure and causes appearance of a 1,2 structure at the point of radical combination.

6.3 Branching in Vinyl Polymers

Consideration of an olefinic unsaturation as equivalent to bifunctionality leads one to ordinarily presume the vinyl and related polymerizations as processes yielding linear macromolecules. But in radical polymerization both the monomers and the polymers being formed or present in the system are subject to attack by propagating radicals through what may be termed as monomer transfer and polymer transfer reactions, thereby leading to formation of some branched polymer molecules in the product, and this aspect has been discussed with some specific examples in Chapter 3. In certain cases, such as in vinyl acetate, chain transfer with the monomer may lead to a polymer molecule with a terminal unsaturation as shown in Eq. (6.6):

$$\sim\sim M_n^{\bullet} + CH_2{=}\underset{}{\overset{XH}{\overset{|}{C}H}} \rightarrow \sim\sim M_n{-}H + CH_2{=}\overset{X^{\bullet}}{\overset{|}{C}H} \xrightarrow{nM} CH_2{=}CH{-}X{-}M_n^{\bullet} \qquad (6.6)$$

$$\sim\sim X{-}CH{=}CH_2 + \sim\sim M_n^{\bullet} \rightarrow \sim\sim X{-}\underset{\wr}{C}H{-}CH_2{-}M_n\sim\sim \qquad (6.6a)$$

The terminal unsaturation may then be picked up by another propagating radical during its growth process, thus yielding a branched polymer molecule, [Eq. (6.6a)]. Similarly, branching may occur through chain-end unsaturation arising from termination by disproportionation detailed in Chapter 3.

$$\sim\sim CH{=}CHX \xrightarrow{\sim\sim M_n^{\bullet}} \sim\sim \underset{\underset{\wr}{M_n}}{\underset{|}{C}H}{-}CHX\sim\sim \qquad (6.7)$$

Polymer transfer during polymerization of a given monomer may sometimes lead to different degrees of branching; and in selected systems extensive chain transfer with the monomer and with the polymer formed may lead to extensive branching and even to limited cross-linking, particularly when the polymerization is allowed to proceed to high conversions. Among the common monomers widely studied, extensive long-chain branching is the characteristic feature of vinyl acetate polymerization. Polymerization of ethylene by the high pressure (free radical) process leads to extensive short-chain branching as a consequence of intramolecular chain transfer as discussed in Chapter 3.

The polymerization of conjugated diolefins and related diene monomers tends to yield partly insoluble or even largely insoluble polymer products due to limited or extensive cross-linking involving the residual or rearranged unsaturation in the repeat units. It is quite obvious that the proportion of insoluble polymer or the gel content increases with degree of conversion. Soluble and industrially useful polymers are usually prepared by limiting the degree of conversion and often by controlling the chain length using calculated molar proportions of a chain transfer agent commonly known as a regulator.

One sure means of producing cross-linked polymers is the use of small or large proportions of a divinyl compound as a comonomer with a (mono)-vinyl monomer, typically exemplified by the copolymerization of styrene with a small amount of divinyl benzene.[7] Examples of other bi-or polyfunctional (olefinic) monomers used as cross-linking agents are ethylene glycol dimethacrylate, triethylene glycol dimethacrylate, glyceryl trimethacrylate, triallyl cyanurate, diallyl sebacate, etc.[8–11]

6.4 Polymer Degradation

By the term degradation, we generally mean breaking down of chemical structure of a given material to smaller fragments, and with respect to polymeric materials it simply implies a decrease in molecular weight or chain length. But sometimes, we may even see a polymer lose its useful or desirable properties and in the process it gains in molecular weight. It is difficult to give a close definition of polymer degradation. In a broad perspective, any deterioration of properties of polymers which make them variously useful as rubbers, plastics or fibres should be considered as a case of degradation. Discolouration or embrittlement of a polymer is a degradation process even though it may not lead to chain breakage or any fall in molecular weight. A degradation process may result in no measurable change or different degrees of decrease or increase in molecular weight, depending on the nature of the polymer and its environments.

In most degradation studies, however, the focus of attention is on: (a) change in molecular weight, (b) change in physical and mechanical properties, (c) evolution of volatiles and their chemical identity, and (d) the chemical structure of the residual polymer vis-a-vis the same of the original polymer. Almost in all practical applications, it is our objective and desire that the polymer performs up to expectations and in tune with its potential without degradation due to the various agencies inherent or infused in its environments, such as heat, light and other radiation, atmospheric oxygen, moisture, acid, alkali, and other chemicals, mechanical forces, microorganisms, etc. Even if a polymer material degrades, the degradation process can be possibly prevented or delayed and the polymer made more resistant to the degrading agent with the use of certain additives only if one has a precise understanding of the mechanism of the degradation reactions. As for example, when the role of oxygen was recognized in the deterioration of polymers, the relevance and importance of antioxidant additives were better realized and appreciated.

The backbone structures of different polymers tend to degrade by two distinct mechanisms by thermal or hydrolytic means, i.e., by a chain mechanism or by a step mechanism. Condensation or step-growth polymers are generally prone to hydrolysis (or thermal splitting) of the condensed interunit linkages, such as the ester and amide linkages and the like, by a step mechanism (much like the mechanism of their formation) to yield low molecular weight degraded products. Addition or chain-growth polymers, on the other hand, are usually prone to degradation under heat, light, high energy radiations or otherwise by a chain-mechanism to yield low molecular weight products.

The hydrolytic degradation of condensation polymers takes place randomly and it follows a pattern similar to that known for related low molecular weight compounds and hence, such degradations of specific condensation polymers are not taken up for discussion here. Drop in molecular weight is quite severe and drastic even for limited hydrolytic degradation of condensation polymers.

6.4.1 Thermal Degradation

Degradation of polymers at elevated temperatures has been most widely studied. The degradation mechanism for different polymers does not follow a uniform pattern. The thermal stability and mode of decomposition of a polymer are largely dependent on the chemical structure of the chain segments or repeat units. It is interesting to note that the degradation characteristics of a polymer cannot be precisely predicted on the basis of those exhibited by low molecular weight model compounds. Even though the mode of decomposition is much the same, the polymers degrade or decompose at temperatures far lower than those of the corresponding model compounds. The degradation may primarily lead to chain scission or involve non-chain scission reactions to begin with. The chain scission or depolymerization reactions result in cut down in chain length or molecular weight due to breaking of main polymer chain backbone. For a degradation of this nature, the products at any intermediate stage of the reaction are very much similar to the parent polymer in the sense that the characteristic monomer units can still be distinctly recognized. The non-chain scission reactions are basically substituent reactions in that the substituents attached to the main polymer chain backbone are partially or totally modified or eliminated. This leads to major changes in the chemical nature of the repeat units in the macromolecular structure and the volatiles liberated in such cases are chemically unlike the monomer.

6.4.2 Depolymerization

Depolymerization brought about thermally may be manifested as depropagation with elimination of monomers only as the volatile or by random breaking of the chain backbone giving rise to low molecular weight volatiles other than the monomer with or without the elimination of monomer in part. Poly(methyl methacrylate) provides an example of one extreme system where depolymerization yields monomer almost quantitatively, and for such a system, decrease in molecular weight proceeds very slowly with the beginning of degradation and the fall in molecular weight is measurable only at some advanced stages of degradation or volatilization. Polyethylene, on the other hand, represents another extreme system that yields little monomer on depolymerization while resulting in a sharp fall in the molecular weight to low values at quite small extents of volatilization.

The depolymerization mechanisms are far from simple and a number of reactions presumably take place simultaneously and consecutively. The initiation of chain degradation process may occur by random chain scission or it may be a case of chain-end initiation. The initiation reaction is followed by a depropagation reaction in selected systems, forming monomer. The overall depolymerization process may be schematically represented as follows:

1. Initiation of depolymerization

   (a) Random chain scission

$$P_n \rightarrow R^{\bullet}_{n-i} + R^{\bullet}_i$$

   (b) Chain-end initiation

$$P_n \rightarrow R^{\bullet}_{n-1} + R^{\bullet}_1 \qquad (6.8)$$

2. Depropagation

$$R_n^{\bullet} \rightarrow R_{n-1}^{\bullet} + M$$

3. Chain transfer involving random chain scission

$$R_n^{\bullet} + P_m \rightarrow P_n + P_{m-i} + R_i^{\bullet}$$

4. Termination

$$R_n^{\bullet} \rightarrow P_n \qquad \text{unimolecular (obscure mechanism)}$$

$$R_n^{\bullet} + R_m^{\bullet} \rightarrow P_{n+m} \qquad \text{bimolecular (radical coupling)}$$

$$R_n^{\bullet} + R_m^{\bullet} \rightarrow P_n + P_m \qquad \text{bimolecular (radical disproportionation)}$$

In scheme (6.8), P, $R^{\bullet}$ and M denote polymer, radical and monomer respectively and the subscripts, as usual, signify chain length or degree of polymerization. The monomer content in the volatiles obtained on depolymerization of several polymers are listed in Table 6.1 from which one can well appreciate how significantly the course of depolymerization is dependent on the structure of the repeat units of the polymer. Substituents modify the reaction course profoundly even if the main chain scission reactions prevail. For ethylene and 1-substituted ethylenes, non-chain scission reactions are predominant, producing either little or up to 50% monomer as volatiles. 1,1-disubstituted repeat units apparently favour 100% monomer yield. Monomer yield is negligible for the presence of chlorine substituent.

6.4.3 Substituent Roles

Substituents control the degradation process in many polymers containing labile α-hydrogen in the repeating units. Substituent reaction or substituent elimination often spares the main-chain C—C links from breakdown in such polymers. However, the substituent reactions have little comparable uniformity; each kind of substituent has its characteristic chemical nature and reactivity. The substituent reactions assume prominence if only they are initiated and accomplished at temperatures lower than that at which main chain bonds are broken. Evidently, then, substituent reactions are normally observed at comparatively low temperatures.

One of the more common polymers showing characteristic substituent reaction is the poly(vinyl chloride), PVC. Had PVC in reality corresponded to the idealized concept of regular repeating units it would have been expected to be a polymer of high inherent stability based on evidences available from studies of low molecular weight model compounds such as 1,3,5-trichlorohexane.[12] Thermal dehydrochlorination of PVC begins slowly at 80–100°C but it becomes quite rapid at >140°C.

Table 6.1 *Depolymerization of several polymers*

Polymer	*Repeat Unit*	*Monomer in Volatiles (%)*
Polyethylene	$-CH_2-CH_2-$	Negligible
Poly(vinyl chloride)	$-CH_2-CH(Cl)-$	Negligible
Poly(trifluorochloroethylene)	$-CF_2-CF(Cl)-$	0
Polystyrene	$-CH_2-CH-$ 	42
Poly(*m*-methyl styrene)	$-CH_2-CH-$ 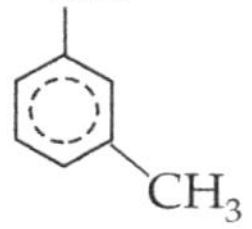	52
Poly(α-methyl styrene)	$-CH_2-C(CH_3)-$ 	100
Poly(methyl methacrylate)	$-CH_2-C(CH_3)(COOCH_3)-$	100
Polytetrafluoroethylene	$-CF_2-CF_2-$	100

Evidences from many studies indicate that the dehydrochlorination of PVC is due initially to a loss of HCl molecule which takes place at a site on the molecule either containing or adjacent to a tertiary or (terminal) allylic chlorine atom (arising in the polymer due to chain transfer by different mechanisms or termination by disproportionation), each of which potentially functions as an activating group or centre. Radical detection in the electron spin resonance spectra of dehydrochlorinated PVC is indicative of a radical mechanism for thermal degradation of PVC.[13,14] However, it is unlikely that a radical chain process of any great propagation length is involved. It is more likely that a radical reaction of short chain length is involved as follows:

$$\sim\sim CH_2—CHCl—CH_2—CHCl—CH=CHCl$$
$$\downarrow$$
$$\sim\sim CH_2—CHCl—CH_2—\dot{C}H—CH=CHCl + \dot{C}l$$
$$\downarrow$$
$$\sim\sim CH_2—CHCl—CH=CH—CH=CHCl + HCl \quad (6.9)$$

The elimination of the first molecule of HCl and consequent formation of a double bond on PVC chain subsequently activates the neighbouring chlorine atom which assumes the nature of an allylic chlorine, causing elimination of another hydrogen chloride molecule, with the process continuing to repeat itself, finally producing a conjugated polyene structure (VIII) by an autocatalytic effect.

$$\sim\sim CH=CH—CH=CH—CH=CH \sim\sim$$
(VIII)

Certain features of thermal dehydrochlorination of PVC, not compatible with a radical mechanism, can be explained and appreciated on the basis of ionic mechanism. Evidences in the studies of HCl "unzippering" from PVC lend support to the currently held theory that both radical and ionic processes may be at work simultaneously.[15] The most effective and widely adopted means of arresting PVC degradation is through incorporation of chemical additives such as basic lead carbonate, tribasic lead sulphate, etc., which are basically HCl scavengers. They effectively react with HCl as and when liberated under processing or service conditions and inhibit the HCl unzippering. Inverse proportionality of PVC decomposition rate to molecular weight stands in evidence for chain-end initiation.

Another good example of substituent reaction is found in the colouration of polyacrylonitrile or polymethacrylonitrile thermally at temperatures near and above 175°C. The colour reactions are mostly due to intramolecular linking of nitrile groups to form conjugated unsaturation of carbon nitrogen sequences (IX), forming ladder-type structure:

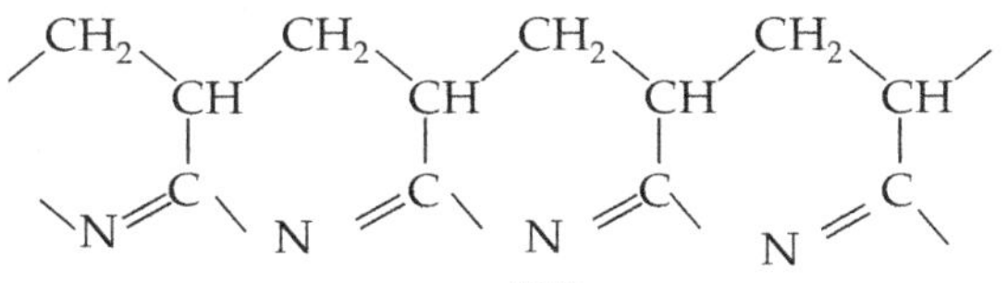

(IX)

The reaction leading to colour formation is catalyzed by carboxylic acid, more particularly for polymethacrylonitrile. For polyacrylonitrile, the intramolecular reaction is occasionally intercepted by intermolecular reaction leading to insolubilization of the product even before development of visible colour. Polymethacrylonitrile, however, remains soluble even when extensively coloured, indicating exclusive or near-exclusive occurrence of intramolecular reaction for this polymer.

6.4.4 Mechanochemical Degradation

Breakdown of molecular chains under shear or mechanical force, often aided by a chemical reaction (oxidation) is called mechanochemical degradation. Mastication of rubber leading to chain breakage and development of plasticity under shear is a good example of mechanochemical degradation of polymers. It is a widespread experience that mastication of natural rubber at ordinary temperatures in an atmosphere of nitrogen brings about no appreciable change in molecular weight and development of plasticity. In air or oxygen, however, degradation is immediate and rapid. Under mechanical shear, the rubber molecules break into radicals, and oxygen, a radical scavenger, readily reacts with them leading to permanent chain breakage. In nitrogen, however, the primary radicals formed under shear immediately recombine to give no effective chain breakage. If benzoquinone, also a radical scavenger, is used in small concentrations (1 phr), rapid degradation even in nitrogen is observed, while little additional effect is exhibited in oxygen.

In natural rubber, the weakest bonds are those linking the different isoprene units as shown below, and scission of each such bond results in generation of two relatively unreactive radicals stabilized notably by allylic resonance.

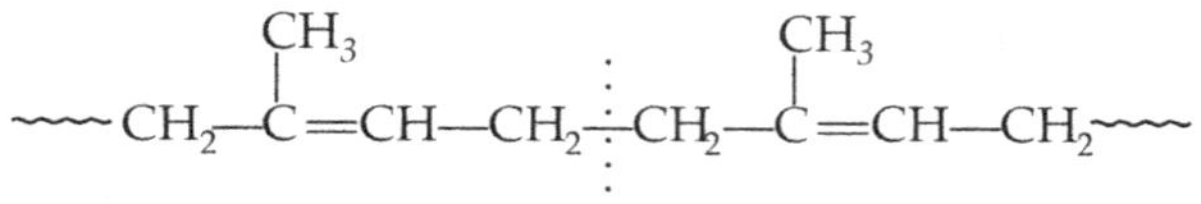

The radicals thus derived are so unreactive as to be practically unable to take part in disproportionation reaction required to lead to permanent chain breakage in absence of radical acceptors. For polymers having no backbone unsaturation, stabilization of primary radicals formed under shear in this manner are remote such that radicals formed due to bond scission under shear are very much more reactive and effective chain degradation takes place even in nitrogen.

6.4.5 Aging or Oxidative Degradation

Almost all polymers are subject to limited oxidative degradation, i.e., chain breakage subsequent to reaction with oxygen, particularly when processed at elevated temperatures, often under high pressure for short time intervals. Saturated hydrocarbon polymers, such as polyethylene, are very much more resistant to aging or oxidative degradation as compared to unsaturated hydrocarbon polymers, such as natural rubber or synthetic diene polymers under ambient conditions. Polymers bearing tertiary hydrogens in the repeat units or methylene groups in the α-position with respect to chain unsaturation are usually more prone to aging. The primary attack by oxygen leads to hydroperoxidation and the hydroperoxides formed subsequently decompose radically under heat/light with the onset of a chain process that ultimately results in chain degradation.

For unsaturated polymers such as natural rubber, the mechanism of oxidation is more complicated and the process turns out to be an autoxidation process. Here, the hydroperoxidation by primary oxygen attack takes place in the α-methylenic position. Subsequent sequence of reactions involve (a) decomposition of the hydroperoxide into radicals and onset of an ill characterized chain degradation or cross-linking process, and (b) epoxidation involving double bonds and the primarily formed hydroperoxides. Subsequent decomposition of the reactive epoxides and further oxidation by oxygen cause chain breakage at olefinic bonds or even saturation of double bonds without chain rupture. The net result is incorporation of oxygenated polar groups such as —OH, —COOH, —CO— and —O— in the polymer. In any event, the loss of unsaturation does not follow oxygen intake quantitatively.

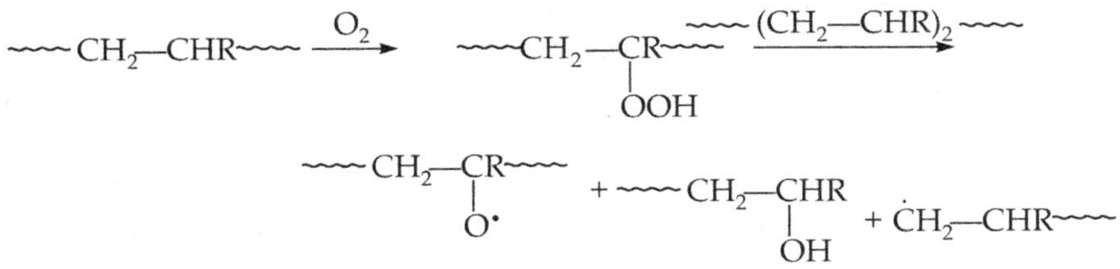

(6.10)

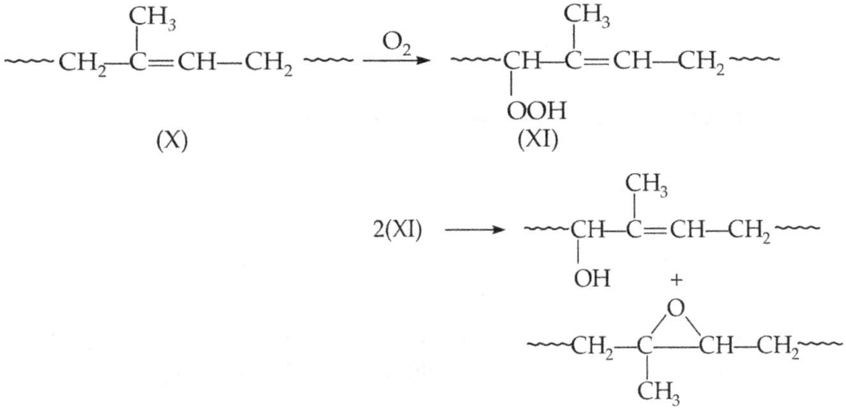

(X) (XI)

(6.11)

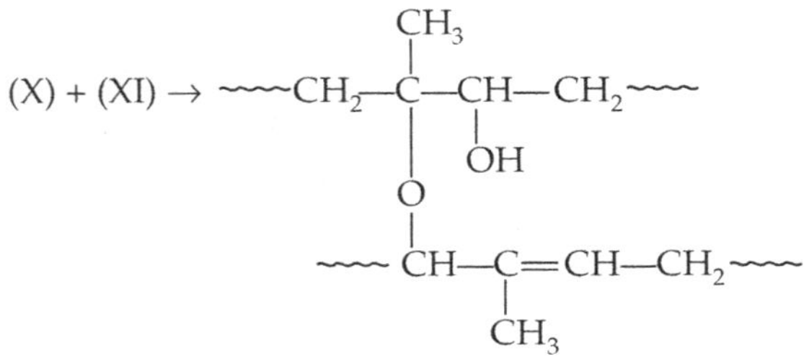

(6.12)

6.4.6 Photodegradation

Polymers are also subject to photodegradation to different extents depending on their chemical structures and certain contaminants that may be present in them. Degradation by light is rather selective in nature. Specific groups would absorb light of specific wavelengths and induce degradation. The power to photodegrade increases as the wavelength of the incident radiation decreases, i.e., as the energy of the quanta increases. Light of wavelength 2800 Å corresponds approximately to a bond energy of 100 kcal/mol, and the wavelength region of 3000–4000 Å corresponding to bond energy of nearly 90–70 kcal/mol is considered as the threshold. For light above this wavelength range, photodegradation scarcely occurs (a C—C single bond has energy in the region 80 kcal/mol and the C—H bond strength is in the region 80–100 kcal/mol depending on environment). Photodegradation is widely recognized to proceed via free radical mechanism. The bonds or groups which absorb energy and those which subsequently break into radicals and the course of radical chain degradation that follows largely depend on the chemical nature of the material and the energy of the incident radiation. High energy radiations such as X-rays, γ-rays, β-rays, high speed electrons, etc., often bring about uncontrolled degradation usually involving simultaneous radical and ionic mechanisms.

6.5 The Concept of Average Molecular Weight

A given polymer material is mostly a mixture of molecules of (nearly) identical chemical structure but varying in chain length or molecular weight. The molecules produced in polymerization reactions have lengths that are distributed in accordance with a probability function which is governed by the mechanism of the reaction and by the conditions under which it has been carried out. The concept of average molecular weight is, therefore, important and relevant and the assignment of a numerical value to the molecular weight of polymer requires the definition of a particular average. An average molecular weight $\overline{M}$ may be generally expressed as

$$\overline{M} = f_1M_1 + f_2M_2 + f_3M_3 + \ldots = \Sigma f_iM_i \tag{6.13}$$

where M_1, M_2, M_3, etc., are the molecular weights of different sizes of molecules and the coefficients f_1, f_2, f_3, etc., are fractions such that their summation, i.e., Σf_i equals to unity. The average molecular weight $\overline{M}$ may also be conveniently expressed as

$$\bar{M} = \frac{\Sigma N_iM_i^a}{\Sigma N_iM_i^{(a-1)}} \tag{6.14}$$

where N_i is the number of molecules with molecular weight M_i and the index 'a' may have any real value. Two most important averages are: (i) number average molecular weight and (ii) weight average molecular weight. Setting $a = 1$, one obtains the expression for number average moleclar weight, $(\bar{M}_n)$ from Eq. (6.14), i.e.

$$\bar{M}_n = \frac{\Sigma N_iM_i}{\Sigma N_i} \tag{6.15}$$

Equation (6.15), in effect, can be expressed as a summation series resembling Eq. (6.13) where the fractional coefficients represent mole fractions of the respective molecular species present in the system, knowing that total weight $W = \Sigma N_iM_i$ and total number of molecules $N = \Sigma N_i$. Thus,

$$\bar{M}_n = \frac{W}{N} = \frac{\Sigma N_iM_i}{\Sigma N_i} = \frac{N_1}{N}\cdot M_1 + \frac{N_2}{N}\cdot M_2 + \frac{N_3}{N}\cdot M_3 + \ldots$$

$$= f_1M_1 + f_2M_2 + f_3M_3 + \ldots \tag{6.16}$$

On the other hand, setting $a = 2$ in Eq. (6.14), one obtains the expression for weight average molecular weight $(\bar{M}_w)$, i.e.

$$\bar{M}_w = \frac{\Sigma N_iM_i^2}{\Sigma N_iM_i} \tag{6.17}$$

Equation (6.17) may also be rearranged and expressed as a summation series as given by Eq. (6.13), but in this case the fractional coefficients correspond to weight fractions of different molecular species present. Thus,

$$\bar{M}_w = \frac{\Sigma N_iM_i \cdot M_i}{\Sigma N_iM_i} = \frac{\Sigma w_iM_i}{\Sigma w_i} = \frac{\Sigma w_iM_i}{W}$$

$$= \frac{w_1}{W}\cdot M_1 + \frac{w_2}{W}\cdot M_2 + \frac{w_3}{W}\cdot M_3 + \ldots$$

$$= f_1M_1 + f_2M_2 + f_3M_3 + \cdots \qquad (6.18)$$

Here w_1, w_2, w_3, etc., stand for weights of different species having molecular weights, M_1, M_2, M_3, etc., respectively and $\Sigma w_i = W$ is the total weight of all the molecules present.

The consequence of definitions given above is that $\bar{M}_w \geq \bar{M}_n$; the equality, however, corresponds to the limiting case of a perfectly monodisperse sample. The deviation from unity of the ratio $\bar{M}_w / \bar{M}_n$, known as the distribution ratio, is taken as a measure of polydispersity of polymer samples, a higher value of the ratio indicating greater polydispersity.

Evaluation of number average molecular weight is useful in understanding the polymerization mechanism and kinetics. The evaluation of $\bar{M}_n$ is also useful in the analysis of kinetic data to examine the effects of many side reactions including chain transfer, inhibition and retardation, and autoacceleration or gel effect in vinyl polymerization. The number average molecular weight also assumes prime importance in determining the solution properties of the polymer, commonly known as the colligative properties. Polymer molecules of lower molecular weight contribute equally and enjoy equal status with those of higher molecular weight in determining these properties.

Weight average molecular weight, on the other hand, is important in relation to bulk properties of polymers that reflect their load bearing capacity. Softening, hot deformation, tensile and compressive strength, modulus and elongation, toughness and impact resistance and some other related bulk properties of polymer are better appreciated on the basis of weight average molecular weight, keeping in mind, however, the influence of chemical nature of the repeat units, degree of branching and cross-linking, thermal or thermomechanical history of the sample and other related factors in this context.

6.6 Viscosity Average Molecular Weight

The viscosity of a polymer solution (η) is higher than that of the pure solvent (η_0) at a given temperature and the gain in medium viscosity on dissolving the polymer in the solvent is a function of both molecular weight and concentration of the polymer solute. Even though the solution viscosity is easy to measure, it does not give a direct and absolute value of molecular weight. If the polymer solution is very dilute and consequently the density change of the solvent due to dissolved polymer is negligible, then the viscosities of the solution and solvent at a given temperature would be proportional to their flow times in a capillary viscometer such that the relative viscosity, η_r expressed as the ratio, η/η_0 would be given by the flow time ratio t/t_0 where t and t_0 are the flow times of the solution and solvent respectively. Both relative viscosity (η_r) and specific viscosity (η_{sp}) defined as $\eta_{sp} = (\eta - \eta_0)/\eta_0$ are dimensionless. If the solute macromolecules do not interfere with one another during flow, the viscosity gain is proportional to their concentration and η_{sp}/c, commonly known as the reduced viscosity, would be a constant. But for polymer solutions, η_{sp}/c is generally found to increase with increase in c. The intrinsic viscosity or the limiting viscosity number, $[\eta]$ for a given polymer–solvent system at a given temperature is given by the intercept of the plot of η_{sp}/c vs c when the linear experimental plot is extrapolated to zero concentration, or more precisely, to infinite dilution condition. The concentration dependence of polymer solution viscosity is conveniently expressed by two empirical equations, known as Huggins' Equation[16] given by Eq. (6.19) and Kraemer's Equation[17] given by Eq. (6.20), i.e.

$$\eta_{sp}/c = [\eta] + k_1[\eta]^2 c \tag{6.19}$$

and

$$\ln \eta_r/c = [\eta] + k_2[\eta]^2 c \tag{6.20}$$

The term $\ln \eta_r/c$ is commonly termed as inherent viscosity. The reduced viscosity, inherent viscosity and intrinsic viscosity are expressed in units of reciprocal concentration and usually in decilitre per gram (c, expressed in g/100 cc). The constants, k_1 and k_2 are known as the Huggins constant and Kraemer constant respectively. For most cases, k_2 is negative and in a general manner $k_1 - k_2 = 0.5$. The slope of each plot, left hand side vs c based on Eqs (6.19) and (6.20) are proportional to square of the intercept, i.e., of the intrinsic viscosity, and the two plots made using common ordinate and abscissa would extrapolate to a common point on the ordinate, thus enabling a precise determination of the $[\eta]$ value (Fig. 6.1).

The relationship between $[\eta]$ and the viscosity average molecular weight $\bar{M}_v$ is given by the semi-empirical Mark-Houwink equation:

$$(\eta_{sp}/c)_{c\to 0} = [\eta] = K\bar{M}_v^a \tag{6.21}$$

where K and a are constants for a particular polymer–solvent system at a given temperature. This equation provides a basis for determining molecular weight from viscosity measurements. The values of $\bar{M}_v$ obtained are not absolute in view of incomplete interpretations of K and a. The values of K and a must normally be

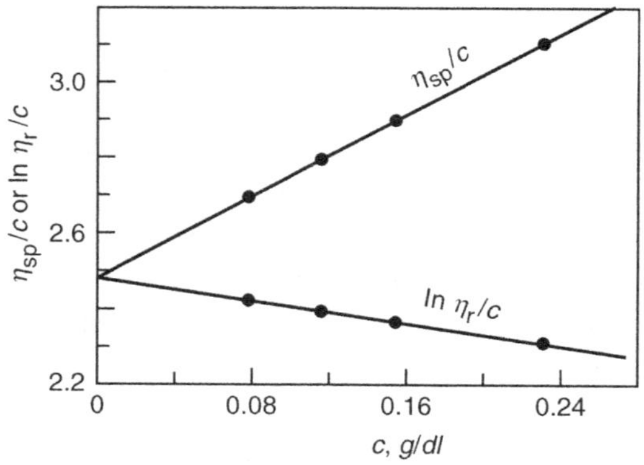

Fig. 6.1 *Typical plots showing η_{sp}/c vs c and ln η_r/c vs c using common ordinate and abscissa for the determination of intrinsic viscosity $[\eta]$ of a given polymer–solvent system at a given temperature*

determined by measuring the $[\eta]$ values of monodisperse polymer samples whose molecular weights have been obtained from one of the absolute methods such as osmometry and light scattering and making use of a plot[18] of log $[\eta]$ vs log M (Fig. 6.2). The exponent a varies with both the polymer and the solvent; its value usually ranges between 0.5 and 0.8. It does not fall below 0.5 in any case and exceeds 0.8 in exceptional cases particularly for polyelectrolytes in the absence of added salts.

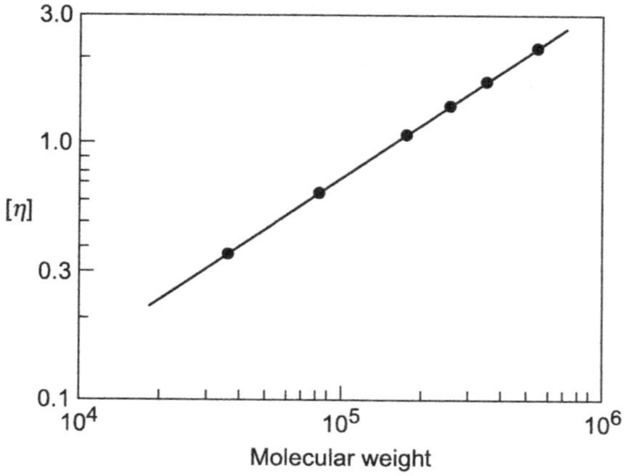

Fig. 6.2 *Intrinsic viscosity–molecular weight relationship at 30° for polyisobutylene in cyclohexane[18] (Krigbaum and Flory, 1953;* **Courtesy,** *American Chemical Society, Washington)*

For most systems, K and a are best understood when $[\eta]$ is determined in a solvent at what is known as the theta (θ) temperature. The value of a is then equal to 0.5 and K depends on the temperature while remaining independent of the solvent, keeping in mind, however, that the solvent fixes the temperature of measurements. At the θ temperature, the chemical potential due to polymer (segment)–solvent interactions is zero and the deviations from ideality just vanish. Therefore, the free energy of interactions of the segments within a volume element is also zero. In fact, θ temperature is the lowest temperature for complete miscibility in the poor solvent used at the limit of infinite molecular weight. The ideality is struck at the θ temperature because the molecular dimensions are unperturbed by intramolecular interactions (*See* Sec. 6.15).

6.6.1 General Expression for Viscosity Average Molecular Weight

For determination of the intrinsic viscosity, it is necessary to extrapolate the η_{sp}/c vs c plot to infinite dilution ($c \to 0$), since it is only under this condition that the polymer molecules in solution contribute to viscosity discretely without mutual interference. Solubilization of a polymer sample is preceded by a large amount of swelling if left undisturbed and the degree of swelling is higher in a better solvent. Likewise, the intrinsic viscosity is also higher in a good solvent than in a poor solvent. What it really means is that in a better solvent, as the polymer goes into solution, a unit mass of polymer expands more to give a higher hydrodynamic volume.

Let us now consider a heterogeneous polymer in dilute solution of concentration c behaving ideally in that the individual molecules contribute to viscosity independently of one another. In that event, if $(\eta_{sp})_i$ is the specific viscosity contribution due to species of size i, then one may write:

$$\eta_{sp} = \Sigma\,(\eta_{sp})_i \tag{6.22}$$

considering c_i and M_i as the concentration and molecular weight of the said species and in view of the ideal specific viscosity component $(\eta_{sp})_i = KM_i^a c_i$, one further obtains

$$\eta_{sp} = K\,\Sigma M_i^a c_i \tag{6.23}$$

and hence,

$$\eta_{sp}/c = [\eta] = K\,\Sigma M_i^a c_i/c \tag{6.24}$$

where, $c = \Sigma c_i$, stands for the total concentration of all polymer species. Taking $c = \Sigma N_i M_i$ and $c_i = N_i M_i$, Eq. (6.24) may be expressed as:

$$[\eta] = K \frac{\Sigma N_i M_i^{(1+a)}}{\Sigma N_i M_i} \tag{6.25}$$

Combining Eq. (6.25) with the Mark–Houwink equation, the general expression for the viscosity average molecular weight becomes:

$$\overline{M}_v = \left\{ \frac{\Sigma N_i M_i^{(1+a)}}{\Sigma N_i M_i} \right\}^{1/a} \tag{6.26}$$

Clearly, for a approaching the value of unity, $\overline{M}_v$ approaches $\overline{M}_w$ and in the limiting case of $a = 1$, $\overline{M}_v = \overline{M}_w$.

The viscometric studies of polymer solutions as a means of molecular characterization of polymers are well recognized and widely practised because of simplicity in terms of experimental approach and the apparatus needed. Dilute solution viscosity is conveniently measured in capillary viscometers of different kinds such as the Ostwald type or the Ubbelohde type (Fig. 6.3). Ubbelohde viscometer is a

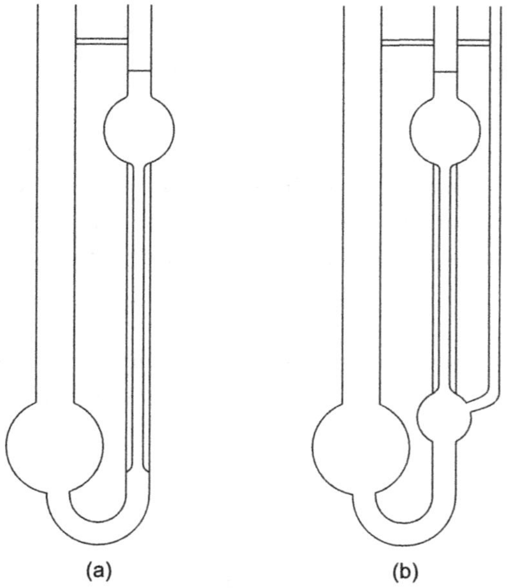

Fig. 6.3 *Capillary viscometers: (a) Ostwald type, (b) Ubbelohde type*

suspended level viscometer and it has the advantage that the flow time measurements are not dependent on the volume of liquid (solution or solvent) in the viscometer and hence, measurements at a series of concentrations can be conveniently made by successive dilution within the viscometer. All flow time measurements for the solvent and solutions of different concentrations are made in a thermostat bath regulated within ±0.1°C. The flow time data should then be treated graphically according to Eq. (6.19) or (6.20) and then extrapolated to infinite dilution ($c \rightarrow 0$) to obtain the value of the intrinsic viscosity, $[\eta]$ as described earlier. $\overline{M}_v$ can then be evaluated using the Mark–Houwink equation and using the appropriate K and a values from the literature, if available, or from an independent determination as described earlier in this section.

6.7 Number Average Molecular Weight

Number average molecular weight can be conveniently evaluated using a polymer solution and taking recourse to ebulliometric, cryoscopic and osmometric measurements. Direct measurement of vapour pressure lowering in dilute polymer solutions lacks precision and hence often leads to uncertain results. However, vapour-phase osmometry permits indirect utilization of vapour pressure lowering and this technique led to development of equipment which can measure a temperature difference that is proportional to the vapour pressure lowering of polymer solution at equilibrium, as can be related through the Clapeyron equation. Temperature differences thus noted are of the same order of magnitude as those to be observed in cryoscopy or ebulliometry; these methods customarily require calibration with low molecular weight standards and usually give reliable results for molecular weight up to 30,000. However, lack of development of equipment for ebulliometric and cryoscopic techniques has turned them much less relevant and useful.

6.8 Membrane Osmometry

Consider a solution of a polymer separated from the pure solvent by a semipermeable membrane; the chemical potential of the solvent, in solution (μ_s) is less than that of the pure solvent (μ_0) and consequently, to keep the system in equilibrium, the chemical potential of the solvent on either side of the membrane must be balanced or made equal. This may be conveniently done by applying an excess pressure, called the osmotic pressure π to the solution side to compensate for the deficiency in chemical potential. Thus, the condition of equilibrium for the chemical potential of the solvent on the two sides of the osmometer membrane gives

$$\mu_0 - \mu_s = \Delta\mu_1 = -\pi\overline{V}_1 \tag{6.27}$$

or,

$$RT \ln f_1 x_1 = -\pi\overline{V}_1 \tag{6.28}$$

where, R is the universal gas constant, T, the absolute temperature, $\overline{V}_1$ the partial molar volume and f_1 the activity coefficient of the solvent in solution. For a very dilute solution, $f_1 \to 1$ and $\overline{V}_1$ may be taken as equal to the molar volume, V_1^0 of the pure solvent. Replacing solvent mole fraction x_1 by $(1 - x_2)$ where x_2 is the mole fraction of the solute in solution, and expanding the logarithm factor, one obtains for a very dilute solution:

$$\pi\overline{V}_1^0 = RT\left(x_2 + \frac{x_2^2}{2} + \frac{x_2^3}{3} + \ldots\right) \tag{6.29}$$

If c is the concentration in gram per unit volume of solution, then x_2 is given by

$$x_2 = \frac{c/\overline{M}_n}{1/V_1^0 + c/\overline{M}_n} \simeq \frac{V_1^0 c}{\overline{M}_n} \tag{6.30}$$

combining Eqs (6.29) and (6.30), one obtains

$$\pi/c = \frac{RT}{\overline{M}_n}\left\{1 + \frac{1}{2}\cdot\left(\frac{V_1^0}{\overline{M}_n}\right)c + \frac{1}{3}\cdot\left(\frac{V_1^0}{\overline{M}_n}\right)^2 c^2 + \ldots\right\} \tag{6.31}$$

Polymer solutions largely deviate from ideality and the activity coefficient f_1 is less than unity even at the lowest concentration at which accurate osmometric measurements can be made. The real coefficients of the concentration terms are somewhat higher in magnitude than those in the equation above. Even then, π/c may be expressed as a power series in c using empirical coefficients, viz.,

$$\pi/c = RT(A_1 + A_2c + A_3c^2 + \ldots) \tag{6.32}$$

or alternatively,

$$\pi/c = \frac{RT}{\overline{M}_n}(1 + \Gamma_2 c + \Gamma_3 c^2 + \ldots) \tag{6.32a}$$

where $\Gamma_2 = A_2/A_1$, $\Gamma_3 = A_3/A_1$ and so on, and $A_1 = (1/\overline{M}_n)$.

The coefficients A_2, A_3, etc. are known as the second, third etc. virial coefficients. In most cases, the term in c^2 and higher powers of c may be neglected. Thus, π/c is measured as a function of c and plotted on a graph paper (Fig. 6.4a) and extrapolation[19] of the plot to $c \to 0$ gives an intercept on the π/c axis which is equal to $RT/\overline{M}_n$. Alternatively, (π/RTc) may be plotted against c (Fig. 6.4b). Evaluation of the number average molecular weight then readily follows from the intercept. The plots are usually linear and in each case, the slope allows evaluation of the second virial coefficient. In good solvents and at high concentrations, the plots may deviate from straight line and tend to become concave upward. In that event, a plot of $(\pi/RTc)^{1/2}$ vs c gives a linear plot. The slopes of plots shown in Fig. 6.4 or the values of the second virial coefficient allow evaluation of the respective polymer solvent interaction. Both slope and curvature are zero at θ temperature. The membrane osmometry is based on the principle illustrated in Fig. 6.5. The membrane is critically important. It should offer high permeability for solvent molecules and practically no permeability to the smallest macromolecules present in the polymer under study. All measurements must be made at a specified and constant temperature. The thermodynamic drive to reach equilibrium causes the liquid level in the capillary in the solution side to rise till the hydrostatic pressure on the membrane in the solution side balances the osmotic pressure on the same in the solvent side. A measure of the difference in liquid levels in the two capillaries after attainment of equilibrium is used to calculate the osmotic pressure. Among the different kinds of membrane, those based on cellulose such as regenerated cellulose (gel cellophane), are used most widely; other suitable membrane materials are collodion (nitrocellulose, 11–13.5% N_2) and denitrated collodion,

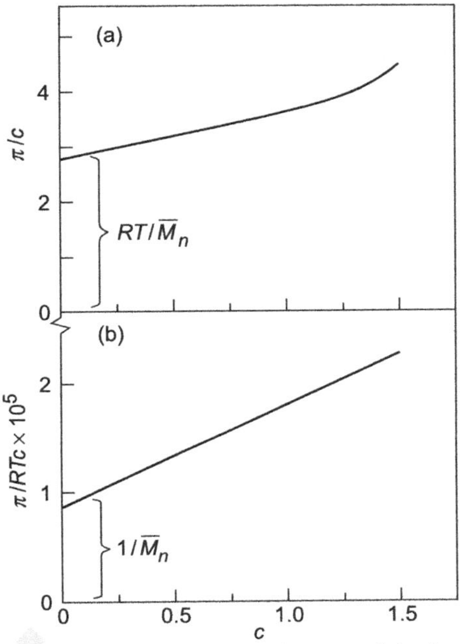

Fig. 6.4 *Typical plots showing (a) π/c vs and (b) π/RTc vs c for determination of $\overline{M}_n$.*

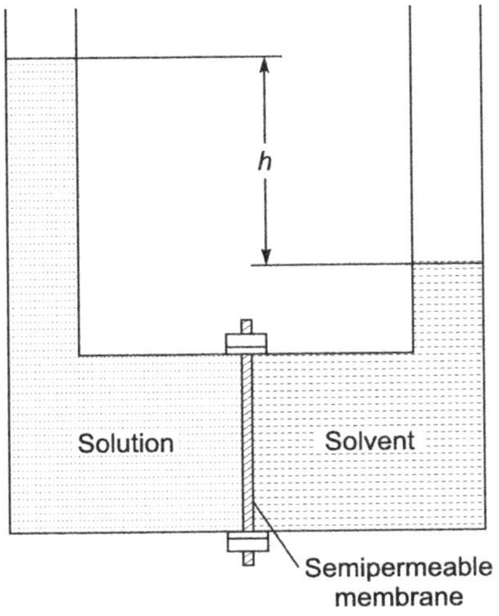

Fig. 6.5 *Operating principle of a membrane osmometer*

poly(vinyl alcohol), poly(vinyl butyral) etc. Among the different types of osmometer cells and assembly, the one based on Zimm-Meyerson design (Fig. 6.6), is more popular because of its simplicity.[20] Typical time period required for the attainment of equilibrium in classical osmometers using dilute polymer solutions is in the range of 10–20 h.

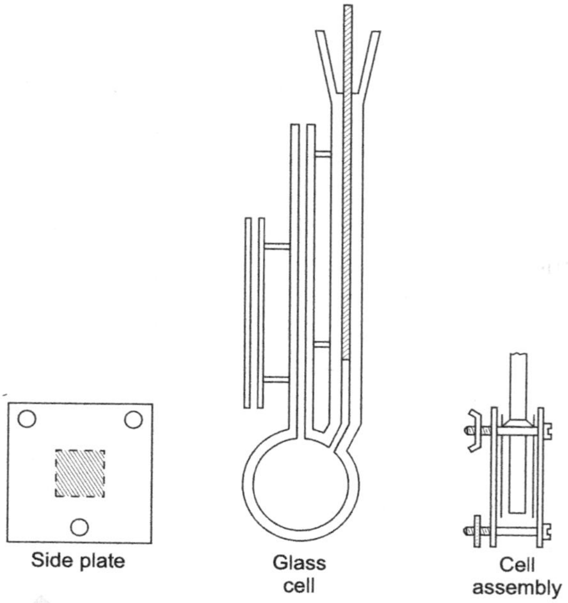

Fig. 6.6 *Sketch of a Zimm-Meyerson osmometer*

However, different models of high speed osmometers are now-a-days available. Their design features closed solvent compartment gadgeted with a sensitive pressure-sensing device without the use of a capillary. The equipment may use a photoelectric or capacitance detecting device using a servo mechanism, or else a strain gauge for direct pressure detection. The high speed equipments allow attainment of equilibrium often in less than 5 min.

6.9 Weight Average Molecular Weight: Light Scattering by Polymer Solutions

The phenomenon of scattering of light[21] by the molecules of a gas (Rayleigh scattering) or by colloidal particles suspended in a liquid medium (Tyndall scattering) is well known. The intensity of the scattered light depends on the polarizability of the molecules or particles compared with that of the medium in which they are contained, i.e. dissolved or suspended. It also depends on the molecular or particle size and on their concentration. If the solution/suspension is sufficiently dilute, the intensity of scattered light is equal to the sum of the contributions from the individual molecules/particles, each being unaffected by the others in the medium.

Light passing through an optically inhomogeneous medium is scattered in all directions and the intensity of the transmitted beam decreases exponentially. The turbidity τ is simply defined by

$$I = I_0 e^{-\tau l} \tag{6.33}$$

where I_0 and I are the intensities of the beam before and after passing through a path of length, l, of the medium.

Let us consider a polymer solution. Thermal agitation of the molecules in solution causes instantaneous local fluctuations of density and concentration. If the solute and solvent have different polarizabilities, the intensity of light scattered by a tiny volume element varies with these fluctuations continuously and arbitrarily. Consequent to these variations in intensity, the rays scattered by two volume elements so situated as to produce interference at a point of observation, will not cancel each other.

In a dilute solution, the work needed to produce a concentration fluctuation is less than that required to produce a density fluctuation corresponding to equal scattering power. Concentration fluctuations are the main origin of scattered light from a dilute polymer solution. The effect due to density fluctuations can be accounted for by subtracting the intensity of the light scattered by the pure solvent from that scattered by the solution.

The work required to produce a given concentration fluctuation is related directly to the free energy of dilution, ΔG_1. The intensity of scattered light can therefore be used to measure the thermodynamic properties. The scattered light intensity from a solution is commonly expressed in terms of its turbidity, which is the fraction by which the intensity of the scattered beam is reduced over 1 cm path length of solution as related by Eq. (6.33). For polymer molecules small in comparison with the wavelength of light used, τ is also expressed as

$$\tau = \frac{32\pi^3 kTn^2 c(\partial n/\partial c)^2 \overline{V_1}}{3\lambda^4 (-\partial \Delta G_1/\partial c)} \tag{6.34}$$

Here, k is Boltzman's constant, n, the refractive index of the medium, $(\partial n/\partial c)$, the change in refractive index with concentration where c is the concentration, λ, the wavelength of the incident beam and ΔG_1 signifies the difference between the molar free energy of the pure solvent and the partial molar free energy of the solvent in solution. Now, $\Delta G_1 = -\pi\overline{V}_1$ where π is osmotic pressure, and using the relation between osmotic pressure and molecular weight one may write

$$-(\partial \Delta G_1/\partial c) = \frac{RT\overline{V}_1}{M}(1 + \Gamma_2 c + \ldots) \tag{6.35}$$

Combining Eqs (6.35) and (6.34) one gets

$$H \cdot \frac{c}{\tau} = \frac{1}{M}(1 + \Gamma_2 c + \ldots) \tag{6.36}$$

where $H = (32\pi^3 n^2/3\lambda^4 N)(\partial n/\partial c)^2$, and $N = R/k$ is the Avogadro number. If τ is determined as a function of c and Hc/τ is then plotted against c, then the intercept on the Hc/τ axis obtained by extrapolation to zero concentration (Fig. 6.7) allows ready calculation of the molecular weigh M, which can be shown to be the weight average molecular weight, $\overline{M}_w$.

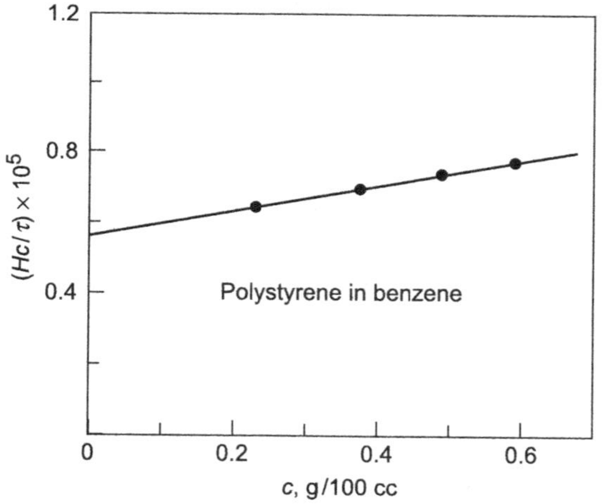

Fig. 6.7 ***A typical plot of Hc/τ vs c for determination of $\overline{M}_w$***

The polarizability of a molecule is a function of the number and kinds of atoms constituting the molecule. It is obvious then that the polarizability of a molecule made up of many identical units, i.e. of a polymer molecule is proportional to the number of units and therefore, the degree of polymerization and hence the molecular weight. The intensity of light scattered by a molecule is proportional to the square of its molecular weight. The intensity of light scattered by N_i molecules of molecular weight M_i in a unit volume of solution is, thus, proportional to $N_iM^2_i$ and the total intensity of light scattered by all the molecules in the unit volume is proportional to $\Sigma N_iM_i^2$ and the total weight of the polymer molecules contained in the unit volume is given by $c = \Sigma N_iM_i$. Hence,

$$\tau/c \propto \frac{\Sigma N_i M_i^2}{\Sigma N_i M_i} \simeq \overline{M}_w \tag{6.37}$$

According to Rayleigh's law, the angular variation of the intensity of unpolarized light scattered by a molecule much smaller than the wavelength, λ of the incident light is proportional to $(1 + \cos^2 \theta)$, where θ is the angle between the incident and scattered beam. It is, however, necessary to measure the scattered intensity i_θ at only one angle, θ for calculation of τ, and according to Rayleigh

$$\tau = \frac{16\,\pi i_\theta r^2}{3I_0(1+\cos^2\theta)} = \frac{16\pi}{3}\cdot R_\theta \tag{6.38}$$

where $(i_\theta r^2)/I_0\,(1 + \cos^2\theta) = R_\theta$ is known as the Rayleigh ratio and it is independent of θ. Here, I_0 is the incident intensity and r is the distance from the scattering particle to the point of observation. Normally, if the scattered intensity at zero angle, i_0 (along the direction of the incident beam) could be measured, one could obtain the value of τ using Eq. (6.38). However, direct observation of i_0 is not practically possible as the much more intense incident beam would also be observed at the same time. R_θ may also be expressed as

$$R_\theta = \frac{Kc}{1/M + A_2c + \ldots} \tag{6.39}$$

and K is an optical constant given by the expression:

$$K = \frac{2\pi^2 n^2 (\partial n/\partial c)^2}{\lambda^4 N} \tag{6.40}$$

Equation (6.39) may be rearranged for the determination of molecular weight in the following form to allow a plot of (Kc/R_θ) vs c.

$$\frac{Kc}{R_\theta} = \frac{1}{M} + A_2c + \ldots \tag{6.41}$$

The intercept at $c = 0$ gives the molecular weight and the limiting slope gives a measure of the second virial coefficient, A_2. The above relationship is particularly applicable for particles much smaller than λ, but it applies to larger solute particles as well for values of Kc/R_θ extrapolated to $\theta = 0°$. The effect of large size may, however, be described by a function, $P(\theta)$:

$$P(\theta) = \frac{\text{Scattered intensity (with intraparticle interference) for large particles}}{\text{scattered intensity without interference}}$$

$$= i_\theta/i_\theta^0 = R_\theta/R_\theta^0 \qquad (6.42)$$

$P(\theta)$ is usually much less than 1 when θ is large and its value increases as θ becomes smaller. For $\theta = 0°$, $P(\theta)$ is unity, and Eq. (6.41) is applicable in this limiting condition. For large particles, equation (6.41) is modified to

$$(Kc/R_\theta) = (1/M \cdot P(\theta)) + A_2c + \ldots \qquad (6.43)$$

In view of difficulty in direct measurement of i_θ for $\theta = 0°$ and in view of the fact that any equation for $P(\theta)$ is valid only at very high dilutions, it is necessary to extrapolate the appropriate scattering data at each angle to zero concentration, before one does the extrapolation to $\theta \rightarrow 0°$ to use $P(\theta)$ data for evaluation of molecular weight. At the same instance, it is required to extrapolate the data at each concentration to zero angle for evaluation of the weight average molecular weight ($\overline{M}_w$) and the second virial coefficient using Eq. (6.41), which for polymer (large) molecules is valid only at zero angle.

The two objectives are combined in one graphical plot as developed by Zimm.[22] The scattered intensity is measured as a function of angle for several solutions of successively decreasing concentrations. A plot is then prepared using Kc/R_θ as the ordinate and $\sin^2(\theta/2) + kc$ as the abscissa where k is an arbitrary constant so chosen as to permit a suitable spread of the data on the piece of graph paper (Fig. 6.8). From such graphical presentation of data, two types of limiting plots are obtained; (i) through extrapolation to $\theta = 0°$, at constant c, and (ii) through extrapolation to $c = 0$, at constant θ. Each set of extrapolated points thus obtained are joined by a curve or line which can be extrapolated to give an intercept on the Kc/R_θ axis, and the intercept gives the value of $1/\overline{M}_w$. The fact that both curves should extrapolate to the same point on the ordinate increases the precision of the determination of $\overline{M}_w$. From the Zimm plot, the slope of the limiting curve corresponding to $\theta = 0°$ gives the second virial coefficient. The mean square radius of gyration, S_z^2 of the particles is given by the slope and intercept corresponding to the other limiting curve in the Zimm plot (for $c = 0$), according to the relationship:

$$S_z^2 = (3\lambda^2/16\pi^2n^2) \cdot (\text{slope/intercept}) \qquad (6.44)$$

Light scattering photometers based on photoelectric measurements are used for measurements of scattering data. A simple measurement principle and approach is

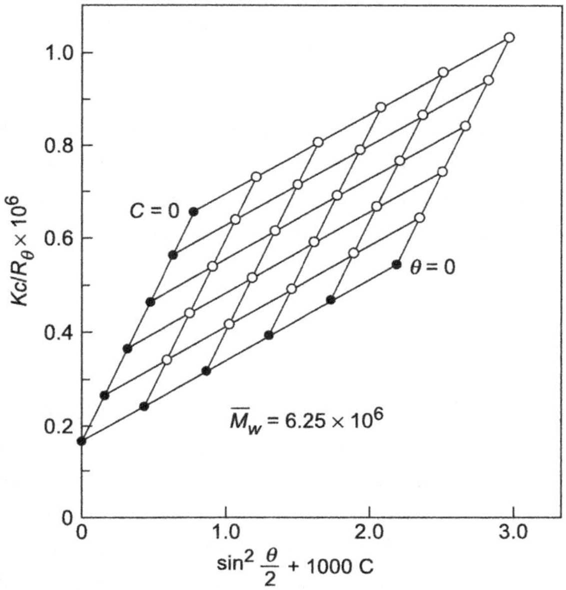

Fig. 6.8 *A typical Zimm plot, Kc/R_θ vs. $\sin^2(\theta/2) + kc$ from light scattering measurements for sodium amylose xanthate in 1 M NaOH solution at 30°C*

shown in Fig. 6.9. It is absolutely necessary to keep the measuring chamber dust free. The scattering glass cell is centred on the axis of rotation of the receiver photomultiplier tube assembly which can be rotated with controlled and measurable angular positions. Besides the measurement of τ or R_θ, it is necessary to determine the refractive index n and the parameter $(\partial n/\partial c)$, whch is a constant for a given

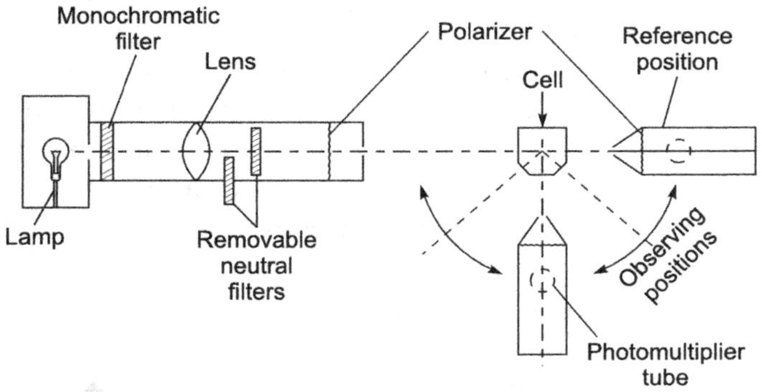

Fig. 6.9 *Principle of light scattering measurements*

polymer–solvent combination at a given temperature. It is measured with the help of a differential refractometer. The choice of solvent is important. The difference in refractive index between the polymer and the solvent should be as large as possible. A solvent of low second virial coefficient allows a more precise evaluation of $\overline{M}_w$ by extrapolation. Dust free solvents and solutions preferably obtained by pressure filtration should be used. The solvent and solutions must be free from all extraneous scattering material. Molecular weight range of 10,000 to 10,000,000 are measurable by this technique.

6.9.1 Dissymmetry

For polymer molecules much smaller than the wavelength of the incident light, the scatterings in the forward and backward directions measured at two angles symmetrical about 90° (say, 45° and 135°) are not appreciably different, but for particles larger than about a tenth of the wavelength of light, the intensity of the scattered light follows a decreasing trend from front to rear. The ratio ($i_{45°}/i_{135°}$), known as the dissymmetry, z of light scattering is unity for small particles and it increases with size. Evaluation of the dissymmetry is conveniently used to estimate particle size which is the effective expanse of the particle with respect to the wave. length. If the particle weight is also known, the measure of dissymmetry gives an indication of the shape of the particles as to whether they are spherical, rod-like, disc-like or random coils. There are, however, limitations to this approach of particle shape determination.[23]

6.10 End-Group Analysis

Analysis of functional groups in polymers, particularly of those incorporated at chain ends is a very important means of polymer characterization. Polymer characterization by this approach is meaningful specially when it is a clear-cut case with well-defined mechanism of polymer growth under a specified condition, and purification from low molecular weight monomers, additives and spurious contaminants has been meticulously accomplished. For high polymers, desired purifications are satisfactorily accomplished by repeated precipitations.[24,25]

In well-defined polymer systems, a knowledge of the number of a specific end-group in a given mass of polymer permits determination of number average molecular weight from the results of end-group analysis, which in effect, gives a count of the molecules in a given weight of the sample. Or else, if the number average molecular weight is determined by an independent technique, end-group analysis gives an idea of the number of a specific end-group per polymer chain which further throws light on the detailed mechanism of polymerization, including initiation and termination. Meaningful information about chain transfer or radical transfer reactions may also be obtained from well-designed experiments.[26–28] Titrimetric methods are particularly useful to estimate such groups as $—NH_2$, —COOH, etc. in relatively low molecular weight polyamide, polyester and related polymers.[29,30]

Spectroscopic methods (*UV*, visible and IR absorption spectroscopy) have been conveniently used for polymer end-group characterization on many occasions.[31–34] Radioactive tracer technique has also been used for some systems (limited though they are) for studies of the mechanism of polymerization and for end-group characterization.[35–38] Sensitive colourimetric methods[39–41] of selective nature have also been extensively used for characterization of end-groups.

The dye-partition technique and the dye-interaction technique introduced by Palit[40,41] and developed for quantitative characterization of polymer microstructure by Palit, Ghosh and coworkers[24,25,42] have been quite suitable for the analysis of —COOH, —OH, $—NH_2$, —Cl, —Br, $—OSO_3^-$, $—SO_3^-$ groups or end groups in high polymer systems.

The general chemical methods of analysis of functional end-groups suited for characterization of condensation polymers are prohibitively difficult for a vinyl polymer because of the latter's much higher molecular weight in general. Introduction of highly sensitive methods is therefore necessary to enable easy detection (and possible estimation) of groups or atoms present as end groups in macromolecules of very high molecular weight. The dye techniques, viz. the dye-partition technique and the dye-interaction technique provide simple solutions to these problems.

6.10.1 Dye-Partition Technique

In the analytical chemistry of detergents and surfactants in general, the colourimetic methods involving the use of ionic dyes of various kinds have been widely employed.[43–45] Large surfactant ions and large dye ions of opposite charge react to produce compounds having very low or little solubility in water. The ionic dyes soluble in water are so selected that they are ordinarily insoluble in such organic solvents as benzene and chloroform. As a result, in a two-phase system involving one such immiscible solvent with an aqueous solution of a dye, little colour is seen in the organic layer. If some detergent bearing an ionic group of a specific kind and charge is present along with excess of a suitable dye of opposite charge, the dye-detergent salt formed, usually being more soluble in the organic solvent, is partitioned preferentially in the organic layer.[46]

According to the principle of dye-partition technique, ionizable groups present in polymers easily undergo metathetic reactions with suitable dyes. In case the dye is water soluble and insoluble in a water immiscible organic solvent and the polymer has the reverse characteristics, i.e. the polymer dissolves in the organic solvent but not in water, in any partition experiment, the dye ion would combine with the polymer counter-ion and as a result, the coloured dye ion would pass into the non-aqueous phase; or in other words, colouration of the nonaqueous phase would indicate the presence of ionizable end-groups in the polymer.

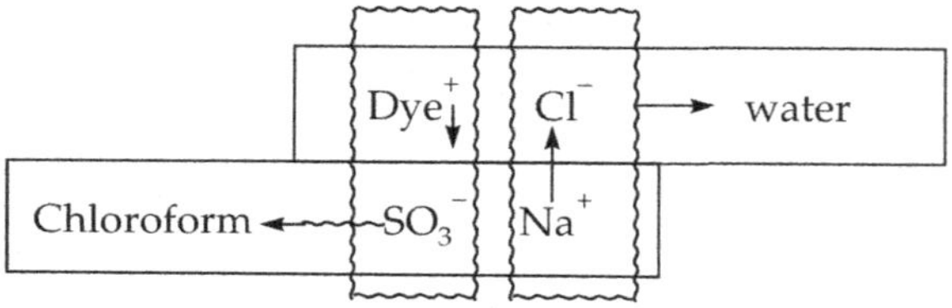

(6.45)

A possible source of complication, namely the adsorption of the dye by the polymer, is practically negligible in view of the fact that a purely thermal polymer such as polystyrene or poly(methyl methacrylate) having no ionizable end group, solubilizes little dye in such partition experiments. Moreover, the polymer in most cases is insoluble in aqueous phase and so there is little complication due to its distribution in both the phases, as in the case of detergents.

Carboxyl (—COO^-) end-group is detected by doing the partition test using the cationic dye pynacyanol (chloride) in neutral or mildly alkaline aqueous solution (pH 8) and the polymer in benzene solution.[47] Strong acid end-groups such as sulphate (—OSO_3^-), sulphonate (—SO_3^-) etc., also respond to this test but they are easily distinguished by carrying out the partition test under acid condition, preferably using methylene blue (chloride) as the basic or cationic dye in aqueous centinormal HCl solution and the polymer in chloroform solution. Under this condition —COOH groups fail to respond to the dye-partition test. Similarly, basic end-groups such as the amino end-groups and quaternary ammonium (cationic) end-groups are readily detected and estimated using an acid or anionic dye such as disulphine blue VN150 or bromophenol blue dye in aqueous acidic (0.01 N HCl) solution and taking the polymer in chloroform solution. For quantitative purposes, equal volumes of the aqueous dye solutions (10 mg/litre) and dilute solution of a polymer in the appropriate organic solvent (0.05–0.5%) are intimately shaken together in a stoppered tube, allowed to settle for distinct separation of layers (centrifuged if necessary) and the colour developed in the organic layer is then measured spectrophotometrically with the help of a calibration curve prepared for a relatively low molecular weight appropriate model compound such as stearic acid, sodium lauryl sulphate, cetyl trimethyl ammonium bromide or lauryl amine.

6.10.2 Dye-Interaction Technique

Solution of a number of basic and acid dyes in aqueous buffer solutions when extracted with benzene, toluene and similar solvents, yields an organo extract of the dye base of dye acid, whose colour depends on the pH of the solution extracted, though the colour of the dye in the aqueous layer may not be pH sensitive. A simple example is the well-known basic dye methyl violet 3B (gentian violet) which, though practically insoluble in benzene is but little extracted by benzene from aqueous solution of pH up to about 7.0. It gives a light violet extract at pH 7.0 and an unstable deep brown extract at about pH 9.0. A strongly alkaline solution produces brown yellow benzene extract of the dye. The dye shows practically no colour change in aqueous solution over the whole pH range. The most sensitive extracts are obtained from more strongly alkaline solutions. Traces of a compound bearing —COO^-, —OSO_3^- or —SO_3^- functional groups sharply change the colour of the benzene extract to violet. Basic dyes of the rhodamine 6G class extracted in benzene from aqueous solution (10 mg/litre) at pH 10–12 also give yellow extracts of high acid sensitivity. Micronormal acid solutions in benzene e.g. formic acid, stearic acid, etc.) can be easily detected by these reagents. The colour change for the sensitive rhodamine dye reagents in the dye interaction test is from yellow to pink [often with fluorescence if the concentration of the (acid) anionic groups is relatively high]. The basic dye reagents (benzene extracts) are quite stable and retain their sensitivity if preserved over NaOH pellets in the dark. For quantitative purposes, equal volumes of the dye reagent and dilute (benzene) solution of an acid or of a polymer purified appropriately are mixed together and the extent of colour change brought about, depending on the nature and concentration of the acid present[41], is measured spectrophotometrically taking the help of a suitable calibration curve.[24]

For the detection of bases such as amines or basic end-groups (amino end-groups) in polymers, almost all phthalein dyes, particularly the halogenated ones are highly sensitive and they can be successfully used after extraction with benzene from their aqueous solution (50 mg/litre) at pH 3–4. Eosin A, eosin saure L neu, erythrosin J or erythrosin pur and rose Bengal give sensitive benzene extracts as dye reagents for detection of basic and cationic end-groups. For quantitative measurements, equal volumes of a selected dye reagent and benzene solution of a polymer are mixed together. The colour change (colourless to pink/yellow to pink/pale brown to pink, as the case may be) brought about in proportion to the concentration of the base or basic end-group present is measured spectrophotometrically using a calibration curve given by a long chain model compound such as lauryl amine or cetyl trimethyl ammonium bromide. For end-group analysis by the dye techniques, tests for acid end-groups are not interfered with by the presence of basic end-groups and vice versa.

The dye techniques suited for the analysis of selected ionic (acidic or basic) end-groups can also be used for the analysis of non-ionizable end-groups after conveniently modifying them to ionizable groups. Before subjecting the polymers to dye tests using the highly sensitive dye reagents, it is necessary that they are scrupulously purified by a process of repeated precipitation to free them from low molecular weight (ionic) acidic or basic contaminants.

Hydroxyl (OH) end-groups in a polymer are not responsive to dye tests. They can be conveniently transformed into dye test responsive carboxyl (—COOH) groups by phthalation[25] (heating 0.2–0.5 g of polymer dissolved in 5 ml pyridine with excess of phthalic anhydride at 90–100°C for 6 h).

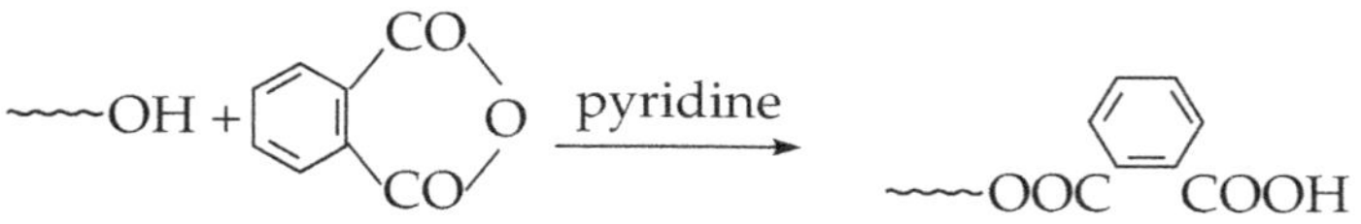

(6.46)

On the other hand, OH end-groups in a polymer can also be transformed into dye test responsive anionic $—OSO_3^-$ (sulphate) end-groups by sulphation[42] (heating the polymer with chlorosulphonic acid in pyridine medium at 90°C for 4–6 h)

$$\sim\!\sim OH + ClSO_3H \xrightarrow{\text{pyridine}} \sim\!\sim O \cdot SO_3H + HCl \qquad (6.47)$$

The sulphur atom of the anionic sulphate end-group is attached to the polymer chain through an oxygen atom (C–O–S linkage); the anionic sulphonate end-group ($—SO_3^-$) is, however, attached to the polymer chain through a C–S linkage. Sulphate end-roups are readily hydrolyzable to —OH groups under acidic conditions and they are distinguishable from sulphonate end-groups as the latter are non-hydrolyzable under similar conditions.[48]

Halogen (Cl, Br, I) end-groups or atoms present in a polymer are not responsive to the dye tests, but they can be conveniently transformed into dye test responsive cationic groups (pyridinium halide) by quaternization using pyridine as the quaternizing base (heating 0.5 g polymer dissolved in 5 ml pyridine at 95°C for 24 h).

C=C unsaturation (as a polymer end-group) may be conveniently traced and detected and possibly estimated by converting the (chain end) unsaturation into an amine (end) group. This is done by heating the polymer in benzene solution with excess of a primary amine (*n*-butylamine) in a sealed tube for 24 h, when addition of the amine takes place at the unsaturation.[49] The polymer is then purified and tested for amine end group by the application of the appropriate dye techniques.

Employing different dye reagents and the two dye techniques in appropriate sequence and following a well-designed approach, one can readily characterize polymers bearing different functional end-groups at the same time. Ordinary chemical or physico-chemical methods of end-group analysis are normally applicable with good degree of precision for polymers of molecular weight up to about 30,000, but highly sensitive microchemical techniques such as the dye techniques for end-group analysis are generally applicable to characterize end-groups of benzene or chloroform soluble polymers of molecular weight up to about 5×10^5.

6.11 The *Z* Average Molecular Weight

The Z average molecular weight $\overline{M}_z$ is expressed as

$$\overline{M}_z = \frac{\Sigma N_i M_i^3}{\Sigma N_i M_i^2} \tag{6.48}$$

The use of the weighting factor $N_i M_i^2$ for the determination of the *Z* average molecular weight clearly indicates that the higher molecular weight molecules are weighted even more heavily in this case than in the case of the weight average molecular weight. For a given distribution, the order of the averages is clearly $\overline{M}_z > \overline{M}_w \geq \overline{M}_v > \overline{M}_n$. The *Z* average molecular weight is conveniently measured by sedimentation equilibrium method in an ultracentrifuge.

The ultracentrifugation techniques are relatively complicated and are much less commonly employed for molecular weight measurements of synthetic high polymers. They are, however, more commonly used for characterization of biological polymers such as proteins.

Using a relatively low speed of rotation with the polymer solution in the cell held in position, and operating the centrifuge under constant conditions over long periods, thus avoiding convection and related disturbances within the cell, a state of equilibrium is reached in which the polymer is distributed in the cell according to its molecular weight and molecular weight distribution. The force of sedimentation on a species in solution is just balanced by its tendency to diffuse out. For a dilute solution closely approaching ideal thermodynamic behaviour and for a monodisperse polymer, the molecular weight, M is given by the expression

$$M = \frac{2RT \ln(c_2/c_1)}{(1 - v\rho)\,\omega^2 (r_2^2 - r_1^2)} \tag{6.49}$$

where c_1 and c_2 are the concentrations at two points corresponding to distances r_1 and r_2 in the cell and ω is the angular velocity of rotation, v, the partial specific volume of the polymer and ρ, the density of the medium. A solvent chosen should preferably be a poor solvent and it should be quite different in density from the polymer so as to allow good sedimentation, and also in refractive index so as to facilitate measurement. If the polymer is polydisperse in nature, then different approaches for measuring the concentration as a function of r yield different molecular weight averages ($\overline{M}_w$, $\overline{M}_z$, etc.). Measurements based on refractive index yield $\overline{M}_z$. Preparative ultracentrifuges are useful in fractionating polymer samples and in separating them from easily sedimented contaminants.

6.12 General Requirement of Extrapolation to Infinite Dilution

The process of dissolution of a polymer is usually slow and it mostly occurs in two characteristic stages: (i) an initial phase of swelling and (ii) gradual formation of a molecular solution if it is not a cross-linked system. The second stage is often aided and accomplished quicker by stirring or agitation and by warming or heating. Solubility features in polymer systems are far more complex than those for low molecular weight substances. The polymer molecules are usually far bigger in size than the solvent molecules, and this factor along with the extent of interactions between polymer segments (intramolecular and intermolecular) and between the polymer and the solvent under a given set of conditions are important considerations to understand and analyze solubility of polymers and their solution properties.

A simple theory considers a polymer chain as an assemblage of a large number of points (chain units or segments) on a lattice work which are joined or tied together by flexible (chemical) bonds of equal lengths. The polymer chain may assume any specific arrangement on the lattice out of many statistically possible arrangements. For ideal solution behaviour, there should be no interaction between segments of different chains, which can be approached and possibly attained only with infinitely dilute solution. But actual measurements of any solution property at such vanishing concentrations are practically impossible to make, thereby necessitating extrapolation of measured properties at finite concentrations to infinite dilution. For finite concentration, howsoever dilute, the interactions between chains cannot be altogether ignored because of chain entanglements.

6.13 Polymer Fractionation and Molecular Weight Distribution

In a very poor solvent, or more appropriately, in a non-solvent, the polymer will not dissolve. On gradual or dropwise addition of a non-solvent to a dilute solution of the polymer in a good solvent, some polymer will be thrown out of solution and will appear as a precipitate at certain point. It has been found that polymer molecules of the highest molecular weight or molecular weight range get precipitated first. On separation of the precipitate, further addition of the non-solvent in a similar manner throws out at a subsequent point a second fraction of polymer with a somewhat lower molecular weight or molecular weight range. The separation into fractions may be made sharper if after adding the requisite amount of the non-solvent the mixture is warmed to make the system just homogeneous and then the polymer fraction is allowed to precipitate as the mixture is cooled. The process is repeated to isolate successive fractions of decreasing molecular weight or molecular weight range. Each of the successive fractions is carefully isolated, dried, weighed and its molecular weight determined by one of the techniques discussed in the preceding sections. It is then possible to draw an integral molecular weight distribution curve as given in Fig. 6.10 showing cumulative weight per cent plotted against molecular weight. Another useful curve based on differential molecular weight distribution shown in Fig. 6.11 is obtained by differentiating the integral distribution curve. The relative positions of $\overline{M}_n$, $\overline{M}_v$, $\overline{M}_w$ and $\overline{M}_z$ are shown on this curve.

Fractionation based on the above approach separates the various molecular species in a given polymer sample mainly on the basis of their solubility characteristics rather than on the basis of their molecular weight. For a given polymer, the solubility characteristics are, however, dependent not only on chain length, but also on branching (including its nature and frequency), cross-linking, end-groups present and on changes in chemical structure on aging.

6.14 Gel Permeation Chromatography

A chromatographic process is one in which the solute is transferred between two phases, one of which is stationary while the other is moving, and the transfer is

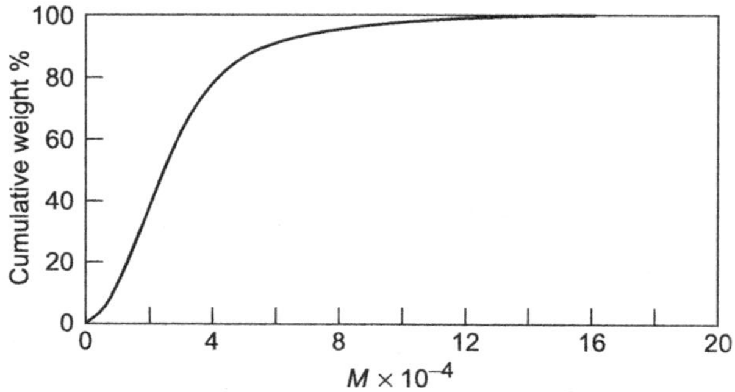

Fig. 6.10 *A typical integral molecular weight distribution curve: plot of cumulative weight % vs molecular weight*

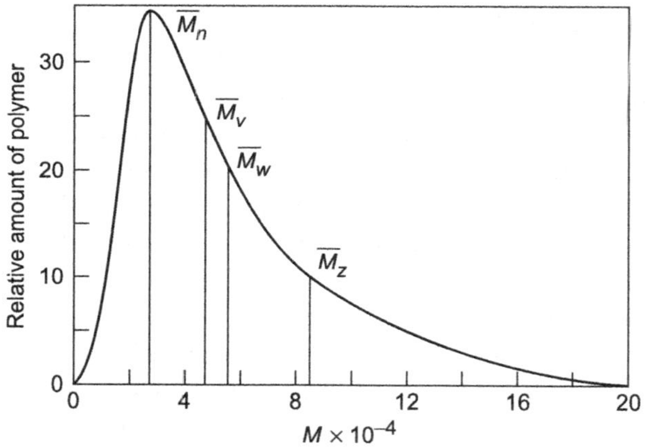

Fig. 6.11 *A typical differential molecular weight distribution curve: plot of relative amount of polymer vs molecular weight*

often allowed to take place in a long column. In gel permeation chromatography the same solvent or liquid is allowed to form the two phases in the column packed with a microporous gel such that the stationary phase is that part of the solvent which is inside the porous gel particles and the mobile phase is made by the solvent remaining outside. The driving force behind the transfer of solute polymer molecules between the two phases is diffusional in nature and it gives rise to a difference in concentrations of solute in the two phases, the transfer process being also largely restricted by the solute molecules' capacity to penetrate or permeate through the pore structure of the gel. The gels used are typically hard, incompressible polymers, those most commonly used being microporous polystyrene (cross-linked with divinyl benzene) prepared by a suspension technique using a suitable inert solvent–non-solvent media. Another material commonly used is porous glass. The pores in the gels used are nearly of the same size as that of the polymer molecules.

A known amount of polymer in a known volume of dilute solution is injected into a solvent stream flowing down the column. The solute polymer molecules flow past the porous beads of the gel and at the same time diffuse into their inner pore structures according to size distribution of the solute polymer molecules and the pore size distribution of the gel. A fractionation of the polymers is achieved in the process in view of the fact that the entry of the larger molecules into the pores of the gel are more restricted or completely hindred due to relatively low pore sizes, and they flow out of the gel column faster, spending less time inside the gel. The smaller molecules follow just the opposite trend as they spend more time inside the gel. The largest among the solute molecules emerge first while the smallest of them emerge last from the gel column. The technique, very commonly known as the "gel permeation chromatography" (GPC), allows separation of polymer molecules by their size. For a properly selected gel, the smallest of the solute polymer molecules find most of the stationary phase accessible.

The method initially requires empirical calibration of a column or a set of columns with gels of graded pore size to yield a calibration curve[50] such as the one shown in Fig. 6.12 relating a molecular size parameter, $[\eta]M$, [see Eq. (6.50) in Sec. 6.15] and retention volume such that with its help a plot of amount of solute versus retention volume of a test polymer known as its chromatogram, Fig. 6.13, can be transformed into a molecular size distribution curve from which a molecular weight distribution curve can be drawn. The GPC is a neat and fast technique for both analytical and preparative

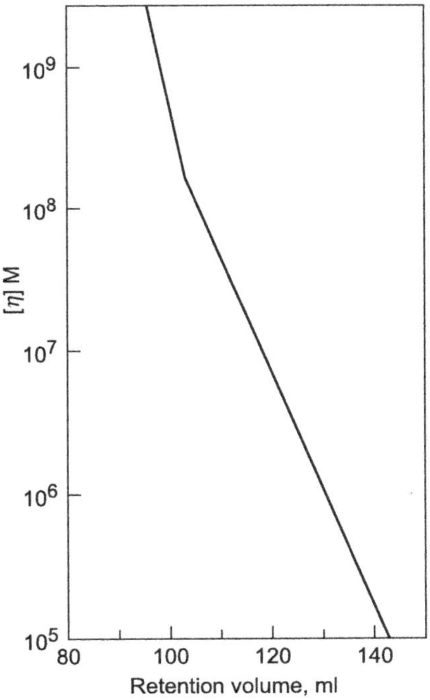

Fig. 6.12 *Calibration curve for gel permeation chromatography: plot of $[\eta]M$ vs retention volume[50] (Grubisic, Remp and Benoit, 1967;* **Courtesy,** *Wiley-Interscience, New York)*

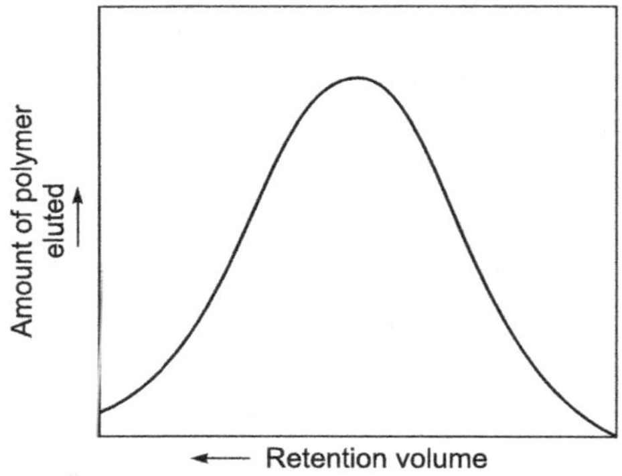

Fig. 6.13 *A typical GPC chromatogram*

work applicable to both linear and branched polymers, requiring a sample size of only a few milligrams and the analysis is usually complete in a time scale of 2–5 h.

Gel permeation chromatography separates molecules in a given polymer sample according to their molecular sizes (or hydrodynamic volumes). Any extraneous physico-chemical factor that contributes to the perturbation of the hydrodynamic volumes of the dissolved polymer molecules and also brings about changes in their rate of elution would not only complicate measurements and interpretations but also may lead to erroneous results and conclusions. For example, in case of non-polar polymers with limited number of charged groups (—SO_3^-, —COO^-, etc.) such as the ionomers or even in case of those having end groups only as the charged groups in their structure, the macromolecules are absorbed on the surface of the microgels as they pass through the columns, thus offering somewhat enhanced resistance to elution as a consequence and then, the size exclusion basis of GPC separation loses its relevance. Such a phenomenon would lead to larger elution volumes and to relatively low molecular weights than actual. The ion-containing polymers tend to aggregate in solvents of low polarity and in case of such macromolecular aggregation, fractionation and molecular weight determination based on separation according to molecular size in solution are largely affected. Analysis of such polymers by GPC can be reliably done only if the charged groups are turned non-ionic or by selecting an eluent solvent system which would prevent adsorption of polymers in the gel columns and would also eliminate macromolecular aggregation. It is important to have a good knowledge about the history of a polymer including its method of synthesis and its microstructure particularly with reference to the presence of charged groups (end groups or branch or repeat units) before attempting to know more about it by employing gel permeation chromatography.

6.15 The Molecular Size Parameter

The molecular size parameter given by $[\eta]M$ is conveniently used in the GPC calibration plot. It is known that the intrinsic viscosity $[\eta]$ of a polymer solution is proportional to the effective hydrodynamic volume of its molecule in solution, $\{(\bar{r}^2)^{1/2}\}^3$ divided by its molecular weight,[51]

$$[\eta] = \phi \frac{(\bar{r}^2)^{3/2}}{M} \tag{6.50}$$

where ϕ, the proportionality constant, was originally considered to be a universal constant, commonly known as the Flory–Fox constant. But analysis of accumulated data later indicated that this was not true,[52] and the value of ϕ was reported to vary between 2.0×10^{21} and 2.8×10^{21}. The linear parameter $\bar{r}$ represents end-to-end distance of the polymer molecule in solution. Equation (6.50) may be expressed by equation (6.51), simply by replacing $(\bar{r}^2)^{1/2}$ by $\alpha(\bar{r}_0^2)^{1/2}$ where α is known as the expansion factor and $K = \phi(\bar{r}_0^2/M)^{3/2}$ is a constant

$$[\eta] = \phi \frac{(\bar{r}_0^2)^{3/2}\alpha^3}{M} = \phi \left(\frac{\bar{r}_0^2}{M}\right)^{3/2} \cdot M^{1/2} \cdot \alpha^3 = KM^{1/2}\alpha^3 \tag{6.51}$$

for a given polymer, independent of its surroundings and molecular weight. At θ temperature, $\alpha = 1$ and

$$[\eta]_\theta = KM^{1/2} \tag{6.52}$$

This expression allows the determination of the unperturbed dimension, $(\bar{r}_0^2)^{1/2}$ of the polymer chain. The value of α depends on the nature of the solvent used; a large value of α arises as a consequence of the use of a thermodynamically 'good' solvent and vice versa. In any solution, a polymer molecule exists as a randomly coiling mass having conformations that occupy many times the volume of all its segments. In a poor solvent characterized by poor solute–solvent interactions, the coils remain relatively contracted, while in good solvents they are relatively expanded or extended, giving a relatively large value for the expansion factor α due to favourable solute–solvent interactions.

6.16 Molecular Weight Distribution in Vinyl Polymers

Molecular weight distribution in vinyl polymers is conveniently calculated by restricting the polymerization to low conversions. This is required to ensure having the same average molecular weight all through the conversion range. Defining p as the probability that a propagating radical will continue to propagate instead of terminating such that $p = \{R_p/(R_p + R_t + R_{tr})\}$, the probability of formation of an n-mer consequent to $(n - 1)$ propagations followed by a termination by disproportionation is p^{n-1} $(1 - p)$ and in analogy with linear step-growth polymerization, the situation is just about the same or similar and the expressions for N_n, W_n, $\overline{X}_n$, and $\overline{X}_w$ as given in Sec. 2.9 are generally applicable. The distribution ratio $\overline{X}_w/\overline{X}_n$ is likewise equal to $(1 + p)$, and for $p \rightarrow 1$, $(\overline{X}_w/\overline{X}_n) = 2$.

For chain termination by coupling, yielding polymer molecules by mutual combination of two kinetic chains, a sharper size distribution results. For this, the distribution is similar to the case of multichain step-growth polymerization with $f = 2$, giving $\overline{X}_n = [2/(1 - p)]$ and $\overline{X}_w = [(2 + p)/(1 - p)]$, such that the distribution ratio or the breadth of the distribution $\overline{X}_w/\overline{X}_n$ is characterized by a limiting value of 1.5 for $p \rightarrow 1$, assuming absence of chain transfer.

6.17 Thermal Analysis

The thermal properties of polymers are conveniently studied by employing such techniques as differential thermal analysis (DTA) and differential scanning calorimetry (DSC).

The DTA technique usually allows detection of thermal response and effects accompanying chemical or physical changes in a material sample when it is heated or cooled in a programmed manner through a zone of transition or reaction. It allows measurements of glass transition temperature (T_g), the crystallization temperature (T_{cr}), crystalline melting point (T_m) and the temperatures of thermal/oxidative degradation, cross-linking or other reactions. A block diagram showing the major components of the equipment and a schematic representation of a DTA thermogram are shown in Figs 6.14 and 6.15, respectively.

In practice, the material sample and a thermally inert reference material held in the respective holder of the DTA cell are heated or cooled in a programmed manner. Any physical or chemical change in the test sample at a specific temperature, which is a characteristic feature of the material under study, is usually associated with a thermal transition leading to an enhanced difference of temperature, ΔT between the test and reference materials held in the same furnace chamber. ΔT is recorded as a function of temperature, T. For no thermal transitions in the test sample, ΔT remains nearly constant. In DTA, the correlation between ΔT and energy changes over a specific transition or reaction is unknown, thus making the conversion of endotherm or exotherm peak areas to energies uncertain. However, the technique is applicable to virtually all polymers and many other material systems, giving mostly qualitative information about thermal effects with clear indications of transition temperatures (Fig. 6.15), but it is normally unsuitable for quantitative measurements of such parameters as heat capacity, heat of fusion or heat of

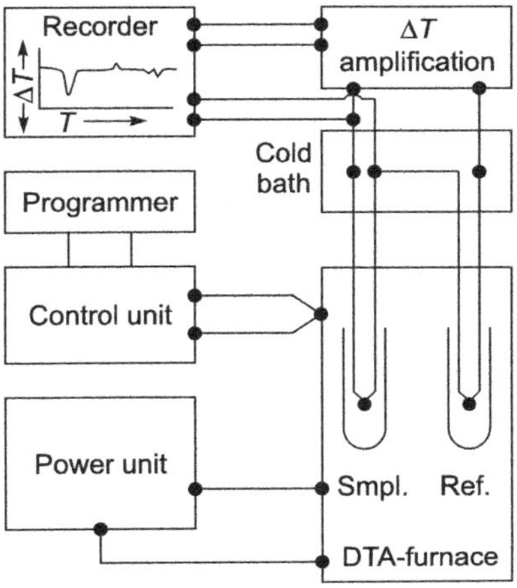

Fig. 6.14 *A block diagram of a DTA apparatus*

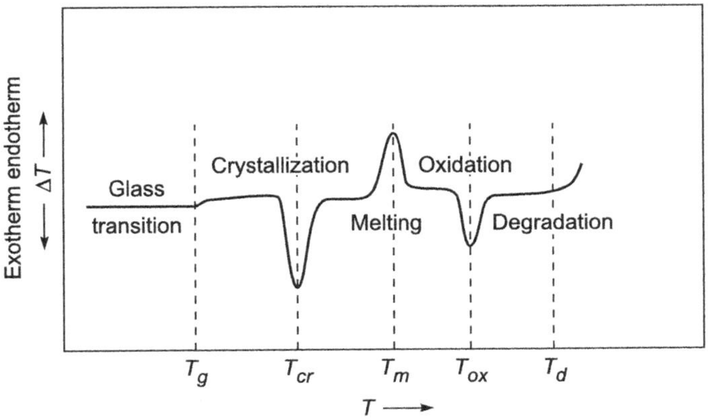

Fig. 6.15 *A typical DTA thermogram showing thermal characteristics of a crystallizable polymer (schematic)*

crystallization of crystalline polymers or change in specific heat at the glass transition for amorphous polymers, which are, however, readily measured employing DSC. In DSC, the test sample and the reference material are heated separately by individually controlled units. The power or electrical energy input to these heaters are so controlled and continuously adjusted consequent to any thermal effect in the test sample as to maintain the two at identical temperatures. The differential power or heat energy required to achieve this state of affairs is recorded against the programmed temperature of the system. For transitions involving latent heat such as fusion, the heat of transition (fusion) is determined by integrating the (heat) energy input over the time interval covering the transition.

Different polymers decompose over different ranges of temperature yielding different proportions of volatiles and residues. Thermogravimetric analysis (TGA) is a useful analytical technique for recording weight loss of a test sample as a function of temperature which may then be used for an understanding of the chemical nature of the polymer. Coupled with analysis of the volatiles and residue, TGA provides information about stability and decomposition of the material in an inert atmosphere or in air or oxygen, and about moisture content and other volatiles and plasticizer content, ash content, extent of cure of cross-linked polymers; in suitable cases, the analysis enables identification of polymers taking help of characteristic thermograms. The test sample is placed in a furnace while it remains suspended from one arm of a precision balance. TGA thermograms are obtained by recording change in the weight of the test sample as it is held at a constant high temperature or it is dynamically heated in a programmed manner. A curve may be drawn to represent weight loss of the sample as a function of temperature (or else in a differential form showing weight change with time as a function of temperature) as shown in Fig. 6.16.

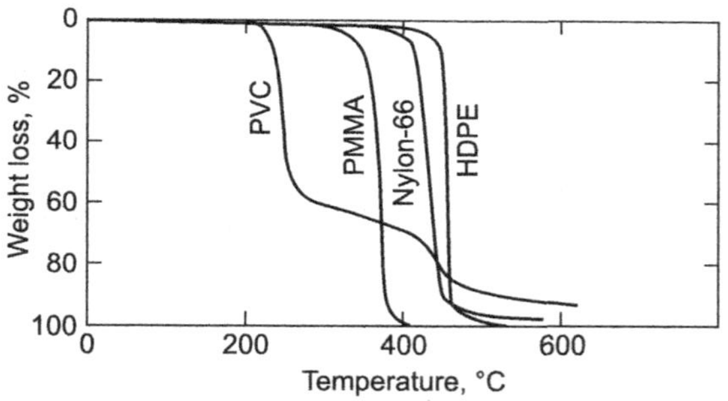

Fig. 6.16 *TGA thermograms of some selected polymers: heating at a rate of 10°C/min under N_2 atmosphere*

6.18 Other Methods and Techniques of Polymer Characterization

Many other destructive or non-destructive methods and techniques are employed in an attempt to characterize polymers or copolymers as obtained in nature or synthetically produced or as subsequently modified following different approaches:

(i) For analysis of volatile or low molecular weight products of deomposition of a polymer such as those produced during heat, light or radiation induced degradation, application of mass spectrometry and gas chromatography permits one to obtain useful and valuable data and information relating to the number and identity of the products and their relative amounts.

(ii) The spectroscopic methods covering ultra-violet, visible and infrared absorption spectroscopy are employed for characterization of chemical structures of polymers, as has already been mentioned earlier in this chapter during discussion of end-group analysis. Photometric methods are commonly used for measurements of the absorbance or optical density, or else the transmittance (%), using samples in dilute solutions. In the infra-red method, samples in their film form prepared by milling or microtoming the sample or casting from solution and solvent evaporation, or in a finely ground form mixed with KBr to form a disc, are commonly used for analysis. Figure 6.17 shows an example of analysis of vinyl polymer (PMMA) employing UV spectroscopy with a view to ascertaining mechanism of chain initiation and identity of initiating radicals for polymerization of methyl methacrylate (MMA) using acetic acid and its chloroderivatives (CX_3 COOH where X stands for H and/or Cl atom) in combination with dimethylaniline (DMA), $\left(C_6H_5-N\begin{matrix}CH_3\\CH_3\end{matrix}\right)$ as photoinitiator. Evidence for incorporation of the DMA moiety presumably as end-group in the test polymers as given by a peak or hump at 255 nm is indicative of chain initiation having taken place, at least in part, by N-methyl anilinomethyl radicals derived from DMA.[53] UV spectroscopy is also useful for the characterization of composition of selected binary copolymers such as those of styrene and methyl methacrylate.[54,55] Figure 6.18 is given as an example of examination of characteristic absorption bands of different polymers and constituent fractions of a graft copolymerization product (polymerization of MMA on aminated PVC)[56] employing IR spectroscopy.

(iii) Polymers are also conveniently characterized by microscopic techniques including both light microscopy and electron microscopy. For light microscopy, techniques utilizing the reflected light allow study of the surface features and texture of solid and opaque polymers. For morphological and related studies, utilizing transmitted light passing through films of samples cast on the microscopic slide, it is convenient to follow the techniques of polarized light microscopy and phase contrast microscopy. The latter technique allows observation and analysis of structural features of molecular aggregates involving differences in refractive index rather than absorption of light, while the former takes advantage of the ability of the crystallites to rotate the plane of polarized light.

Electron microscopy has, however, emerged as a very powerful tool in the morphological studies of crystalline polymers. Direct technique is convenient

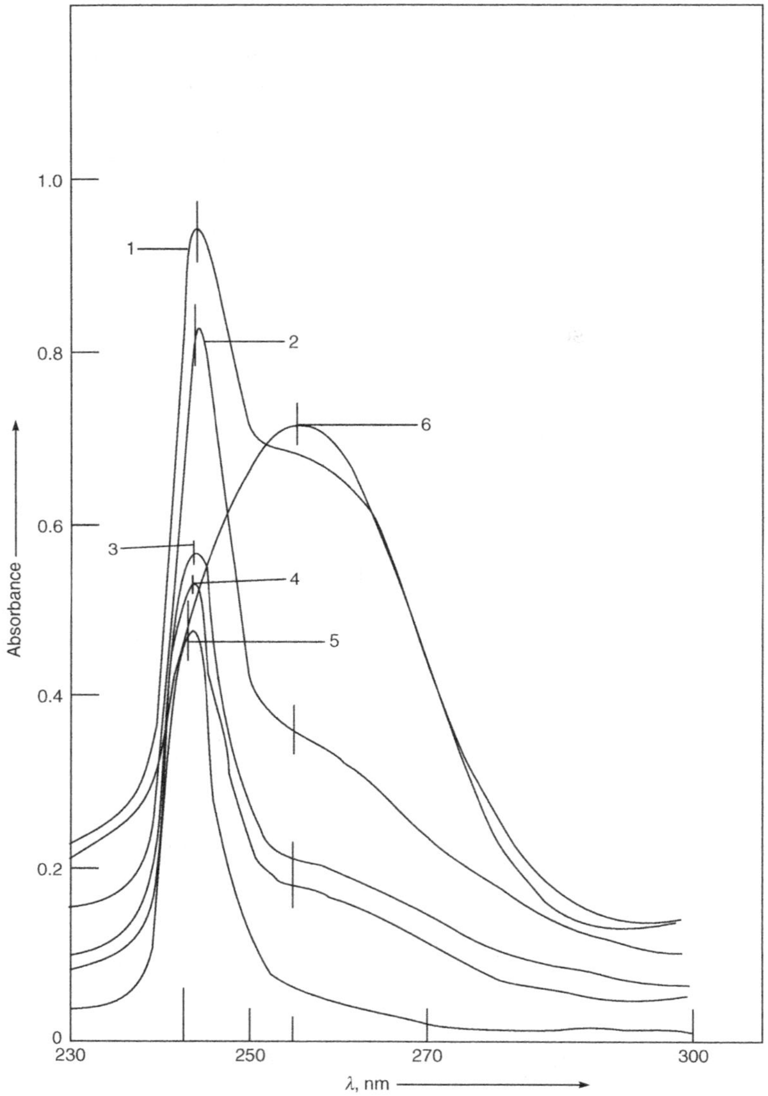

Fig. 6.17 *UV absorption spectra of solutions of poly(methyl methacrylate)*[53]*, PMMA (0.75%), (Curves 1-5) and of dimethylaniline, DMA (6.5×10^{-5} mol 1^{-1}), (Curve 6) in $CHCl_3$. Data given for Curves 1-5 are initiator system (each initiator component, 5×10^{-3} mol 1^{-1}), condition of polymerization including temperature and average degree of polymerization, $\overline{X}_n$: (1) CCl_3—COOH/DMA; photo, 40°C; 2.81×10^3; (2) $Cl_2CHCOOH$/DMA; photo, 40°C; 7.17×10^3; (3) $ClCH_2 \times COOH$/DMA; photo, 40°C; 7.36×10^3; (4) $CH_3 \cdot COOH$/DMA; photo, 40°C; 9.27×10^3, and (5) no initiator; dark, 21.0×10^3. (Ghosh and Mukherjee 1986;* Courtesy. *Pergamon Journals, Devon, England)*

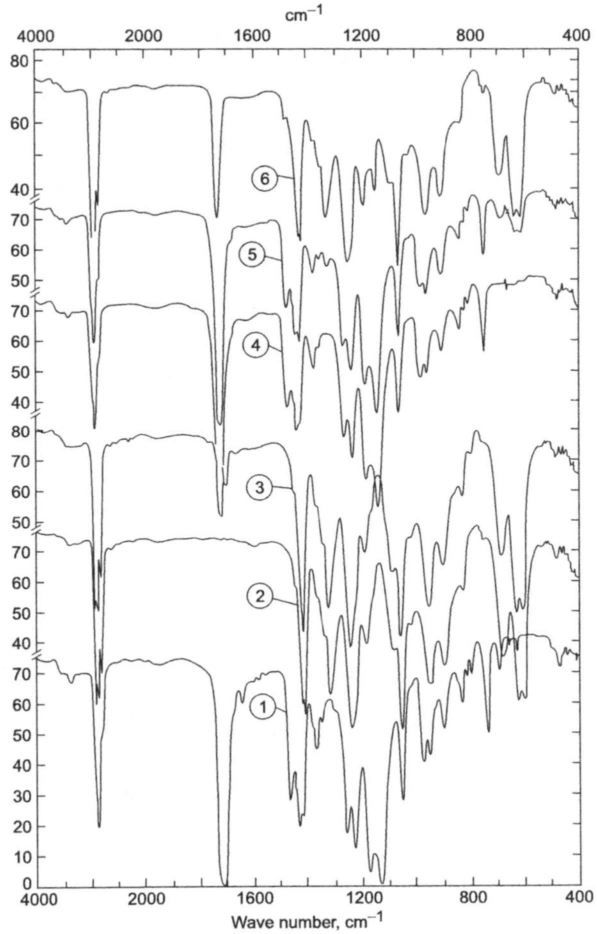

Fig. 6.18 *IR Spectra of several polymers[56]: Curve 1, PMMA; Curve 2, poly(vinyl chloide), PVC and Curve 3, n-butylamine (n-BA)-treated PVC; and those of three fractionated polymer species form a gross product from graft copolymerization of MMA on (n-BA)-treated PVC: Curve 4, the first fraction corresponding to unbound PMMA (comparable with curve 1); Curve 5, the second fraction corresponding to the graft copolymer of PMMA on (n-BA) treated PVC, and Curve 6, the third fraction corresponding to unreacted (n-BA) modified PVC (comparable with Curve 3); (n-BA) modification of PVC (2.5 g) was done by heating the polymer with n-BA (5 ml) in cyclohexanone solution (50 ml solvent) at 90°C under a pressure of 100 psi for 24 h. Fractionation of the gross product from graft copolymerization was done at 25°C using tetrahydrofuran as the solvent and petroleum ether as the precipitant. (Ghosh, Bhattacharya and Maitra, 1988; Courtesy, Huthig & Wepf Verlag, Basel)*

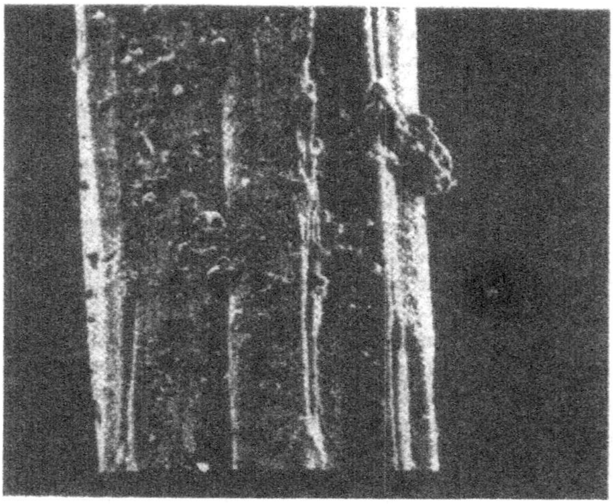

(a)

(b)

Fig. 6.19 *Scanning electron micrograph showing morphological patterns of: (a) the surface of a mature jute fibre and (b) fractured section of the mature jute fibre, revealing multi cellular structure of the fibre* for observation of the structure of thin specimens such as polymer single crystals. But more commonly, the heavy metal shadowing techniques or those of replication or solvent etching are used for electron microscopic studies and observations. Scanning electron microscopy (SEM) has been more lately developed as a technique in which a fine electron beam is scanned across the surface of an opaque (non-conducting) polymer specimen coated with a conducting film of materials such as gold or copper. The scanning technique results in images with great depth of field and excellent three dimensional appearance (Fig. 6.19). SEM technique as applied to polymer systems is primarily employed in the studies of (a) phase morphology of polymer blends, and of graft and block copolymers, (b) surface features of natural and synthetic fibres, and (c) fracture and failure mechanisms of plastics, rubbers and composites under tear, tension and abrasion, etc.

(iv) X-ray diffraction method is another useful and powerful analytical technique employed for understanding and elucidation of the structure of crystalline, drawn or oriented polymers. Polymer single crystals are far too small for X-ray diffraction experiments and the crystal structure of a polymer is usually determined from analysis of X-ray patterns of a fibre drawn from the polymer. The periodicity of molecular structure of a polymer is linked to the existence of a repeat unit and a characteristic repeat distance is associated with the periodicity of its crystal. In view of the crystalline regions being aligned longitudinally along the fibre axis, the X-ray pattern of the oriented polymer is essentially identical with the rotation pattern to be obtained from a single crystal. For such a pattern, diffraction maxima appear in rows perpendicular to the fibre axis, commonly called layer lines. Greater the repeat distance, closer together appear the layer lines. Measurement of the distance between the layer lines allows direct determination of the repeat distance.

X-ray diffraction studies also reveal a great deal of information relating to disorder in the crystal structure, though a quantitative measurement of the disorder is difficult. Disorder in the crystal structure leads to broadening of the diffraction maxima and this is caused by both small crystallites as well as distortions within large crystals. For calculation of the relative amounts of crystalline and amorphous material in a given sample, it is necessary to resolve the contributions of the two types of structures to the X-ray diffraction pattern.[57]

Unoriented polymers usually give X-ray diffraction patterns characterized by rings, but on orientation, the rings split into arcs or even into spots, Fig. 6.20. For a non-crystalline or amorphous polymer such as polystyrene, considerable molecular orientation can be imparted if the material is drawn or put

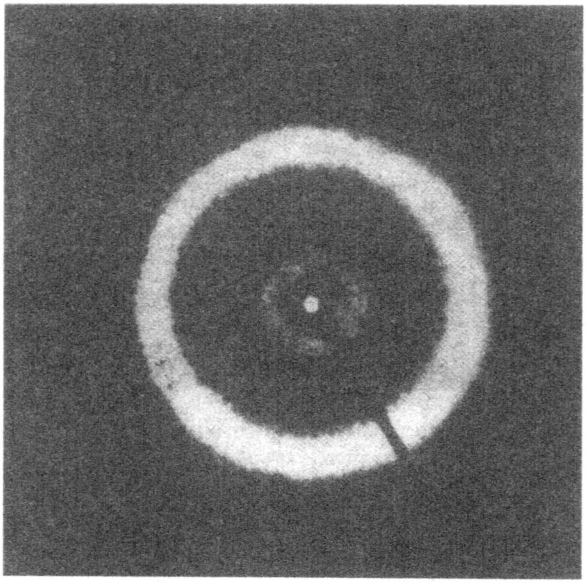

(a)

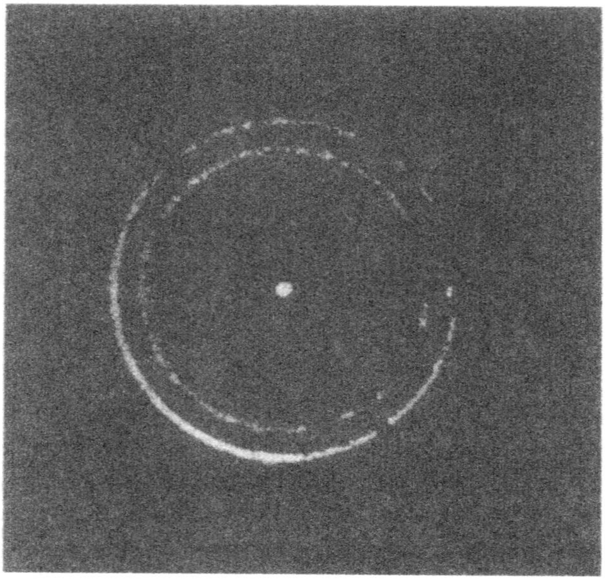

(b)

Fig. 6.20(a) (b)

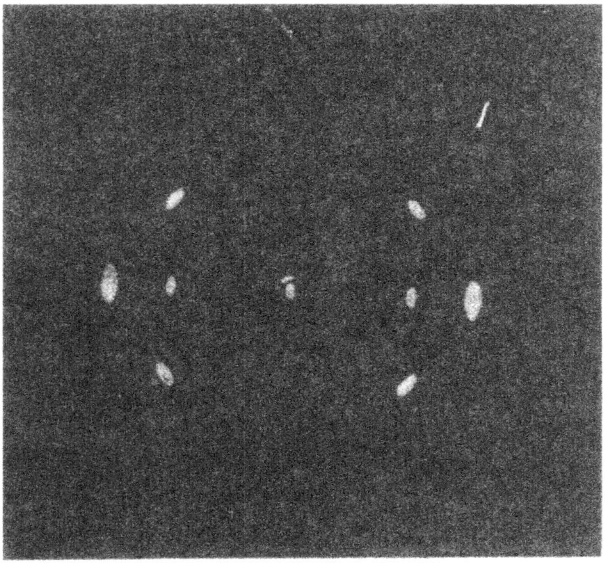

(c)

Fig. 6.20 *X-ray diffraction patterns for: (a) unstretched (ambient) (b) unstretched (frozen) and (c) stretched (600% elongation) Hevea rubber*

under stress at temperature closely above T_g. If it is then quickly cooled to below T_g, the imparted orientation will be stabilized. This type of orientation can be measured by birefringence or change in the index of refraction with direction as well as by the enhanced strength along the direction of applied stress or by the force with which the specimen snaps or retracts on heating above T_g.

(v) Studies based on nuclear magnetic resonance (NMR) spectroscopy are made on polymers with the objective of ascertaining and establishing chain configurations and microstructure. NMR technique is particularly advantageous in that it allows study of the motion and location of protons which are not that readily accomplished by other approaches of study and analysis. As a distinctive and powerful analytical tool, this technique allows independent confirmation of isomeric structures (isotactic and syndiotactic sequences and their random combination giving atactic structure) and evaluation of copolymer composition including the composition of selected graft copolymers.[58,59]

Figure 6.21 shows the 60 **MHz NMR** spectra of three poly(methyl methacrylate) samples labelled according to the predominant tacticity and prepared using different appropriate catalysts.[60]

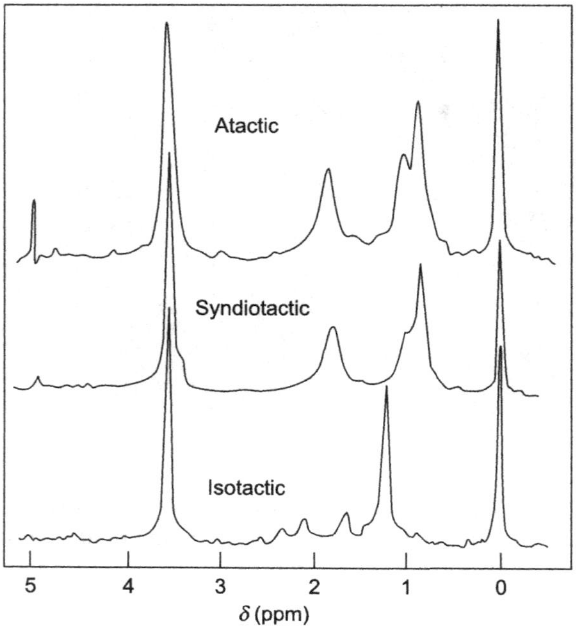

Fig. 6.21 *Nuclear magnetic resonance spectra[60] of three poly(methyl methacrylate) samples with prominence of isotacticity, syndiotacticity and atacticity (McCall and Slichter, 1964;* **Courtesy,** *Wiley-Interscience, New York)*

References

1. Staudinger, H. and A. Steinhofer, *Ann.* **517** (1935), 35.
2. Marvel, C.S., J.H. Sample and M.F. Roy, *J. Am. Chem. Soc.*, **61** (1939), 3241.
3. Flory, P.J., *J. Am. Chem. Soc.*, **61** (1939), 1518.
4. Wall, F.T., *J. Am. Chem. Soc.*, **62** (1940) 803; **63** (1941), 821.
5. Marvel, C.S. and C.L. Levesque, *J. Am. Chem. Soc.*, **60** (1938), 280.
6. Marvel, C.S. and C.E. Denoon, *J. Am. Chem. Soc.*, **60** (1938), 1045.
7. Staudinger, H. et al., *Ber.*, **67** (1935) 1164; **68** (1935), 1618.
8. Walling, C., *J. Am. Chem. Soc.*, **67** (1945) 441.

9. Miller, A.A. *J. Appl. Polym. Sci.*, **5** (1961) 388.
10. Pinner, S.H., *Nature*, **183** (1959), 1108.
11. Dannis, M.L. and F.L. Ramp, *Encyclopedia of PVC*, L.I. Nass, ed., Marcel Dekker, New York, 1976, Vol. 1. p. 237.
12. Staudinger, J.J.P. *Plastics Progr.* (1953), p. 9.
13. Ouchi, J., *J. Polym. Sci.*, **A3** (1965), 2685.
14. Miller, A.A., *J. Phys. Chem.*, **63** (1959), 1755.
15. Baum, B., *J. Soc. Plast. Eng.*, **17** (1961), 71.
16. Huggins, M.L., *J. Am. Chem. Soc.*, **64** (1942), 2716.
17. Kraemer, E.O., *Ind. Eng. Chem.*, **30** (1938), 1200.
18. Krigbaum, W.R. and P.J. Flory, *J. Am. Chem. Soc.*, **75** (1953), 1775.
19. Masson, C.R. and H.W. Melville, *J. Polym. Sci.*, **4** (1949), 337.
20. Zimm, B.H. and I. Meyerson, *J. Am. Chem. Soc.*, **68** (1946), 911.
21. Lord Rayleigh, *Phil. Mag.*, **41** (1871), pp. 107, 274 & 447.
22. Zimm, B.H., *J. Chem. Phys.*, **16** (1948), 1099.
23. Benoit, H., *J. Chem. Phys.*, **65** (1968), 23.
24. Palit, S.R. and P. Ghosh, *J. Polym. Sci.*, **58** (1962), 1225.
25. Ghosh, P., A.R. Mukherjee and S.R. Palit, *J. Polym. Sci.*, **A2** (1964), 2807 & 2817. p. 196.
26. Nandi, U.S., P. Ghosh and S.R. Palit, *Nature*, **195** (1962), 1197.
27. Mitra, B.C., S.C. Chadha, P. Ghosh and S.R. Palit, *J. Polym. Sci.*, **A1, (4)** (1966), 901.
28. Saha, M.K., P. Ghosh and S.R. Palit, *Bull. Chem. Soc.* (Japan), **29** (1966), 1336.
29. Waltz, J.E. and G.B. Taylor, *Anal. Chem.*, **19** (1947), 448.
30. Pohl, H.A. *Anal. Chem.*, **26** (1954), 1614.
31. Dainton, F.S., *J. Chem. Soc.*, (1952), 1533.
32. Simha, R., and L.A. Wall, *J. Research*, NBS, **41** (1948), 521.
33. Bawn, C.E.H., A. Ledwith and R. Mathies, *J. Polym. Sci.*, **34** (1959) 104.
34. Bro, M.I. and C.A. Sperati, *J. Polym. Sci.*, **38** (1959), 289.
35. Bevington, J.C., H.W. Melville and R.P. Taylor, *J. Polym. Sci.*, **12** (1954), 449; **14** (1954), 463.
36. Smith, W.V., *J. Am. Chem. Soc.*, **71** (1949), 4077.
37. Berry, K.L., and J.H. Peterson, *J. Am. Chem. Soc.*, **73** (1951), 5195.
38. Ghosh, P. and K.F.O'Driscoll, *J. Macromol. Sci. Chem.* **A1** (1967), 1393.
39. Rosenthal I., G.J. Frisone and J.K. Coberg, *Anal. Chem.*, **32** (1960), 1713.
40. Palit, S.R., *Makromol. Chem.*, **36** (1959), 89; **38** (1960), 96.
41. Palit, S.R., *Chem. Ind.*, (1960), 1531 S.R. Palit and P. Ghosh, *Microchem. Tech.*, **2** (1961), 663.
42. Ghosh, P., P.K. Sen Gupta, and A. Pramanik, *J. Polym. Sci.*, **A3** (1965), 1925.
43. Auerbach, M.E., *Ind. Eng. Chem. Anal. Ed.*, **15** (1943), 492.

44. Colichman, E.L., *Anal. Chem.*, **19** (1947), 490.
45. Mukherjee, P., *Anal. Chem.*, **28** (1956), 870.
46. Hollis, H.H. and W.R. Stone, *Analyst*, **82** (1957) 166.
47. Palit, S.R., B. Mukherjee and R.S. Konar, *J. Polym. Sci.*, **50** (1961), 45.
48. Ghosh, P., S.C. Chadha and S.R. Palit, *J. Polym. Sci.*, **A2** (1964), 4441.
49. Entwistle, E.R., *Trans. Faraday Soc.* **56** (1960), 284.
50. Grubisic, Z., P. Remp and H. Benoit, *J. Polym. Sci.*, **B5** (1967), 753.
51. Flory, P.J. and T.G. Fox, Jr., *J. Polym. Sci.*, **5** (1950), 745. *J. Am. Chem. Soc.*, **73** (1951), 1904.
52. Krigbaum, W.R. and D.K. Carpenter, *J. Phys. Chem.*, **59** (1955), 1166.
53. Ghosh, P. and G.S. Mukherjee, *Eur. Polym. J.*, **22** (1986), 103.
54. Eisenberg, A., K.F.O'Driscoll, and A.V. Tobolsky, *Anal. Chem.*, **31** (1959), 203.
55. Ghosh, P. and K.F.O'Driscoll, *J. Polym. Sci.*, **B4** (1966) 519 and P. Ghosh *J. Polym. Sci.*, **B7** (1969) 293.
56. Ghosh P., A.S. Bhattacharya and S. Maitra, *Angew. Makromol. Chem.*, **162** (1988), 135.
57. Alexander, L.E., *X-ray Diffraction Methods in Polymer Science*, Wiley-Interscience, New York, 1969.
58. Bovey, F.A., *High resolution NMR of Macromolecules*, Academic Press, New York, 1972.
59. Ghosh, P. and P.K. Sen Gupta, *J. Appl. Polym. Sci.*, **11** (1967), 1603.
60. McCall, D.W. and W.P. Slichter, *Newer Methods in Polymer Characterization*, B. Ke, Ed., Wiley-Interscience, New York, 1964.

CHAPTER 7

Polymer Rheology and Polymer Morphology

7.1 Polymer Rheology—Introduction

A material body in any physical state is subject to deformation under external stress. Deformations are primarily of two kinds: (i) ideal elastic deformation, which takes place instantaneously on application of stress and disappears instantaneously upon release of the applied stress (temporary and reversible deformation), and (ii) "flow", characterized by irreversible and permanent deformation. The study of deformation and flow characteristics of a material aimed at understanding and interpretation of the mechanical behavioural pattern in terms of the applied stress and consequent strain or strain-rate developed at a given temperature is commonly known as Rheology.

Under processing conditions and during performance, the polymeric materials (plastics and rubbers etc.) are subjected to different forms of stress and they undergo flow and other deformations. Hence, studies of polymer rheology is significant and very much relevant for understanding and control of their formulation with compounding ingredients and subsequent processing and for evaluation of the scope and limitations of their uses and applications. In fact, rheology is the science relating to the study of the inter-relationships between the flow and other deformations of matter and the factors causing and controlling the flow and deformation.

7.2 Stress and Strain

Distortion or deformation of a material of a given dimension are generally caused by three types of forces when applied on it—(a) tensile force, (b) compressive force and (c) shear force, and the nature of deformation that results in each case is shown in Fig. 7.1.

In general, stress is defined as the force applied, F, divided by the area, A, on which the force acts. For tensile and compressive stress, the area considered to define the stress is the cross-sectional area, A, perpendicular to the direction in which the force acts [Fig. 7.1(a) and (b)]. For calculation of shear stress, however, the area considered is the surface area, A which is parallel to the direction of the applied force, F [Fig. 7.1 (c)]. In either case the stress concerned is equal to F/A. The strain developed is expressed as the ratio of absolute distortion or deformation, ΔL in length divided by the original length of the unstressed sample, L_0, i.e. strain = $\Delta L/L_0$.

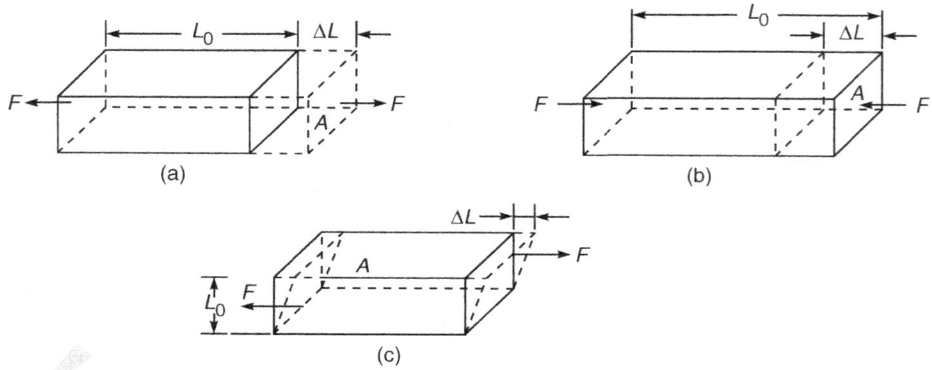

Fig. 7.1 ***Deformation of a rectangular body under (a) tension, (b) compression and (c) shear***

7.3 Ideal Elastic Solid

Let us now consider an ideal elastic solid and its deformation characteristics. A solid is considered as ideally elastic in which deformation under a specific kind of stress takes place instantaneously on application, and disappears completely and instantaneously on withdrawal of the deforming stress. Here, strain developed is time-independent and is directly proportional to the applied stress, i.e.

$$(F/A) = \text{constant} \cdot (\Delta L/L_0) \tag{7.1}$$

When the ideal elastic body is subjected to tensile (or compressive) stress, the proportionality is expressed as

$$\tau = E \cdot \gamma \tag{7.2}$$

where τ is the applied stress (tensile/compressive) in lb · in^{-2}, kg cm^{-2} or other appropriate units of force per unit cross-sectional area, γ is the axial strain ($\Delta L/L_0$), and E is the modulus of elasticity, commonly known as the Young's modulus. The proportionality law as defined above is known as Hooke's Law.

Likewise, if the ideal solid is subjected to a shear stress (τ_s), then the shear strain (γ_s) developed as a function of the stress applied is given by the expression

$$\tau_s = G\gamma_s \tag{7.3}$$

Here G, the proportionality constant is known as the shear modulus.

A constant stress of any kind applied to an ideal elastic solid, immediately deforms it to a fixed and constant level of strain in equilibrium with the applied stress and it will not deform or strain further with time. Hooke's Law usually holds for strain within 0.1%. Hookian (ideal) elastic deformation is characterized by: (a) strain–time and (b) stress–strain curves shown in Fig. 7.2.

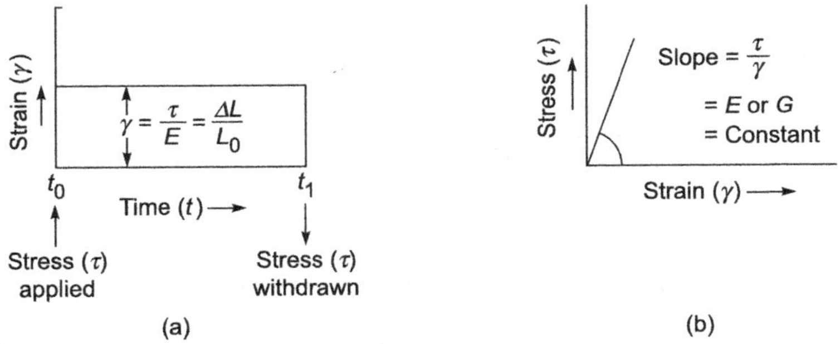

Fig. 7.2 ***Deformation characteristics of an elastic body: (a) strain vs time; (b) stress vs strain***

Just beyond the Hookian zone, characterized by the yield point, increase in stress on a solid leads to permanent deformation or flow, and on release of the stress, the deformed solid no longer snaps back to its original dimensions, having suffered a permanent set. A solid (elastic) body is, therefore, characterized by a finite yield stress. This is in sharp contrast with a fluid or liquid (viscous) body for which the yield stress value is zero. By mechanical analogy, the deformation behaviour of an ideally elastic body is that of a steel spring.

Elongation under tensile stress is simultaneously accompanied by a lateral contraction and the ratio of lateral to longitudinal strain, known as "Poisson's ratio" is expressed by the symbol ν. The value of ν is equal to, greater or less than 0.5 for density of the material remaining unchanged, increasing or decreasing on deformation under stress.

7.4 Ideal or Newtonian Fluid

When a fixed and constant stress is applied to a liquid or fluid body, it undergoes continuously increasing amount of strain or deformation which is non-recoverable on withdrawal of the stress. A liquid is thus a material of zero yield value in which the strain is a function of stress as well as time.

Let us now consider the liquid to be confined between two plates, the lower plate AB being stationary while the upper plate CD moves with a specific and constant velocity, V as shown in Fig. 7.3 under the influence of a constant shearing force F.

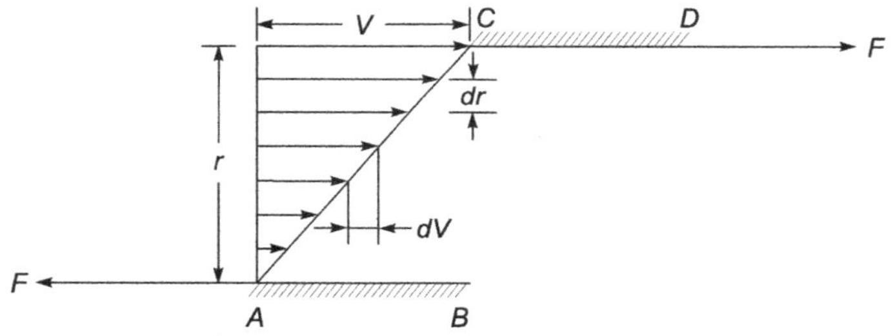

Fig. 7.3 *Flow of a liquid under shear*

If the liquid is ideal, the applied force will distribute itself uniformly across the layers or thickness of the material such that the shear stress τ_s at any point within the fluid mass maintained at a constant temperature is linearly and directly proportional to the velocity gradient such that one may write:

$$\tau_s = \eta\left(\frac{dV}{dr}\right) = \frac{1}{\phi}\left(\frac{dV}{dr}\right) \tag{7.4}$$

or

$$\eta = \frac{1}{\phi} = \frac{\tau_s}{\left(\frac{dV}{dr}\right)} \tag{7.5}$$

where η is the coefficient of viscosity or simply the viscosity or internal friction of the liquid and ϕ, the reciprocal of viscosity is its fluidity. The relationship given by Eq. (7.4) or (7.5) is known as Newton's law and liquids which behave in this manner are called Newtonian fluids or ideal fluids.

Further, the differential shear strain ($d\gamma_s$) in the material recurring across the differential fluid layer of thickness (dr) over the differential time period (dt) is given by

$$d\gamma_s = \left(\frac{dV}{dr}\right) dt; \qquad \text{or,} \quad \left(\frac{d\gamma_s}{dt}\right) = \left(\frac{dV}{dr}\right) \tag{7.6}$$

and hence

$$\tau_s = \eta\left(\frac{d\gamma_s}{dt}\right) \tag{7.7}$$

Equation (7.7) states that the shear stress (i.e., the tangential force per unit area of the surface on which the force acts) required to shear a Newtonian fluid is linearly and directly proportional to the shear strain rate. The variations of deformation or flow with time and of the shear strain rate or simply the shear rate developed with shear stress are shown in Fig. 7.4.

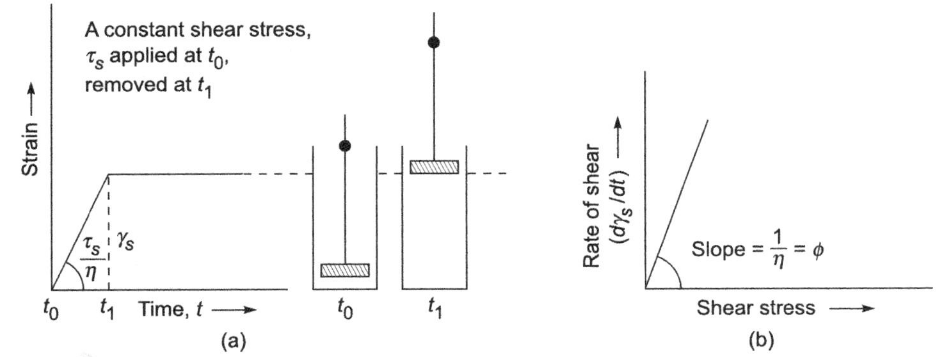

Fig. 7.4 *Flow of a Newtonian fluid: (a) strain vs time; (b) rate of shear vs shear stress*

A Newtonian liquid, being completely devoid of elasticity, is considered as a purely viscous material, and by mechanical analogy, its deformation or flow under stress is represented by a weightless piston moving in a cylinder or a dash pot filled with a medium that offers some resistance [Fig. 7.4(a)]. The amount of flow, γ_s (i.e., strain) is a linear function of time, t, i.e.

$$\gamma_s = \frac{\tau_s}{\eta} \cdot t = \phi \cdot \tau_s \cdot t \tag{7.8}$$

The coefficient of viscosity η is a true constant of the fluid body independent of the shear stress and shear rate or rate of flow (without turbulance) at which it is measured and does not depend on the prior shear deformation history of the fluid body. Plot of shear rate vs shear stress shows a straight line passing through the origin for which the slope represents fluidity or reciprocal of viscosity [Fig. 7.4(b)]. Hence, higher the slope, higher is the flow rate or shear rate and lower the viscosity.

7.5 Non-Newtonian Fluid

The flow behaviour of a real fluid is best understood by graphical comparison of the respective shear stress vs shear (strain) rate curves with the straight line through origin that a Newtonian (ideal) fluid produces using common axes (Fig. 7.5). From a look at the flow curves shown in the figure, one can observe and appreciate how the flow behaviours of some real or non-Newtonian fluids (polymer melts, solutions, pastes and dispersions) differ from that of an ideal or Newtonian fluid. In general, one finds that the relationships for plastic or polymeric fluid systems are not perfectly linear, but they produce curves which are concave upward or concave downward or more complex in nature. For such anomalous or non-Newtonian systems, the coefficient of viscosity depends on the stress, strain rate and previous shear history of the sample.

A Bingham body represents the case of an ideal plastic material which supports a finite stress before a strain is initiated. Thus, after the material reaches a definite stress level known as the yield stress, $(\tau_s)_y$, a finite strain rate is generated; the stress–strain rate relationship is assumed to be linear, as in Newtonian fluids, but

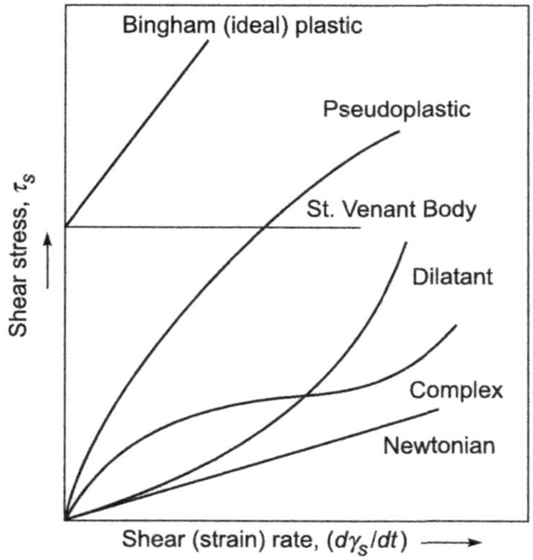

Fig. 7.5 *Flow curves, τ_s vs $(d\gamma_s/dt)$, for different types of fluid material*

only after attainment of the yield stress $(\tau_s)_y$. The behaviour of most real plastics is, however, characterized by non-linear relationships which are concave downward as shown by the pseudoplastic behaviour in Fig. 7.5. Most plastic melts, solutions or dispersions exhibit pseudoplastic flow patterns. At rest, long chain molecules of a plastic melt, solution or dispersion are believed to engage each other into stable associations as a consequence of chain entanglements and due to molar cohesion. At low shear rates, the stress required to overcome or undo the effect of chain entanglements is relatively high, but proportionately lower stress is required to allow the melt, solution or dispersion flow at a higher strain rate once the molecules are set in motion with respect to each other. The entanglement effect becomes virtually non-existent and almost Newtonian behaviour assumes prominence at high shear rates depending on the exact nature of the fluid materials. If allowed to stand at rest thereafter, the fluid materials set again. Such isothermal reversible sol–gel transformations are known as *thixotropy*.

An altogether different kind of flow behaviour is characterized by a shear stress vs shear rate curve that is concave upward. The fluid system of this nature is known as dilatant. Such behaviour is exemplified by some highly concentrated suspensions of certain fine powders in appropriate liquids and by some melts in rare cases where crystallization or structure formation occurs at high rate of shearing. Such systems usually resemble a Newtonian fluid at low shear rates, but beyond a specific shear rate, they sharply gain in viscosity which may finally reach a very high value. When set at rest again, they return to the flow behaviour of a near-normal viscous liquid. This reversible flow pattern is called *dilatancy*.

Pseudoplastics are *shear-thinning* while dilatants are *shear-thickening* fluid systems.

Yet another type of flow behaviour is that exhibited by what is known as the St. Venant body, Fig. 7.5. Such a material is characterized by some yield stress value $(\tau_s)_y$ before the attainment of which the material is assumed to undergo no strain, but after which it cannot support any further stress and it will then shear at any rate from zero to infinity to disallow the stress on the material to rise beyond the critical value.

A more anomalous flow pattern for some non-ideal fluid systems may combine pseudoplastic, dilatant and Newtonian characters at different shear rate zones and it is dependent on the structure of the fluid in equilibrium with a given stress and strain rate condition. The flow behavioural pattern, being dependent on the structure of the material, will also change with the temperature of the system.

Diverse flow behaviours or requirements are encountered or met during processing of plastics and rubbers at different temperatures and pressures by various approaches and techniques, such as extrusion, pressing and moulding by compression and injection techniques, calendering, blow moulding, thermoforming, lamination and coating, brushing and spraying, etc. Each process or technique involves, at different stages of the operation, a critical and delicate relation between temperature, applied stress and inherent fluid property such as the viscosity of the material. A thorough evaluation and understanding of the rheological characteristics are necessary prerequisites for satisfactory operation and control of such processes.

The reduction of viscosity with increasing rate of shear is taken advantage of in achieving desirable and optimum viscosity for polymers in conversion machineries and equipments without raising the temperature to detrimental levels, simply by controlling the shear rate to as high a level as economically and otherwise permissible. A reduction in viscosity with increasing shear rate is also taken advantage of in brushing and spraying of paints which are polymer solutions/ dispersions containing pigments. Shearing with a brush at high rates reduces viscosity, thus facilitating adequate spreading and coverage and ensuring recovery of much higher viscosity when left at rest after brushing to prevent running down or dripping and consequent surface defects under very much smaller gravitational forces.

7.6 Apparent Viscosity

For a better description and appreciation of flow behaviour, it is sometimes useful to plot the apparent viscosity of the fluid materials as a function of the applied shear rate. The apparent viscosity at a specific shear rate is given by the ratio of the shear stress to the shear rate. The general differences in flow behaviours of Newtonian, pseudoplastic and dilatant fluid systems are quite clearly seen from a plot of $\tau_s/(d\gamma_s/dt)$ vs $(d\gamma_s/dt)$ (Fig. 7.6).

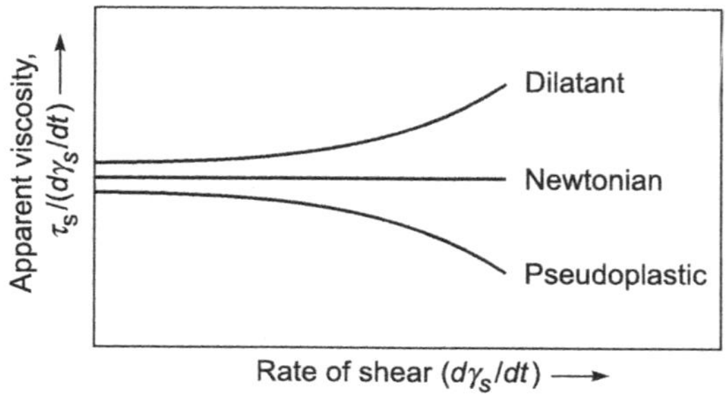

Fig. 7.6 *Plot of apparent viscosity vs rate of shear (schematic)*

7.7 The Power Law

Different mathematical relationships have been put forward to describe the flow behaviour of different kinds of viscous solutions, melts, pastes and other dispersions quantitatively. But very few of them are useful in rheological studies and calculations because of limitations. A simple equation, commonly referred to as Ostwald de Waale or Power Law equation, has, however, been accepted to be of general relevance and applicability.

The equation relates shear stress τ_s and shear rate $(d\gamma_s/dt)$ in the manner given by:

$$\tau_s = K\left(\frac{d\gamma_s}{dt}\right)^n \tag{7.9}$$

Here, K is a constant and the power or exponent n is a number which may be equal to, greater or smaller than unity. For a Newtonian fluid, the expression reduces to the form as given by Eq. (7.7):

$$\tau_s = \eta\left(\frac{d\gamma_s}{dt}\right)$$

Thus, for a Newtonian fluid, the power or exponent n is exactly equal to unity and the constant K is given by the coefficient of viscosity, η. The value of n for a real fluid gives a relative measure of how its behaviour deviates from that of an ideal (Newtonian) fluid characterized by an n value of unity. Thus, n is commonly referred to as the flow behaviour index of a fluid which can be considered as an inherent property of the fluid, just like the modulus or coefficient of viscosity of material systems. Equation (7.9) may be rewritten to give Eq. (7.10).

$$\log \tau_s = \log K + n \log (d\gamma_s/dt) \tag{7.10}$$

Equation (7.10) gives a linear relationship between $\log \tau_s$ and $\log (d\gamma_s/dt)$ and the slope of the experimental plot (Fig. 7.7) gives the value of n. For the analysis of flow

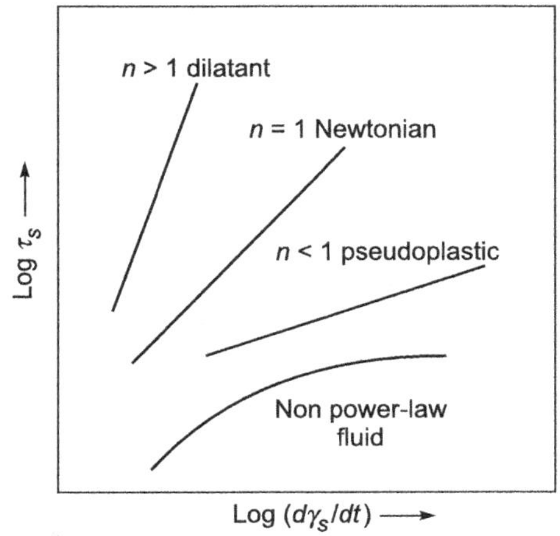

Fig. 7.7 *Power Law plot showing $\log \tau_s$ vs $\log (d\gamma_s/dt)$ for different types of fluid material (schematic)*

behaviour of many systems, the Power Law relationship has been useful and from a plot of data as in Fig. 7.7 and measuring the slope, one can readily get an idea of just how non-Newtonian the fluid is. A value of $n > 1$ indicates dilatant behaviour while $n < 1$ indicates pseudoplastic behaviour.

7.8 Free Volume or Molecular Hole Concept

The flowability or fluidity of a material is considered[1] to be a direct function of the free volume or space in the liquid. The free volume may be calculated from volume difference by comparing the volume occupied by the solid material from the volume of the same amount of liquid and the molecular holes or free volumes are caused by thermal expansion. Larger availability of space for movement is manifested in lower resistance to flow for many systems and the general relation: Fluidity = (Constant × Free volume) immediately follows. According to the free volume theory, any factor that reduces the free space increases the viscosity of a liquid. Thus, external pressure reduces the amount of free space by forcing the molecules together and hence reduces the fluidity or increases the viscosity. Rise in temperature leads to volume expansion, i.e., increase in free space and hence lowering in viscosity.

For materials of low molecular weight, the dependence of viscosity on absolute temperature is given by the relationship:

$$\eta = A \exp (b/T) \tag{7.11}$$

where A and b are constants. Molecules of a liquid of low molecular weight are considered to be roughly spherical and according to Eyring, the molecules in the liquid will pack together in a fashion much similar to that in the solid or the crystalline state though with somewhat lower degree of regularity. According to Eyring,[2] a liquid is visualized as a mass of regular array of molecules with certain holes, vacant sites or free space between molecules here and there, as shown in Fig. 7.8.

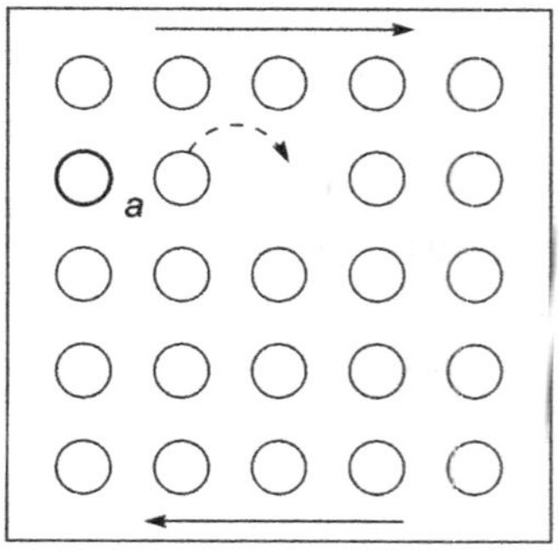

Fig. 7.8 *Model of a liquid as a mass of regular array of molecules with occasional vacant sites or holes*

The molecules under prevalent thermal condition normally remain confined to fixed mean positions under the influence of forces exerted by the surrounding molecules. However, once a while, a molecule adjacent to a vacant site such as the molecule 'a' in the figure will overcome the influence of its neighbouring molecules to jump into the hole or vacant site, thus creating a new hole in its previous position. Thus, the structure and relative position of vacant sites undergo constant change as a consequence of jumping of molecules into holes. The changes would be of random nature in the absence of any external or applied stress and no overall flow would result in view of a lack of direction of jumping. When a shear stress is applied, the random effect gives way to a directional effect and jumping to neighboured vacant sites favoured in the direction of applied stress is manifested into a resultant bulk movement or flow in that direction.

This molecular model to depict flow considers that flow arises as a consequence of independent jumping of individual molecules from one equilibrium position to another in a stepwise manner, but not due to the simultaneous sliding of all the molecules in the layer. To break away from the binding forces of the neighbours a jumping molecule must acquire certain amount of energy, called the activation energy. Higher temperature raises the probability of a given molecule acquire the required energy. Equation (7.12) is an equation of the Arrhenius type relating the viscosity (η), to the frequency of jumping (A) at the absolute temperature (T), E being the activation energy and R the universal gas constant. From measurements of η at different temperatures for the melt of a given material and subsequent graphical treatment of data as a plot of log η vs $1/T$ (Fig. 7.9), one can calculate the jump energy or the activation energy E and thus, a measure of the magnitude of the cohesive forces holding the molecules in their normal positions in the liquid state may be obtained.

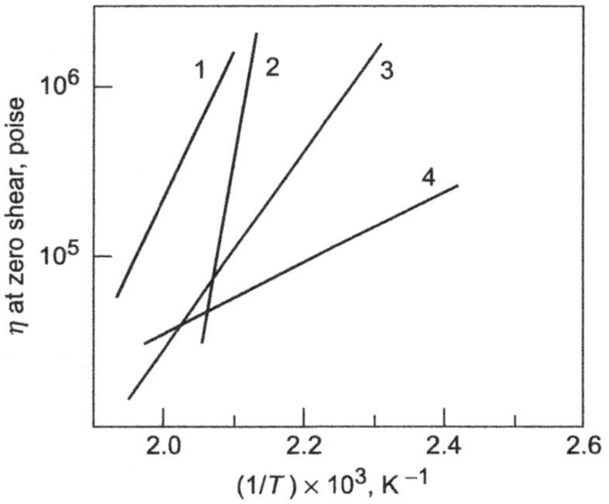

Fig. 7.9 *Effect of temperature on the viscosity of some polymers; (1) PMMA, (2) ethyl cellulose, (3) polystyrene, and (4) polyethylene*

$$\eta = A \exp (E/RT) \tag{7.12}$$

Extension of the Eyring theory to polymers should be considered keeping in mind that polymer molecules are not spherical. They are rather very long (high length to diameter ratio). Experiments on temperature dependence of melt viscosity, however, show that the E values for polymer melts, are not much higher than for low molecular weight liquids or melts of comparable chemical nature and composition, indicating that the flow units in the two systems are of comparable dimensions.

In fact, in the case of a polymer system, it is not the whole molecule but a short segment of the chain molecule that moves at a given instant. It is a natural consequence of the concept of chain-like structure of randomly joined links for polymer molecules that a specific chain cannot move as a single entity but that separate short chain segments move practically independently of each other.

The temperature dependence of melt viscosity and the activation energy obtained from this study are both by and large independent of chain length but the melt viscosity as such is largely dependent on chain length as the viscosity value is related to the mobility of the chain molecule as a whole.

7.8.1 Williams–Landel–Ferry (WLF) Equation

In polymer systems, the glass transition corresponds to onset of long-range liquid like mobility of segment of molecules characterizing the rubbery state. The enhanced mobility at or above T_g necessitates utilization of more free space or volume than that required and utilized for short-movement of atoms in the glassy state. The enhancement in the relative free volume with rise in temperature (T) above the glass transition temperature, T_g (see Sec. 7.20.2) leads to a much enhanced volume expansion coefficient in this region. The fraction f of the "free" volume may be expressed as:

$$\text{(i) } f = f_g, \text{ for } T < T_g, \text{ and (ii) } f = f_g + (T - T_g)\,\Delta\,\alpha, \text{ for } T \geq T_g$$

Thus, f is a constant at its minimum value for all temperatures below T_g which then becomes a reference point; above T_g, additional free volume created by the thermal effect becomes dependent on the increase ($\Delta\,\alpha$) in the expansion coefficient such that $f = f_g + (T - T_g)\,\Delta\,\alpha$ for $T \geq T_g$.

WLF equation relates log of relative viscosity (η/η_g) with the temperature difference ($T - T_g$) by an empirical expression as shown below:

$$\log \frac{\eta}{\eta_g} = \frac{-a(T - T_g)}{b + (T - T_g)} \tag{7.12a}$$

Analysis by fitting available data on η–T behaviour of many glass forming polymers fixes the values of the numerical constants a and b in the WLF equation as 17.44 and 51.6 respectively. It is thus implied that the viscosity of the polymer should become infinite at $T = (T_g - b) = (T_g - 51.6)$, indicating that for polymers, all molecular motions should become totally frozen at $T < (T_g - 51.6)$. What actually happens is that altogether new mechanism of deformation takes over or sharply becomes overwhelmingly prevalent as the temperature falls below or much below T_g.

7.9 Viscosity as a Function of Molecular Weight

The relationship between viscosity and molecular weight for most polymers examined by a graphical logarithmic plot as shown in Fig. 7.10 is characterized by a sharp change at a particular value of molecular weight. The slope of the two linear portions of the plot below and above this critical molecular weight, M_c are about 1.75 and 3.4 respectively. The value of M_c varies from one polymer to another. The critical molecular weight corresponding to the transition in the viscosity behaviour at low rates of shear or zero shear rate points to additional hindrance to flow from this point onwards due to chain entanglements. Below this point the molecules usually move independently as in low molecular weight liquids, but above this point mutual entanglements of chain molecules is so prominent that movement of one involves dragging of others along with it; as a consequence, very high rate of viscosity increase follows with increasing molecular weight.[3–5]

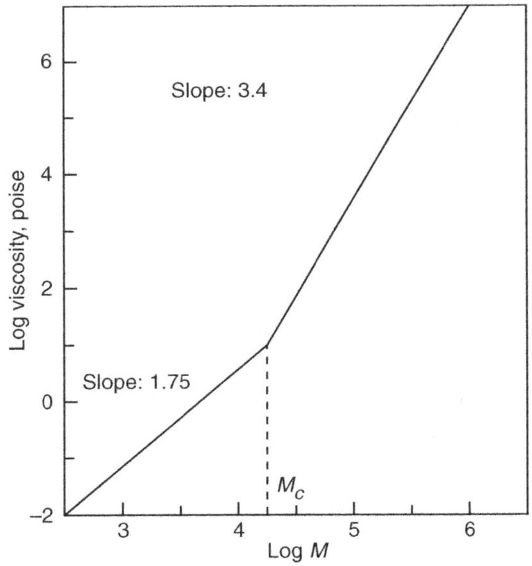

Fig. 7.10 *Dependence of polymer (melt) viscosity on molecular weight (M): a typical plot of log viscosity against log M*

7.10 Weissenberg Effects

An elastic after-effect is generally found in high molecular weight fluid materials after extrusion under high shear stress through an orifice or die and this is seen to happen within fraction of a second after extrusion. Herzog and Weissenberg[6] observed the existence of a "normal" force in polymers subjected to shear stress. In the polymeric melt systems, the entangled molecular structure gets deformed elastically during flow and a different kind of stress is generated within the flowing melt in addition to the shear stress applied. The additional stress generated is manifested as tensile stress acting on planes perpendicular to and along a direction parallel to the direction of sliding.

The effect, known as the Weissenberg effect is easily demonstrated by subjecting a viscous liquid mass to shear between a coaxial cylinder system by rotating one while keeping the other fixed Fig. 7.11. It is most easily demonstrated by rotating the inner cylinder. On rotation, the liquid climbs up the rotating inner rod or cylinder to a significant height. The tensile stress generated on shearing action cannot be directly observed; but consequent to its action round the circumference of a circle, it gives rise to an internal pressure within the liquid under the shear, as if the liquid were encompassed by stretched elastic bands. This pressure forces the liquid up the surface of the inner cylinder.

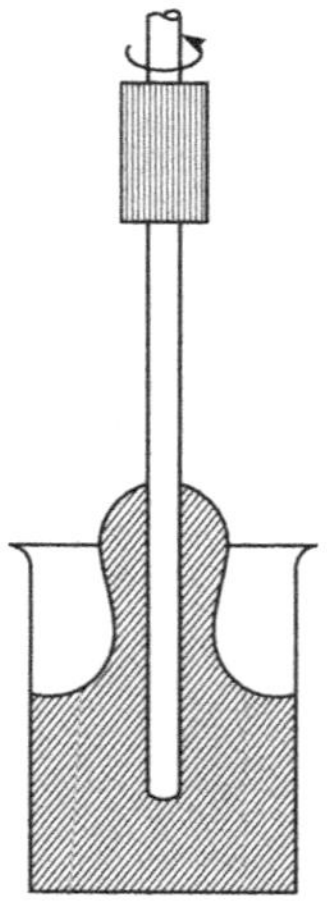

Fig. 7.11 *Demonstration of liquid climb in a coaxial cylinder system on rotation*

The Weissenberg effect is clearly manifested in the increase in diameter of extruded profiles of a variety of molten polymers. In certain cases the spontaneous increase in diameter is favoured with an increase in temperature while in certain other cases, it may not be so. For high polymers, higher molecular weight and lower temperature lead to higher viscosity of melts and there will be an intermediate molecular weight range and temperature for maximum expansion or ballooning of extruded profiles.

The extrusion swelling, more commonly known as "die swell" arises probably due to a combination of normal stress effects and of a possible elastic recovery consequent to prior compression before the melt or liquid enters the die. However, expansion or swelling takes place even if the elastic recovery effect is eliminated by using a long tubular die such that the flow at the exit point is practically unaffected by the stresses due to compression before entry into the long tube used as the die.

A practical aspect of the Weissenberg effect or die swell is that on extrusion or calendering, the melt coming out is larger than the die or nip size. The control of size is done by partly using a smaller aperture and by partly having an increased draw down. In extrusion, the swelling may be largely minimized by having a conical die design with the narrow end towards the interior and by lowering of temperatures, extrusion rates or molecular weight of the polymer.

7.11 Measurement of Flow or Rheological Properties of Fluids

Instruments or equipments for measurement of rheological properties of fluids must necessarily accomplish two things: (a) the fluid must be sheared at measurable rates, and (b) the stress developed must be known. Two kinds of instruments having simple geometry and widely in use are: (i) rotational viscometer and (ii) capillary or extrusion rheometer.

A rotational viscometer (using cylinders, cones, spheres and discs) simply resembles the co-axial cylinder system described in Fig. 7.11. The fluid is sheared at a given temperature in the annular or enclosed space due to rotation of the inner cylinder or the like device while the outer cylinder or device is kept stationary or vice versa; and in either case, the difference in torque required for the rotation is a measure of the shearing stress and the speed of rotation gives a measure of the rate of shear. The principle is used in the design and functioning of the Mooney viscometer used in the studies of the flow or rheological characteristics of rubber and rubber compounds. Brabender plastograph or rheotron based on similar design and working principle is useful in the studies of melt and solution rheology of thermoplastics and elastomers.

Capillary or extrusion rheometers are useful in the studies of melt viscosity of thermoplastics and selected elastomer systems if selection of appropriate ranges of temperature and shear rate is permissible. A sample in powder, granule or other form is loaded into an extrusion cylinder or chamber and heated to a specific temperature using thermostatic control. On attainment of temperature equilibrium, the melt is forced by a plunger through a cylindrical or capillary die or orifice attached to the bottom. The pressure on the plunger is a measure of the shear stress. The plunger is allowed to move at constant pre-set speed covering the desired range of shear rate. The amount of fluid extruded per unit time is a measure of the rate of shear. A very simplified and elementary model of this instrument is used for gradation of polyolefins and related thermoplastics by what is known as the melt flow index (MFI) given by the amount of polymer collected under a given condition of pressure and temperature through a specified cylinder and capillary over a specified time period. (*See also* Sec. 10.5.)

7.12 Melt Fracture and Irregular Flow

Above a certain shear stress considered as the critical shear stress, instabilities in the flow of many polymer melts exhibited by an abrupt change in the shape of the molten stream extruded out of a capillary rheometer are found to occur. The irregular flow pattern can be conveniently photographed using a glass rheometer. The irregularities may suggestively be due to "Reynolds Turbulence" or structural turbulence and thermomechanical breakdown of the polymer.[8,9] It was, however, noted that for a given polymer and a given die or orifice, the critical shear stress was the same for different temperatures. Also, the adhesion of the polymer melt to the capillary wall suffers adversely and it gets more sinewy at very high shear stress as a consequence of flow orientation. Above the critical shear stress, the material near the capillary wall relaxes very much faster than the core material leading to the flow irregularities, which, according to Tordella, is caused by a fracture of the melt before its entry into the capillary tube.[10] Melt fracture depends on die geometry, molecular weight and molecular weight distribution and chain branching. A linear polymer such as high density polyethylene is characterized by a higher critical shear stress than the corresponding branched polymer (viz., low density polyethylene) of comparable molecular weight.

The origin of flow irregularities and the site of melt fracture is at the entrance to the die, since this is the zone of greatest shear stress. The stress enhancement is so great compared to the relaxation time that the polymer melt fractures much like a solid.

7.13 Time-Dependent Flow

In addition to the time-independent flow of some fluids discussed so far, there are some pseudoplastic or other kinds of non-Newtonian fluid materials which display time dependency in their flow behaviour.

For requirement of a material that must stay true to dimension during use, considerations of its yield value (yield stress) are of primary practical importance. Rheologically, a solid is distinguished from a liquid as a substance having a finite yield value. Flow in solid systems occurs by movements of blocks of molecules because of their so-called 'fixed' structures, characterizing them by a finite yield value. On the other hand, liquid or fluid systems permit movement of individual molecules and thus they are characterized by a zero or infinitesimal yield stress.

7.14 Viscoelastic Behaviour

Elastoviscous liquids of viscoelastic solids are liquids or solids, respectively, which exhibit a combination of properties of elastic solids and truely viscous liquids. Their flow behaviour is time-dependent in nature and is characterized by creep and relaxation phenomena.

Creep or *cold flow* of a viscoelastic material is the long-time continuous deformation or strain that occurs in it under the influence of a constant stress (less than the short-time yield value). Creep or creep strain may be either fully recoverable with time or involve permanent deformation with part recovery.

Relaxation is a flow process that takes place in a viscoelastic material in which the stress needed to keep the material extended or strained to a constant amount of deformation or strain decreases or decays with time. The stress on the test piece may slowly drop to zero as it undergoes a permanent set to the stretched or extended length.

Metals and low-molecular weight crystalline solids undergo elastic deformation and can be strained to merely less than 1% before the elastic limit is attained. The deformation results mainly from distortion of normal bond lengths and bond angles. Polymeric solids, however, undergo much larger elastic deformations. Their chain molecules remain in a highly coiled or folded form. If a specific coiled position corresponds to the position of minimum free energy, application of external tensile stress causes temporary uncoiling or unfolding of the chain molecules leading to elongation of the test specimen, followed by a return to he original dimension as the stress is withdrawn. The retraction is aided by chain entanglements, limited cross-linkage and thermal agitation (*See also* Sec. 1.6.7). Reversible elongation as high as 1000% are exhibited by some rubbers. The *kinetic* or *long-range elasticity* of rubbers involves a change in the structure of the molecular aggregates leading the amorphous mass of unstretched rubber to a crystalline, fibrous mass of distinctly different properties on stretching to high elongations (~1000%), (*See* also Fig. 6.20 and explanation thereof).

7.15 Mechanical Model of a Viscoelastic Material

A viscoelastic body is assumed to be composed of an elastic element represented by a steel spring and a viscous element represented by a piston and dashpot. A simple model shown in Fig. 7.12(a) combines the pair placed in series (Maxwell unit), and another, shown in Fig. 7.12(b), combines them in parallel (Voigt unit).

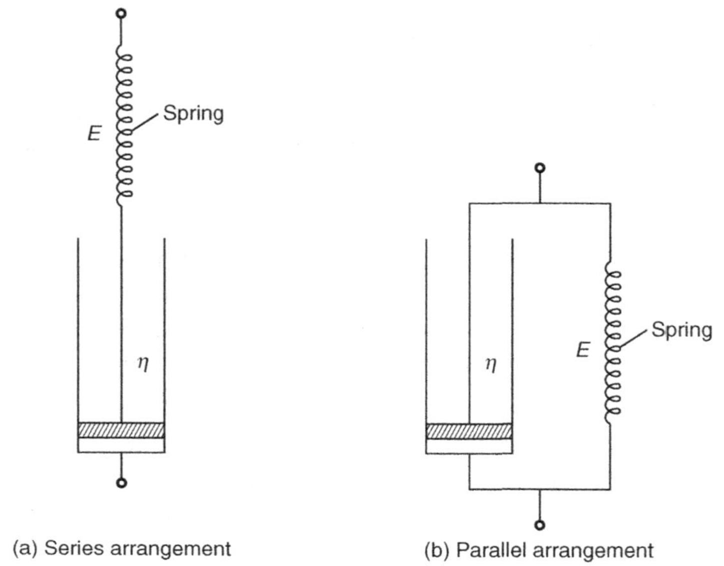

Fig. 7.12 ***Mechanical model of relaxation using an elastic component (spring) and a viscous component (dashpot with a frictionless piston) placed in (a) series, and (b) in parallel***

For a Maxwell unit exemplified by a fibre containing crystalline and amorphous regions, the overall rate of flow combining the contribution of the elastic and viscous elements, is expressed as:

$$\frac{d\gamma}{dt} = \frac{1}{E}\cdot\frac{d\tau}{dt} + \frac{\tau}{\eta} \tag{7.13}$$

The corresponding equation for shear stress τ_s is

$$\frac{d\gamma_s}{dt} = \frac{1}{G}\cdot\frac{d\tau_s}{dt} + \frac{\tau_s}{\eta} \tag{7.13a}$$

On rearrangement, Eq. (7.13) may be written as:

$$\frac{d\tau}{dt} = E\cdot\frac{d\gamma}{dt} - \frac{E}{\eta}\tau$$

$$= E\cdot\frac{d\gamma}{dt} - \frac{\tau}{\lambda} \tag{7.14}$$

where $\lambda = \eta/E$ is a characteristic constant of the selected material, and is commonly called the *relaxation time*.

7.16 Decay of Stress under Constant Strain (Stress Relaxation)

On application of an initial stress τ_0 on a viscoelastic material represented by a Maxwell unit, the deformation will instantly occur entirely in the spring (i.e., the elastic element). If held at a constant strain, giving $(d\gamma/dt) = 0$, flow takes place in the viscous element (causing the piston to move) and consequently progressive decay of stress takes place. The residual stress at any time t after application of the stress is obtained by integrating Eq. (7.14), given, $(d\gamma/dt) = 0$, and one obtains

$$\tau = \tau_0 e^{-t/\lambda} \tag{7.15}$$

The relaxation time λ is the time over which the stress decays to $1/e$ of its original value.

For a viscoelastic material combining an elastic and a viscous element in parallel, as in Fig. 7.12b, stress required to maintain a fixed strain would depend on the rate of the deformation. The stress reaches a constant value as the desired strain is achieved, and thus, the system is not associated with a relaxation effect under this condition.

7.17 Relaxation or Strain Enhancement under Constant Stress

For a material having the elastic and viscous elements in series (Fig. 7.12a), subjected to a constant stress, one obtains on integration of Eq. (7.13):

$$\gamma = \frac{\tau}{E} + \frac{\tau}{\eta}t = \frac{\tau}{E} + \frac{\tau}{E}\cdot\frac{t}{\lambda} \tag{7.16}$$

Equation (7.16) describes the time dependent flow behaviour under the given condition and (τ/η) or $(\tau/E\lambda)$ is taken as the creep rate. The flow curve is shown in Fig. 7.13(a).

For a material with parallel positioning of the elastic and viscous elements [Fig. 7.12(b)], the viscous element starts deforming (i.e., the piston moves) at a rate given by τ/η, but as the stress is progressively transferred to the spring, the viscous flow rate decreases to zero asymptotically as shown in Fig. 7.13(b). At any time, the stress on the viscous element is given by the difference between the external stress and the stress transferred to the elastic element, and the flow rate is given entirely by the rate of flow of the viscous element. The overall deformation observed is thus a retarded elastic deformation γ_R which is expressed as

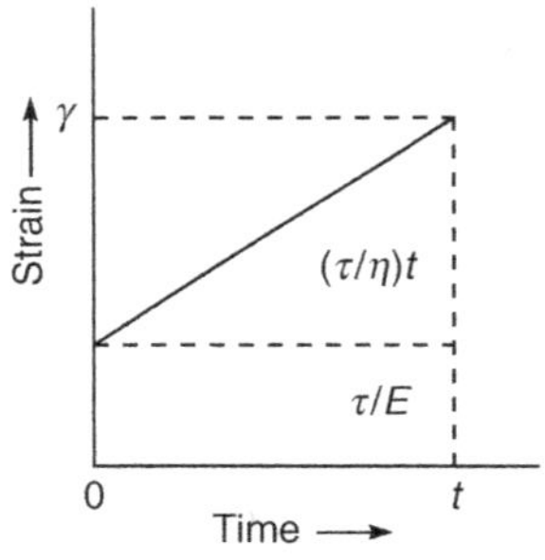

τ/E = Hookian elastic deformation
(τ/η)t = Deformation due to viscous flow
γ = Total deformation
(a)

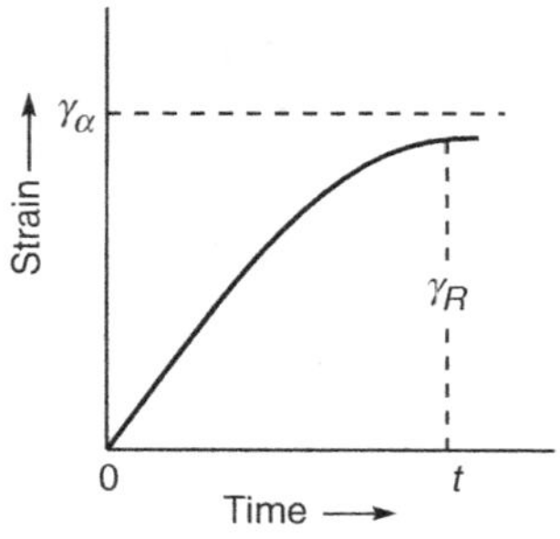

γ_R = Retarded elastic deformation
$\gamma_\alpha = \tau/E$
$\gamma_R = (1 - e^{-t/\lambda})\gamma_\alpha$
(b)

Fig. 7.13 *Constant stress flow curves for (a) a Maxwell unit, and (b) a Voigt unit*

$$\gamma_R = \frac{\tau}{E}[1 - e^{-(E/\eta)t}] = \gamma_\infty(1 - e^{-t/\lambda}) \tag{7.17}$$

Here, $\gamma_\infty = \tau/E$ corresponds to the elongation or strain at infinite time.

In practical polymers—rubbers, plastics, fibres and the like, showing viscoelasticity and characterized by creep, relaxation time and fatigue—the yield value is a function of time and rate of loading. The relaxation effect is more prominent in less crystalline materials. The larger the t/λ ratio [Eqs (7.15)–(7.17)], the more viscous the material behaviour; and the smaller the ratio, the less the relaxation and the more elastic the material behaviour.

7.18 Hysteresis

For all viscoelastic (solid) materials, a measurable part of the work expended in deforming the solid is dissipated as heat (*See also* Sec. 1.6.6). It is a common experience that the load-elongation curve followed during loading or extension is not the same as that during unloading or retraction, and the loading–unloading cycle is schematically shown in Fig. 7.14. The work of stretching during loading is measured by the area under the curve of extension or loading. Similarly, the area under the unloading curve gives a measure of the work of retraction. The area encompassed between the loading and unloading curves, called the hysteresis loop gives a measure of the mechanical energy dissipated as heat. The hysteresis loss for an ideally elastic (Hookian) body is zero as the loading and unloading curves are one and the same and there is no loop. For non-Hookian, i.e., imperfectly elastic or viscoelastic bodies, the hysteresis loop is measurable and the area of the loop is dependent on the degree of structural changes brought about during extension due to uncoiling or unfolding of the molecular chains. The loop shown schematically in Fig. 7.14 is negligible for steel and very prominent and significant for vulcanized rubbers. For a permanent set, the unloading curve would somewhere cut the strain axis. The hysteresis loop shown in Fig. 7.14 indicates no permanent set or deformation.

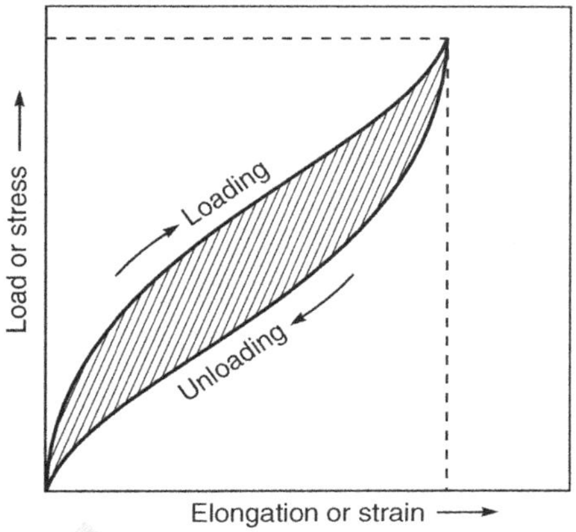

Fig. 7.14 *Hysteresis loop (schematic representation) for vulcanized rubber*

7.19 Creep and Relaxation of Typical Plastics

A typical experimental curve for initial flow and relaxation of many plastics over a cycle of loading and unloading is shown in Fig. 7.15.

On loading or stress application, there appears an instantaneous (elastic) deformation given by OA. Then a retarded deformation or creep follows up to the point B. On removal of the stress, as the point B is reached at time t_1, an instantaneous and partial recovery BC_1 takes place followed by a time dependent recovery (creep recovery) from C_1 to D over the time period t_1 to t_2: further recovery is negligible such that DE is the permanent deformation left after the loading-unloading cycle. The creep AB consists of two components—recoverable and non-recoverable—which become apparent during the unloading (creep recovery) stage. The primary creep (C_1C_2), representing the elastic component, is recoverable, but being time-dependent in nature, it is different in nature from the instantaneous

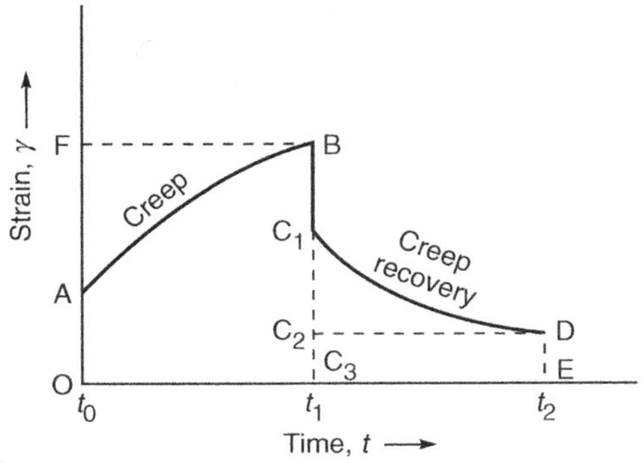

Fig. 7.15 *Creep and relaxation phenomena of a typical plastic material*

(Hookian) recovery. The non-recoverable creep C_2C_3 which is equal to DE is commonly known as secondary creep. The three components of the overall deformation (OF) may be identified as: (i) BC_1 (equal to OA), corresponding to elastic deformation, (ii) C_1C_2, corresponding to retarded elastic deformation or primary creep, and (iii) C_2C_3, corresponding to viscous deformation, flow or a blockade retarded deformation or secondary creep. The overall deformation may be understood and interpreted using mechanical models comprising spring and dashpot in appropriate series and parallel combinations with or without a blockade.

7.20 Transition Phenomena in High Polymers

Raising or lowering of temperature, just as application or withdrawal of stress, greatly influences the physical structure and properties of polymers. As temperature is changed, a high polymer material usually passes two transitions characterized by (i) melting point or first order transition corresponding to the melting temperature denoted by T_m, and (ii) the second order transition or the glass transition corresponding to the temperature denoted by T_g.

7.20.1 Melting Point and First Order Transition

Melting of a crystalline solid or boiling of a liquid is associated with change of phase. Many high polymers possess enough molecular symmetry and/or structural regularity that they crystallize sufficiently to produce a solid–liquid phase transition phenomenon or melting, i.e., they have a crystalline melting point, T_m. The melting is quite sharp for a few polymers such as, the nylons, while in most other cases as for polystyrene, different rubbers, etc., the phase change takes place over a range of temperature. Phase transitions of this kind, particularly in low molecular weight materials, being associated with sharp discontinuities in some primary physical properties such as the density or volume, V, [$V = (\partial G/\partial P)_T$] and entropy, S, [$-S = (\partial G/\partial T)_p$], which are first derivatives of free energy, are commonly called first order transitions, (Fig. 7.16). Although we observe

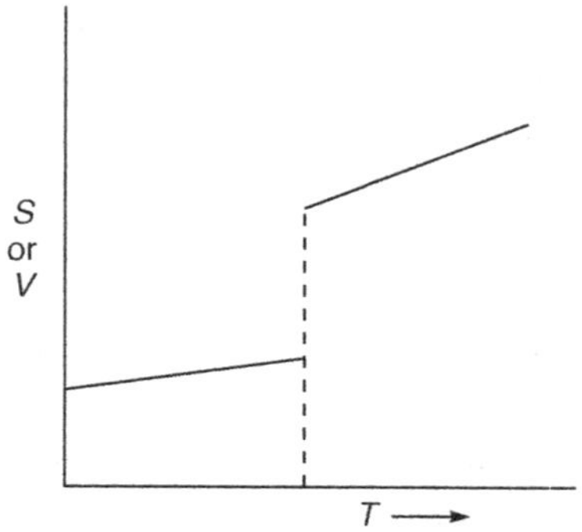

Fig. 7.16 *First order transition showing an idealized phase transition: Trend of change of entropy (S) or volume (V) with temperature (T)*

melting, a true first order transition or ideal melting in high polymers is frequently absent primarily in view of the distribution of molecular weight and entanglements of chain molecules in the material system.

7.20.2 Glass Transition or Second Order Transition

Glass transition or second order transition is not a phase transition and almost each polymeric or high polymeric material is characterized by a specific glass transition temperature (T_g), appearing well below its (crystaline) melting point T_m.

At T_g, the variables S, V and H merely undergo change of slope when plotted against temperature without showing sharp changes or discontinuities as observed in case of first order transitions. The properties which undergo discontinuities at the glass transition temperature are: heat capacity C_p, $[C_p = (\partial H/\partial T)_p]$, coefficient of thermal expansion α, $\left[\alpha = \frac{1}{V}(\partial V/\partial T)_P = \frac{1}{V}\frac{\partial}{\partial T}\{(\partial G/\partial P)_T\}_P\right]$ and isothermal compressibility K, $\left[K = -\frac{1}{V}(\partial V/\partial P)_T = -\frac{1}{V}(\partial^2 G/\partial P^2)_T\right]$ which are second derivatives of free energy, and it is for this reason that the glass transition temperature, T_g is commonly referred to as the second order transition temperature (Fig. 7.17).

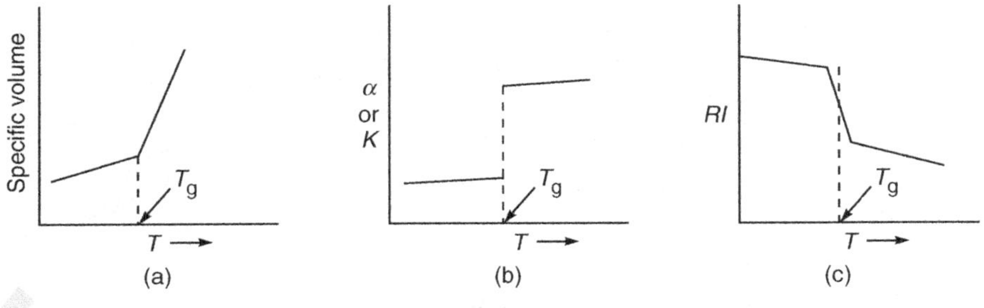

Fig. 7.17 ***Trend of change of (a) specific volume, (b) coefficient of thermal expansion (α) or isothermal compressibility (K), and (c) refractive index (RI) with temperature indicating the glass transition point (T_g)***

The glass transition is not a phase transition and so, it involves no latent heat. Below this temperature normally rubber-like polymers become rigid, hard and dimensionally stable and they are considered to be in a glassy state while above this temperature, all normally rigid, hard and stiff polymers become soft, and readily susceptible to cold flow or creep, and they turn into a rubbery state. The difference between the glassy and rubbery states of the polymer lies not in their geometrical structure, but in the state and level of molecular motion. Below the glass transition temperature, T_g, the molecular segments or groups constituting the chain backbone can undergo only limited degrees of vibration and they do not possess the energy required to rotate about bonds and change their positions with respect to segments of neighbouring chains. At or slightly above T_g, rotation sets in and it is conceivable that only the molecular segments rather than the entire high polymer molecule would rotate at this stage of transition. The much higher coefficient of thermal expansion just beyond T_g points to the greater degree of freedom of rotation.

At their respective glass transition or second order transition temperatures, different polymers may be considered to be in an isoviscous state, and as such, T_g is a common reference point for polymers of diverse nature, below which they behave as rigid plastics (glassy polymer) and above which they are rubbery in nature. For us, a useful rubber is provided by a polymer whose T_g is well below the ambient temperature and a useful plastic is one whose T_g is far above the ambient temperature. Table 7.1 gives a list of some common polymers and their T_m and T_g values.

Molecular weight, external tension or pressure, plasticizer incorporation, copolymerization and cross-linking are some of the more important factors that influence the glass transition temperature and the melting point of polymers. The comparative lowering of T_m and T_g for polymer modification by plasticizer incorporation and by copolymerization (comonomer incorporation) is shown in Fig. 7.18. In general, a comonomer incorporation, i.e., copolymerization is more effective than plasticization in lowering the melting point, while plasticization is normally more effective than copolymerization in lowering the glass transition point. Cross-linking tends to increase the glass transition temperature as the cross-links inhibit the freedom of the chain segments to rotate, necessitating a higher temperature for inception of rotation. Similarly, higher molecular weight of linear chains restricts scope for segmental rotation and

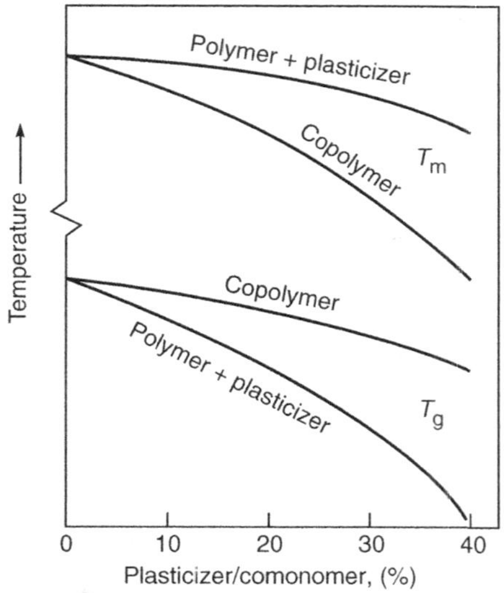

Fig. 7.18 *Schematic plots showing comparative lowering of T_m and T_g of a polymer by separately incorporating: (i) a plasticizer and (ii) a comonomer by copolymerization*

Table 7.1 *T_m and T_g values of some polymers**

Polymer	*Repeat Unit*	T_m °C	T_g °C
Polyethylene	—CH_2—CH_2—	137	–115, –60
Polyoxymethylene	—CH_2O—	181	–85, –50
Polypropylene (isotactic)	—CH_2—$CH(CH_3)$—	176	–20
Polyisobutylene	—CH_2—$C(CH_3)_2$—	44	–73
Polybutadiene (1, 4-*cis*)	—CH_2—CH=CH—CH_2—	2	–108
Polyisoprene (1, 4-*cis*) (natural rubber)	—CH_2—$CH(CH_3)$=CH—CH_2—	14	–73
Poly(dimethyl siloxane)	—$OSi(CH_3)_2$—	–85	–123
Poly(vinyl acetate)	—CH_2—$CH(OCOCH_3)$—		28
Poly(vinyl chloride)	—CH_2—CH·Cl—	212	81
Polystyrene	—CH_2—$CH(C_6H_5)$—	240	95
Poly(methyl methacrylate)	—CH_2—$C(CH_3)(COOCH_3)$—	200	105
Polytetrafluoroethylene	—CF_2—CF_2—	327	126
Poly(Caprolactam)	—$(CH_2)_5CONH$—	215, 223	50
Poly(hexamethylene adipamide)	—$HN(CH_2)_6$—NHCO—$(CH_2)_4CO$—	264	53
Poly(ethylene terephthalate)	—$O(CH_2)_2$—O·CO—$(C_6H_4)CO$—	254	69
Poly(ethylene adipate)	—$O(CH_2)_2$—OCO—$(CH_2)_4CO$—	50	–70

**See also* Table 1.3

hence causes rise in T_g, with a levelling off effect in the higher molecular weight range, however.

7.20.3 Measurement of Glass Transition Temperature

The glass transition temperature, T_g, is conveniently measured in the laboratory by taking help of dilatometry (Fig. 7.19). The polymer appropriately confined in the bulb at the bottom is kept immersed in a suitable liquid, usually mercury so as to give a column of the liquid in the capillary up to a convenient height for measurement. The positioning of the glass plug, as shown, enables heating the test specimens avoiding overheating. The dilatometer placed in an outer bath may be heated at a preset rate and pattern. From the rise of the liquid in the capillary on heating and consequent rise in temperature, the change in volume of the specimen (taking into account the volume expansion of the liquid, etc.) may be conveniently obtained. A typical plot showing variation of specific volume with change of temperature for polystyrene is shown in Fig. 7.20.

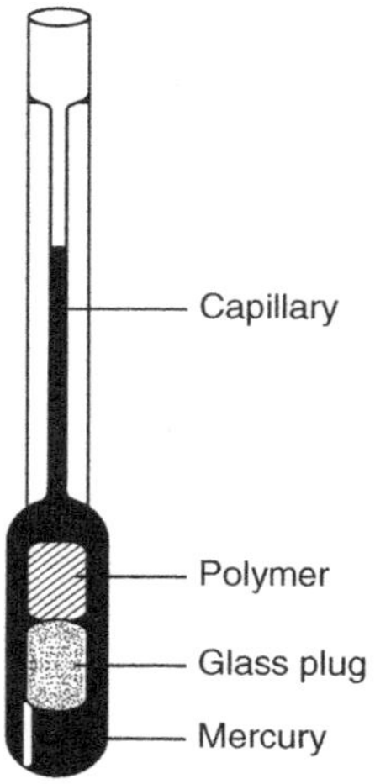

Fig. 7.19 *A dilatometric set up for measurement of change in volume of a polymer with change in temperature*

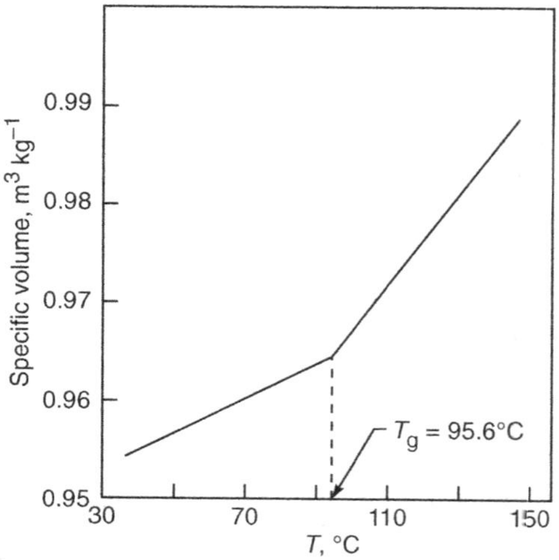

Fig. 7.20 *Volume expansion of polystyrene with rise in temperature: the inflection point corresponds to* T_g

7.20.4 The Brittle Point

A polymer is usually characterized by a temperature called the brittle point which is either close to or somewhat higher than the glass transition temperature. As the temperature of a rubber is lowered, the rubbery properties are gradually lost and the polymer hardens; finally, a temperature called the brittle point is attained at or below which the polymer sample becomes brittle and breaks or fractures on sudden application of load.

To enable comparison of brittle points of different polymer samples, it is necessary to do the testing under specified conditions of testing, including specified thickness for test specimen, degree and rate of loading, etc., in view of the test being empirical in nature. The brittle point corresponds to a temperature at which the time interval of load application in the test method just matches or equals that required by the test specimen to undergo the necessary deformation. At lower temperatures the specimen fails to deform as rapidly and hence it breaks while above the brittle temperature, the time of load application is more than adequate for the specimen to absorb the applied energy and deform without breaking. Lower molecular weight limits the scope for long range molecular interactions and chain entanglements and hence leads to a higher brittle temperature. Variation of brittle temperature and glass transition temperature with molecular weight, schematically depicted in Fig. 7.21, reveals that the trend of change for the two parameters are just the opposite. The difference between the two is relatively narrow in the higher range of molecular weight but it becomes much larger as the molecular weight decreases.

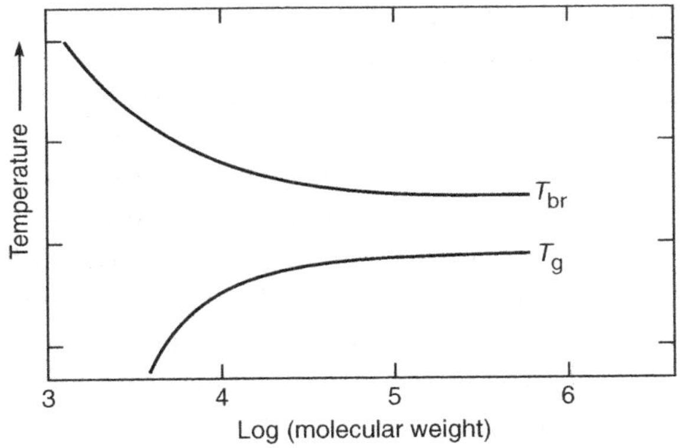

Fig. 7.21 *Typical plots showing influence of molecular weight on glass transition temperature (T_g) and brittle temperature (T_{br}) of a polymer (schematic)*

7.21 Polymer Morphology—Introduction

Studies of physical form, arrangement and structure of the molecules of a material system relates to its morphology. Polymer morphology covers the study of the arrangement of macromolecules into amorphous and crystalline regions and the overall physical structure of the molecular aggregates.

In molecules including those of polymers, fixed arrangements of atoms that remain normally unaltered and which can be altered only by breaking and reforming of primary valence bonds, are commonly referred to as *configurations*. On the other hand, arrangements of molecular segments that can be altered by rotation of segments or groups of atoms around single bonds are referred to as *conformations*. Different modes of chain growth (and sometimes chain termination) give rise to different configurations including head-to-tail, head-to-head or tail-to-tail arrangements (Sec. 6.2), stereospecific or random arrangements given by isotactic, syndiotactic and atactic structures (Sec. 1.5.6) in vinyl polymers, and structures arising from 3, 4, or 1, 2 addition and *cis* and *trans* isomers from 1, 4 addition in diene polymers (Sec. 5.9).

7.22 Development of Crystallinity

Morphological studies about polymers are primarily related to the molecular pattern and physical behaviour of the crystalline regions of crystallizable polymers. Amorphous, semicrystalline and highly crystalline polymers are known, but it is difficult or even impossible to attain 100% crystallinity in bulk polymers. It is also difficult, if not impossible, according to the latest studies based on electron microscopy and other approaches, to obtain solid amorphous polymers completely devoid of any degree of molecular order or crystallinity. A whole spectrum of structures ranging from total disorder, and different kinds and degrees of order to very high degree of (if not total) order may describe the physical state of a given polymeric system depending on test environment, thermomechanical treatment to which the polymer has been subjected and the chemical environment from which the polymer has been isolated. Moreover, the collected data for the degree of crystallinity may vary according to the employed method of estimation. The crystallinity data given in Table 7.2 must, therefore, be regarded as approximate.

Table 7.2 *Approximate degree of crystallinity (%) for different polymers*

Polyethylene (low density)	60–80
Polyethylene (high density)	80–98
Polypropylene (fibre)	55–60
Nylon (polyamide fibre)	55–60
Terylene (polyester fibre)	55–60
Cellulose (cotton fibre)	65–70
Cellulose (viscose rayon fibre)	35–40
Gutta percha	50–60
Natural rubber (crystallized)	20–30

Polymers having crystallites in excess of 50% are generally recognized to be crystalline. Cellulose derivatives used as fibres have crystallinity lower than that shown by cellulose. The predominantly linear chains of high density polyethylene

(HDPE) exhibit crystallinity much higher than any other polymer known, even substantially higher than that exhibited by the low density variety; and for the HDPE, the attainable crystallinity is very much close to the upper limit (100%). On the other hand, atactic polymers in general (including those of styrene and methyl methacrylate bearing bulky side-groups) having irregular configurations fail to meaningfully crystallize under any circumstances.

Figure 7.22 gives an idea of the rate of crystallization of a highly crystallizable polymer, polyethylene, at different temperatures, studied by following volume changes.[11] As the temperature is lowered, the rate of crystallization rapidly increases and well below the melting point (<127°C), it soon becomes so rapid as to make observations practically impossible. The practical consequence of the very high rate of crystallization in polyethylene is that it is not possible to obtain the polymer in the amorphous state at room temperature or under ambient conditions. Sudden quenching of the melt to below room temperature yields a material which is still crystalline (though usually with the likelihood of a lower degree of crystallinity than that developed on normal cooling of the melt). This is because the time required for crystallization is far shorter than the time taken in cooling the test sample.

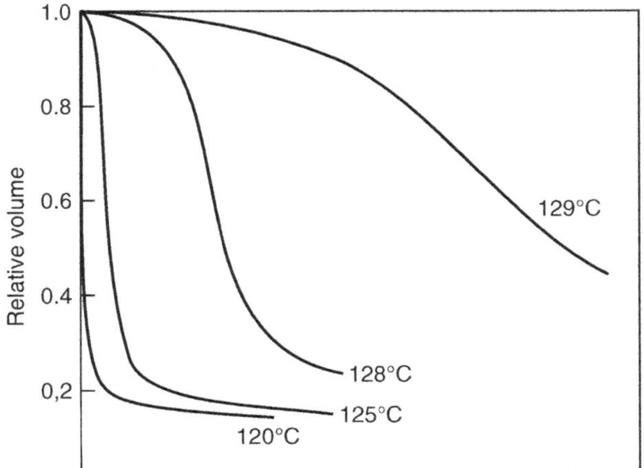

Fig. 7.22 ***Progress of densification or development of crystallinity with time in polyethylene held at different temperatures. The ordinate is the ratio of actual volume change to the ultimate change in volume after an indefinitely long time period (Mandelkern, 1964; Courtesy, McGraw Hill, New York)***

For practical reasons, therefore, the phenomenon of polymer crystallization is conveniently studied and measured using an amorphous polymer, such as natural rubber. The advantage of using rubber as a kind of model material for the study of crystallization is that crystallization process is quite slow in the substance and it takes place in a convenient range of temperature. The manipulation of the crystallization process for experimental purposes is done with much greater ease with rubber than with a crystalline polymer which crystallizes at rates too high for easy and precise measurements, and that also at much higher temperatures.

All rubbers are, however, not crystallizable. Only those built up of chains characterized by chemically identical and regular repeating units, such as natural rubber, are capable of crystallization.

Natural rubber may be conveniently crystallized by cooling below room temperature, preferably below 0°C or by simply stretching at room temperature.

7.23 Crystallization of Rubber on Cooling

When unvulcanized natural rubber is held at a fixed low temperature such as 0°C, it slowly gets somewhat stiffened and hard and loses flexibility proportionately. However, the material still retains a good degree of flexibility and toughness. The change is also associated with some gain in density or lowering of volume, and it is understood to be due to development of crystallinity in the material.

Crystallization in ordinary low molecular weight liquid on cooling to the freezing point or below takes place very rapidly by the required molecular rearrangement from a disordered to a very regular state of packing. A polymer melt system is somewhat more complicated due to complex chain entanglements restricting free mobility of the chain segments and thus hindering and delaying the desired rearrangement process on cooling. For polymers the time scale for crystallization is usually much longer than for liquids of low molecular weight.

Trends of change in relative volume of rubber taking place during crystallization at several temperatures are shown[12] in Fig. 7.23. The maximum attainable crystallinity and the time required for this is very much dependent on the temperature. In each case the rate of volume reduction is initially quite slow; the rate gradually builds up, passes through a steady zone and then finally decays giving a limiting maximum development of crystallinity. Lowering of temperature leads to enhancement of the steady rate of crystallization till about –25°C at which point the steady rate passes through a maximum and on further reduction in temperature the crystallization rate follows a decreasing trend. The dependence of rate on temperature is shown[12] in Fig. 7.24.

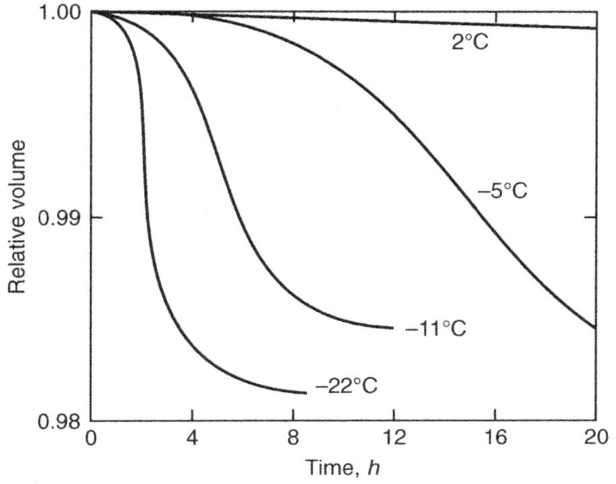

Fig. 7.23 *Crystallization of natural rubber; plot showing trends of change in relative volume vs time at different temperatures*[12]

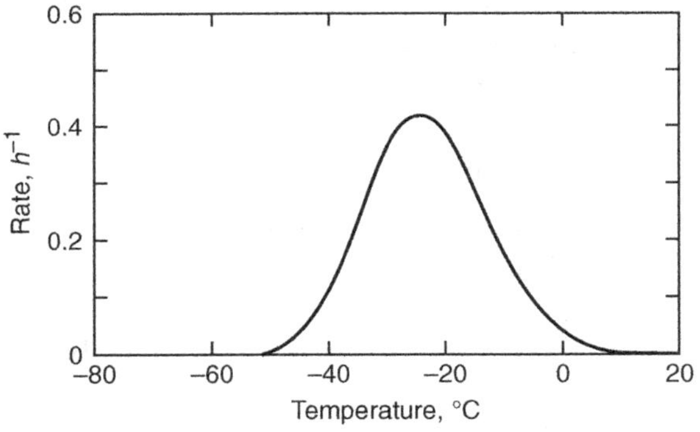

Fig. 7.24 *Plot showing variation of rate of crystallization with variation of temperature for natural rubber*[12] *(Wood, 1946;* **Courtesy,** *Wiley-Interscience, New York)*

The crystallization is completed in about five hours at –25°C. In natural rubber, the extent of crystallinity developed does not exceed 30%.

7.24 Mechanism of Crystallization

As the polymer melt is held at a temperature close to or slightly above its melting range, the initial delay in crystallization rate build-up is linked with the initial process of *nucleation*. Growth of crystallites is contingent upon the development and existence of a certain number of very small growth centres or nuclei for the deposition of oriented chain segments. The centres are initially formed on extended cooling or holding of the melt at the given temperature by coming together of a small number of molecular segments in the course of their random motion (micro Brownian motion) under the prevalent thermal condition. Nucleation, however, is common to all processes that turn an initially homogeneous medium into a heterogeneous system due to deposition of a separate phase.

The tiny nuclei formed initially are rather unstable and probably many of them will disappear soon after formation. But they become more stable if they survive and grow in size and finally may reach a size that assures them a permanent life and permits further growth by more deposition of the material on their surface. Larger the surface area of a nucleus, faster is the rate of deposition of the oriented segments on the surface. This explains the rapid rise in rate of crystallization just beyond the initial stage of development of crystallinity (Figs 7.22 and 7.23).

As the growth continues, the opposing effect of entanglements between molecular chains attached to neighbouring crystallites becomes more and more serious and ultimately critical, putting severe restrictions on the mobility of chain segments and thus making it difficult for them to get to a position for attachment to any one of the crystallites formed. Beyond this stage, the rate of crystallization sharply diminishes and finally the process dies down.

Lowering of temperature favours nucleation and lower thermal energy of the chain segments makes it less likely that a nucleus once formed will again disappear, the overall result being a gain in the number of nuclei and an increase in the overall rate of crystallization as the temperature is progressively lowered. However, at a very low temperature, the overall energy of the system and that available to a chain segment are so much diminished that the segments lose much of their mobility and hence their deposition on a nucleus generated is hindered more effectively. For natural rubber the crystallization process is practically frozen out below –50°C (Fig. 7.24).

7.25 Stress-Induced Crystallization

It is well known and a matter of wide experience that stretching of a strip of vulcanized rubber makes it develop a temporary crystallinity which disappears instantly on removal of the stretching force. A strip of raw or unvulcanized rubber also develops crystallinity when subjected to high extensions on application of a stretching force, but it remains in the extended state without retraction to its original state on stress release. However, when heated carefully in the subsequent stage, such as by dipping the strip into hot water, the crystals melt and the strip reverts to its unstrained state. The cross-links in the vulcanized material are responsible for accumulation of the strong restoring or retracting force which breaks the stress-induced orientation or the crystalline structure when the stress is withdrawn. In the unvulcanized form, the absence of cross-links allows varied degrees of chain slippage on extension and whatever elastic restoring force accumulates is far too insufficient to break the crystalline structure. Raising the temperature to about 30°C or to a slightly higher level to melt the crystallites allows the rubber strip to return to its original state. Time dependency of crystallization of unvulcanized rubber at a low temperature, 0°C by stretching to different fixed extensions can be studied by following density changes with time and the trends of change are shown in Fig. 7.25.

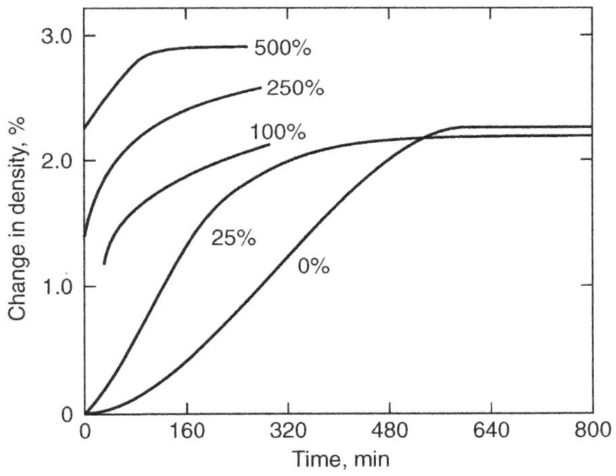

Fig. 7.25 *Time dependency of stress-induced crystallization (densification) of unvulcanized rubber held at a low temperature (0°C) for different orders of fixed extensions as indicated*

Moderate extensions produce effects comparable to those produced by lowering of temperature. However, for extensions greater than 100%, the crystallization rates are very high and only the final stages of the process can be practically observed.

7.26 Melting of Rubber

Just as crystallization, melting processes in polymer systems too are distinctive, more complex and diverse in nature than in low molecular weight material systems.

Figures 7.26 and 7.27 show the melting curves of a typical rubber (natural rubber) of low degree of crystallinity and of a typical polymer of very high degree of crystallinity (high density polyethylene) respectively. For rubber, the curve begins with volume data at 0°C as in Fig. 7.26. The beginning of melting is indicated by a relatively steep rise in volume with temperature which continues over a range of about 10°C for rubber crystallized at or slightly below 0°C till the melting is complete. Beyond this point further rise in temperature leads to the usual volume expansion of the amorphous rubber. The melting curve of polyethylene is characterized by a relatively sharp volume change and the temperatures of the beginning and completion of the melting process are usually limited well within a range of 10°C. If after melting, the temperature is lowered again, the linear volume contraction for the amorphous rubber (Fig. 7.26) continues to much lower temperatures and the melting curve is not retraced in the reverse direction simply because measurable recrystallization does not occur in the time span of the experiment. For polyethylene, however, the process, is reversible in a practical sense and the recrystallization curve is, by and large, a retrace of the melting curve.

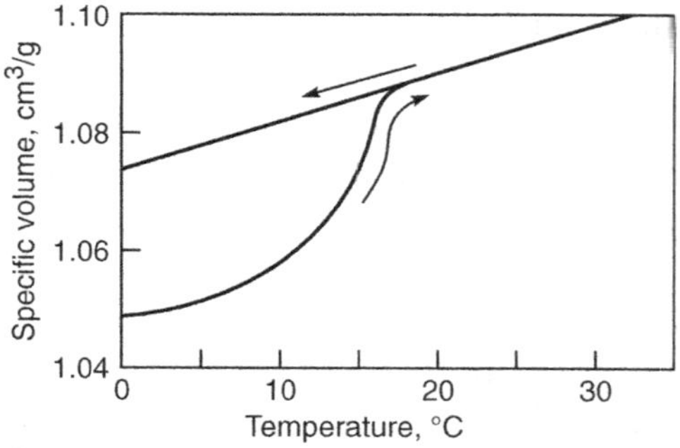

Fig. 7.26 *'Melting' curve showing a plot of specific volume vs temperature for (crystalline) natural rubber*[13] *(Bekkedahl and Wood, 1941;* **Courtesy,** *American Chemical Society, Washington)*

For rubber, whereas melting takes place over an extended range of temperature, the beginning of melting and the range of temperature over which it is accomplished are also largely dependent on the temperature at which the preceding crystallization had been achieved. Usually, melting begins at a temperature which is four to six degrees higher than the temperature of crystallization, as illustrated in Fig. 7.28. In a polymer crystallizing from the melt, a crystallite does not remain in equilibrium with the amorphous

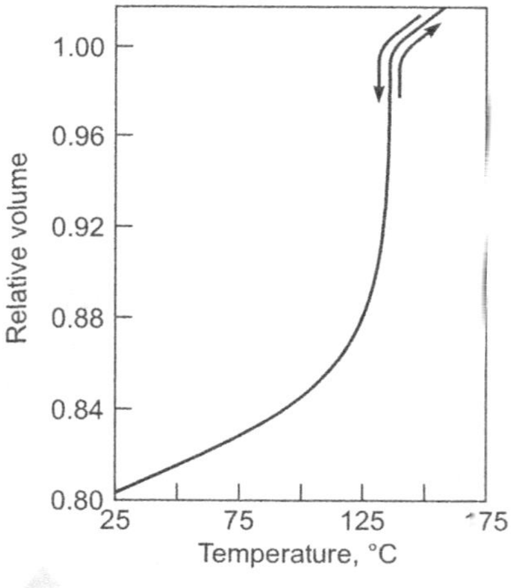

Fig. 7.27 *'Melting' curve showing a plot of relative volume vs temperature for polyethylene.*[11] *(Mandelkern, 1964;* **Courtesy,** *McGraw-Hill, New York)*

phase, as the chain segments of molecules in the liquid or amorphous phase are physically locked into the crystallites and as the same molecule may pass from one crystallite to another via the intervening amorphous zone. This feature of lack of true equilibrium between the solid (crystalline) and liquid (amorphous) phases in the rubber leads to some unusual but interesting effects. Figure 7.28 shows that for rubber crystallized at very low temperatures, say, –40°C, melting is complete at temperatures close to but < 0°C. If after completion of melting, the temperature of the amorphous rubber is not allowed to rise and the sample is held at the upper melting temperature for a long period, a second crystallization would set in. Thus, it is possible to have simultaneous or consecutive melting and recrystallization in a given piece of rubber as it is heated over the melting range after initial crystallization and held at a specific temperature within that temperature range.

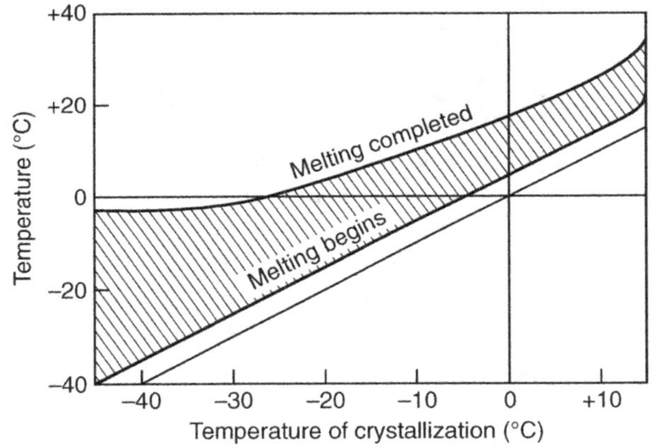

Fig. 7.28 ***Plot showing dependence of melting range of natural rubber on temperature of crystallization[12], the diagonal line below the melting zone indicating temperature of crystallization (Wood, 1946;*** **Courtesy,** ***Wiley-Interscience, New York)***

7.27 Polymer Single Crystals

Single crystals of different polymers can be grown by slow precipitation from very dilute solutions. They appear in the form of very thin plates or lamellae, usually diamond-shaped with spiral growth pattern and showing step-like formation on the surface (Fig. 7.29).

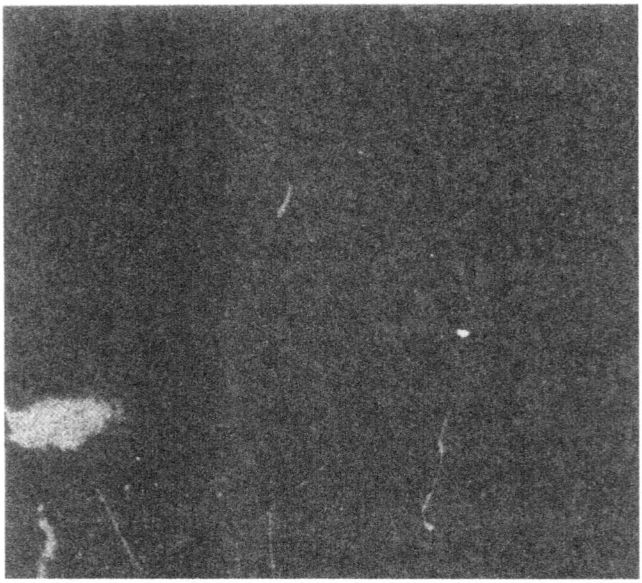

Fig. 7.29 *Electron micrograph of single crystals of polyethylene.*[14] *(Reneker and Geil, 1960; Courtesy, American Institute of Physics, New York)*

The single crystals are very small in size and cannot be examined by X-ray diffraction. However, they are readily and conveniently studied by electron microscopy. Electron diffraction pattern and electron micrographs reveal certain interesting features about the single crystals. The thickness of the lamellae is very small (100–200 Å) compared to the length of the polymer chain. The diffraction pattern indicates without uncertainty that the chain axis is perpendicular to the plane of the lamellae. The structural pattern of the single crystal is then understood well on the basis of what is known as the folded chain theory which envisages that a single molecule of the polymer must bend or fold forwards and backwards many number of times across the thickness of the lamellae. Such folded chains can be stacked in the crystal lattice without much difficulty. It is widely believed that the single crystal comprises an array of folded chains packed individually and successively between the top and bottom surfaces, or planes, and on the growing edges of the lamellae (Fig. 7.30).

Crystallization of this nature involving whole individual polymer molecules discretely without interference of other molecules is made possible due to large distances that separate the individual molecules in the very dilute solution (Fig. 7.31). This wide separation ensures near elimination of chain entanglement. Thus, when one segment of a polymer molecule gets attached to one of the thin edges of

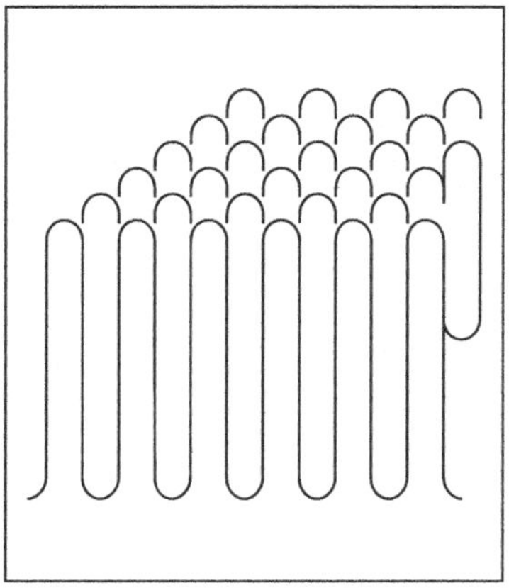

Fig. 7.30 *Schematic representation of chain folding in the formation of a single crystal*

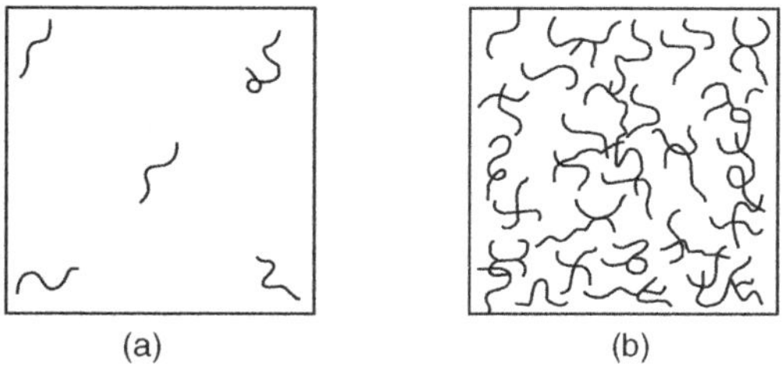

Fig. 7.31 *Separation between polymer molecules in (a) very dilute solution, and (b) concentrated solution (schematic representation)*

the growing crystal, it faces no competition from other far away molecules for occupation of the adjacent lattice site. There will be no hindrance to the successive occupation of immediately adjacent sites by the segments of the same molecule by a chain folding mechanism until the whole molecule is used up.

7.28 Structure of Bulk Polymers

Crystalline polymers obtained on cooling of their melts also produce electron micrographs showing the lamellae structure for the crystallites[15] and giving little direct evidence for the presence of major amorphous regions (Fig. 7.32). An idealized model representation of the lamellae structure shown in Fig. 7.33(a), however, may not be applicable to all types of polymers. Most polymers (other than the different types of polyethylenes) contain amorphous regions to the tune of 20–50% distributed in the material along with the crystalline zones. In the structural model, a provision has to be made to accommodate the amorphous material. In a fringed-micelle or fringed-crystallite model [Fig. 7.33(b)], the amorphous material is interspaced between the randomly distributed crystallites. This model explains well the morphological features in such materials as rubber and some cellulosic polymers. For other polymers of intermediate orders of crystallinity, different structures combining the fringed micelle and regularly stacked lamellae model may represent the overall structural pattern. These structural concepts allow for imperfections commonly encountered, such as the interlamellar entanglements, molecular loops of

Fig. 7.32 *Electron micrograph of the surface of polyoxymethylene crystallized by melt cooling*[15,16] *(Geil, 1960;* **Courtesy,** *Wiley-Interscience, New York)*

varying dimensions, irregular fold lengths, and interconnecting chains passing through different lamellae.

Any model consisting of stacks of lamellae interspaced with and connected by amorphous regions may be referred to as the interlamellar amorphous model [Fig. 7.33(c)]. This model provides the most useful approach to understanding the mechanical properties of bulk crystallized polymers of moderate to high degrees of crystallinity. The ductility and cohesive character is a direct consequence of the interlamellar ties. Much like stacks of bricks without mortar, stacks of lamellae (crystals) without measurable presence of tie molecules, such as those obtained by slow cooling of a very dilute solution, would prove relatively brittle. The tie molecules reduce brittleness and induce ductility.

7.29 Spherulites

The most common and prominent feature in bulk crystallized polymers is the spherulite or spherical crystallite. A spherulite is usually constituted of a symmetrical

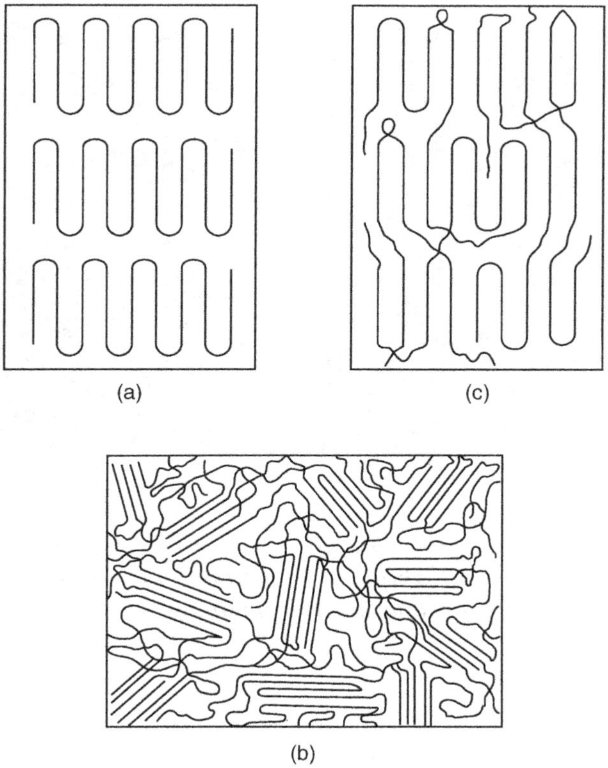

Fig. 7.33 ***Schematic representation of: (a) ideal stacking of lamellar crystals, (b) fringed-micelle model of randomly distributed crystallites, and (c) interlamellar amorphous model***

structure arising as a consequence of the cooperative growth of crystallites radially outward from a core or nucleus in three dimensions (Fig. 7.34). Bulk crystallized polymers are not merely a series of stacked lamellae separated and interconnected by amorphous regions. The lamellae units are organized radially within the spherulites (Fig. 7.35). The crystallization process through which spherulites are formed follows sequential steps of nucleation of spherulite growth. Nucleation may be aided by intentional addition of a foreign substance, called the nucleating agent. Nucleating agents reduce the size of spherulites by increasing the number of nuclei. Growth of large spherulites contributes to brittleness.

Most polymers continue to increase in density long after spherulite growth is complete. The post-primary crystallization densification takes place both in the inter- and intra-spherulitic regions. The densification due to secondary crystallization

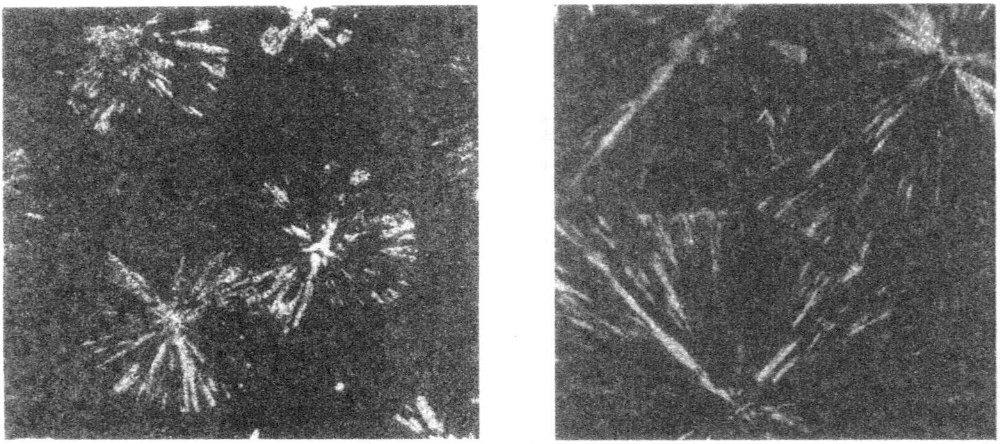

(a) (b)

Fig. 7.34 *Stages of spherulite growth in melt crystallized polypropylene: (a) the spherulites begin to grow from point sources, (b) their gain in size may continue until they impinge on one another*

Fig. 7.35 *Spherulitic structure shown schematically: (1) radial growth and branching of the lamellae, (2) portion of a lamellae enlarged to indicate a folded chain structure showing chain axis perpendicular to the radius of the spherulite*

taking place after the primary process of spherulite growth, leads to thickening of the lamellae as chain segments are gradually pulled in from the amorphous regions. Another consequence of secondary crystallization may be increase in brittleness. However, the process of secondary crystallization and its effect on mechanical properties are recognized to be complex and they depend largely on many factors including the rate and span of cooling, annealing and cold drawing or orientation by stretching.

References

1. Batschinski, A.J. *Z. Physik. Chem.*, **84** (1913) 643.
2. Kauzmann, W. and H. Eyring, *J. Am. Chem. Soc.*, **62** (1940), 3113.
3. Flory, P.J., *J. Am. Chem. Soc.*, **62** (1940), 1057.
4. Fox, T.G. and P.J. Flory, *J. Phys. Chem.*, **55** (1951), 221.
5. Fox, T.G., S. Gratch and S. Loshack, *Rheology—Theory and Application*, F.R. Eirich, Ed., Academic Press, New York, 1956, Vol 1, Ch. 12.
6. Herzog, R. and K. Weissenberg, *Kolloid Z.*, **46** (1928), 227.
7. Spencer, R.S. and R.E. Dillon, *J. Colloid Sci.*, **3** (1948), 163.
8. Gilmore, G.D. and R.S. Spencer, *Modern Plastics*, **27** (1951), 117.
9. Nason, H.K. *J. Appl. Phys.* **16** (1945), 338.
10. Tordella, J.P. *Rheology—Theory and Applications*, F.R. Eirich, Ed., Academic Press, New York, 1969, Vol. 5, Ch. 2.
11. Mandelkern, L. *Crystallization of Polymers*, McGraw-Hill, New York, 1964.
12. Wood, L.A. *Advances in Colloid Science*, H. Mark and G.S. Whitby, Eds., Wiley-Interscience, New York, 1946, Vol. 2, pp. 57–95.
13. Bekkedahl, N. and L.A. Wood, *Ind. Eng. Chem.*, **23** (1941), 381.
14. Reneker, D.H. and P.H. Geil, *J. Appl. Phys.* **31** (1960), 1916.
15. Geil, P.H. *J. Polym. Sci.*, **47** (1960), 65.
16. Geil, P.H. *Polymer Single Crystals*, Interscience, New York, 1963.

CHAPTER 8

Plastics—Materials and Processing Technology

8.1 Plastics Materials—Introduction

Material systems which can be conveniently made to flow or deform into a desired shape or form, usually under heat and/or pressure, are commonly known as plastics. The binding material in a given plastic product is invariably a polymer or resin. A resin is defined as a material, organic in nature and of either synthetic or natural origin, which has the capacity to produce a thin coherent film either from its melt or from its solution. A resin may be polymeric (characterized by high molecular weight and a characteristic repeat (chemical) unit in the structure) or non-polymeric (characterized by low molecular weight) in nature. According to the resin-concept as above, almost all polymers may be considered as resins, but all resins are not polymers. Conventionally, however, resins are brittle materials showing different degrees of hardness under ambient conditions. Rosin and Shellac are two important natural resins, and among the relatively low molecular weight ($M < 2000$) synthetic resins, the coumarone-indene resins and the so-called petroleum resins are important. These resins and their derivatives find widespread use as tackifiers and as improvers of adhesion, gloss, hardness, etc., in many adhesive and coating formulations based on different polymers. Most commercially important structural resins are, however, thermosetting polymers covering such products as the phenolic (phenol–formaldehyde) resins, amino(urea–formaldehyde and melamine–formaldehyde) resins, alkyd (polyester) resins, epoxy resins, etc.

The technology of a polymer begins with the polymer itself, which is the base material of a plastic, rubber, adhesive, coating or related item that we see around us. The synthetic polymer industry can be either coal-based or petroleum-based.

Ethylene, an important monomer and an intermediate for some other monomers may also be obtained by dehydration of alcohol. The developing trends in the decades after 1950 switched the technology of monomer synthesis for almost all kinds of polymer in favour of petroleum as the basic raw material, so much so that today's monomers and polymers are widely recognized as petrochemicals. Salient features of polymerization relating to some important polymers finding plastic applications in the main and their properties, processing technology and applications will be discussed in the present chapter.

8.2 Polyethylene

Polyethylene of different density ranges can be obtained by chain polymerization of ethylene, the simplest olefinic monomer. The observed chemical nature and physical properties of the polymer depend largely on the nature of the catalyst used and conditions employed during polymerization. Polyethylenes obtained by polymerization of ethylene are mostly branched polymers, the chemical nature and frequency of branches being largely dependent on reaction conditions and the mechanism involved. Polyethylenes of commerce are available in two major density ranges and the classified products are commonly known as: (a) low density polyethylene, LDPE (density range, 0.915–0.94 g/cm^3), and (b) high density polyethylene, HDPE (density range, 0.945–0.96 g/cm^3).

8.2.1 Low Density Polyethylene (LDPE)

LDPE is produced by free radical polymerization of ethylene under very high pressure (1500–3000 atm), using oxygen, peroxides or azo-compounds as initiators. The development of a commercial process of this kind was associated with numerous technical difficulties. Highly exothermic nature of the polymerization requires strict control of reaction variables (pressure, catalyst concentration, temperature, monomer purity, flow velocity in the reactor, etc.) in order to ensure safety and to prevent run-away reactions and explosion. The early ICI process, commercialized in 1939, employed oxygen (air) as the catalyst. The polymerization reaction, simply, written as

$$nCH_2 = CH_2 \rightarrow —(CH_2 - CH_2)_n— \qquad (8.1)$$

is expected to be favoured under high pressures and low temperatures from thermodynamic considerations. On kinetic considerations, however, a temperature range of 170–250°C is usually employed. At a lower temperature, the radical generation process is far too slow for commercial production of the polymer.

To get high molecular weight, ethylene of high purity is required. Traces of oxygen catalyzes the polymerization process. The allowable oxygen at a pressure of 2000 atm and 165°C is about 0.075%. If this limit is crossed, explosive decomposition reaction occurs yielding hydrogen, methane and carbon in the main. The conditions are so chosen as to permit and favour only polymer formation at commercially viable rates. High pressure increases local concentration of the gaseous monomer in the vicinity of the transient chain radicals, and thereby brightens the prospect of obtaining high molecular weight for the polymer product. Higher temperature, however, tends to lower the molecular weight.

A pressure of the range of 1500–2000 atm is obtained by compressing refined ethylene, usually in two or three successive stages. For continuous operation, the reactants are passed through narrow-bore tubular reactors or stirred reactors, fitted with efficient heat-transfer system. Conversion in the range of 10–30% is normally achieved after which the unreacted monomer is flashed out and recovered and the viscous molten polymer is extruded out in the form of strands or ribbons, cooled and granulated.

In the conventional activation with oxygen, the formation of some unknown peroxide/hydroperoxide must be considered as the source that furnishes the radicals for chain initiation on subsequent decomposition, which apparently takes place fast enough only above 150°C. The heat of polymerization of ethylene[1] (800–1000 cal/g) is considerably higher than that of other monomers, e.g., styrene (164 cal/g). In addition to many difficulties inherent in high pressure reactions, the control of temperature consequent to this high heat of polymerization is one of the most important problems in the polymerization of ethylene by the high pressure technique.

Chain branching (frequency, branch-length and distribution) as well as molecular weight and molecular weight distribution can be conveniently varied and controlled by varying the temperature, pressure, nature of initiator and its concentration and other additives such as solvents/non-solvents or suitable chain transfer agents, and by injecting the initiators at different points in the reactor.

8.2.2 High Density Polyethylene (HDPE)

Use of coordination catalysts or supported metal oxide catalysts allows polymerization of ethylene at a much lower pressure as well as at a relatively low temperature; the polymers thus produced are by and large linear, having much fewer branches than found in the product obtained by the high pressure free radical process. The HDPE synthesis is conveniently done by (i) the Ziegler process using an aluminium trialkyl–$TiCl_4$ complex as the catalyst dispersed in a hydrocarbon solvent such as *n*-heptane and employing a slightly elevated pressure and a temperature in the range of 70–100°C, (ii) the Phillips process, using CrO_3 (5%) supported on finely divided silica-alumina (75–90% silica) base as the catalyst; the catalyst is activated by heating to about 250°C, dispersed in a hydrocarbon solvent such as cyclohexane and polymerization of ethylene is done employing a pressure of 15–30 atm and a temperature in the range of 130–150°C, or (iii) the Standard Oil (Indiana) process using a dispersion of molybdenum oxide catalyst containing sodium and calcium metal or their hydrides as promoters in a hydrocarbon solvent, employing a pressure in the range of 49–80 atm and a temperature range of 220–250°C. The Ziegler process yields polymers of density of about 0.945 g/cm^3 while the other processes using (supported) metal oxide catalysts yield polymers of density as high as 0.96 g/cm^3. On completion of reaction in the Ziegler process, the catalyst is destroyed using water or alcohol with traces of an acid and the polymer is collected by filtration or centrifugation, which is then washed and dried. The processes using metal oxide catalysts may be either fixed-bed, moving-bed or slurry processes. On completion of the reaction, the polymer formed and evenly distributed in the paraffin or cycloparaffin diluent in the high reaction temperature is separated from the catalyst fragments and then recovered and isolated by cooling and/or solvent evaporation.

8.2.3 Structure and Properties of Polyethylenes

The differences in the density range of LDPE and HDPE as mentioned earlier have their origin in the inherent difference in the mechanistics which are consequential in giving rise to significant structural differences in these polymers. The structural differences relate to differences in molecular weight and molecular weight distribution and differences in degrees of chain branching in the main. Due to the high chain transfer (polymer transfer—"back biting" mechanism, Sec. 3.15.4) in the radical polymerization effected under high pressure, the resultant polymer, LDPE has nearly 20–50 branches (both long and short branches) per 1000 linear carbon atoms in the chain molecules. The loss of molecular symmetry due to the high degree of branching results in lower density range (0.915–0.94 g/cm^3) and lower softening or melting temperature (110–117°C) for LDPE. The low pressure polymers, HDPE, are by and large linear, having only 2–5 short branches or side chains per 1000 carbon atoms in the main chain, thus having a higher density range (0.945–0.96

g/cm^3) and higher melting temperature (125–130°C) compared to LDPE. The theoretical linear (unbranched polymer) polymethylene, —$(CH_2)_n$—as obtained from diazomethane $CH_2\!<\!\overset{N}{\underset{N}{\|}}$ using BF_3 – Et_2O catalyst at about 0°C through elimination of N_2 on polymerization [$nCH_2N_2 \rightarrow$ — $(CH_2)n$ — + nN_2] is, however, characterized by high molecular weight with a density as high as 0.98 g/cm^3 and a melting temperature of 137°C.

Polyethylenes and other polyolefins are graded by what is known as the melt flow index (MFI), i.e., the weight in grams extruded by piston action through a standard opening under a standard load in a standard plastimeter at 190°C over a specified time period (*See also* Sec. 10.5). MFI gives a rough guide to processability of different grades of polymers of a given kind or monomeric origin. For a given kind of polymer, a higher MFI indicates lower molecular weight and vice versa. LDPE comes in a wide MFI range (0.3–70), but for HDPE, the useful MFI range is much lower (0.02–2.5). The tensile strength of LDPE ranges between 1500–2500 psi and that for HDPE is in the range of 3000–5000 psi. The Izod impact strength for LDPE is as high as 10 ft lb per inch of notch, while that for HDPE is much lower (1.5–5 ft lb per inch of notch).

Polyethylenes are chemically paraffinic in nature, inert to non-oxidizing acids, alkalies, water and many aqueous solutions and organic solvents. HDPE is usually more resistant to chemicals than LDPE. There is no singular solvent for polyethylenes at room temperature. LDPE dissolves in toluene at or above 60°C, while HDPE dissolves at temperatures about 20–30°C higher in the same solvent. With larger amorphous regions, LDPE is more permeable to gases than HDPE. Both LDPE and HDPE are prone to aerial oxidation, which is quite severe at processing temperatures. LDPE is more prone to oxidation than HDPE (if free from metallic impurities) in view of higher concentrations of tertiary hydrogen (at branch sites) in LDPE. Uncontrolled oxidation leads to measurable incorporation of carbonyl and peroxy groups which make the polymer more prone to degradation and embrittlement due to absorption of heat and UV light. Polyethylenes generally show poor environmental stress cracking resistance, LDPE being poorer in this respect than HDPE. However, improved processing conditions and better design result in low stresses being imparted to products, thus ensuring desired long service life and performance for them. The stress cracking problem is less critical with relatively high molecular weight (relatively low MFI) and high density materials.

Polyethylenes are very good insulating materials and in this respect they compare favourably with other dielectric materials. Being adequately flexible, LDPE is extensively used as an insulator in wires and cables.

8.2.4 Uses and Applications of Polyethylenes

Polyethylenes, being thermoplastic in nature, are conveniently processed by injection moulding, blow moulding, rotational moulding, extrusion and thermoforming techniques. They find extensive uses and applications as moulded or formed objects, films, sheets, bottles and containers, pipes and tubes, and in wire insulation and cables. Films, commonly made from LDPE, find application in packaging, water-proofing, irrigation and water management including canal lining and mulching, and in coating and lamination. Various additives such as fillers, pigments and colouring matters, antioxidants, slip additives, anti-block and antistatic agents, UV-absorbers, blowing agents and flame retardants are mixed or blended with polyethylene, polypropylene and other polyolefins for various reasons. For high voltage cables application and for heat resistant films or moulded objects, polyethylene is preferably cross-linked using selected chemical cross-linking agents, such as peroxides or by irradiation with high speed electrons or γ-rays. Factors contributing to widespread application of polyethylenes are low cost, easy processability by variety of techniques, excellent chemical inertness and electrical insulation properties, low temperature toughness and flexibility, good clarity of thin films, heat sealability, non-toxic nature and low water vapour permeability. The major limiting factors in this respect are low softening point or heat distortion temperature, proneness to environmental stress cracking, poor scratch resistance, low tensile strength and high gas permeability.

8.2.5 Chlorosulphonated Polyethylene

The high molecular symmetry of polyethylene (LDPE) may be broken by treating the polymer with chlorine in presence of small quantity of sulphur dioxide, thereby introducing some —SO_2Cl groups and —Cl atoms along the polymer chain as shown in the reaction scheme (8.2).

$$\sim\sim CH_2—CH_2—CH_2—CH_2 \sim\sim + SO_2 + Cl_2$$

$$\longrightarrow \left[\left[(CH_2)_n—\underset{\displaystyle Cl}{\underset{|}{CH}}\right]_x CH_2—\underset{\displaystyle SO_2Cl}{\underset{|}{CH}}—\right]_y \qquad (8.2)$$

Random introduction of chlorine atoms in the chain to the extent of one chlorine atom for at least seven carbon atoms ($n \geq 6$) gives a useful rubbery product. For an average product, $x = 12$. The SO_2Cl groups provide sites for cross-linking or vulcanization. The vulcanization is conveniently done with certain metal oxides such as PbO as shown in the reaction scheme (8.3). This polyethylene-derived rubber is popularly known as 'Hypalon' in the trade.

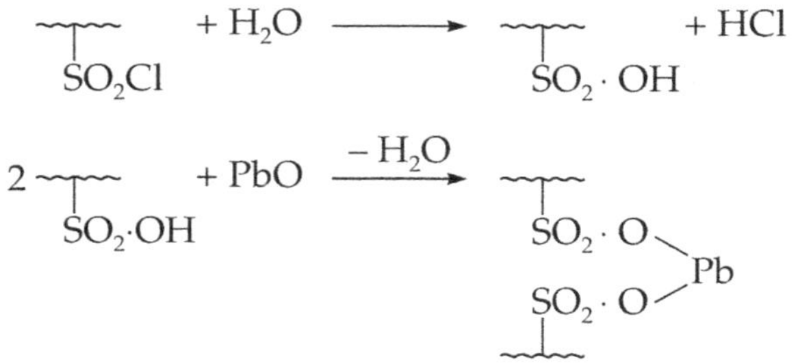

(8.3)

The chlorosulphonated polyethylene (CSP) rubbers are superior to many hydrocarbon (diene) rubbers in ozone resistance. They are also among the best chemical-resistant rubbers and have good resistance to heat-aging and flexing.

Chlorination of polyethylene to different degrees (up to about 40% chlorine) is also practised to obtain different products—thermoplastic or elastomeric. The reaction is carried out at an elevated temperature in solution or in suspension using either LDPE or HDPE. With HDPE, chlorination in aqueous suspension at or above 120°C gives a convenient elastomer that can be vulcanized by peroxides.

8.3 Linear Low Density Polyethylene (LLDPE)

The understanding of a link between different kinds and degrees of branching and different ranges of densities as exemplified by LDPE and HDPE has led to the more recent development of a mechanistically linear polyethylene of a low density range by a copolymerization approach. The product, a linear copolymer of ethylene and a minor proportion of an α-olefin such as 1-butene, 1-hexene or 1-octene, appears as a branched polyethylene having controlled proportions of uniform short chain branching along the chain. Uniform short branches appear in the chain at each point of the α-olefin comonomer incorporation. This linear grade of polyethylene, having a low density range (0.915–0.934 g/cm^3) is commonly known as Linear Low Density Polyethylene (LLDPE) and it can be synthesized employing the Ziegler process or the Phillips process.

Structurally, LDPE and HDPE may also be considered, as copolymers of ethylene and minor proportions of α-olefins, in view of the presence of significant long and short branches in LDPE and very limited short branches in HDPE. While LDPE and HDPE, if so considered as copolymers, are copolymers by incidence and not by definition or design, LLDPE has the distinction of being a copolymer primarily by definition and design.

LLDPE displays a superior balance of toughness, stiffness, stress cracking resistance and electrical properties over LDPE. With favourable production economics and progressive acceptability in the market, LLDPE may soon take over up to 30–50% of the traditional LDPE market throughout the world.

There are significant differences between the rheological behaviours of LDPE and LLDPE which pose some resistance towards acceptance of LLDPE by many LDPE end product manufacturers. It is, however, most likely that on grounds of superior performance, most non-film applications of LDPE will finally switch over to LLDPE. Judicial blends of LDPE and LLDPE are also being developed for various applications. The melting points of commercial grades of LLDPE are in between the melting points of commercial LDPE and HDPE.

8.4 Polypropylene

The polypropylene of commerce is isotactic polypropylene. Its synthesis became possible through the commercial utilization of coordination polymerization during 1957–60. Propylene, the monomer, is obtained from the cracking of petroleum products (natural gas, light oil or naphtha). The separation of propylene from propane in the C_3 fraction is difficult and is achieved by fractional distillation in a carefully designed distillation plant. Propylene of high purity is required for polymer synthesis.

8.4.1 Synthesis of Polypropylene

The synthetic approach for polypropylene is much the same as that for the production of HDPE by the Ziegler process. The same reactor can be used to polymerize ethylene, propylene or other α-olefins with some modifications, if necessary. For propylene, the reaction condition should be so chosen as to overwhelmingly favour formation of the isotactic polymer. The catalyst may be prepared from $TiCl_4$ and aluminium triethyl combination dispersed in naphtha. Propylene is charged into the reactor under pressure. Nearly 80–85% conversion is achieved in 8–10 h at 60°C. The reaction variables are molar ratio of catalyst components, catalyst concentration, reaction temperature and monomer pressure. Hydrogen may be used as a chain terminating or transfer agent for control of molecular weight.

About 3–10% atactic polymer is also formed along with the formation of the isotactic polymer in the main. After the reaction reaches 80–85% conversion stage, the ingredients are transferred into a flash drum to purge off the unreacted monomer which is then recycled. The atactic product is soluble in the solvent naphtha. The solution is decanted off after centrifugation whereby most of the atactic material is removed. The residue is then treated with methanol acidified with traces of HCl to decompose and dissolve the catalyst. The polymer is then centrifuged, washed, dried and collected.

8.4.2 Structure and Properties of Polypropylene

Polypropylene (PP) is the lightest homopolymer known, its density being 0.90 g/cm^3. With a rigid, short methyl group attached to every alternate carbon atom in the linear molecular chains and the highest order of steric regularity (isotactic), polypropylene, as commercially produced, is comparatively more rigid, stiff and stronger than HDPE. Its melting point is nearly 45–50°C higher than the melting point of HDPE. Having a much higher concentration of tertiary carbon atoms in its chains, PP is much more prone to oxidation or aging than HDPE and LDPE.

Commercial PP is nearly 90–97% isotactic. The degree of isotacticity or the isotactic index is given by the percentage of the polymer that is insoluble in *n*-heptane. Isotactic PP is much less crystalline than HDPE. The high melting point (≥ 170°C) of PP allows the moulded objects to be steam sterilized. Its high strength (tensile strength: 4500–5500 psi, elongation at break 100–300%), high melting point and low density is advantageous in many applications. PP, however, becomes brittle close to 0°C. In this respect, PP is inferior to polyethylene. But PP has a much superior hinge property. The problem of brittleness is very well overcome by block copolymerization of propylene with minor proportions (5–15%) of ethylene. PP is generally free from environmental stress cracking problems. Its specific heat is lower than that of polyethylene but higher than that of polystyrene. PP melt is more non-Newtonian than polyethylene melt. The melt viscosity of PP is also more sensitive to change of temperature. It is also graded by melt flow index for processing purposes. Generally, PP shows lower moulding shrinkage than polyethylene. If free from catalyst fragments, PP and polyethylene are similar in electrical properties.

PP exhibits higher clarity (less opacity) than HDPE, mainly because the amorphous and crystal densities are closer for PP than for polyethylene. Biaxially oriented polypropylene (BOPP) shows a high degree of clarity and biaxial stretching also produces polymers of much improved tensile strength. PP is used in the form of moulded and formed objects, sheets and films, bristles, monofilaments and fibres, covering such items as luggage, frames containers and different packaging items, ropes, textiles, tows and nets, pipes and tubes, etc. Typical mouldings also include components of automobile and home appliances industry, machine parts and components, sterilizable health-care and hospital equipment, kitchenware, furniture, etc.

8.4.3 New Generation Polyolefins

Polyolefins, viz. polyethylene (PE), polypropylene (PP) and higher poly(α-olefins) and olefin copolymers are commercially produced by chain polymerization of selected monomers or monomer combinations. The exact mechanism of polymerization, rate of polymerization, polymer microstructure, molecular weight and molecular weight distribution very much depend on the nature of the olefin monomer, catalyst, co-catalyst, temperature and pressure and also on the nature of other additives such as hydrogen as chain transfer agent or chain length regulator, solvents and comonomers. Free radical catalysts such as organic peroxides on the one hand and selected (supported) metal oxide catalysts (CrO_3, MoO_2 etc.) or some transition metal halide ($TiCl_4/TiCl_3$) catalysts along with a selected aluminium alkyl acting as a co-catalyst forming what is known as the Ziegler–Natta catalysts on the other, are most commonly used for olefin polymerization so far.

Today, metallocene catalysts are drawing the attention of all concerned with production and use of polyolefins, in view of overwhelming advantages that they offer in terms of much higher rate of polymerization, uniform structure and uniform polymer melt-rheology in view of the polymer having substantially narrower molecular weight distribution. Metallocene catalysts are based on Zr/Ti complexes with organic ligands such as the cyclopentadienes (Cp) in conjunction with aluminoxanes (oxyaluminium compounds). Characteristically, metallocenes are single site catalysts (SSC) providing nearly 100% active sites against only 1–3% active sites for the conventional Ziegler–Natta catalyst systems, indeed providing a tremendous improvement in catalyst activity. Metallocene catalyzed processes much like the Ziegler–Natta catalyzed processes, are also low-pressure polymerization of ethylene, propylene and other α-olefins, though of higher sophistication in many respects.

The mechanism of formation of the active species in the metallocene catalyst system comprising, say, cyclopentadienyl (Cp) metal (M = Ti or Zr) halide (X), structure I, and methyl aluminoxane, structure (II) may be schematically shown as:

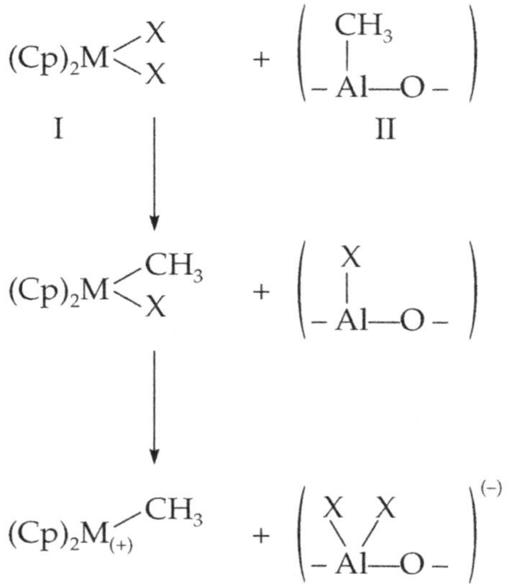

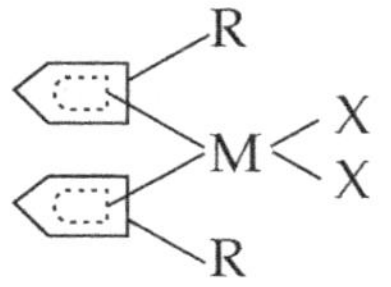

$[(Cp)_2 MX_2]$

I

Gas phase, slurry and solution processes have been in practice for industrial polymerization of ethylene and higher olefins. The gas phase processes are usually more versatile. Low density polyethylene (LDPE) of density range 0.91–0.94 g/cm^3 are produced by the high pressure process of two types, viz. tubular process and autoclave process; the tubular and autoclave processes are almost similar except the reactor type and pressure requirement. Pressure requirement in the autoclave reactor (continuous stirred tank reactor, CSTR) is usually lower than that employed in the tubular reactor. Depending on the particular product to be manufactured, measured doses of selected free radical (peroxide) catalysts, assorted comonomers and modifiers such as hydrogen, if required, are also to be fed to the high pressure rectors along with ethylene monomer.

Slurry phase polymerization encompasses the largest group of HDPE technologies. Excellent temperature control is a major advantage of the slurry process. However, for the production of LLDPE, by using an α-olefin comonomer (3–6%) along with ethylene as the bulk monomer, swelling of the copolymer produced in the slurry medium may pose some problem causing lowering of polymer production rate and in its density range. Long jacketed loop reactors and continuous stirred tank reactors are most widely used for slurry polymerization that is commonly operated at 30–40 atm pressure and at 85–110°C temperature.

Solution phase processes have some distinctive advantages over slurry processes in that the molecular weight distribution (MWD) can be better controlled along with easy control of the process variables in the inherent homogeneous phase. High polymerization temperature also permits high reaction rates and high polymer throughputs in the reactor. The solution process tends to produce polymers of relatively wide density range and high melt index. A pressure of 100 atm and a temperature of 150°C are commonly maintained in the solution (CSTR) process. Because of the short residence time and high polymer concentration, reactors of smaller size, including tubular reactors put in series or parallel to the CSTR, may be used for control of MWD of the relevant HDPE or LLDPE products.

The distinctive advantage of the gas phase polymerization process is the elimination of polymer separation, diluent recovery or handling and polymer drying steps. The gas phase process is also not constrained by solubility, swelling and viscosity characteristics but it suffers from the disadvantages like limitation of reaction operating temperature to resin softening point, agglomeration of polymer particles due to generation of local hot spots and heat transfer limitation etc.

Polyolefin synthesis is slowly changing from a semi-empirical art to an exact science due to improved understanding of nature of active centres, stereoregulating mechanisms, control of polymer architecture including monomer sequence distribution and process parameter–polymer composition–product properties relationship.

Bimodal and broad MWD LLDPEs are commonly produced in a low pressure gas phase process at a much lower cost than that of the high cost high-pressure LDPE process. Often LLDPE penetration into LDPE market is constrained by its performance deficiencies in respect of clarity and in extrusion coating. The gas phase higher α-olefin (Octene-1) LLDPE holds high promise in respect cost-performance benefit. These new LLDPEs will not only expand the total market of low-pressure, polyethylene (HDPE) but also take a larger market share of the high pressure product, LDPE.

Metallocene based new (third) generation polyolefin technology has the potential to provide a much better control over chain length, chain branching, tacticity, compositional homogeneity and MWD, all resulting in improved/controlled rheology and morphology of the product polymers and copolymers so produced. The new generation polyolefin manufacturing processes (slurry, bulk or solution and gas phase) using high performance catalyst systems have improved the reactor productivities by two to three fold (showing scope for even higher productivities) and substantial increase in manufacturing capabilities for specific product grades. Binary random copolymers and di, tri or quadri block or random copolymers providing wide scope for improvements in product properties and processing characteristics have been developed and further refinements are in the process of development. Metallocene based technology had its beginning in the early nineties and it is still in the process of development and is expected to change the polyolefin scenario in the new millennium. It is likely to be further extended to polymerization of vinylic and diolefinic monomers too.

Newly developed metallocene based LLDPE has a characteristically narrow molecular weight distribution and the potentiality of higher comonomer incorporation compared to LLDPE produced by the Ziegler–Natta and Chromium catalysts. The metallocene LLDPE exhibits a lower and sharper melting point, better hot tack and improved heat-seal, clarity and impact properties. Narrow MWD, however, may pose some processing difficulties which are meticulously overcome by controlled incorporation of a higher α-olefin (providing longer branches).

Metallocene based polypropylene (PP) synthesis offers option to change each property of the polymer on molecular level, independent of each other, and also enabling synthesis of syndiotactic PP that is not obtained by supported titanium based Ziegler–Natta catalysts. Elastomeric PP consisting blocks of atactic and isotactic stereosequences can be synthesized through unprecedented control over the micro-structures. The liquid-pool (using liquefied propylene) and gas-phase PP processes are not only advantageous on account of productivity and production economy, they are also advantageous due to elimination of catalyst residue and atactic removal steps.

Reactor-made polyolefin alloy is an important area of commercial relevance and significance due to its better performance characteristics compared to blends/alloys made by post-reactor blending processes. The reactor PP alloys are made by repeated sequential introduction of selected comonomers during polymerization of propylene. The properties of such polyolefin alloys are controlled by the chemical structure and physical state of the different phases provided by the sequential blocks. These products are commercially produced on small scales by what is known as the Catalloy process. These novel flexible, tough and impact resistant clear PP based alloy resins are developed to compete with non-olefinic (plasticized) polymers, such as poly(vinyl chloride). Use of styrene in sequential monomer addition in the Catalloy process along with propylene may produce polymer alloys with higher heat distortion temperatures, stiffness and chemical resistance. Such polyolefinic resins are viewed as suitable competitors of polyamides, polyacetals polyphenelene oxide and polycarbonates. Future developments are aimed at working of computer controlled swing reactors enabling production of HDPE, LLDPE and random or block copolymers of ethylene, propylene and higher olefins using a single reactor by appropriately changing feed monomer, catalyst, solvent and other additives and manipulating physical process parameters.

8.5 Copolymers of Ethylene

Among the copolymers of ethylene, those between ethylene and propylene and between ethylene and vinyl acetate have reached a stage of widespread development and application. Other important copolymers involving ethylene and some other α-olefin such as 1-butene, include the wide range of linear low density polyethylene (LLDPE). The α-olefin copolymers are produced by the Ziegler or Phillips process, while copolymers of ethylene with vinyl acetate, maleic anhydride and acrylate monomers are industrially produced by radical polymerization.

Stereo block copolymers of propylene and ethylene (4–15% ethylene) show improved impact properties. Though copolymeric in nature, they are essentially considered as polypropylenes of minor chemical modifications. When the two monomers are used in nearly equal or 60 : 40 proportions, products commonly known as ethylene-propylene rubbers (EP rubbers) are formed on copolymerization. EP rubbers are more aging resistant than diene rubbers, being devoid of unsaturations and they are conveniently vulcanized by peroxides, e.g., dicumyl peroxide. Carbon black is used with EP rubber to develop moderate to high tensile strength and modulus. Low dose of sulphur (0.3 phr) is often used along with a peroxide (2–3 phr) as the curative and Ca-stearate is preferred to stearic acid as the activator.

Ethylene vinyl acetate (EVA) copolymers are available in different grades varying in vinyl acetate (VA) content (3–48%). Those containing low VA content (3–5%), are considered as modified low density polyethylenes. Their high surface tack, disadvantageous in conventional film applications, is taken advantage of in meat packaging as stretch films as well as for cling-wrap purposes. In higher vinyl acetate compositions, EVA copolymers are used as wax additives, wax coatings, hot melt adhesives and elastomers. Selected grades of hydrolyzed EVA, commonly designated as EVOH, offer excellent barrier properties with respect to gas (oxygen) permeability.

Copolymerization of ethylene with carboxylic monomers such as maleic anhydride, methacrylic acid, etc., by the high pressure process produces materials which are used as thermosetting materials but processed like thermoplastics. When treated with metal salts, oxides or alkoxides (Zn, Ca, Mg, etc), ionic (metal carboxylate) cross-links, stable under ambient conditions, are readily established. The ionic cross-links get weakened with rise in temperature and their influence disappears at the processing (melting) temperatures, thus permitting reshaping. On cooling after reshaping, the ionic cross-links are regrouped and reformed and the thermosetting character reappears. Such polymers with "meltable" cross-links are known as "ionomers".

8.6 Polystyrene

8.6.1 Monomer Synthesis

Polystyrene is another high tonnage, general purpose thermoplastic polymer. The monomer, styrene, or vinyl benzene, $CH_2 = CH \cdot C_6H_5$ is made from benzene and ethylene in two steps. The first step involves alkylation of benzene (formation of ethyl benzene) through reaction with ethylene in the presence of a Friedel–Crafts Catalyst such as $AlCl_3$.

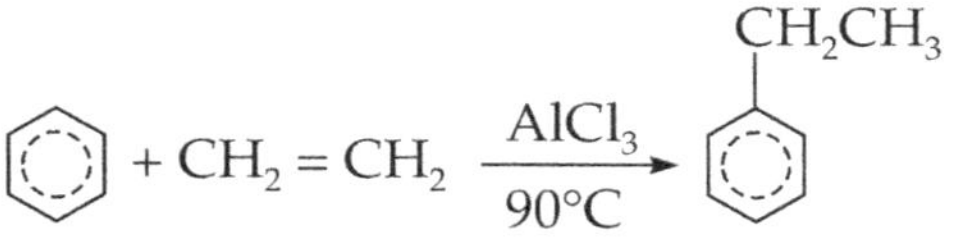

(8.4)

In a subsequent step, ethyl benzene is dehydrogenated to styrene by heating its stream over iron oxide or magnesium oxide catalyst at about 600°C. Styrene, isolated from the dehydrogenation product by distillation, is stabilized by addition of traces of a phenolic stabilizer, such as t-butyl catechol and then stored in a cold storage for longer shelf-life.

8.6.2 Polymerization of Styrene

Polymerization of styrene is mostly done following the bulk polymerization or suspension polymerization techniques. The bulk polymerization is done in two stages. In the first stage, the stabilizer-free monomer containing dissolved initiator is allowed to polymerize at about 80°C in one or more "prepolymerizer" batch reactors, each fitted with an efficient stirrer and an outer jacket for circulation of heat transfer fluid until a conversion of about 30% is reached. The viscous monomer–polymer mixture is then forced into the top of a tower reactor (nearly 25–40 ft in height and 10–15 ft in diameter) which is fitted with heating and cooling jackets and coils for external and internal control of temperatures respectively. The temperature down the tower changes from 100°C at the top to nearly 180–200°C at the bottom. The melt containing polystyrene and traces of the monomer passes through an extruder at the bottom from which it emerges as narrow strands or filaments which are then cooled, granulated and packed.

8.6.3 Structure and Properties of Polystyrene

Commercial polystyrene is a radically initiated linear atactic polymer. It is amorphous and transparent. It comes in different grades such as general purpose grade, high molecular weight grade and heat resistant grade. Removal of last traces of unreacted monomer (from 5–0%) improves the softening point or heat distortion temperature from nearly 70°C to about 100°C. The polymer is hard, brittle and transparent at room temperature. It is soluble in aromatic hydrocarbon solvents, cyclohexane and chlorinated hydrocarbons. The polymer has a density of 1.054 g/ cm^3, a refractive index of 1.59–1.60 and a T_m value of 240°C. It sounds like metal when dropped. It decomposes at elevated temperature into a mixture of volatiles, a significant part of which is the monomer which can be identified from its characteristic smell (*See* Table 6.1).

Polystyrene, being hydrocarbon in nature, has very low moisture absorption; this feature coupled with its good mouldability, dimensional stability and low moulding shrinkage makes polystyrene an ideal polymer for injection moulding and for thermoforming. As a rigid thermoplastic, it has moderate tensile strength (6000–8000 psi) and very low elongation at break (1–3%). The major limitations of the polymer are its brittleness and very low impact strength (Izod 0.2–0.35 ft lb in^{-1} notch) and low heat distortion temperature (~80°C).

Polystyrene has poor outdoor weathering resistance and a tendency to yellowing and crazing on long use. It possesses very good electrical insulation characteristics and a low dielectric loss factor at moderate frequencies.

8.6.4 Modification to High Impact Grades

There have been many attempts to improve impact properties of polystyrene, but those achieved through use of rubbery additives and through copolymerization approach have been commercially successful and widely practised. Rubber modified polystyrenes are used to nearly the same extent as straight polystyrene. Rubbers commonly in use for the purpose are styrene–butadiene rubber (25–30% styrene content) or simply polybutadiene rubber. Simple blending of the rubbery polymer with polystyrene done on two-roll mills or in the latex form brings only marginal improvements in impact strength. Much better improvements are achieved by polymerization of styrene in an emulsion of the rubbery polymer or in solution containing the rubbery polymer, through what may be termed as a graft copolymerization approach. To derive the best effect, the rubber used should be semicompatible with polystyrene and should have limited degrees of cross-linking. The rubber should be dispersed in fine discrete droplets within the polystyrene matrix, the rubbery phase comprising only 5–20% of the total volume of the resin–rubber composite. Such high impact grade polystyrenes may have two to five times the impact resistance of straight polystyrene, but they exhibit nearly half the tensile strength and a much lower softening point. They also show inferior aging resistance and require stabilization through incorporation of adequate doses of antioxidant stabilizers.

8.6.5 Styrene–Acrylonitrile (SAN) Copolymers and ABS Resins

To obtain a styrene-based polymer of higher impact strength and higher heat distortion temperature at the same time, styrene is copolymerized with 20–30% acrylonitrile. Such copolymers have better chemical and solvent resistance, and much better resistance to stress cracking and crazing while retaining the transparency of the homopolymer at the same time. In many respects SAN copolymers are also better than poly(methyl methacrylate) and cellulose acetate, two other transparent thermoplastics.

ABS resins are terpolymers of acrylonitrile, butadiene and styrene, prepared by interpolymerization (grafting) of styrene and acrylonitrile on polybutadiene or through blending of SAN copolymers with butadiene–acrylonitrile (Nitrile) rubber. Impact improvement is far better if the rubber in the blend is lightly cross-linked. The impact resistance of ABS resins may be as high as 6–7 ft lb. per inch of notch.

8.6.6 Processing, Uses and Applications of Polystyrene

Polystyrene and related copolymers, terpolymers and blends are widely processed by injection moulding, blow moulding, thermoforming and extrusion techniques. Expanded polystyrene finds extensive use in packaging and shock absorbing applications, in thermal insulation and as acoustic improvers in halls and auditoria. Largest outlet of polystyrenes is in the packaging field. High impact grades are suitable for use as toys, games and sports articles, casings and cabinets for electrical/electronic gadgets and equipments and inner liners of refrigerators. ABS resins have emerged as engineering plastics for toughness, strength and durability and are being increasingly used as housewares and as housings for domestic and industrial electrical equipment, automobile components, etc. Another major outlet of polystyrene is in the making of ion-exchange resins. For this, the polymer is produced in the cross-linked form by suspension polymerization technique using divinyl benzene (1–5%) as the comonomer. Modifications of the suspension resin beads by controlled sulphonation with concentrated H_2SO_4 produces cation-exchange resin. Chloromethylation of the resins with chloromethyl ether and their subsequent transformation into quaternary ammonium salts by treatment with tertiary amines followed by alkali treatment lead to the formation of anion-exchange resins.

8.7 Acrylic Plastics: Poly(Methyl Methacrylate)

Plastics based on acrylate and methacrylate monomers are commonly known as acrylic plastics. Most important among the acrylic plastics is poly(methyl methacrylate) prepared by the polymerization of methyl methacrylate. Monomer methyl methacrylate is prepared from acetone through the cyanohydrin route. The cyanohydrin is treated with 98% sulphuric acid in a cooled kettle to yield methacrylamide sulphate intermediate which on esterification with methanol produces methyl methacrylate. The synthetic route is shown in the reaction scheme (8.5).

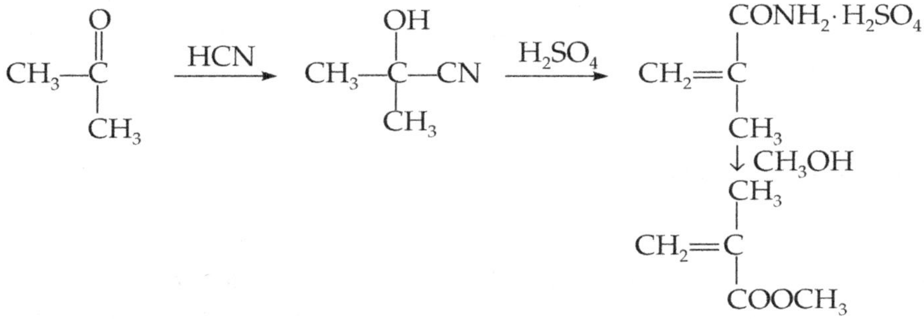

(8.5)

Moulding grade poly(methyl methacrylate) is made by bulk or suspension polymerization of the monomer, using peroxide or azonitrile initiators at about 100°C, preferably in the absence of air. For the production of cast sheets, the monomer is first converted into a prepolymer syrup of convenient viscosity and the syrup is then poured into the casting cell consisting of heat resistant, parallel, polished glass plates separated by rubber gaskets round the edges, and held together by spring loaded clamps or spring clips. Precautions must be taken to obtain sheets free of sink marks and voids. The cell is then gradually heated in a heating tunnel for 15–20h over a temperature zone of 50–100°C to complete the polymerization. Rods are made by polymerizing the syrup under pressure in vertical aluminium tubes, while tubes are prepared by polymerizing a calculated amount of syrup in an aluminium tube under nitrogen pressure, with its both ends closed. The assembly is heated as the tube is rotated horizontally at a constant rate, and the syrup polymerizes on the wall of the rotating tube. Natural shrinkage on polymerization enables easy removal of the casting.

A combination of excellent optical clarity (refractive index, 1.49), total lack of colour, some toughness and outstanding outdoor weathering behaviour makes poly(methyl methacrylate) very useful as automotive tail lamp and signal light lenses, jewellery, lenses of optical equipment and contact lenses. The polymer is very popular in display and advertisement applications; use in the building industry is also notable. Perspex, Plexiglas, Lucite and Acrylite are common trade names.

Minor proportions (5–25%) of different monomers such as *n*-butyl acrylate, 2-ethyl hexyl acrylate, methacrylamide, dimethyl aminoethyl acrylate, etc., are used to make coating grade poly(methyl methacrylate), usually by solution or emulsion techniques, for paints and enamel applications. Production of dentures is another important outlet of poly(methyl methacrylate). Poly(methyl acrylate) and poly(methyl methacrylate) may be saponified to form the polymeric acids which correspond to polymers of acrylic acid and methacrylic acid. They show characteristic/polyelectrolytic behaviour in aqueous solution. They are not suitable for use as formed or moulded objects, films, sheets, etc. As such or as sodium salts they are useful as thickening agents.

8.8 Acrylic Fibres

Fibres primarily based on polyacrylonitrile are known as acrylic fibres. Acrylonitrile monomer is produced by addition of HCN to acetylene or by dehydration of ethylene cyanohydrin formed by the addition of HCN to ethylene oxide. Acrylonitrile is a monomer which is a non-solvent for its polymer. Bulk polymerization turns out to be a case of heterogeneous polymerization and is associated with high heat generation and consequent to that, with hazards of violent explosion. The monomer is soluble in water to the extent of nearly 8% at room temperature and polyacrylonitrile of commerce is conveniently formed by heterogeneous aqueous polymerization of acrylonitrile using redox catalysts at or near room temperature. The polymer is obtained in the form of a fine precipitate.

Polyacrylonitrile, i.e., the homopolymer of acrylonitrile is unsuitable for many conventional thermoplastic applications as it suffers discolouration and infusibility on heating. It starts yellowing on heating and finally turns black, insoluble and infusible (Sec. 6.4.3). It is ordinarily insoluble in most common solvents. However, dimethyl formamide, dimethyl sulphoxide and tetramethylene sulphone are good solvents for the polymer. It also dissolves in aqueous concentrated solution of some salts. The polymer is dissolved in one of these solvents and then processed by wet or dry spinning techniques to produce the fibre.

To improve certain properties such as dyeability, feel and texture of the acrylic fibre, the polymer is prepared by copolymerizing 5–15% of one or more monomers, such as vinyl pyrrolidone, vinyl acetate, acrylates, acrylic/methacrylic acid, along with acrylonitrile. The incorporation of the comonomer units makes the fibre more suited for dyeing with many common dyes for textiles. Fibres from copolymers having much higher proportions (20–35%) of a comonomer, usually vinyl chloride and vinylidene chloride, are commonly known as modacrylic fibres.

The commercial acrylic fibres come in different colours and shades. They have outstanding weathering resistance. They also exhibit good resistance to chemicals, stains, insects and microorganisms. In bulkiness, feel and warmth, the acrylic fibres are very much similar to wool. They are widely blended with other fibres, particularly wool, to form various textile items.

8.9 Poly(Vinyl Acetate)

Vinyl acetate monomer is most conveniently prepared by catalytic addition of acetic acid to acetylene, which may be accomplished by a liquid-phase or a vapour-phase process. The vapour-phase synthesis is more in commercial practice and is accomplished by passing a mixture of acetylene and acetic acid through a reaction tube packed with such catalysts as cadmium acetate or zinc acetate at a temperature of 220–250°C. The monomer formed is purified by distillation.

$$\mathrm{CH{\equiv}CH + CH_3COOH \rightarrow CH_2{=}\underset{\displaystyle OCOCH_3}{\underset{|}{CH}}} \tag{8.6}$$

Because of soft and gummy nature of poly(vinyl acetate), the polymer has poor dimensional stability. It rarely finds an application as moulded or formed objects and is seldom processed like common thermoplastics. The polymer is widely used in an emulsion form and hence, emulsion polymerization of vinyl acetate is widely employed for commercial purposes. The commercial emulsion polymerization is usually accomplished at a temperature of 50–70°C in a reactor with all provisions for temperature control. For desirable control of particle size and particle size distribution in the emulsion on latex and to regulate chain branching, a measured stream of the monomer and initiator is slowly and steadily added over a period of 3–4 h into the polymerization system and the polymerization is usually stopped after about 4–6 h.

Poly(vinyl acetate) as formed commercially is atactic in nature and hence amorphous and it is transparent if free from contaminants such as emulsifiers. It is soluble in benzene, acetone and chloroform. Poly(vinyl acetate) is also widely miscible with methanol. It finds widespread use as adhesives for wood, decorative laminates and related items and as a sizing material for paper and textiles; gum application (chewing gum) is also popular.

8.9.1 Polymers Derived from Poly(Vinyl Acetate)

Poly(vinyl acetate) is also used as the intermediate for the commercial production of (i) poly(vinyl alcohol) and (ii) poly(vinyl acetals). The monomer, vinyl alcohol, does not exist in the free state and all attempts to make it result in the formation of the stable tautomer, acetaldehyde.

$$CH_2{=}\underset{\displaystyle OH}{\underset{|}{CH}} \rightarrow CH_3{-}\underset{\displaystyle O}{\underset{\|}{CH}} \qquad (8.7)$$

(i) *Poly(vinyl alcohol)* Poly (vinyl alcohol) is prepared by the alcoholysis of poly-(vinyl acetate) as shown in reaction (8.8). The process is often inappropriately termed as "hydrolysis". Methanol is preferred over ethanol for the alcoholysis which is better carried out with alkaline catalysts (NaOH, $NaOCH_3$, etc.) than with acidic catalysts. The process is accomplished much quicker (< 1 h) at a higher temperature (say, 60°C) than at or near room temperature (6–8 h at 20–25°C). The average degree of polymerization of poly(vinyl alcohol) is considerably lower than that of the parent poly(vinyl acetate) because the alcoholysis process splits the branch units formed via the acetate groups.

$$\left[CH_2{-}CH(OCOCH_3) \right]_n + n\,CH_3OH \longrightarrow \left[CH_2{-}CH(OH) \right]_n + n\,CH_3{-}\underset{\displaystyle O}{\underset{\|}{C}}{-}OCH_3 \qquad (8.8)$$

Polymers with a degree of alcoholysis close to or higher than 85% (85% or more of the acetate groups of poly(vinyl acetate) transformed into alcohol (OH) groups) are readily soluble in water. For polymers with >90% of the substituents transformed into OH groups by alcoholysis, water-solubility follows a falling trend with increasing degree of alcoholysis.

Poly(vinyl alcohol) is normally unsuitable as a material for thermoplastic applications. It tends to decompose before melting, liberating moisture and forming chain unsaturations. Extended heating leads to discolouration and insolubilization.

Readily water soluble poly(vinyl alcohols) are made by keeping the degree of alcoholysis in the range of 80–90% and they find widespread use as thickening agents for aqueous suspensions and emulsions. Films obtained from them are useful as water-soluble packages. Tubular film forming technology has been developed for poly(vinyl alcohol). As a polyol, it behaves as a nonionic surfactant and is extensively used as a stabilizer in suspension and emulsion polymerization. It is also variously useful as a binder, adhesive and sizing agent in ceramic, paper and textile industry.

Poly(vinyl alcohol) in the form of a warm aqueous solution may be wet spun into a concentrated aqueous salt (Na_2SO_4) solution containing known proportions of H_2SO_4 and formaldehyde whereby it is insolubilized through formation of formal groups and drawn into a stable fibre. Establishment of limited cross-linking through formaldehyde helps in optimizing the shrinkage property of the fibre. The fibre is useful as a textile material with washability and quick drying properties.

(ii) ***Poly(vinyl acetals)*** Poly(vinyl acetals) are products of interaction between poly(vinyl alcohol) and various aldehydes. Products based on reactions involving formaldehyde, acetaldehyde, propionaldehyde and butyraldehyde are separately known as poly(vinyl formal), poly(vinyl acetal), poly(vinyl propional) and poly(vinyl butyral) respectively. Reactions usually involve condensation between 1,3-hydroxyl groups of poly(vinyl alcohol) and the aldehyde molecules. Besides containing the acetal structures formed,

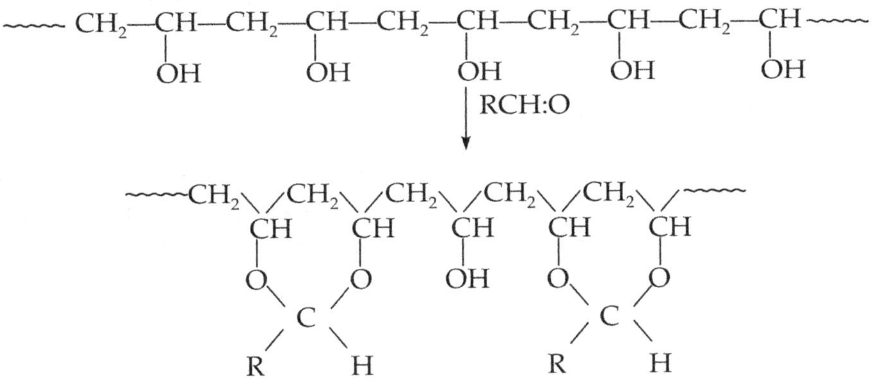

(8.9)

the derived polymer molecules also contain some residual hydroxyl groups as well as some residual acetate groups retained in the polymer due to incomplete alcoholysis of the original poly(vinyl acetate) to poly(vinyl alcohol).

Among the poly(vinyl acetals), only poly(vinyl butyral) is important considering commercial aspects. It finds extensive use as safety glass interlayer, particularly because of its outstanding clarity, toughness, photostability, moisture insensitivity and exceptional adhesion to glass. The polymer may be prepared by the direct process from poly(vinyl acetate); but for safety glass purposes it is most conveniently prepared from poly(vinyl alcohol). The catalyst most commonly used is H_2SO_4. If the reaction is done using aqueous solution of poly(vinyl alcohol) and butyraldehyde to give a homogeneous system to begin with, poly(vinyl butyral) precipitates out as it is formed. On the other hand, if one begins with an initial inhomogeneous system, i.e. a suspension of poly(vinyl alcohol) in water-ethanol mixture and requisite proportions of butyraldehyde and H_2SO_4 catalyst, one obtains a homogeneous solution at the completion of the reaction. Poly(vinyl butyral) is precipitated from the solution by addition of water under stirring conditions, washed and dried. For safety glass application, about 25% of the hydroxyl groups are usually left unreacted, and about 30–40 wt.% of a plasticizer, such as dibutyl sebacate or triethylene glycol di-2-ethyl butyrate, are commonly used.

8.10 Poly(Vinyl Chloride)

Poly(vinyl chloride), commonly named as PVC, is the most important of the vinyl thermoplastics considering volume of production and fields of application, the commercial products ranging from very rigid to very flexible items. The polymer is highly unstable when thermally treated at the processing temperatures and it is due to the discovery of a variety of heat stabilizers that the prospect of PVC technology became very bright and the technology reached its present height of success.

8.10.1 Preparation of Vinyl Chloride

Vinyl chloride monomer ($CH_2 = CH \cdot Cl$) is being synthesized throughout the world on commercial scales mainly according to the following processes employing (1) acetylene, and (2) ethylene as the immediate organic raw materials. Main processes developed and in practice are:

(i) hydrochlorination of acetylene,

(ii) chlorination of ethylene to ethylene dichloride (EDC) and thermal cracking of the latter to vinyl chloride and hydrogen chloride,

(iii) the byproduct hydrogen chloride in process (ii) can be utilized in: (a) hydrochlorinating acetylene to produce more vinyl chloride straight away, or (b) in oxychlorinating more ethylene to produce EDC, and

(iv) mixed gas process starting with a dilute mixed stream of acetylene and ethylene.

Historically, vinyl chloride is definitely an acetylene derivative and as late as 1965, more than 45% of all vinyl chloride produced was still based on acetylene derived from coal via calcium carbide or from petroleum sources. However, the technology later shifted overwhelmingly in favour of ethylene.

The earliest route to vinyl chloride (VC) was from acetylene and HCl. In a typical synthesis

$$CH \equiv CH + HCl \rightarrow CH_2 = CH \cdot Cl \qquad (8.10)$$

dry hydrogen chloride free from chlorine is mixed with an equimolar proportion of dry acetylene and the mixture is then passed through a multitubular reactor packed with mercuric chloride catalyst on an activated carbon support. Temperature is maintained at 90–100°C. Vinyl chloride monomer formed is then purified and stored under nitrogen in stainless steel tanks.

As ethylene became more abundant, technology based on chlorination of ethylene to ethylene dichloride (EDC) and then cracking of EDC to vinyl chloride was developed. The byproduct HCl was recovered and utilized to make more vinyl chloride by reaction with acetylene.

$$CH_2{=}CH_2 \xrightarrow{Cl_2} Cl \cdot CH_2{-}CH_2 \cdot Cl \xrightarrow[500°C]{-HCl} CH_2{=}CHCl \tag{8.11}$$

The next development was the so-called 'oxychlorination' process involving reaction of HCl with ethylene in presence of air (oxygen) to produce EDC. In a subsequent step,

$$CH_2{=}CH_2 + 2HCl + \tfrac{1}{2}O_2 \xrightarrow{CuCl_2} Cl \cdot CH_2{-}CH_2 \cdot Cl + H_2O \tag{8.12}$$

the EDC is cracked to yield vinyl chloride and HCl. The byproduct HCl is recycled in the oxychlorination step.

Mixed gas processes based on cracking of naphtha to equimolar proportions of ethylene and acetylene have also been developed with the objective of total chlorine utilization. The mixture of the two is first reacted with HCl to form vinyl chloride, the reaction taking place between the acid and acetylene contained in the cracked gases. Vinyl chloride formed is separated from ethylene which is then chlorinated to EDC and then cracked to vinyl chloride and HCl. The mixed gas processes are competitive with the oxychlorination process.

8.10.2 Polymerization of Vinyl Chloride

Commercial polymerization is done by using free radical catalysts and employing bulk, suspension and emulsion techniques. Suspension and emulsion techniques are, however, most commonly employed.

Bulk polymerization is heterogeneous in view of insolubility of the polymer in the monomer. Peroxy dicarbonates are conveniently used as initiators in the bulk or suspension polymerization. Bulk polymerization may be done in rotating cylindrical reactors with tumbling steel balls inside to facilitate removal of heat of polymerization from the monomer–polymer heterogeneous system. The bulk polymerization may also be done in two stages as in the Pechiney–St. Gobain process to overcome the difficulties of heat removal and to have better control on molecular weight and particle size of the polymer. Suspension and emulsion polymerizations are done in stirred tank jacketed pressure vessels following approaches as described in *See* Secs 4.11.3 and 4.11.4. For injection moulding, extrusion and calendering purposes, particularly for clear objects and for electrical insulation purposes, bulk and suspension grade PVC resins are used. Emulsion grades are utilized in organosols and plastisols and in some other areas, but they are unsuitable in insulation (wire and cable) industry because of the presence of traces of detergent or soap and other ionic components of the polymerization recipe in the isolated polymer (PVC).

8.10.3 Structure and Properties of PVC

Poly(vinyl chloride), PVC is by and large a linear polymer, colourless and thermoplastic in nature, and having a chlorine content of about 56.8%. The polymer is thermally unstable and extensive heating transforms it into a dark coloured residue resembling polyacetylene and liberating HCl as the volatile (*See* Sec. 6.4.3). PVC is insoluble in all hydrocarbon solvents. Two of its important solvents are cyclohexanone and tetrahydrofuran. It is also soluble in ethylene dichloride and nitrobenzene. It possesses flame retardation and self-extinguishing characteristics. The polymer as produced commercially, is substantially amorphous in nature. Structural irregularities in the polymer arise due to occasional branching effect during polymerization and due to chain-end unsaturation consequent to termination by disproportionation. Technically, the polymer is graded on the basis of a solution viscosity parameter known as the K-value but not by melt viscosity or melt flow index, in view of its poor thermal stability.

Commercial polymers differ not only in molecular weight and molecular structure (degree of branching) but also in their particle characteristics such as porosity, shape, size and size distribution. The processing behaviour of the polymer is largely linked with these particle characteristics. In its massive form, PVC is a hard, horny, rigid material with a characteristic tendency to stick to metallic surfaces at elevated temperatures.

8.10.4 Compounding and Processing of PVC

PVC can be conveniently processed to rigid items if only it is compounded with stabilizers and lubricants. It is variously compounded with many other compounding ingredients such as fillers, plasticizers, extenders, and other process aids, impact improvers, colouring matters, etc.

Stabilizers of PVC protect it from measurable degradation at processing temperatures. Among the common stabilizers are basic lead salts such as basic lead carbonate, tribasic lead sulphate, dibasic lead phosphate, etc. Organo compounds of other metals used as stabilizers include those of cadmium, barium, calcium, zinc and tin, mostly in the form of phenates, octoates, benzoates and laurates. Some of them produce much improved effects in presence of phosphite antioxidants, such as trisnonyl phenyl phosphite. Organo-tin compounds are specially useful for producing (crystal) clear compounds and products.

All flexible applications of PVC depend on significant use of plasticizers; notable among them are the high boiling phthalates such as, dibutyl phthalate, dioctyl or diisooctyl phthalate, etc., and the phosphate plasticizers, such as trioctyl phosphate and tricresyl phosphate. Chlorinated paraffin and the phosphate plasticizers are used in fire retardant compounds. Aliphatic esters such as dibutyl sebacate and dioctyl sebacate or adipate are specially useful for having compounds with high resilience and a low cold flex temperature.

Impact modification of rigid PVC is variously accomplished by blending with it different proportions of such polymers as nitrile rubber, ABS graft terpolymers, chlorinated polyethylene, selected polyacrylates, etc.

Compounded PVC is converted into moulded or formed objects by melt processing or by processing of PVC pastes and latices. Widespread processing techniques include injection moulding, extrusion, calendering, blow moulding and thermoforming. The processing of unplasticized (rigid) PVC is much more difficult and critical than that of plasticized PVC, primarily because of much higher temperature needed for the processing of the former at which measurable decomposition of the polymer occurs. However, trouble-free processing of rigid PVC can be done through judicious blending and compounding of lubricants, stabilizers (organotin compounds) and other process aids. Blending or compounding of different ingredients is conveniently accomplished by dry blending of powders, thus avoiding unnecessary heating in mills and mixers.

8.10.5 Applications of PVC

Both rigid and flexible applications of PVC have been developed. Rigid applications include chemical plants and equipments, storage tanks, building items, pipes, sheets, specific moulded objects and containers. PVC guttering and rain water piping, window frames and transparent roof sheeting are some of its building applications. Floor tiles and wall linings from plasticized PVC are also worthy of mention. Other flexible or semi-rigid applications include toys, packaging items, tubes, pipes and hoses, leather cloths, moulded objects, sheets, films, containers, footwear, belting, wire insulation and cables.

8.10.6 Copolymers of Vinyl Chloride

The processing temperature of PVC can be substantially lowered by plasticizer incorporation (external plasticization) or a comonomer incorporation (internal plasticization). Vinyl chloride–vinyl acetate (VC–VA) copolymers containing about 15% vinyl acetate can be processed at about 130–140°C. Gramophone records moulded from VC–VA copolymers are made from compositions containing only stabilizer, lubricant, pigment and sometimes an anti-static agent as the additives. The copolymer loaded with large proportions of filler and other essential ingredients finds significant use in flooring. Vinyl chloride copolymers with vinyl acetate (15%) or acrylonitrile (40%) as comonomers can be drawn into fibres. Copolymers of vinyl chloride (15%) and vinylidene chloride (85%) known as Saran in the trade have excellent chemical resistance including resistance to acids and alkalies and prominent self-extinguishing properties. Saran films (biaxially oriented) have excellent clarity, strength, toughness and water and gas impermeability. The copolymer can be conveniently drawn into a fibre. Saran fibres are suitable as filter cloths for filtering acidic, alkaline or corrosive chemicals.

8.11 Polytetrafluoroethylene (PTFE)

Among the fluorocarbon polymers, the use of polytetrafluoroethylene is relatively widespread and it is popular in certain specific areas of application. The monomer tetrafluoroethylene (bp – 76.3°C) is obtained by pyrolysis of monochlorodifluoromethane in contact with platinum at about 700°C. Monochlorodifluoromethane (bp 40.8°C) is obtained by reacting chloroform with hydrofluoric acid:

$$CHCl_3 + 2HF \rightarrow CHClF_2 + 2HCl \tag{8.13}$$

$$2CHClF_2 \rightarrow CF_2{=}CF_2 + 2HCl \tag{8.14}$$

Commercial polymerization of tetrafluoroethylene is essentially aqueous polymerization accomplished by using free radical initiators (persulphate or hydrogen peroxide). An elevated pressure is maintained during polymerization. Polymer is obtained either in a granular form or in the form of a fine aqueous dispersion.

PTFE is essentially a linear polymer having a density of about 2.2 g/cm^3. The properties of the polymer depend on its particle size and molecular weight. As manufactured, the polymer has a crystallinity > 93%. There are no solvents for it at room temperature. The degree of crystallinity of processed items depends much on the rate of cooling. Polymers of lower molecular weight are usually more crystalline. PTFE has a waxy feel and it is self-lubricant in nature, having a coefficient of friction lower than any other solid. Aerosol dispersions of low molecular weight PTFE are effective dry lubricants. PTFE shows a moderate tensile strength (2500–4000 psi) and excellent heat resistance, its melting point (T_m) being 327°C. Its electrical insulation property, weathering resistance and chemical resistance are excellent.

Above melting point, the melt viscosity of PTFE is very high, so much so that the normal processing technology for thermoplastics is not applicable. For moulding, techniques similar to those employed in powder metallurgy or ceramic processing are used for PTFE. The process involves preforming the powder usually at room temperature under a pressure of 2000–8000 psi, sintering at 370°C and finally cooling. The dispersions can be used for casting films, dip coating, etc.

Major applications of PTFE are as seals, films, gaskets, laboratory equipments or components thereof, packings in pumps and valves, stopcocks, machine components, kitchenwares (non-stick PTFE coated pans), etc. and in electrical insulation and electronics. Its high volume cost is a constraint for its use for making large objects. Outstanding weather resistant and chemical resistant polymers are also obtained from polymerization and copolymerization of chlorotrifluoroethylene, vinyl fluoride, vinylidine fluoride and hexafluoropropylene (*See* also Chapter 9).

8.12 Coumarone-Indene Resins

Coumarone and indene are two major constituents of the fraction boiling at 165–180°C obtained from fractionation of coal tar naphtha. Being aromatic in nature and bearing vinylic kind of structure attached to the aromatic ring, they closely resemble styrene. For resinification, they are polymerized in the crude naphtha without separation using sulphuric acid as the (ionic) catalyst. The commercial polymers or resins are obtained in the molecular weight range of 2000–2500, giving an average degree of polymerization of 20–25. The resins of different grades, from sticky and soft to hard and brittle, are commercially available. They are variously used as process-aids and property improvers in plastics and rubbers and in the formulation of coatings and adhesives. They are also useful as binders of different fibrous or particulate fillers, waxes, etc., for making hard thermoplastic floor tiles.

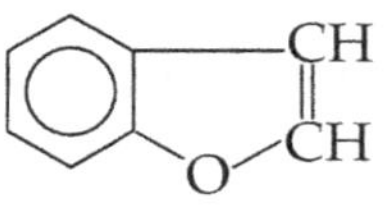

coumarone
bp 168-172°C

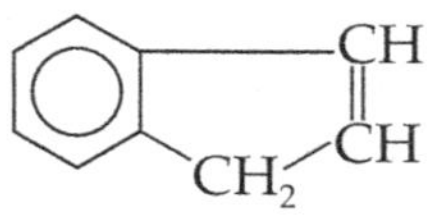

Indene
bp 182°C

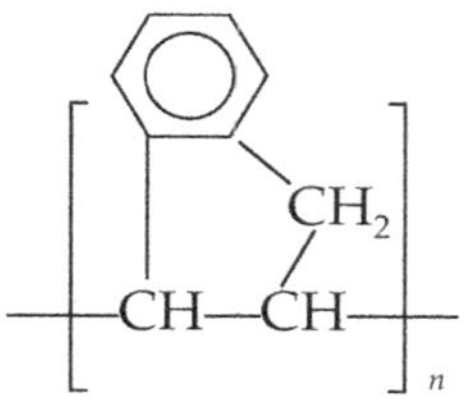

polyindene

8.13 Polyacetals and Polyethers (Acetal Resins)

Hydrocarbon chain segments when interlinked with one another by (—O—) ether linkages give polymeric structures commonly known as polyethers. Polyoxymethylenes $-(-CH_2—O-)_n-$ or polyformaldehyde resins derived from formaldehyde are commonly referred to by the generic term acetal resins.

For polymerization purposes, formaldehyde of very high purity is required. Polymerization may be done by ionic mechanism in a dry inert solvent such as *n*-hexane or *n*-heptane over a wide range of temperature (–40 to +60°C). The polymer is formed under rapidly stirring conditions and is removed as a slurry. Catalysts used include Lewis acids, amines, phosphines, arsines and stibines. Heat stability of the polyoxymethylenes or the acetal resins is much improved by transforming OH end groups to acetate groups using acetic anhydride as the acetylating agent and sodium acetate as the catalyst (0.01% based on the anhydride) at about 140°C. Under these conditions acetylation is substantially complete in 5 min.

$$\sim\sim CH_2—O—CH_2—OH + (CH_3CO)_2O \rightarrow \sim\sim CH_2—O—CH_2OCOCH_3 + CH_3COOH \quad (8.15)$$

Processes based on copolymerization of small amounts of ethylene oxide, 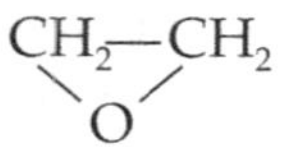 or 1,3-dioxolane 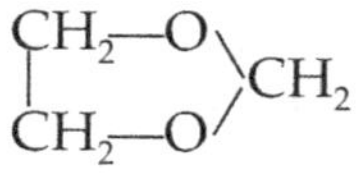 with formaldehyde have also been commercially developed to produce thermally stable polyoxymethylenes or acetal resins. Occasional incorporation of molecules having two successive methylene groups make the polymers much less prone to degradation.

Due to their simple structures and linearity of the chains, the polyoxymethylenes are capable of crystallization just as the polyethylenes. In commercial polymers, copolymerization leading to thermal stability tends to reduce crystallinity. The density of acetal homopolymer is 1.425 g/cm^3 while that of acetal copolymer is nearly 1.41 g/cm^3, their crystalline melting points are nearly 180° and 160°C and tensile strengths are 10000 and 8000 psi respectively. They are considered as engineering polymers because of their high melting point and high mechanical properties and resistance to chemicals and common solvents at room temperature. They are processable like all other thermoplastics. Typical applications of acetal resins cover machine parts and components, automobile instrument panels, in the construction of pumps and pump housings, etc. The polymers are not generally suitable for use in strong acid, alkali or oxidizing chemicals. They are commonly used with the incorporation of small quantities of an antioxidant and a UV stabilizer. Their ease of fabrication, low coefficient of friction and good corrosion resistance and resistance to fatigue also make them suitable as engineering plastics.

Linear polyethers known as poly(ethylene oxides) are conveniently made by controlled ring opening polymerization of ethylene oxide or through polycondensation of ethylene glycol. Low molecular weight poly(ethylene oxides) of the molecular weight range 1,000–20,000 are greasy to waxy

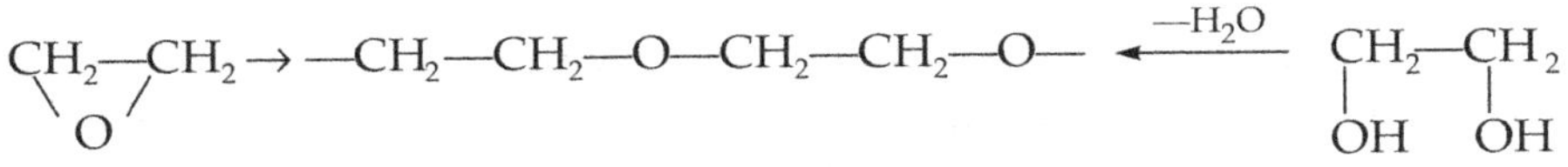

(8.16)

in appearance depending on molecular weight and they are readily soluble in water. High molecular weight grades (molecular weight >50,000) have melting points close to 66°C and tensile strength in the range of 1500–2000 psi. They can be processed like many thermoplastics. Poly(ethylene oxides) or poly(ethylene glycols) are particularly suitable for application requiring good water solubility together with very good to excellent film forming property. They are also used as textile sizes and thickening agents. Copolymers of epichlorohydrin and ethylene oxide are elastomeric in nature. Polyether from epichlorohydrin 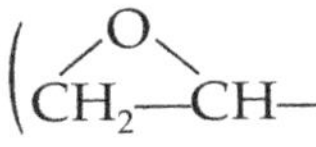 is also elastomeric. These elastomers are conveniently vulcanized with the help of diamines or polyamines or by using lead oxides. They have good heat resistance, excellent ozone resistance, and good weathering resistance. The polyepichlorohydrins are noted for their good flame resistance.

8.14 Polyamides

Synthesis of linear polyamides can be made by polycondensation of bifunctional acids and amines or by polycondensation of amino acids of the type $\mathrm{HOO}\underset{\mathrm{H}}{\underset{|}{\mathrm{C}}}\mathrm{RNH_2}$

or by ring opening polymerization of lactams of the type $(CH_2)_x\left<\begin{matrix}CO\\ |\\ NH\end{matrix}\right.$. The synthetic polyamides are commonly known in the trade by the generic term "nylon". Nylons from diacids and diamines are designated by code of two numbers, the first indicating the number of carbon atoms in a molecule of the diamine and the second indicating the number of carbon atoms in a molecule of the diacid (thus giving identity of the monomeric constituents). Polymers from amino acids or lactams are designated by a code of single number indicating the number of carbon atoms in a molecule of the amino acid or the lactam. Thus, the polyamide from caprolactam is known as nylon 6, that from hexamethylene diamine and adipic acid as nylon-66 or from hexamethylene diamine and sebacic acid as nylon 610. The polymer from ω-aminoundecanoic acid with 11 carbon atoms in its molecule is known as nylon 11.

8.14.1 Preparation of Poly(Hexamethylene Adipamide): Nylon 66

Nylon 66 or poly(hexamethylene adipamide) is synthesized by polycondensation of adipic acid and hexamethylene diamine. Adipic acid can be prepared by oxidation of cyclohexane or cyclohexanol either directly or via cyclohexanone. Benzene is used as the starting material for the production of cyclohexane and cyclohexanol. The different routes of formation of adipic acid are shown in the reaction scheme (8.17). Hexamethylene diamine is conveniently prepared from adipic acid via adiponitrile as shown in the reaction sequence (8.18).

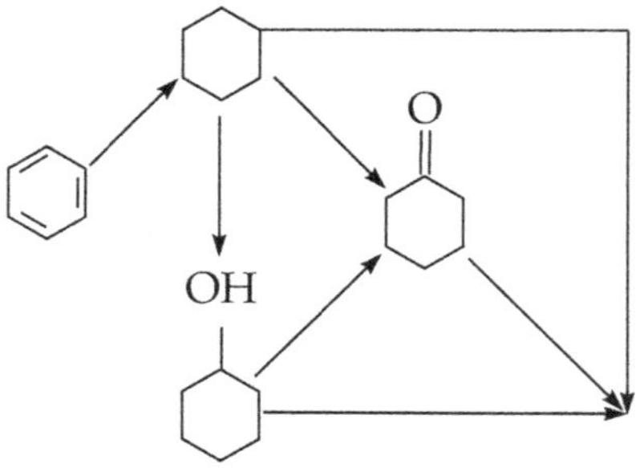

$HOOC\,(CH_2)_4\,COOH$ (8.17)

Adipic acid

$$HOOC \cdot (CH_2)_4 \cdot COOH \xrightarrow{NH_3} NC \cdot (CH_2)_4 \cdot CN \xrightarrow{H_2} \underset{\text{hexamethylene diamine}}{H_2N \cdot (CH_2)_6 \cdot NH_2} \quad (8.18)$$

The first step during the synthesis of nylon 66 or poly(hexamethylene adipamide) is the formation of the nylon 66 salt. The adipic acid and the hexamethylene diamine are reacted in boiling methanol to the insoluble nylon 66 salt (mp 190°C) which precipitates out. The salt is then dissolved in water,

$$HOOC \cdot (CH_2)_4 \cdot COOH + H_2N \cdot (CH_2)_6 \cdot NH_2 \longrightarrow \begin{matrix} {}^{-}OOC \cdot (CH_2)_4 \cdot COO^{-} \\ H_3\overset{+}{N} \cdot (CH_2)_6 \cdot \overset{+}{N}H_3 \\ \text{nylon 66 salt} \end{matrix} \quad (8.19)$$

mixed with about 0.5–0.1 mole percent of acetic acid to act as the viscosity stabilizer, i.e., to limit the molecular weight, and pumped into an autoclave. The autoclave is closed and the temperature is raised to nearly 220°C. The steam generated purges the air and a pressure of nearly 250 psi is allowed to develop. After about 2 h the temperature is raised to 270–280°C and steam is bled off to maintain the pressure at 250 psi. Pressure is then reduced over a period of 1–2 h to the normal level. The molten polymer is then extruded out under nitrogen pressure through a valve at the bottom of the autoclave onto a water-cooled casting wheel forming ribbons which are then chipped into granules and stored. Nylon 610 is prepared by a similar technique via formation of the appropriate salt (mp 170°C).

8.14.2 Preparation of Nylon 6

Caprolactam is made from cyclohexanone oxime through reactions involving Beckmann rearrangement. Nylon 6 or polycaprolactam may be made by a batch or continuous process. Caprolactam, water (acting as the catalyst) and traces of acetic acid (chain length regulator) are charged into a reactor and heated at 250°C under blanket of nitrogen for 10–12 h. Polycaprolactam formed remains in equilibrium with unreacted monomer (about 10%) which may be removed from the product by washing with water. For making large cast objects in particular, polymerization of caprolactam by ionic (chain) mechanism developed later[2] may be employed.

8.14.3 Preparation of Nylon 11 and Nylon 12

Nylon 11 is prepared by self-polycondensation of ω-amino undecanoic acid under melt condition at about 220°C. The equilibrium ring-opening reaction of caprolactam during preparation of nylon 6 is easily catalyzed by water. In case of synthesis of nylon 12 from dodecyl lactam, temperature higher than 260°C is necessary for ring opening and almost 100% yield of high polymer is achieved as the condensation is not an equilibrium reaction.

8.14.4 Properties, Uses and Applications of the Nylon Polyamides

The nylons are polar, crystalline materials with excellent resistance to hydrocarbons. There are only a limited number of solvents for them, the most common ones being formic acid, glacial acetic acid, phenols and cresols. Dilute mineral acids are usually not much active on the nylons, but concentrated acids attack them to different extents depending on the nature of the nylon and the acid concentration. HNO_3 is mostly active at all concentrations. Resistance to alkalies is very good at room temperature.

The nylons are polymers of low specific gravity (1.02–1.14). The specific gravity of nylon 66 and nylon 6 are very close (1.14 and 1.13 respectively). The properties of the nylons are considerably affected by the amount of crystallization. The glass transition temperatures (T_g) of the nylons are below room temperature, thus making them somewhat flexible and reasonably tough despite high order of crystallinity. They have fairly sharp melting points and the homopolymers show low melt viscosities above T_m. Nylons are hygroscopic; nylon 66 and nylon 6 show an equilibrium moisture content of about 8–10% at 100% relative humidity; moulded objects from them exhibit tensile strength of nearly 12,000 psi.

For better processing, the nylon granules should be dry; if necessary, the granules should be dried in an oven at 70–90°C prior to processing. Because of high crystallizing tendencies, a high mould shrinkage is generally observed, which may be reduced using high injection pressure. The shrinkage depends on the melt temperature, mould temperature, injection speed, mould design and, of course, on the type of the nylon.

High cost of the nylons restrict their use as general purpose plastics. They are particularly used where their toughness, rigidity, oil resistance, heat resistance, abrasion resistance and self-lubricating properties are of special advantage. In plastics application, their largest outlet is in mechanical engineering. They are prone to embrittlement on long exposure to direct sunlight. They tend to discolour or turn yellowish on aging. Widespread applications include gears, bearings, cams, bushes, etc. Nylon films feature low odour transmission and are useful in packaging for food stuffs, drugs, and pharmaceuticals.

Nylon 6 and nylon 66 are melt-spun into fibres or filaments and the fibres and cords made from them are extensively used as reinforcing agents for plastics and rubbers (in the construction of composites including hoses and beltings and as tyre cords). Nylon 11 and nylon 610 find application in brush tufting, wigs, outdoor upholstery, etc., because of their flexibility. Other applications of the nylons include textiles, ropes, tows, nets, pipes, tubes, rods, bottles and containers, toys, electrical components, cable sheathings, (chemical, solvent and abrasion) resistant coatings, etc.

8.14.5 Liquid Crystalline Polymers[3]

Liquid crystals are highly anisotropic fluids that exist between the boundaries of solid and conventional isotropic liquid phase. A number of high polymeric substances do exist normally as liquid crystals or they form liquid crystals under specific conditions. Cases of polymeric mesomorphism that cover thermotropics, i.e., liquid crystalline polymer melts and lyotropics meaning liquid crystalline polymer solutions are known. A material of this kind is commonly referred to as a mesophase. Configurational restrictions prevalent in a polymeric mesophase allow formation of structures that readily accommodate the long range orientational order that commonly characterizes liquid crystals. For a polymeric mesophase, not only are the molecules locally parallel to each other to begin with but also the frequency of reversals of chain direction in melt or solution are expected to be low.

For structure formation in melt or solution, it is required to incorporate rigid and polarizable segments in the polymer chain. This may at the same time result in enhanced thermal stability and poor solubility making them ultimately meltable with difficulty and often with decomposition or make the polymers intractable in solvents. An optimum balance of rigidity, nature of interunit linkages and molecular weight is required to enable the polymer attain useful thermal and/or solubility characteristics.

Poly(γ-benzyl-1-glutamate) (PBLG) provides a liquid crystalline polymer that has been most extensively studied:

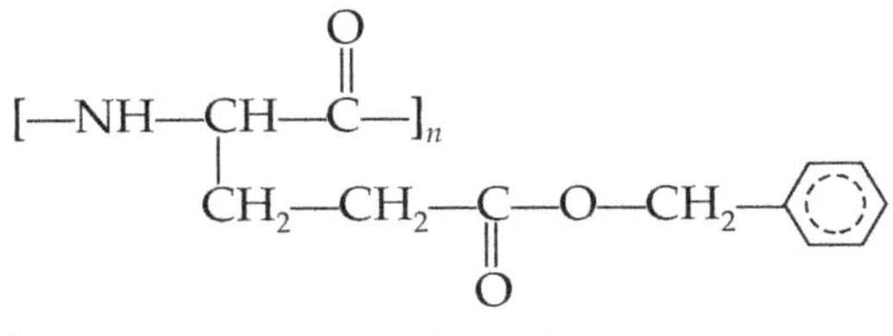

PBLG

PBLG assumes the α-helical conformation in a number of solvents as a result of intramolecular hydrogen bonding and favourable orientation and stacking of the pendant side chains. The polymer molecules readily assume an extended and rigid geometry and may turn into oriented and ordered clusters at sufficiently high concentrations. The formation of a lyotropic liquid crystal occurs at a critical volume fraction of polymer which is inversely proportional to the length to diameter ratio of the polymer. The random copolyester based on residues of hydroxy benzoic acid 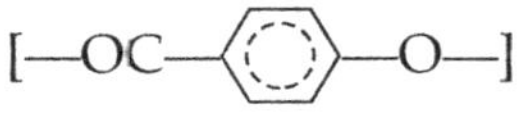 and hydroxy naphthoic acid 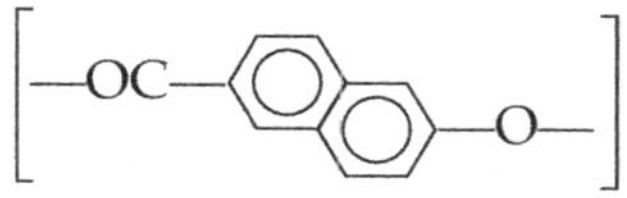 has a melting point of about 300°C. Above this temperature, the polymer can be processed as a thermotropic mesophase while below it, the beneficial mechanical properties are realized (high strength and modulus).

The rigid units in the molecular structure of liquid crystalline polymers (LCP) are primarily responsible for their high performance and liquid crystalline behaviour. The said rigid units are commonly referred to as mesogens and the phases with positional and/or orientational long range order in one or two directions as provided by the thermotropic or lyotropic LCP's are termed as mesophases.

The mesophase structures may be of different types depending on the nature of the mesogens, interunit linkages and the flexible chain segments; the mesophases may be schematically viewed as (a) cholesteric (array of layered uniaxially oriented patterns of the mesogens with different sets of parallel alignments from layer to layer caused by rotational effects, thus giving a helical arrangement of the mesogens through the layers). (b) Smectic (array of layered uniaxially oriented structures of mesogens showing them in parallelly set positions, upright or inclined with respect to the preferred direction of alignments, (c) discotic (cluster of disc-like array of main-chain or side-chain mesogens) and in this type of mesophase, the disc units in clustered stacks are linked with one another along the axial direction of the long chains or in a direction perpendicular to the axial direction, and (d) nematic (uniaxial random spacing of the parallelly aligned mesogens, giving practically no layered arrangement (Fig. 8.1).

One may have (i) main-chain LCP's having mesogenic groups incorporated end-to-end into main chain and interspaced by small flexible chain units or spacer groups, and (ii) side-chain LCP's in which the mesogenic groups are linked to the main flexible chains at regular intervals or chain unit spacings, as pendent groups or branch units, giving a comb-like physical feature (Fig. 8.1). If the segments linking the mesogenic units along the main chain or in the side chain are also rigid, then the LCP is difficultly processable due to high melting point; on the other hand, if the said segments are relatively flexible, the LCP shows lower melting point and easier processability.

Space groups such as —CH = N—, —C = C—, —C : O—O—, or $—N = \overset{\overset{O}{\uparrow}}{N}—$ act as stiff polar linkages; segments such as $—(CH_2)_n—$, $—(CH_2—CH_2—O)_n—$, $—[(CH_3)_2$

SiO]—, however, act as flexible spacer groups. As few as two para-linked aromatic rings (axial ratio about three) are adequate to provide an effective mesogenic unit.

Most if not all LCPs including those main-chain polyaromatics listed in Table 1.3 (bottom part) are high performance polymers, showing high thermal and mechanical resistance properties. Some of the polyaromatics and related polyheterocyclics with favourable structures and having no flexible spacer groups between the main chain mesogenic units (see Table 12.1) also exhibit electrical conductivities of the semi-conducting range that can be enhanced hundred to million fold or even more by doping and for some, the conductivities may even be raised to the metallic range, thereby making it possible to obtain what is known as synthetic metals. LC displays have shown vast potential for application in digital or graphic display systems in watches, calculators, TV, computer screens and other such electronic devices. High performance easy processable LCPs are usually high cost polymers at the current stage, thus justifying their limited use in low volume special applications as thermo-plastics (moulded objects and films), fibres and cords and special purpose cans, casings, containers, etc. for engineering applications and in optoelectronic devices.

8.14.6 Aromatic Polyamides

A number of aromatic polyamides are known to form a mesophase in concentrated solutions.[4,5] Aromatic polyamide fibres, commonly known as aramid fibres, are relatively recent developments. Polycondensation between appropriate diamines (*m*- or *p*-phenylene diamine) and acid chlorides (isophthaloyl or terephthaloyl chloride) leads to the production of industrial aromatic polyamides. The reaction product between *p*-phenylene diamine and terephthaloyl chloride is known by the name Kevlar (Dupont) in the market. Kevlar fibres are nearly one-fifth in density compared to steel but are nearly equal to steel in tensile strength. The high tenacity, high modulus aramid fibres are excellent in dimensional stability. Kevlar has a $T_g >$ 300°C and a melting point of 500°C which is above its decomposition temperature in air. It is wet spun from a solution of concentrated sulphuric acid, methyl pyrdine or a mixture of hexamethylene phosphoramide, *N-N*-dimethyl acetamide and LiCl. The aromatic polyamide solutions initially show a rise in viscosity with increase in volume fraction of polymer followed by a reversal in the trend and an abrupt decrease in viscosity indicative of mesophase formation when a critical volume fraction of polymer is exceeded.

The aromatic polyamides have important practical applications. The exceptionally high degree of orientation, achieved when the lyotropic mesophase is spun or extruded, imparts exceptional high strength and moduli to aramid fibres and films.

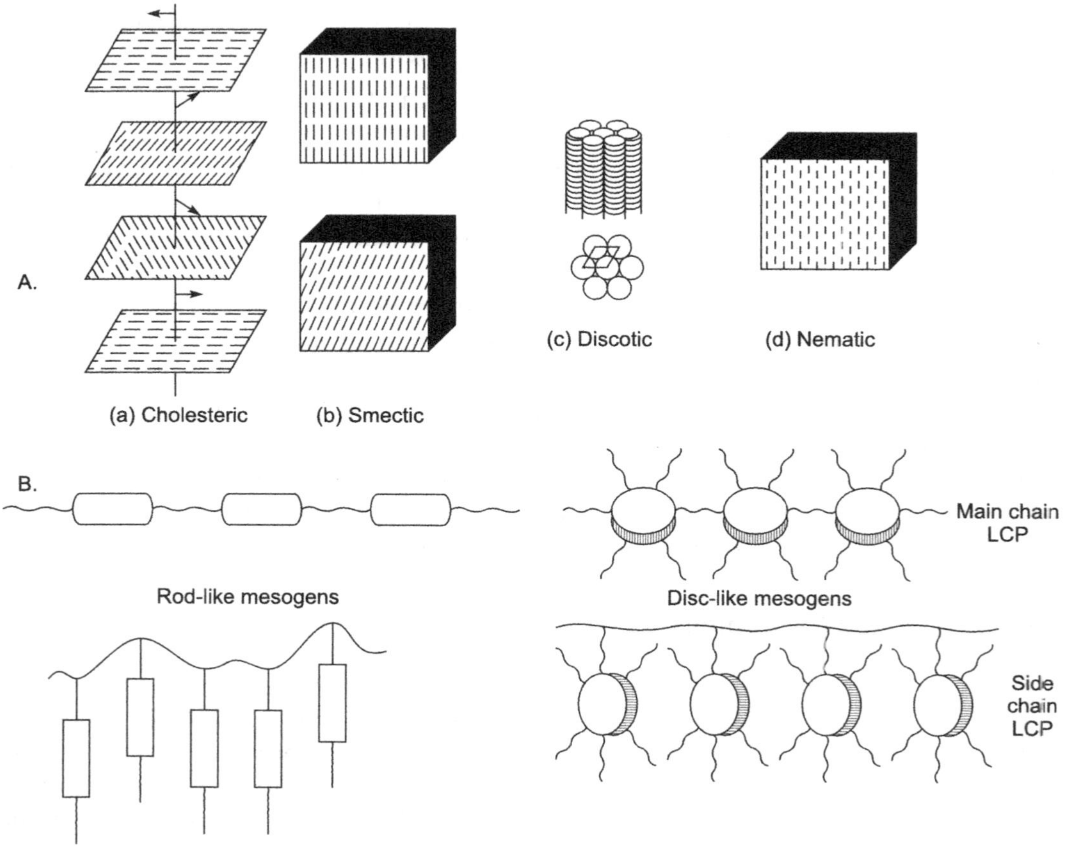

Fig. 8.1 *Schematic representation of : A. mesophasic structures—(a) Cholesteric, (b) smectic, (c) discotic, (d) nematic; and B. liquid crystalline polymers (LCP): main-chain LCP and side-chain LCP*

They are useful in cables for stiffness, corrosion resistance and light weight. They are in demand in fibre reinforced composites where their price and properties lie between those of glass fibre and carbon fibre. Bullet resistant fabrics are another important area of use of the aramid fibres.

8.15 Polyimides

Polyimides which have reached the stage of commercial exploitation are those based on reactions between pyromellitic dianhydride and diamines, particularly aromatic diamines. The sequence of reactions involved in the two stage preparation may be written as:

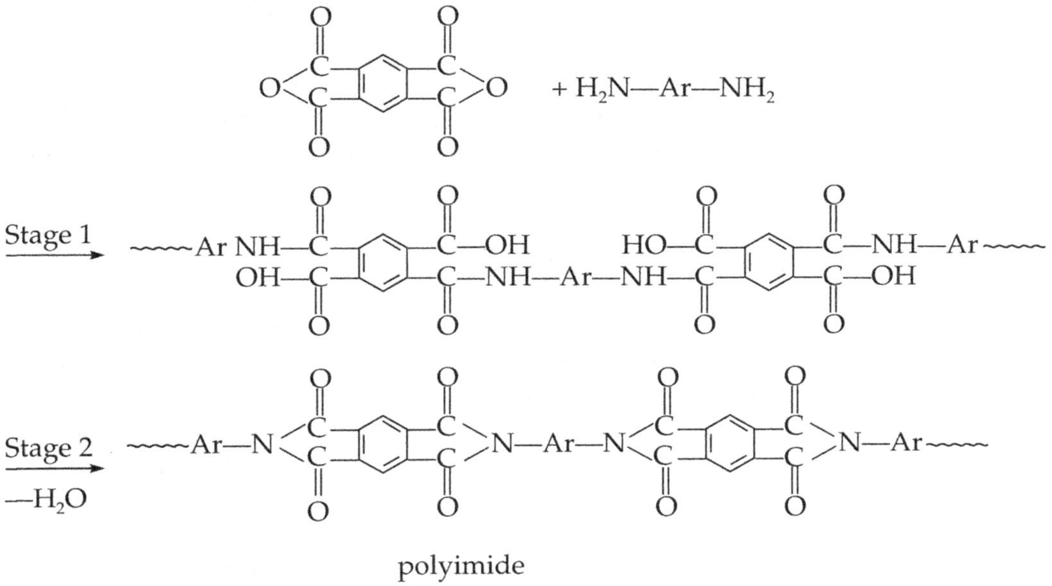

(8.20)

The polymer formed in stage 1 is soluble and it is converted into the thermally stable polymer, polyimide in the second stage. Premature gelation or the second stage reaction can be prevented by using dry reactants in an anhydrous system free from pyromellitic acid and keeping the temperature low (< 50°C). The stage 1 polymer is soluble in dimethyl formamide. The diamines used to make useful polyimides are *m*-phenylene diamine, benzidine and 4,4-diaminodiphenylether.

The polyimides have excellent resistance to thermal and oxidative degradation, chemicals excepting strong bases, and to high energy radiations. Their applications as wire enamels, insulating varnishes and heat resistant coating finishes have been developed. The resins are also useful in the production of heat stable laminates by impregnation of glass or carbon fibres with them and then pressing to form.

8.16 Polyesters

The early developments in polyesters were almost entirely based on resins meant for surface coatings industry. These resins are commonly known as "alkyd" resins, the term having been coined by combining the first part of the word alcohol and the last part of the word acid. Of the different alkyds, the glyptals based on glycerol–phthalic anhydride condensates received most attention in the early stages. A later development is the unsaturated polyester resins for use as laminating resins particularly for the production of glass fibre reinforced structural items. Another significant development in linear polyester subsequent to the development of fibre-forming polyamides is the discovery of polyethylene terephthalate as an excellent fibre forming polyester.

8.16.1 Alkyds for Oleoresinous Varnishes

The polyester resins used for surface coating applications are mostly glyptals usually modified with a drying oil. The resins are produced by extended condensation reactions involving glycerol, phthalic anhydride (acid) and (the fatty acids of) various oils such as linseed oil, soya bean oil, dehydrated castor oil and tung oil. A sizable part of the constituent C_{18} fatty acids of these oils bear conjugated unsaturations in their structures which make them "drying", i.e., they lose unsaturation, gain in viscosity and become progressively less soluble or insoluble on aging or aerial oxidation. The drying is, in fact, a case of oxidative cross-linking often highly catalyzed by such compounds as lead soaps, cobalt or manganese naphthenates, etc., commonly known as driers. The useful properties of the alkyds may be varied by varying the acid, alcohol or the oil. Other polyhydric alcohols commonly used are ethylene glycol, propylene glycol, pentaerythritol and sorbitol, and other acids used are maleic anhydride, isophthalic acid and terephthalic acid.

High or low proportions of drying oils may be used for cooking out the oleoresinous compositions. The oil length of the varnish is given by the number of gallons of drying oils per 100 lb of resin. Long oil compositions are more tough and durable and particularly more suited for outdoor applications while short oil compositions are more hard and glossy and take a higher polish.

In the initial stage of the cooking or resin making, the temperature is kept relatively low (~225°C) for a few hours and extensive ester interchange reaction is accomplished at this stage along with some degree of chain extension involving the dicarboxylic acid giving rise to increased molecular weight and higher viscosity. The ester interchange reactions lead to uniformity of molecular composition. Polymerization of the uniform mixed glycerides is better accomplished by raising the temperature to above 250°C over the final hours of the cooking. The structure of a segment of a molecule of an oil-modified alkyd (glyptal), I is shown below:

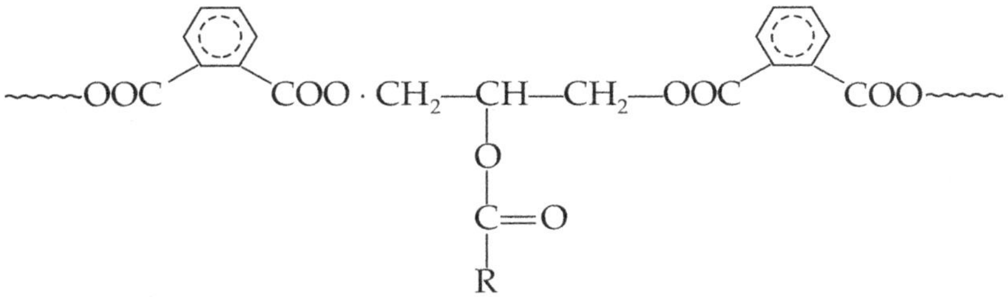

($R = C_{17}$ chain with conjugated unsaturation)

(I)

Segment of an oil-modified alkyd molecule

The alkyd resins are useful because of good combination of properties and a favourable cost structure. They are also widely modified with rosin, phenolic resins, urea resins, epoxy resins and such monomers as styrene to suit different property needs for different applications. Most of the hard resins are based on maleic anhydride or maleic–rosin adducts. The maleic–rosin resins are made by allowing the anhydride and glycerol to react in presence of excess of rosin.

8.16.2 Polyester Resins for Making Laminates and Composites

(i) *The Resin System* The laminating resins and those meant for reinforced composites are widely made by polycondensations involving varied proportions of a saturated acid (phthalic anhydride, isophthalic acid, adipic acid, etc.), an unsaturated acid, which is most invariably maleic anhydride and a glycol (ethylene glycol, diethylene glycol or 1,2-propylene glycol), whose chemical structures are given below:

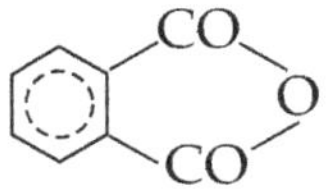

phthalic anhydride

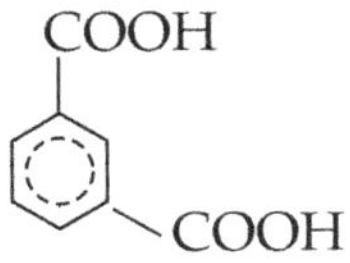

isophthalic acid

$$HOOC—(CH_2)_4—COOH$$

adipic acid

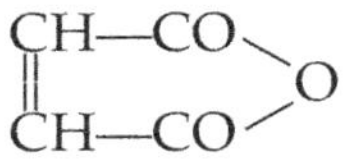

maleic anhydride

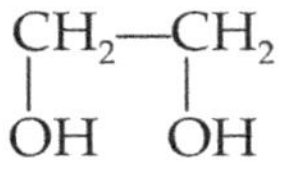

ethylene glycol

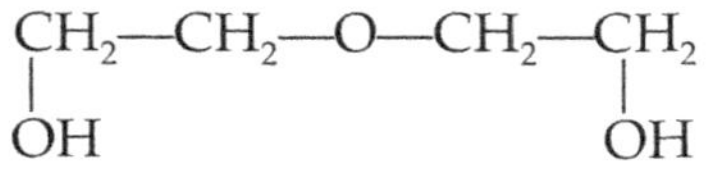

diethylene glycol

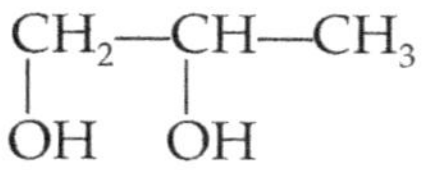

1,2-propylene glycol

The unsaturated polyester resins are prepared by heating the component acids and glycols for 10–12 h at about 160–200°C, under stirring conditions usually under a blanket of nitrogen. An inert atmosphere is necessary to obtain light colour for the resin and to prevent premature gelation. A slight molar excess of glycol is usually taken to allow for evaporation losses. The progress of reaction is followed by measurement of fall in acid number with time of reaction. Xylene is generally used to facilitate removal of water formed in reaction steps through azeotropic distillation. When the acid number drops in the range 25–50, xylene is allowed to boil off and heating is discontinued. The resin is then cooled to below the boiling point of styrene and pumped into a blending tank when it is appropriately blended with right proportions of styrene containing suitable inhibitors. The cooled blend is then ready for storage and shipping.

(ii) ***The Curing System*** The success of the laminating resins is related to the last-stage curing process when the resin blend has been applied to the glass fibre or other reinforcing agents. Benzoyl peroxide dispersed in dimethyl phthalate is often used in appropriate doses for pressure moulding at elevated temperatures (~100°C). In presence of dimethyl aniline, the peroxide forms a redox combination and readily brings about satisfactory room temperature curing. Methyl ethyl ketone peroxide (m.e.k.p.) and cobalt or copper naphthenate or octoate also combine to give an efficient room temperature curing system. Depending on the nature and concentration of the curing agents and the prevalent temperature, the curing may be accomplished in a few minutes to a few hours. Mixed with fillers/glass strands, the resins are processed to finished items either by hand lay up or by compression moulding.

The cured resins formed by a graft copolymerization approach have a cross-linked structure; the polystyrene chains form the cross-links connecting different polyester chains at the point of maleic unsaturations during the curing process, which is induced by a radical mechanism. The major utility of the saturated acid used during the resin-making is to space out the chain unsaturation and as a consequence appropriately reduce or control the cross-link density.

8.16.3 Film- and Fibre-Forming Polyester: Poly(Ethylene Terephthalate)

Poly(ethylene terephthalate) provides very good mouldable, film-forming and fibre-forming polyester because of its high glass transition temperature (70–80°C) and high crystalline melting point (>250°C). The fibres and films must be drawn out with ease and convenience as the spun or extruded melt is cooled.

The polymer is essentially looked upon as the linear polycondensation product of terephthalic acid and ethylene glycol. The terephthalic acid is commonly prepared from *p*-xylene by an oxidation process. In view of difficulties in isolating *p*-xylene in pure form consequent to the closeness of its boiling point to the boiling points of *o*- and *m*-xylenes and strong sublimation tendencies of terephthalic acid at poly-condensation reaction temperatures, it is convenient to produce the polymer poly(ethylene terephthalate) by an ester interchange reaction between the more readily purified dimethyl terephthalate (DMT) and ethylene glycol. The overall reaction is accomplished in two stages. Ester interchange takes place in the first stage with liberation of CH_3OH as the byproduct in presence of Sb_2O_3 with cobaltous acetate as the ester interchange catalyst. A reduced pressure is maintained to facilitate removal of CH_3OH. In the subsequent stage, temperature is raised and taken to a level of 250–280°C towards the end while maintaining a reduced pressure whereby the desired condensation to high molecular weight polymer with removal of the byproduct of condensation, ethylene glycol is accomplished. The molten polymer is then cooled, granulated and stored.

The commercial films [Melinex (ICI) and Mylar (DuPont)] are made by quenching the extruded sheets or films to the amorphous state and subsequently by controlled reheating and stretching of the sheet/film nearly three-fold in machine and transverse directions at 80–100°C. Controlled annealing of the biaxially oriented films at 180–200°C improves crystallinity (~40%) and reduces heat shrinkage. The films are suitable in electrical applications and packaging, and as magnetic recording tapes. The fibre is made by a melt spinning process. Poly(ethylene terephthalate) is the most important synthetic fibre to have found widespread textile applications either alone or more commonly as different blends with cotton or wool. The fibre is widely known as Terylene (ICI trade name) or "Dacron" (DuPont trade name). The polyester fibres possess good "crease resistance" and "wash and wear" properties. A sizable fraction of the polyester fibre is used as the reinforcing cord in the tyre, belting and related industry.

8.16.4 Microbial Thermoplastic Polyesters

(a) ***Background*** Polyesters of the microbial poly(hydroxyalkanoate) family are thermoplastic in nature and they have biodegradable and biocompatible properties. They can be produced from selected carbon substrates by a variety of microorganisms. Prokaryotic organisms including gram-positive and gram-negative bacteria are known to produce poly(3-hydroxybutyrate) [P(3HB)], structurally expressed as a polymer of 3-hydroxybutyric acid:

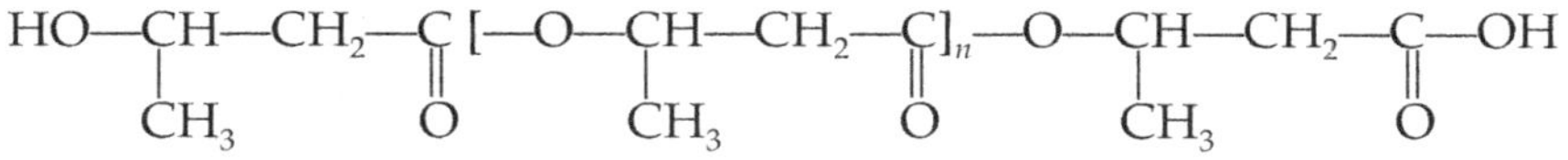

Some bacteria have further been found to accumulate poly(3-hydroxy alkanoates), [P(3HA)] by incorporating into the polymer structure repeat units other than 3-hydroxybutyrate, so that the copolyester derived by the microbial process may be structurally written as:

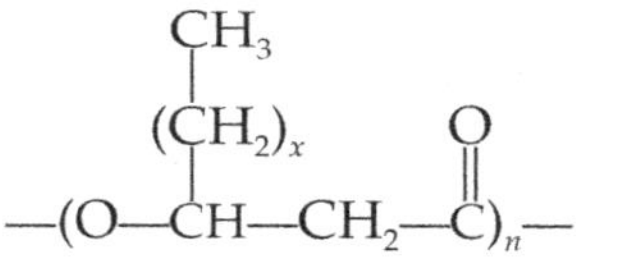

($x = 0, 1, 2, 3$)

A wide variety of P(3HA) copolymers have been reported to be isolated from different environmental sources[6–8] such as fresh water and marine cyanobacteria, marine sediments and sewage sludges. For use of different alkanes and alkanoic acids as the carbon sources, microbial synthesis of P(3HA) copolyesters having alkyl chain lengths from methyl to nonyl or even higher has been reported.

The physical properties of the microbial copolyesters can be regulated by controlled variation of molecular structure, covering copolymer composition and chain length. Copolyesters from hard crystalline thermoplastics to soft elastic rubbers can be made. The microbial thermoplastic polyesters having melting temperatures of 55–180°C can be processed by common extruder and injection moulding equipments.

The microbial copolyester products such as films, fibres, moulded objects and extruded profiles can be degraded in our environment—soil, sludge and sea water. The degradation is very fast under optimal conditions. Some microorganisms such as bacteria and fungi excrete extracellular P(3HA) depolymerases to degrade the microbial polyesters utilizing the intermediates as nutrients. Extracellular P(3HB) depolymerases have been isolated from *Pseudomonas lemoiguei* and *Alcaligenes faecalis*. The enzymatic degradation commonly occurs at the surface of the microbial polyester item and the rate of surface erosion is strongly contingent upon both the molecular structure and copolymer composition of the microbial polyesters.

(b) ***Production and Isolation*** The production of microbial polyesters is done by batch or continuous fermentation using selected bacterial cultures in the presence of a variety of carbon substrates such as glucose, gluconates or methanol as the feed and a salt medium containing dissolved oxygen, ammonium, sulphate, phosphate, potassium or iron ions as nutrients.

The next step is to isolate and purify the produced polymers (microbial polyester/copolyester). They can be extracted from bacterial cells using an organic solvent such as chloroform, methylene chloride, dichloroethane or propylene carbonate. Various lipids are coextracted together with the polyesters from the cells. To remove lipids, the polyesters are precipitated slowly by adding diethyl ether, hexane or lower alcohol as non-solvents. The polyesters can be further purified by reprecipitation from chloroform solution.

Alternatively, the bacterial cells are selectively destroyed and dissolved by treatment with dilute sodium hypochlorite solution for 30–60 min; longer time may result in degradation of the polyesters as well. The isolated polyester/copolyester granules are purified and made free of lipids by washing with diethyl ether or methanol. Poly (3-hydroxybutyrate), [P(3HB)] granules can also be isolated by treating bacterial cells with enzymes that solubilize all of the cell components except P(3HB).

(c) ***Application Prospects*** Man's life-style at the dawn of the twenty first century is so much dependent on use of plastics in packaging, construction, health care and sanitation, agriculture, furniture, food and most industrial sectors that man is confounded with the massive problem of disposal of millions of tons of plastic wastes. The concept of biodegradable plastics is the key to solving the waste disposal problem; thus developing plastics that are biodegradable and that do not pollute the environment has become an extremely important global objective. About a million marine animals face death every year by choking on floating plastic waste items that are mistaken by them as food items or by otherwise getting entangled in non-degradable plastic debris.

Fibres and films of biodegradable microbial polyesters get completely decomposed in about eight weeks in sea water and their use or disposal in marine environments would go a long way in reducing harm and injuries to fishes, turtles, seals and other big or small marine animals caused by synthetic plastic wastes. Other important applications of biodegradable polyesters are in the making of agriculture aids including mulching films, and for making films for packaging and other related applications and of bottles and containers. Microbial (biodegradable) polyesters have high potential for use as a matrix for controlled or sustained release plant-growth regulators such as fertilizers and other nutrients or pesticides. The chemicals are gradually released as the encapsulant polyester is degraded in soil. Blends of biodegradable polyesters or other natural polymers such as starch, gelatin, dextrin and alginates with many synthetic polymers[9,10] such as the nylons, poly(ethylene oxide), poly(vinyl chloride), poly(vinyl acetate) etc hold high prospect of making biodegradable plastic items of diverse property ranges. Further, rigorous and systematic evaluation of effects of incorporation of plasticizers in and blending other polymers with the microbial polyesters hold high prospect toward making biodegradable polymer items of diverse application potentials. Possible medical applications of P(3HB) include controlled drug release, bone plates and surgical sutures. P(3HB) and other microbial polyesters show good bio-compatibility and slow resorption in biological environments (*in-vivo*). P(3HB) exhibits negligible oral toxicity and its ultimate biodegradation product, D(-)-3-hydroxybutyric acid, is a normal metabolite in human blood.

Some synthetic polyesters/copolyesters having biodegradable cross links and backbone links based on trimellitic acid or citric acid and such alcoholic systems as glycerol, castor oil, and 1,2,6 hexane triol have been recently studied and characterized and they are reported to be promising as encapsulants for control release of drugs, fertilizers and pesticides.[11–13]

8.17 Polyurethanes

Polymers formed through interactions of isocyanate groups of di- or polyisocyanates and hydroxyl groups of monomeric or polymeric diols or polyols are commonly known as polyurethanes. The reaction between hexamethylene diisocyanate and 1,4-butanediol yields a fibre-forming polyurethane (Perlon U). The reaction is better considered as rearrangement (rather than condensation) polymerization characterized by a step-growth mechanism:

$$n\text{OCN}\cdot(\text{CH}_2)_6\cdot\text{NCO} + n\text{HO}\cdot(\text{CH}_2)_4\cdot\text{OH} \rightarrow -[\text{OOCNH}\cdot(\text{CH}_2)_6\cdot\text{NHCOO}\cdot(\text{CH}_2)_4]_n- \quad (8.21)$$

Different kinds and grades of polyurethanes have been developed for applications as fibres, resins, rubbers, adhesives and surface coatings. However, their applications as flexible or rigid foams are of special importance.

The reaction (8.21) for the production of *fibre-forming polyurethane* is carried out by controlled heating of the mixture of the diisocyanate and the diol at about 190–200°C, usually under a blanket of nitrogen with agitation, till the reaction is complete. The reaction is exothermic and good care must be taken to avoid overheating in the kettle. The molten polymer is then passed through metal gauge screen and extruded and cut into chips. The melt may be fed directly into the spinneret to form the filaments.

The fibre formation technology of the polyurethane (Perlon U) through melt spinning is much the same as for the nylons. However, the polyurethane begins to decompose close to and above 220°C and overheating should be avoided. The polyurethane fibre is less prone to colour development (yellowing) in air, and has a lower moisture absorption than the nylon fibres. They are comparable in tensile strength, but the polyurethane fibre has a much lower softening temperature which limits its use in the textile field. The fibre or filament finds some use as bristles, monofilaments, nets, sieves and filter cloth. The polymer is otherwise useful as a *thermoplastic resin* and is readily processed by injection moulding and extrusion. The moulded and formed objects have very good dimensional stability and good retention of electrical properties in humid conditions. Like the nylons, the polyurethane is readily dissolved by formic acid and phenols.

The *polyurethane rubbers* are made from chain extension reactions involving linear aliphatic polyesters with terminal hydroxyl groups and diisocyanates (toluene diisocynate, 1,5-naphthalene diisocynate, etc.) as described in Chapter 4 (Sec. 4. 1). The urethane linkages formed during chain extensions may further react with excess diisocyanates or through terminal isocyanate groups of the growing chains at a higher temperature in a mould to establish cross linking. The polyurethane rubbers have good tensile strength (~5000 psi), higher than any other rubber, and excellent tear and abrasion resistance, and they exhibit excellent resistance to oxygen and ozone. They are, however, less resistant to acids and alkalies and steaming (hydrolytic decomposition) than the conventional rubbers. They find application as oil seals, footwear soles and heels, conveyor belts, printing rollers, fabric coatings, waterproof items, tank lining in chemical plants, cable sheathing, pump impellers and diaphragms, industrial hoses, etc.

Polyurethane foams are very popular. They may be rigid, semirigid or flexible. They may be either polyester-based or polyether-based. The foam technology requires concurrent chain extension with diisocyanates, cross-linking and gas evolution. Gas evolution may be achieved by reaction between —COOH end-groups of polyesters and isocyanate groups (—COOH + OCN— → —CONH— + CO_2) or more readily by reaction between isocyanate and water:

$$\begin{matrix} \sim\sim NCO \\ \\ \\ \\ \sim\sim NCO \end{matrix} \quad + H_2O \longrightarrow \quad \begin{matrix} \sim\sim NH \\ | \\ C{=}O \\ | \\ \sim\sim NH \end{matrix} \quad + CO_2$$

The rate of CO_2 evolution should be controlled. The cross-linking should be rapid so as to trap gas evolved in small bubbles as soon as it is formed. Loss of gas would result in poor foams. The polyurethane foams commonly known as "U" foams are extensively used as cushioning materials, soft backing or padding, sponges, paint rollers and in quilting and packaging. Rigid foams are mostly used in the field of thermal insulation. As adhesive and surface finishes, hard polyurethane resins are suitable for metals while the soft elastomers for rubbery or leathery products. A two pack system is usually used, the curing agent being used just prior to application. Single component finishes cure in contact with moisture from air.

8.18 Polycarbonates

Polycarbonates are characterized by repeat units having carbonate $\left(\text{—O—}\overset{\displaystyle\text{O}}{\overset{\|}{\text{C}}}\text{—O—}\right)$ inter-unit linkages. Linear polycarbonates are formed either by reaction between dihydroxy compounds and phosgene or by means of ester interchange reaction involving dihydroxy compounds and organic carbonates (R—O · CO · O—R). The polycarbonate based on bisphenol A and diphenyl carbonate is the only one that has been developed into a commercial product of special importance.

Bisphenol A is readily produced by the condensation of acetone and phenol under acidic condition at temperatures below 70°C. An excess of phenol is used in order to achieve a high yield. Phosgene used (in the direct process or for preparing diphenyl carbonate) is prepared commercially from chlorine and carbon monoxide. Diphenyl carbonate (C_6H_5—O · CO · O—C_6H_5) is conveniently obtained by reacting phenol with phosgene in aqueous NaOH solution in presence of a tertiary amine catalyst. The diphenyl carbonate is purified by redistillation.

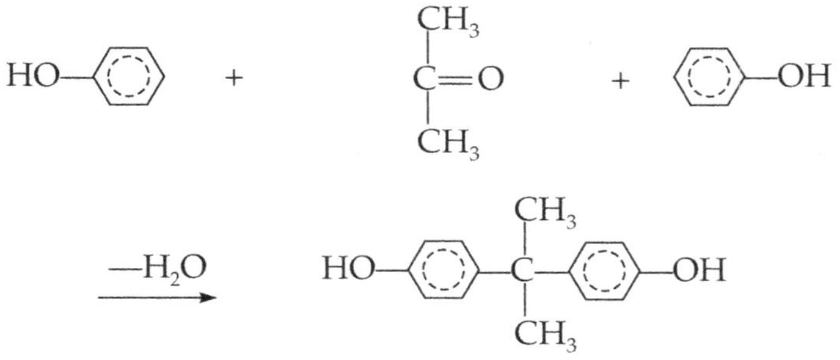

bisphenol A

(8.22)

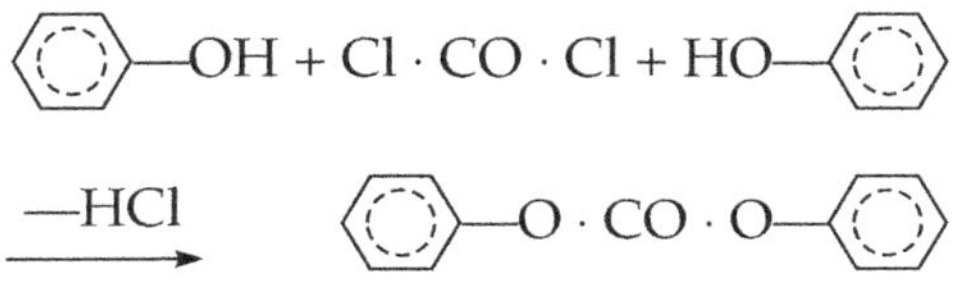

diphenyl carbonate

(8.23)

The ester interchange reaction leading to polycarbonate formation is shown in reaction (8.24). The reaction is normally accomplished at nearly

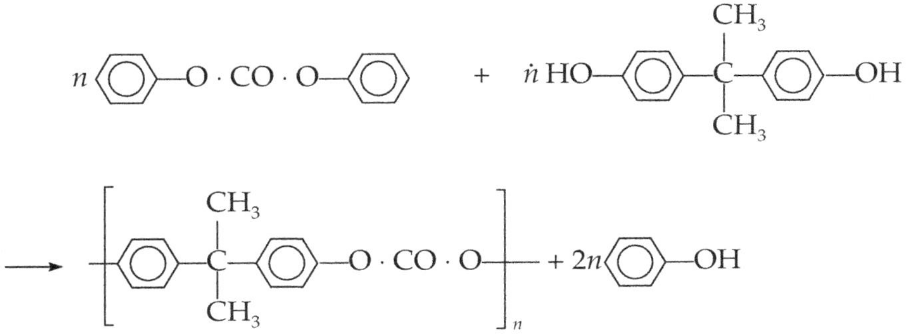

(8.24)

200°C at a reduced pressure (25 mm Hg), until about 90% of the phenol liberated has been removed. The temperature is then raised to about 290–300°C and the pressure further reduced to below 1 mm Hg. Further chain extension takes place and the melt viscosity increases significantly at this stage. The reaction is arrested at an appropriate melt viscosity and the melt is forced out of the kettle under nitrogen pressure. For convenience, the reaction is done using very high molar proportions (≥2) of the diphenyl carbonate so that the initial reaction product is the bis(phenyl carbonate) of bisphenol A. On further heating, polymerization proceeds by splitting out of the diphenyl carbonate thus producing the polycarbonate resin. Direct phosgenation of bisphenol A may be used for the production of high molecular weight polycarbonates:

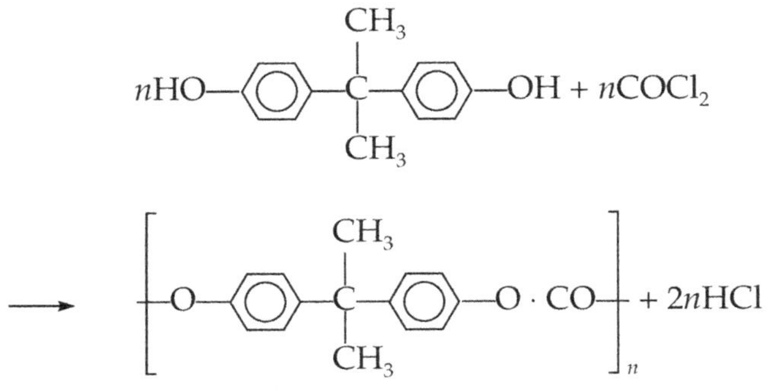

(8.25)

The reaction is better done in presence of such solvents as pyridine (which act as HCl acceptors) diluted with some other solvents as chloroform, methylene chloride or tetrachloroethane. The polymer is isolated by precipitation with methanol. The ester-interchange method is commercially more favoured as it allows avoidance of the handling of toxic and flammable solvents and thus allowing avoidance of costs of solvents and solvent recovery.

Bisphenol-A polycarbonates have densities of nearly 1.2 g/cm^3. They exhibit tensile strengths of the order of 9,000-10,000 psi. They are characterized by very good impact resistance (12–18 ft.lb/in notch). Their creep resistance is also very remarkable and this makes them much superior to polyamides and acetal resins. The polycarbonates are however inferior in respect of stress-cracking or crazing.

The polycarbonates appear in crystal-clear grades, if prepared from high purity ingredients. The refractive index at 25°C is 1.58. Their low water absorption, good insulation properties (low dielectric constant, 2.9–3.2, and high volume resistivity, 2 $\times 10^{16}$ ohm cm), transparency and toughness coupled with heat and flame resistance (T_m = 225–250°C, T_g = 145°C), make them suited for many areas of application including electrical insulation.

Polycarbonates can be processed by all the usual thermoplastic processing techniques—injection moulding, extrusion, blow moulding, roto-moulding, themo-forming, etc. Major applications include electric and electronic appliances, helmet, automotive lighting, protective shields, safety glass/sun glass, bottles, domestic wares, structural foams, components of camera, projector housings, etc. Factors retarding their growth are high cost, limited chemical and UV resistance and stringent processing requirements. For processing, the resin must be thoroughly dried for several hours in an oven at 110-120°C and used for processing immediately thereafter.

8.19 Epoxy Resins

Epoxy or epoxide resins are those resins or polymers which are produced by condensation reactions using an epoxy compound as an ingredient in the monomeric reactant mixture. The most common and widely used epoxy resins are made by allowing polycondensation involving epichlorohydrin and bisphenol A. Epichloro-hydrin, more expensive of the two, is derived from propylene via allyl chloride according to the sequence of reactions given in reaction scheme (8.26). The material, a colourless liquid, is commercially available at 98% purity.

$$CH_2{=}CH\cdot CH_3 + Cl_2 \longrightarrow CH_2{=}CH\cdot CH_2\cdot Cl + HCl$$

$$CH_2{=}CH\cdot CH_2\cdot Cl + HOCl \xrightarrow{H_2O/Cl_2} Cl\cdot CH_2{-}CH(OH)\cdot CH_2\cdot Cl$$

$$Cl\cdot CH_2{-}CH(OH)\cdot CH_2Cl + NaOH \rightarrow \underbrace{CH_2{-}CH}_{O}{-}CH_2\cdot Cl \qquad (8.26)$$

The commercial liquid resins are essentially diglycidyl ether of bisphenol A mixed with small proportions of some high molecular weight polymers. Two to three moles of epichlorohydrin are used for one mole of bisphenol A in order to obtain higher yield of the diglycidyl ether. Commercial resins show molecular weights in the range of 340–400 and seldom their molecular weights exceed the range 3000–4000. The mixture of bisphenol A and epichlorohydrin in the desired molar proportion is heated under stirring condition to about 110°C under an atmosphere of nitrogen while a 30% aqueous solution of NaOH is added dropwise continuously to neutralize the HCl liberated. After the reaction, the organic layer is removed, dried and then fractionally distilled under vacuum.

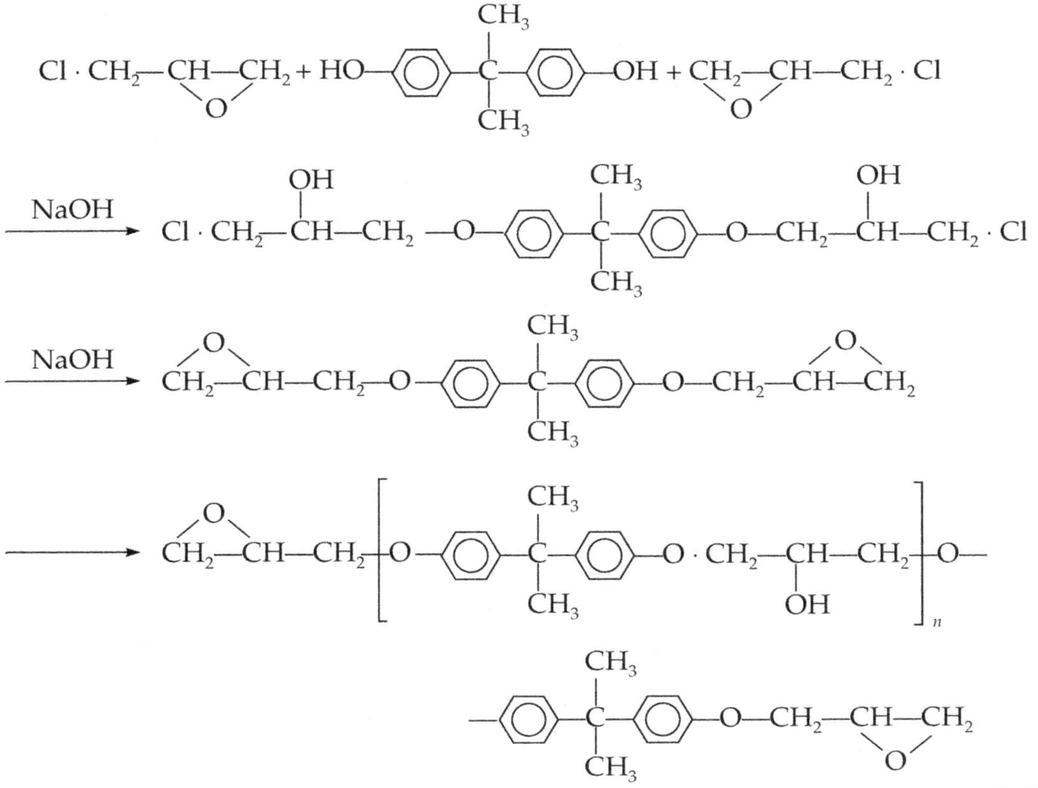

(8.27)

The quality of the epoxy resin is determined by its viscosity, hydroxyl equivalent and epoxy equivalent, and molecular weight. The hydroxyl equivalent is the weight in gram of the resin containing one gram equivalent of hydroxyl group and is normally determined by reacting the resin with acetyl chloride. Likewise, the epoxy equivalent is the weight of the resin in gram containing one gram equivalent of epoxy group and it is normally determined by reacting a known quantity of the resin with HCl and then estimating the unreacted acid. HCl reacts with the epoxy group according to Eq. (8.28).

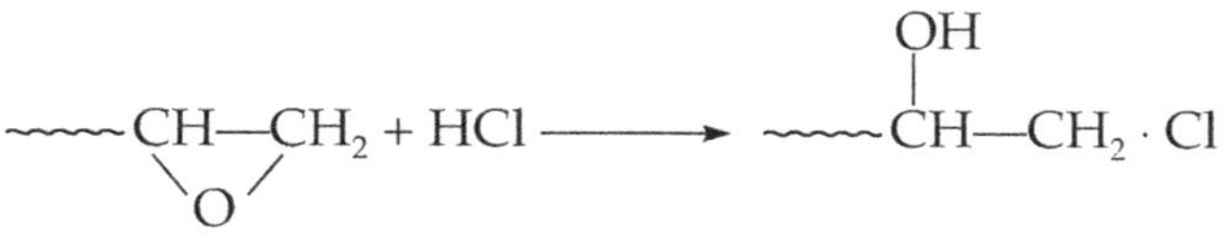
(8.28)

The cross-linking of epoxy resins is done through the epoxy or hydroxyl groups. The curing or cross-linking agents are commonly termed as hardeners or hardening agents. The hardeners commonly in use are monoamines, diamines and polyamines, acids and anhydrides [phthalic, hexahydrophthalic, maleic, pyromellitic (dianhydride), chlorendic, etc.]. With amines or acids, the reaction involves opening of the epoxide rings giving β-hydroxy amino or β-hydroxy ester linkages:

(a)
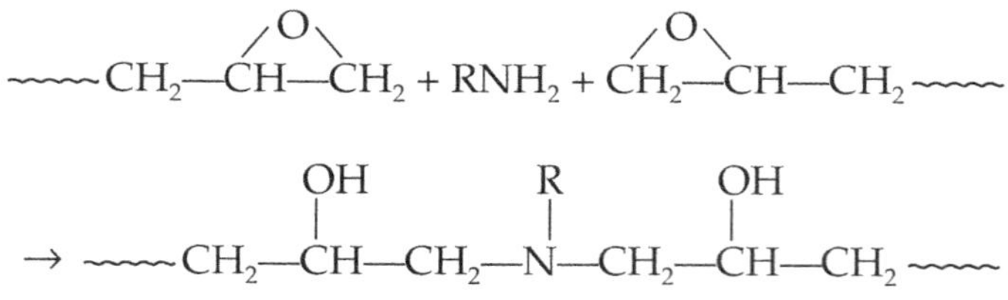
(8.28a)

(b)
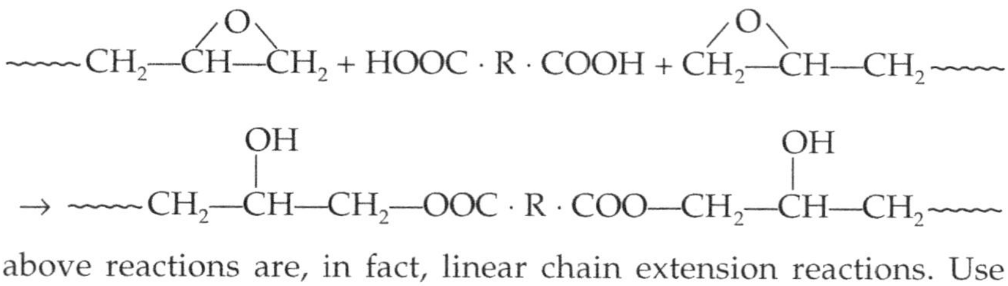
(8.28b)

The above reactions are, in fact, linear chain extension reactions. Use of diamines and polyamines with more than two active hydrogen in the molecule leads to three dimensional networks and cross-linkages. With acids and anhydrides, reactions involving OH groups initially present or those formed on opening of the epoxide ring lead to cross-linking.

$$\left|\text{—OH} + \text{HOOC}\cdot\text{R}'\cdot\text{COOH} + \text{HO—}\right| \xrightarrow{-H_2O} \left|\text{—OOC}\cdot\text{R}'\cdot\text{COO—}\right| \quad (8.29)$$

The cross-linking may also be done using low molecular weight phenolic (novolac) and urea resins where the reactions may involve condensation of methylol groups with the secondary hydroxyl groups of the epoxy resin or epoxide ring opening through reaction with the phenolic hydroxyls or the amino groups of the urea resins.

Amines are fast curing even at room temperatures, and they incorporate good chemical resistance. They are, however, toxic and skin-sensitive. The anhydrides are much less toxic and give cured products of good thermal stability and high heat distortion temperature. They are effective only at elevated temperatures. Better pot life is obtained using Lewis acid hardeners, such as BF_3 and its complexes.

The epoxy resins are suitable for both moulding and laminating applications. They also find special applications as high strength adhesives and cementing agents, in the building industry and road surfacing, in surface coating and in potting and encapsulation of electronic components. The major structural and engineering applications include glass fibre reinforced cast objects and laminates and in these applications, the epoxy resins excel the polyester resins in view of their far superior heat resistance, chemicals and alkali resistance, mechanical properties and aging resistance. The epoxy's excellent adhesion, very good cohesion, low creep and low shrinkage on cure and absence of volatiles have developed the resin into an excellent adhesive, bonding or cementing material for metal-to-metal, metal-to-ceramics, metal-to-plastics and various other combinations of substrates.

The so-called phenoxy resins or the phenoxies are the extended polycondensation products of bisphenol A and epichlorohydrin of molecular weight close to 25,000, represented by a polymer structure given in the final step of reaction sequence (8.27), where n is nearly 100. The phenoxies are usually slightly branched and are readily cross-linked through the hydroxyl groups in the repeating units by diisocyanates.

The phenoxies exhibit very good resistance to alkalies and quite low gas permeability. Their rigidity, creep resistance, high toughness and ductility, high heat resistance and low moulding shrinkage (0.003 cm $\cdot$ cm^{-1}) make them suitable for special engineering applications.

Another class of polymers of related structure are the polysulphones, usually having —O— (ether) and —SO_2— (sulphone) linkages between aromatic rings, e.g.,

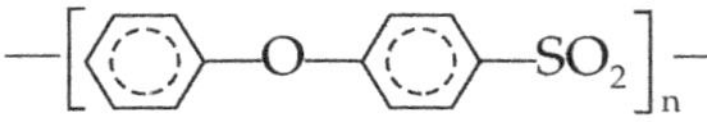

prepared by self condensation of Ar—O—Ar—SO_2Cl or polycondensation involving (Ar—O—Ar) and (ClO_2S—Ar—O—Ar—SO_2Cl). Commercial polysulphones have T_g > 200°C, and they are usually aromatic polyethersulphones. Polysulphones with no interunit ether links may be obtained by reacting bissulphonylchlorides with an aromatic hydrocarbon using Lewis acid catalysts:

$$\begin{aligned} &ClO_2S\text{—}Ar\text{—}SO_2Cl + H\text{—}Ar\text{—}H \\ &\xrightarrow{-HCl} \text{—}[O_2S\text{—}Ar\text{—}SO_2\text{—}Ar]_n\text{—} \end{aligned} \tag{8.30}$$

The processing temperatures of commercial polysulphones are above 300°C. They are rigid, tough, high tensile strength/high modulus thermostable polymers with non-burning behaviour, and they can also withstand ionizing radiations without much degradation or cross-linking and deterioration of other properties. They are, however, notch sensitive, like the polycarbonates. Their high mechanical strength, heat resistance, chemical resistance, oil resistance and good burning behaviour make them suited for many engineering applications.

8.20 Cellulose Plastics

Cellulose is the most abundant constituent of the vegetable or plant kingdom. The cellulose content, however varies from plant to plant. Cellulose content is over 90% in cotton and nearly 50% in an average wood, the most convenient sources of cellulose for chemical treatments and transformations are cotton linters and wood pulp.

Of the very large number of cellulose derivatives that have been prepared, only the esters and ethers are generally important. The three hydroxyl groups of the repeating anhydro-β-glucose units are transformed into the ester or ether groups during esterification and etherification reactions, giving a theoretical degree of substitution (DS) equal to 3.0.

Cellulose esters find widespread applications as moulded plastics, artificial fibres, films, sheets, and in coatings and lamination. Among the important cellulose esters are cellulose nitrate, cellulose acetate, cellulose acetate-butyrate and cellulose propionate.

8.20.1 Cellulose Nitrate

Cellulose nitrate is the oldest cellulosic or cellulose derivative known and it is the only inorganic ester of cellulose of commercial importance. Nitration of cellulose is carried out by using a mixture of HNO_3 and H_2SO_4 over a specific time period under controlled conditions of temperature and mixed acid composition. The nitration can be done by batch or continuous process. The three distinct stages in the preparation of cellulose nitrate are: (i) pretreatment of cellulose, (ii) nitration, and (iii) after treatment and dehydration.

Wood pulp or cotton linters are shredded by a set of rotating blades and then dried to a very low water content (<0.5%). Water content is an important factor in the preparation of mixed acid which may vary in composition from 0–20% water, 15–50% HNO_3 and 30–60% H_2SO_4. A very low water content tends to produce high degree of nitration; H_2SO_4 plays the role of a condensing agent. In a typical batch process, 500 kg of the mixed acids are taken into a jacketed reaction vessel fitted with agitators and about 12 kg of dried cotton linters or shredded wood pulp are added keeping the temperature in the range 25–40°C. The nitration is allowed to proceed for 20–25 min. Even after nitration, the fibrous form of cellulose is retained. The nitrated product is dumped through the bottom of the reactor into a centrifuge and much of the acid mixture is removed. The ester is washed profusely with water in what is known as the drowning tank and pumped as a slurry into a vat that holds products of several nitrations. Due to minor degrees of sulphation through reaction with H_2SO_4, the nitrated product at this stage is unstable as the sulphate groups tend to split off reforming H_2SO_4 and setting off an autocatalytic decomposition process. During stabilization process, the product is boiled in water under controlled pH condition whereby removal of sulphate groups is accomplished. The cellulose nitrate is then washed, bleached if necessary, centrifuged and then dehydrated with alcohol. The transformation of cellulose (R—cell · OH) into the nitrate may be written as:

$$\text{R—cell} \cdot \text{OH} + \text{HO} \cdot \text{NO}_2 \rightarrow \text{R—cell} \cdot \text{O} \cdot \text{NO}_2 + \text{H}_2\text{O} \qquad (8.31)$$

Commercial cellulose nitrates meant for plastics and coating applications have N_2 contents in the range of 10.9–12.2% corresponding to degrees of substitution (DS) 1.8–2.3. Complete nitration leading to DS of 3.0 corresponds to a N_2 content of 14.14%. Nitrates with N_2 content $\geq$ 12.5% are used as explosives.

Cellulose nitrates for plastics and coating (lacquor) applications have N_2 content in the range 11.5–12.2%; they are available in a large number of viscosity grades and they are soluble in esters, ketones, ether–alcohol mixtures and glycol ethers. They are compatible with many other resins. Cellulose nitrates with 10.9–11.2% N_2 content find use in flexographic inks, lacquer coatings for paper, foils and other flexible substances. For plastics applications cellulose nitrate is invariably plasticized with nearly 25% by weight of camphor. The camphor-plasticized product is known as 'celluloid'. Cellulose nitrate, alcohol and camphor are mixed in a steam-heated dough mixer until homogeneous, whereby the fibrous nature of cellulose is lost. The dough is then filtered under pressure, reduced to a low solvent content (12–16%), first by putting under vacuum and then milling on a calender. It is then sheeted into slabs and pressed into a homogeneous block of a specific size. Sheets may be sliced out from the blocks. They are then 'seasoned' to remove last traces of solvent and subjected to surface finishing operations. The high inflammability and somewhat poor chemical and solvent resistance of cellulose nitrate stand in the way of its widespread industrial application. It is no longer considered suitable in many of its earlier areas of application such as photographic and cine film, machine guards or covers, toys, etc. Curent applications include tool or knife handles, table-tennis balls spectacle frames, lacquers and printing ink; in some of these fields, celluloid is facing growing challenge from other synthetic polymers or other cellulosics such as cellulose acetate.

8.20.2 Cellulose Acetate

The most important organic ester of cellulose is cellulose acetate. Cellulose acetate is prepared by acetylation of cellulose using a mixture of acetic anhydride and glacial acetic acid as the acetylating agent under a controlled condition. Small proportion of H_2SO_4 is used as catalyst. In most instances cotton linter and wood pulp are used for acetylation. Acetic acid used acts as a swelling agent for the cellulose and as a homogenizing solvent or diluent for the acetate derivative formed. Three stages are associated with the homogeneous acetylation process: (a) pretreatment of cellulose, (b) acetylation and (c) hydrolysis.

Pretreatment makes cellulose swell whereby it becomes readily accessible to acetic anhydride, the actual acetylating agent. Glacial acetic acid is most commonly used for pretreatment. Cotton or wood pulp is steeped in 30-40% of its own weight of glacial acetic acid for 1–2 h.

The pretreated cellulose is then transferred into the acetylating tank and a requisite amount of the acetylating mixture consisting of acetic anhydride, glacial acetic acid and the catalyst, H_2SO_4 previously cooled to 15–20°C is poured into the tank. As the acetylation proceeds, the physical appearance of the substrate slowly changes from a mass of shredded fibre through an opaque dough like mass to finally a clear viscous solution. Temperature is controlled carefully within 45–50°C. Esterification is usually complete in 5–6 h and the primary product, characterized by a degree of substitution of 3.0, is called the primary cellulose acetate; theoretical acetic acid yield of the primary acetate is 62.5% which corresponds to an acetyl content of 44.8 percent.

The primary acetate is soluble in chlorinated hydrocarbons (such as chloroform, trichloroethane or methylene chloride), dioxane or nitromethane and insoluble in acetone. The primary acetate obtained under the conditions detailed above contains some sulphate groups, and it exhibits a viscosity not much suited for spinning. The primary acetate or the triacetate obtained as a viscous solution in the acetylating mixture is then diluted with aqueous acetic acid and all the sulphate groups as well as part of the acetate groups are hydrolyzed. The hydrolysis is accomplished over several hours to several days depending on temperature employed and degree of hydrolysis desired. As the desired degree of hydrolysis or degree of substitution is reached, the cellulose acetate is precepitated by addition of water to the stirred solution. The product is then filtered, washed and dried. It is commonly known as secondary cellulose acetate or the diacetate. The theoretical acetic acid yield corresponding to the diacetate is 48.8% which is equivalent to 35.0% acetyl content. The cellulose diacetate or the secondary acetate is soluble in acetone and insoluble in chloroform.

Even though the secondary acetate is soluble, it is thermally unstable and decomposes below its softening point. It can be conveniently processed and moulded if only it is plasticized with a suitable plasticizer such as, diethyl phthalate and triphenyl phosphate. Other compounding ingredients used are UV-stabilizers, extenders, lubricants and colouring matters. A major outlet of cellulose acetate is in the area of sheets, films and membranes. Its exceptional clarity makes it suited for photographic films. Injection moulded items include tooth brush handles, combs, etc. Films are useful in packaging and wrapping. The toughness, low flammability (compared to cellulose nitrate), good clarity and favourable cost structure of cellulose acetate are advantageous. On the other hand, its high water absorption, poor solvent and chemical resistance and limited heat resistance and dimensional stability are its limitations.

The diacetate is dry spun from acetone solution into fibre, and it is less prone to wrinkling than regenerated cellulose (viscose) fibres. It also soils less and is known as the acetate rayon. The acetate fibre is less prone to creasing and it is washable more readily than regenerated cellulose fibres. The primary acetate can be dry spun from methylene chloride solution and the relevant fibre is suitable for apparels.

Mixed organic esters of cellulose have also been developed commercially. The important mixed esters are cellulose acetate–butyrate and cellulose acetate-propionate. The cellulose acetate–butyrates meant for plastics application (injection moulded objects) have nearly 12–20% acetyl and 30–40% butyryl content. Cellulose acetate–butyrate is more suitable than cellulose acetate for movie films. Its excellent appearance and clarity, toughness and ease of mouldability are of special advantage. It is also of lower density than cellulose acetate. The mixed ester is competitive with cellulose acetate in most of its thermoplastic applications including eye-glass frame on property considerations. However, the polymer emits smell of butyric acid under processing conditions. The mixed ester is used in the making of automobile parts, tool handles, toys, telephone housings and pipes. Other cellulose esters that have been commercially exploited are cellulose propionate and cellulose acetate–propionate.

8.20.3 Cellulose Ethers

Cellulose ethers are conveniently prepared by allowing selected alkyl chlorides to react with alkali cellulose in presence of alkali at a specific temperature over a specific time period.

$$\text{R—cell}\cdot\text{OH} + \text{R}'\text{Cl} + \text{NaOH} \rightarrow \text{R—Cell}\cdot\text{O}\cdot\text{R}' + \text{NaCl} + \text{H}_2\text{O} \qquad (8.32)$$

For making ethyl cellulose, the alkyl chloride used is ethyl chloride. The reaction is done at 60°C for several hours and the temperature is raised to 125–140°C towards the end. Ethyl alcohol and ethyl ether are formed as byproducts and are removed by distillation. Ethyl cellulose is precipitated by adding hot water; it is then centrifuged, washed and dried.

The completely substituted product (DS = 3.0) exhibits an ethoxyl content of 54.88%. With a DS value of 0.8%–1.2, the ethyl cellulose is water soluble. Highly or fully substituted product is, however, soluble in non-polar solvents (aromatic hydrocarbons). Ethyl cellulose is used in hot melt strippable coating formulations, and surface coating formulations (paints, varnishes and lacquers). Water soluble grades were developed for use in textile and paper sizing and the alkali soluble grades were developed as permanent textile finishes. The cellulosic thermoplastics have been gradually losing their place and importance in many areas of application due to progressive availability of and developments in superior synthetic resins.

Other cellulose ethers of some importance include methyl cellulose (DS 1.6–2.0) which is employed as a thickening agent, and an emulsifying and paper sizing agent being soluble in cold water. It has also limited use in leather tanning operation and ceramic processing. Hydroxy ethyl cellulose prepared by reaction of alkali cellulose with ethylene oxide also finds applications similar to those for methyl cellulose.

Sodium carboxymethyl cellulose is another important cellulose ether. It is produced by reacting alkalicellulose with sodium salt of chloracetic acid:

$$R{-}Cell \cdot ONa + Cl \cdot CH_2 \cdot COONa \rightarrow R{-}Cell \cdot O \cdot CH_2COONa + NaCl \quad (8.33)$$

On acidification the acid form known as the carboxy methyl cellulose (CMC) is obtained. Na–CMC of DS in the range 0.5–1.0 are water soluble and are commercially useful. They find use in synthetic detergents (soil suspending agent) and as stabilizers or viscosity modifiers in aqueous suspensions and emulsions. They also find applications as textile or paper sizing agents, surface active agents and polyelectrotyles. Pure Na–CMC is physiologically inert and is commonly employed as thickening agent in food, pharmaceutical and cosmetic products.

8.20.4 Regenerated Cellulose

One simple way of solubilizing cellulose is to transform the alkali cellulose or soda cellulose (R—Cell · ONa) into sodium cellulose xanthate by reacting it with CS_2. The xanthate made appropriately is soluble in dilute aqueous alkali, which on acidification decomposes; the xanthate structure is broken and cellulose structure is regenerated. This is the basic approach to the production of regenerated cellulose. The main reactions involved may be shown as:

$$R{-}Cell \cdot OH + NaOH\ (18\%) \rightarrow R{-}Cell \cdot ONa + H_2O \quad (8.34)$$

$$R{-}Cell \cdot ONa + CS_2 \rightarrow R{-}Cell \cdot O{-}\overset{\overset{\displaystyle S}{\|}}{C}{-}SNa \quad (8.35)$$

$$R{-}cell{-}O{-}\overset{\overset{\displaystyle S}{\|}}{C}{-}SNa + H^+ \rightarrow R{-}Cell{-}OH + CS_2 + Na^+ \quad (8.36)$$

The solution of sodium cellulose xanthate in dilute alkali is commonly known as the viscose solution. Fibre made from wet spinning of viscose solution in an acid/salt bath is known as the viscose rayon.

In the initial step, cellulose (cotton linter or wood pulp) is steeped in 18% aqueous NaOH at room or lower temperatures. This helps appropriate swelling of the cellulose which is at the same time converted into soda-cellulose. The excess alkali is then pressed out and the soda-cellulose formed is shredded and allowed to age at 25–30°C for 2–3 days to accomplish desirable extents of oxidative degradation of the cellulose substrate. The aged soda-cellulose crumbs are then reacted with liquid or gaseous CS_2 at about 25–30°C for 2–3 h whereby the mass becomes sticky and somewhat gelatinous, assumes a deep orange colour and emits a bad odour. After the desired extent of xanthation is accomplished, excess CS_2 is removed by applying vacuum and the mass is dissolved in dilute aqueous NaOH solution (4%). The odour and colour of these viscose solution is attributed to some sulphurous byproducts formed during xanthation.

The viscose solution is then allowed to stand at a low temperature (15–20°C) for a few days and a process known as ripening follows. The unstable sodium cellulose xanthate undergoes decomposition and a complex sequence of reactions and physical changes take place during ripening at the end of which coagulation of viscose becomes easy and almost spontaneous. The ripened viscose is then filtered and spun into fibre or made into transparent film. The coagulation or regeneration bath is made up of an appropriate aqueous solution of H_2SO_4, Na_2SO_4 and glucose. The fibre made from viscose is known as viscose rayon and the film made is known as cellophane.

The low or normal tenacity viscose rayon (2–2.5 g/denier) was once developed almost solely as an apparel fibre. But it is no longer useful in this field because of subsequent developments of much better fibre forming synthetic polymers. High tenacity viscose rayons (3–6 g/denier) are produced by high degrees of stretching of the fibre after spinning. Because of high order of crystallinity, very low elongation (poor flexibility) and low moisture absorption, the high tenacity rayon is almost exclusively used as tyre cords but not as a textile material. In tyre cord application too, the viscose rayon is facing stiff challenge from the synthetic fibres, particularly polyesters and nylons.

For the production of cellophane films, the viscose solution is extruded as a film which is then immersed in the Na_2SO_4/H_2SO_4 bath. Xanthate-splitting and regeneration of cellulose takes place in the bath and the film is passed through washing, bleaching and desulphurizing baths. Plasticization and finishing of the film is achieved by passing it finally through a bath containing ethylene glycol or glycerol. Cellphane or regenerated cellulose films or foils are popular in wrapping applications and useful as membranes for dialysis.

8.21 Phenolic Resins

The reaction between different phenols and aldehydes leads to the formation of what is known as phenol–aldehyde resins or simply phenolic resins or the phenolics The simplest of these reactants, viz. common phenol and formaldehyde are most commonly employed in the making of different grads of phenolic (phenol-formaldehyde) resins. Other phenols used are cresols, resorcinol and *p*-tertbutyl phenol. Among aldehydes other than formaldehyde, only furfural has been used commercially to a limited degree. Resins from phenol and formaldehyde are also commonly known as p–f resins.

8.21.1 Chemistry of Resin Formation

For resinification with aldehydes, common phenol is recognized to function as a trifunctional compound (the 2 ortho-positions and the para-position are the three reactive sites). In most resinification reactions involving formaldehyde, formalin solution (nearly 37% aqueous formaldehyde) is used and the aqueous formaldehyde is considered to remain in this solution as methylene glycol ($CH_2 = O + H_2O \rightarrow HO \cdot CH_2 \cdot OH$). Thus, phenol–formaldehyde resinification system is a tri-bi functional system and the resinification that proceeds by a polycondensation process invariably leads to the ultimate formation of a space network, i.e. three dimensional cross-linked polymer. The first step in the resinification reaction producing monomethylol phenols may be written as:

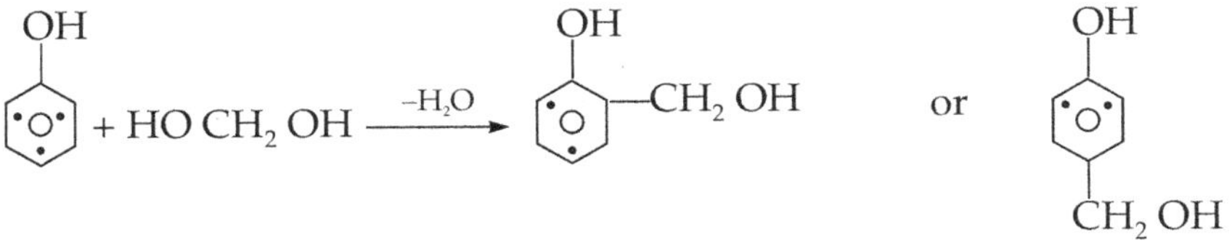

(8.37)

The reaction is catalyzed by both acids and bases. Basic catalysts (Na_2CO_3, NaOH, NH_4OH, etc.) and higher molar proportions of formaldehyde favour formation of di- and trimethylol phenols in the subsequent steps, e.g.

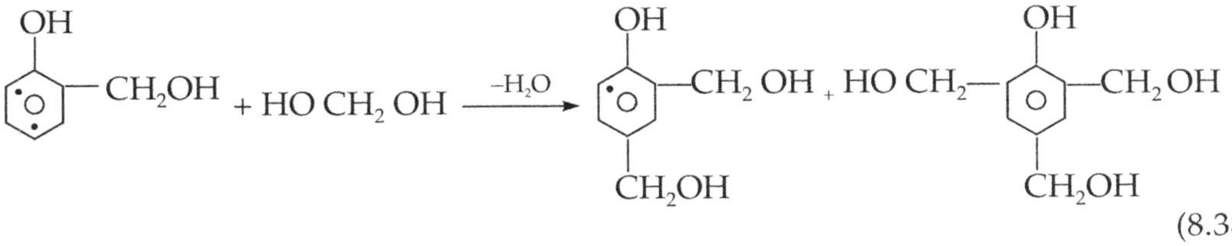

(8.38)

The initial reaction products or the resinoids and the reactants further react among them leading to chain extension and resin formation giving progressively increasing molecular weight. Acids are, however, more powerful catalysts. Acid catalysts (mineral or organic acids) and higher molar proportions of phenol favour immediate reaction involving two phenol molecules for each formaldehyde molecule resulting in the production of 2,4-dihydroxydiphenylmethane as the primary isolable product in the early stages. The early stage resinoids readily react with additional formaldehyde to form larger monomethylol derivatives which immediately

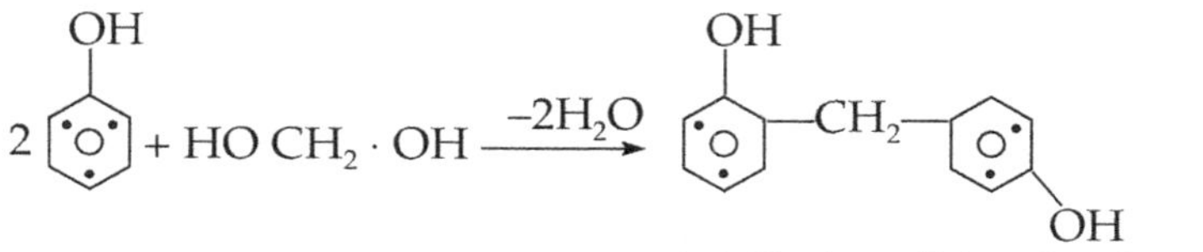

2, 4-dihydroxydiphenyl methane (8.39)

react further with an additional phenol molecule. Repetition of this reaction sequence leads to the formation of practically linear polymers with average molecular weight of nearly 600–700, which are commonly known as novolacs. A typical novolac contains about 6 aromatic rings linked through methylene linkages and having practically no methylol group in the resin molecule, (II). Novolacs are thus permanently soluble and fusible.

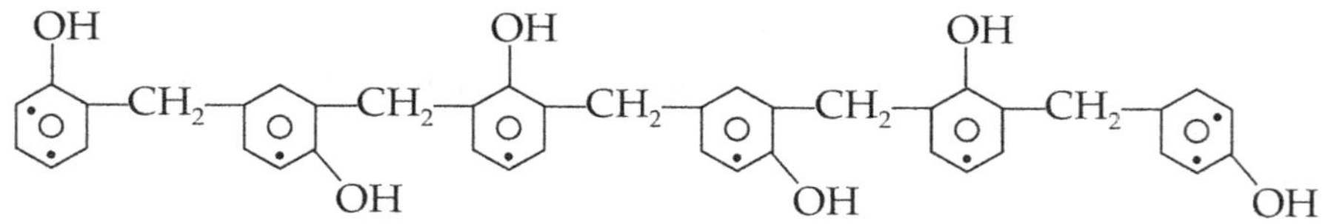

(II)
Average novolac molecule

If additional formaldehyde is made available and the resin is thermally treated, cross-linking reactions between novolac chains via the formation of methylol groups along with chain extensions take place ultimately rendering the resins insoluble and infusible. A segment of a typical cross-linked structure, (III) is shown below:

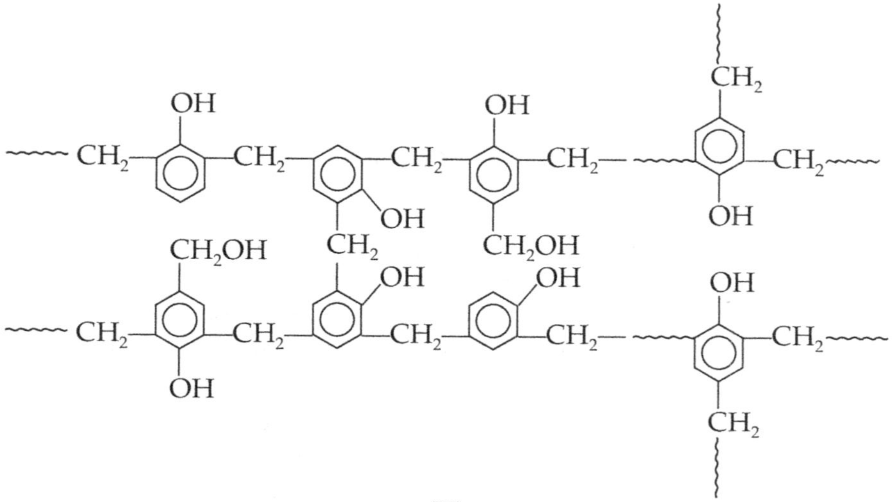

(III)
Segment of a cross-linked phenolic resin

The cross-linking of the novolacs in the second step in presence of additional formaldehyde is better accomplished using a basic catalyst. The overall technology of the novolacs is a two step process; the first step corresponds to the formation of the linear, soluble and fusible resins using phenol and formaldehyde, usually in 1 : 0.8 molar proportions and the second step corresponds to heating the linear resin in presence of what is known as a hardener or cross-linking agent which decomposes under heat and pressure to liberate formaldehyde and bring about cross-linking through establishment of methylene bridges. Hexamethylene tetramine or simply hexa, most commonly used as the cross-linking agent, decomposes under heat and using the moisture present to generate formaldehyde (the curing agent) and ammonia (the curing catalyst).

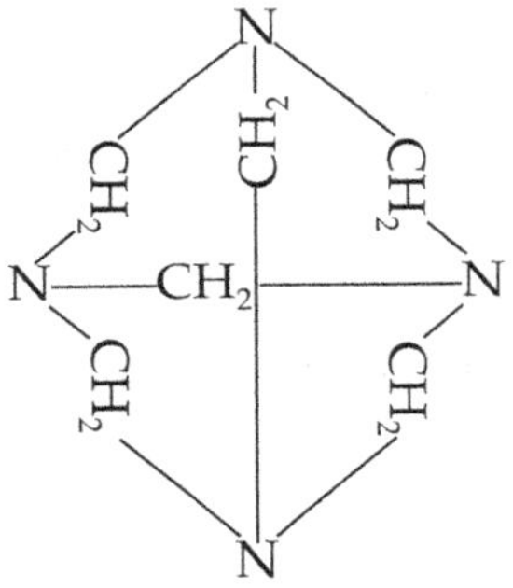

hexamethylene tetramine

$$\xrightarrow{+\,6H_2O} 4NH_3 + 6CH_2O \qquad (8.40)$$

By contrast, the resinification using basic or alkaline catalysts ultimately leading to the production of insoluble, infusible, cross-linked polymer products is a one step process. The resinoids or chain molecules formed initially and at intermediate stages of growth carry appreciable number of methylol groups. If the reaction is stopped in the early stages, a low molecular weight product soluble in alkali or alcohol is obtained. It is commonly known as *resole* or the A-stage resin. A typical resole molecule, contains 3–4 aromatic rings and nearly 3–5 methylol groups and it may be either a linear chain or may be lowly branched (IV):

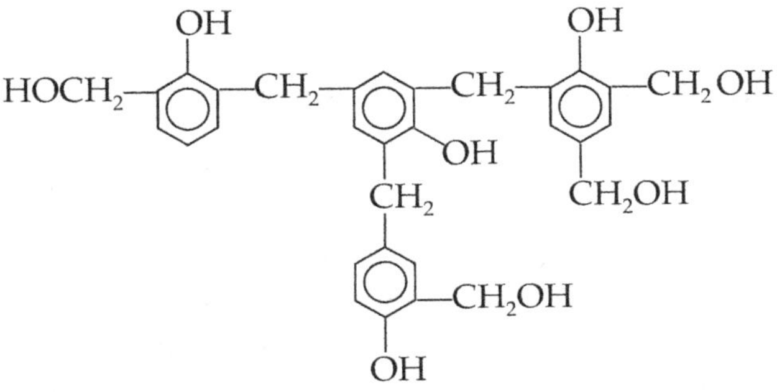

(IV)
Resole or A-stage phenolic resin

Resoles may have molecular weights in the range 350–500. On heating, resoles interact among each other and with other reactive species present in the system producing larger and mildly cross-linked resin molecules known as *resitol* or the B-stage resin. The resitol is insoluble in alkali and alcohol, but is partly or completely soluble in acetone. Resitols are still heat-fusible, and on further heating they turn into fully cured or cross-linked resin known as *resite* or the C-stage resin. The C-stage resin is totally insoluble and infusible, and is segmentally represented by structure (III) shown earlier in this section. There are also evidences that under conditions of higher molar proportions of formaldehyde and alkaline catalysts, reaction between phenol alcohols initially formed leads to measurable formation of methylene ether bridges at temperatures as low as 70–100°C. Heating of the ethers in subsequent step or during cross-linking results in transformation of the methylene ether bridges to methylene bridges by splitting off formaldehyde, at least in part,

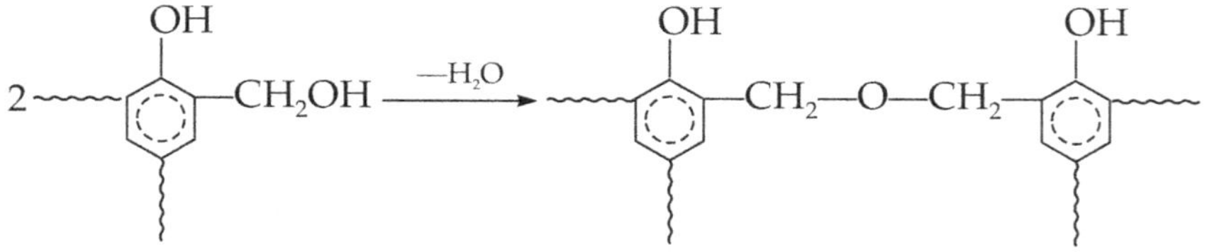

(8.41)

which may then react at other reactive sites ultimately leading to the establishment of additional cross-linkages.

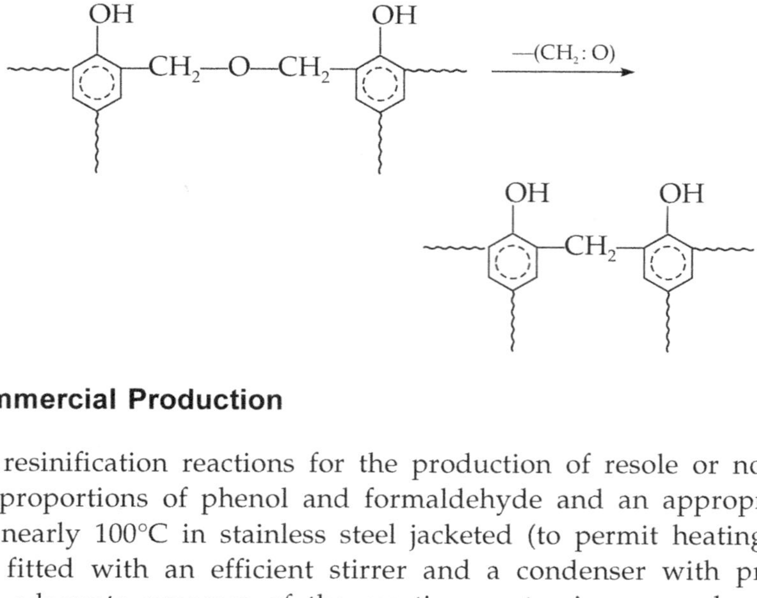

(8.42)

8.21.2 Commercial Production

Commercial resinification reactions for the production of resole or novolac, using appropriate proportions of phenol and formaldehyde and an appropriate catalyst are done at nearly 100°C in stainless steel jacketed (to permit heating or cooling) kettles each fitted with an efficient stirrer and a condenser with provisions for reflux. After adequate progress of the reaction, water is removed usually under reduced pressure and under controlled temperature conditions and the molten or high viscous resin is discharged by opening the main discharge outlet valve at the bottom of the kettle.

8.21.3 Phenolic Moulding Powders

Novolacs are commonly used for preparation of moulding powders by judicious blending or mixing with the resin a host of compounding ingredients such as filler (usually wood flour or saw dust, 40–60 phr), hardener (hexa 10–12 phr), accelerators or curing catalysts (CaO, MgO, 2–5 phr), lubricant (stearic acid or Ca/Zn stearate, 1–2 phr) and pigments (2–5 phr). During curing at high temperatures (160–165°C or higher) the phenolics darken considerably due to additional reactions involving the methylene ether linkages and phenolic OH groups leading to the formation of quinone methide and related structures. The pigments used in the phenolic moulding-powders are limited to blacks, browns and dark greens, blues, reds and oranges. Fillers are used to impart useful properties and to produce good mouldings. Saw dust has the additional advantage of reducing the cost. The phenolic moulding powders are usually moulded by compression moulding techniques using a pressure of 1500–3000 psi and a temperature of 160–170°C. The bulk factor of the moulding powder is given by the ratio of the density of the moulding to the apparent density of the powder fed into the mould. A convenient bulk factor for efficient moulding is normally obtained by what is known as preforming or pelletizing. Both preforming and preheating help in optimizing the moulding cycle. Preheating also reduces shrinkage and required moulding pressure. The curing is usually accomplished in 30–60 s for small and thin objects.

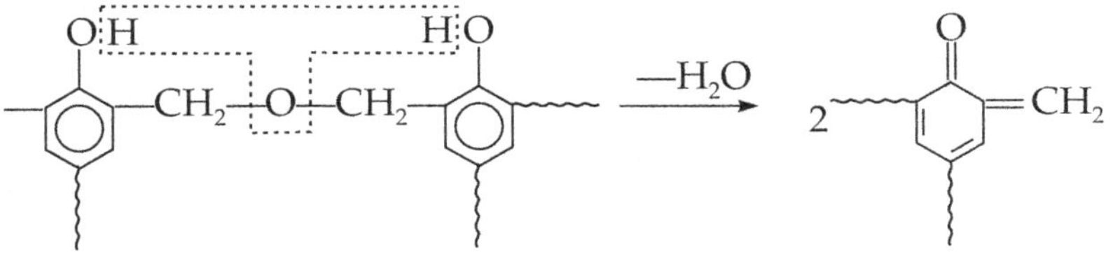

(8.43)

quinone methide structure

The most well-known applications of the phenolics are in areas of electrical insulation such as switches and plugs. The phenolics have better heat and moisture resistance than the urea-formaldehyde mouldings. Phenolic mouldings are also used to make knobs, handles, telephones, instrument cases, automobile components, caps and closures of containers, etc.

8.21.4 Phenolic Laminates

Laminating phenolic resins are usually the resoles. Caustic soda catalyzed resoles are suitable for mechanical and decorative laminates while ammonia-catalyzed resoles are suitable for electrical laminates. The reinforcements in the laminates are either paper, fabric or wood vineer. Resoles having a high methylol content made by using high formaldehyde to phenol ratio are water-soluble and a high degree of impregnation is possible using the aqueous solution. Spirit soluble resoles are also used for impregnation wherever necessary, convenient or advantageous. The continuously impregnated sheets are dried in a vertical or horizontal drying oven; the solvent evaported may be recovered if necessary and the resin advances to a desirable high degree of polymerization. The resin impregnated dry sheets are then cut into pieces and laminates are prepared by plying up the pieces and pressing in (multi-daylight) compression press between metal plates at 150–160°C and under pressure of 1000–2000 psi.

The phenolic (paper) laminates are extensively used for high voltage electrical insulation purposes. The laminates properly made through the right selection of the resin and reinforcement exhibit high strength, stiffness rigidity and good machinability and they are used in many engineering applications such as gears, bearings, bushings, etc. Plywoods and particle boards find use in the building and furniture industry. The laminates are also used in chemical plants and in the construction of transport vehicles and aircrafts.

8.21.5 Cast Phenolics

The casting grade phenolic resins are different from those meant for other applications. They are prepared from highly pure grades of phenol and formaldehyde (formalin) using more than 2 moles of formaldehyde (typically 2.2–2.5 moles) per mole of phenol and employing alkaline catalyst. To minimize colour formation, a stainless steel resin kettle is used and a milder temperature (70–75°C) is maintained. Highly methylolated hydrophilic resin molecules are formed and they remain uniformly dispersed in the water without separation or layer formation. The nearly water-white resin solution is then mixed with lactic acid to neutralize the alkaline catalyst and then the water content is reduced to nearly 5–10% by applying vacuum at 80–85°C. The viscous resin, light straw in colour, is then mixed with glycerol, a mould lubricant and colouring matter and poured into appropriate moulds and allowed to cure slowly at a temperature of 70–75°C over a period of 2–10 days. With growth in sizes of the resin molecules, ultimately leading to the formation of the C-stage resin, water present tends to separate out in fine droplets.

Glycerol used helps to disperse the entrapped water and thus helps minimize loss of transparency. Post-baking improves the dimensional stability of the cast phenolics. The curing time may be significantly lowered by using small amounts of an acidic catalyst into the resin mix just before casting. For their attractive colours and transparency self-lubricating properties and good machinability, the cast phenolics are used as objects of art and jewellery and as handles of tools and umbrella, knobs, etc. Cast phenolics have, however, become of much less commercial relevance in view of the widespread developments of cheaper and much improved materials in the form of vinyl and related thermoplastics, over the decades following the Second World War.

8.21.6 Miscellaneous Applications of Phenolic Resins

Foamed objects based on phenolic resins have gained importance, particularly in structural applications in the building industry. Resins based on para substituted phenols are used in surface coating formulations; *p*-tertbutyl phenol and *p*-phenyl phenol are reacted with formaldehyde to produce oil-soluble grade phenolics. Resorcinol–formaldehyde resins are extensively used as suitable bonding or coupling agents aimed at achieving good adhesion between reinforcing cords and rubber in the tyre, hose and belting industry. Other applications, taking advantage of the binding or adhesion capacity of the phenolics include brake linings, abrasives, sand core bonding and shell moulds for metal castings, grinding wheels, high performance refractory items etc. Another important application of the phenolics is in the area of making ion-exchange resins via sulfonation (cation exchanger) and chloromethylation followed by quaternization by treatment with ter-amines followed by alkali treatment (anion exchanger) (see Sec. 8.6.6.).

8.22 Amino Resins

Resins or plastics based on products of interactions between amines or amides and aldehydes are commonly known as amino resins or aminoplastics. Of the different amino resins prepared and used, the urea–formaldehyde (u–f) resins are the most important in all respects. Next in importance are the melamine–formaldehyde (m–f) resins.

8.22.1 Urea–Formaldehyde Resins

Urea–formaldehyde resins are prepared by reactions between urea and formaldehyde, usually carried out in two stages. The resinification is usually carried out in the first stage under a mildly alkaline condition in the temperature range of 30–45°C leading to the formation of monomethylol urea, and dimethylol urea. They are colourless and extensively soluble in water.

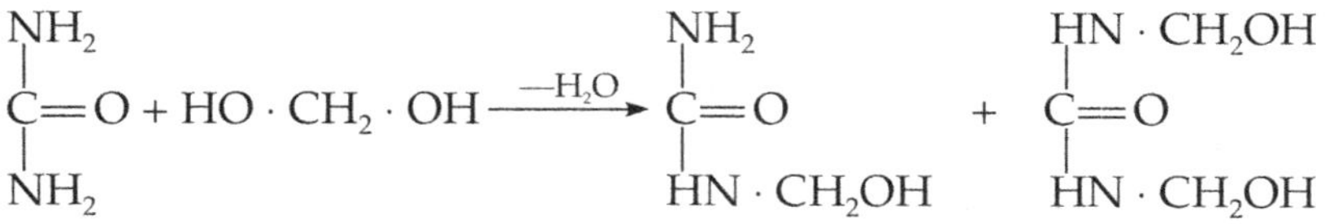 (8.44)

The relative proportions of mono- and di-methylol ureas would depend on the molar ratio of urea to formaladehyde (u : f). Urea is potentially tetrafunctional and use of nearly 1.5 to 2 moles of formaldehyde for each mole of urea has been found to yield useful resins. There should be enough formaldehyde to permit measurable and significant formation of dimethylol urea. Both monomethylol urea and dimethylol urea are isolable in the pure form. They further react amongst each other and with higher condensation products leading to chain extension and ultimately to the formation of high polymers. The polymerization is conveniently carried out in the second stage under acidic condition, the acid acting as a strong catalyst for the polymerization, leading to the formation of insoluble, infusible cross-linked products. The chain extension and cross-linking are accompanied with loss of water and formaldehyde. A segment of a typical cross-linked species is shown below (structure V).

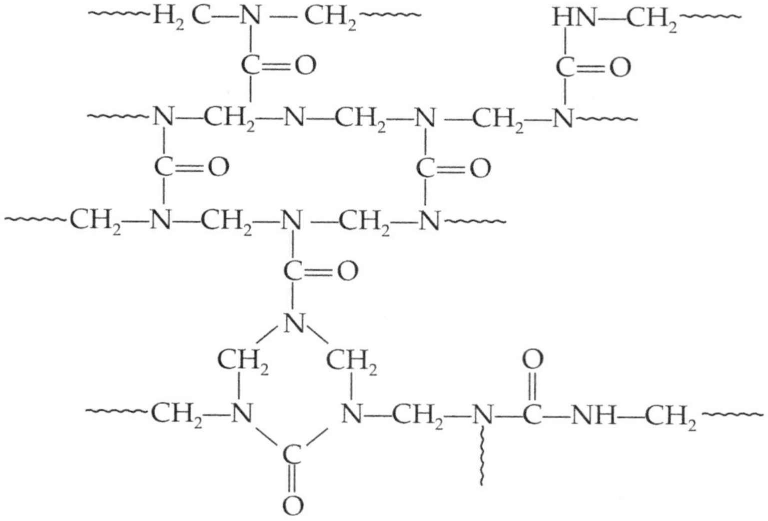

(V)
Segment of a typical cross-linked structure of urea-formaldehyde resins

The first stage resinification using appropriate molar proportions of urea and formaldehyde (37% formalin, adjusted to a *p*H of nearly 8.0 with NaOH) is done in a stainless steel reactor for about 30–90 min at a temperature of 30–45°C. Any drop in *p*H is usually adjusted by the addition of small amounts of ammonia or hexamine, The first stage or the A-stage resin is a water soluble mixture of monomethylol and dimethylol urea and some unreacted urea and formaldehyde.

The aqueous solution is then used straight away for the preparation of moulding powder. Cellulosic fillers in the form of bleached wood pulp are almost universally used to obtain widest range of bright colours or white objects. The pulp, resin solution and other additives are mixed together and heated at 60–70°C for 2–3 h to allow impregnation and further growth in size of the resin molecules. The resin impregnated pulp is then dried in a rotary drier for nearly 2 h at 100°C to a water content of about 5% and the resin advances to the B-stage in the process. The mixture is then milled in a ball mill and the pigments or colouring matters are added at this stage. For efficient and rapid curing of the resin to C-stage at moulding temperatures, a latent catalyst (1–2%) such as ammonium sulphamate, trimethyl phosphate etc., is mixed with the resin-filler systems. The catalyst decomposes at moulding temperatures to liberate an acid that accelerates the cross-linking reactions. To improve the shelf-life, the moulding powders are stored in a cool place. Incorporation of a small amount of a stabilizer such as hexamine in the moulding powder further improves the shelf-life. About 1% of a metal stearate (Zn, Al, Mg) is used as the lubricant in the moulding powder.

Moulding is commonly done by the compression, transfer or screw injection technique. Compression moulding is done employing a temperature of 150–160°C and a pressure of 1000–3000 psi over a time period of 20–60 s depending on the size and thickness of the moulding following the procedure described for the phenolic moulding powders. Typical end-uses are housewares (cups, bowls, saucers, etc.) housings for domestic appliances, electrical plugs and switches, telephones sets, caps and closures, vacuum flask cups and jugs, buttons, trays, knobs, lampshades, kitchen and bathroom fittings, etc.

The cured urea resins are hard and rigid and resistant to stain and scratching. They are non-inflammable but they char around 200°C. They exhibit good impact properties and excellent electrical insulation properties under dry conditions. They (u–f) resins do not impart taste and odour to food and drinks with which they may come in contact.

Urea–formaldehyde resins also find extensive use as wood adhesives and in the manufacture or plywood and laminates. The adhesive resins are prepared in the water soluble form, stabilized by maintaining the *p*H at 7.5 and a solid content of nearly 50–70% according to requirement by allowing partial dehydration using vacuum distillation. The resin may also be spray dried to a stable powder. The high formaldehyde resins exhibit greater clarity, but they are more expensive due to the higher cost of formaldehyde. The resins are hardened under acidic conditions. In plywood manufacture, the more common hardening agent used is ammonium chloride (1.5 parts per 100 parts of resin solution) which acts as an acid donor. The resin/hardener mixture is used to impregnate the wood veneers, which are then plied together and pressed at nearly 100–110°C using a pressure of 200–1000 psi. The plywoods are extensively used indoors but are unsuitable for outdoor work. A major outlet of the u–f resin bonded plywood is in the packaging field.

Urea–formaldehyde and related resins also find widespread use in the finishing of cotton textiles. The fabric is passed through the aqueous solution of methylol urea containing metal salts as hardeners and the resin-soaked fabric is heated in an oven at 140–160°C whereby the resins are hardened *in situ*. The process imparts good crease resistance to the fabric. Mercerization (steeping in NaOH solution) before resin treatment imparts improved combination of properties, particularly with respect to tear and tensile strengths along with crease resistance. Butylated urea resins show a good degree of thermoplasticity and they are soluble in organic solvents. They are used along with alkyd resins in the formulation of stoving enamels or air-drying lacquors. Simultaneous condensation of urea, formaldehyde and a monohydric alcohol such as butanol in presence of an acidic catalyst involve etherification and resinification at the same time. The alkoxy groups formed improve the solubility of the resin in surface coating solvents (make the resin oil-soluble) and by blocking the methylol groups they also reduce the tendency of the resin to get cross-linked.

$$\begin{array}{c}-N-CH_2-\\ |\\ C{=}O\\ |\\ HN-CH_2OH\end{array} + ROH \rightarrow \begin{array}{c}-N-CH_2-\\ |\\ C{=}O\\ |\\ HN-CH_2OR\end{array} + H_2O \qquad (8.45)$$

8.22.2 Melamine–Formaldehyde Resins

The other amino resin of commercial importance is melamine–formaldehyde (m–f) resin. Melamine may be obtained by heating urea under pressure, or more conveniently by heating dicyandiamide,

$$\underset{\text{urea}}{6CO(NH_2)_2} \xrightarrow[\text{400–600 atm}]{\text{400–600°C}} \underset{\text{melamine}}{C_3H_6N_6} + 3CO_2 \qquad (8.46)$$

preferably in presence of ammonia at about 300°C under pressure. Dicyandiamide is produced by heating cyanamide (NH_2CN) which in turn is obtained from calcium cyanamide, $CaCN_2$.

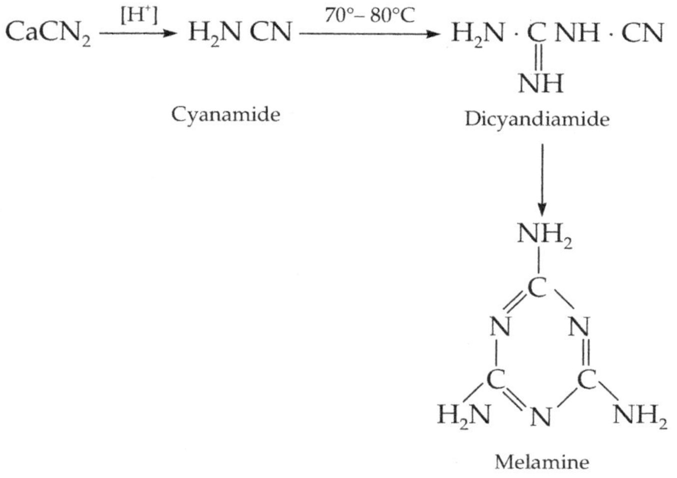

(8.47)

Resinification reaction between melamine and formaldehyde is very much dependent on the melamine to formaldehyde molar ratio and on the *p*H of the medium. Initial reactions lead to the formation of a mixture of water soluble methylol melamines. Depending on the prevalent reaction conditions, methylol derivatives up to six methylol groups per molecule (melamine is potentially hexa-functional) may be obtained. (Eq. 8.48 shown on next page).

Methylol—methylol condensation may subsequently lead to some methylene ether linkages:

$$\sim\sim\underset{|}{N}\text{—}CH_2OH + HOCH_2\text{—}\underset{|}{N}\sim\sim \longrightarrow \sim\sim\underset{|}{N}\text{—}CH_2\text{—}O\text{—}CH_2\text{—}\underset{|}{N}\sim\sim + H_2O \qquad (8.49)$$

However, major reactions leading to formation of methylene bridges are:

$$\sim\sim\underset{|}{N}\text{—}CH_2OH + H\underset{|}{N}\sim\sim \xrightarrow{-H_2O} \sim\sim\underset{|}{N}\text{—}CH_2\text{—}\underset{|}{N}\sim\sim \qquad (8.50)$$

$$\sim\sim\underset{|}{N}\text{—}CH_2\text{—}O\text{—}CH_2\text{—}\underset{|}{N}\sim\sim \xrightarrow{-CH_2:O} \sim\sim\underset{|}{N}\text{—}CH_2\text{—}\underset{|}{N}\sim\sim \qquad (8.51)$$

Commercial resinification reactions are done in much the same manner as employed for urea resins; pH is adjusted to 8–8.5 using Na_2CO_3 and an initial molar ratio of 1 : 3 for melamine to formaldehyde is used. The methylolation is done at a temperature of 80–85°C. The aqueous solution of the methylol melamines may be

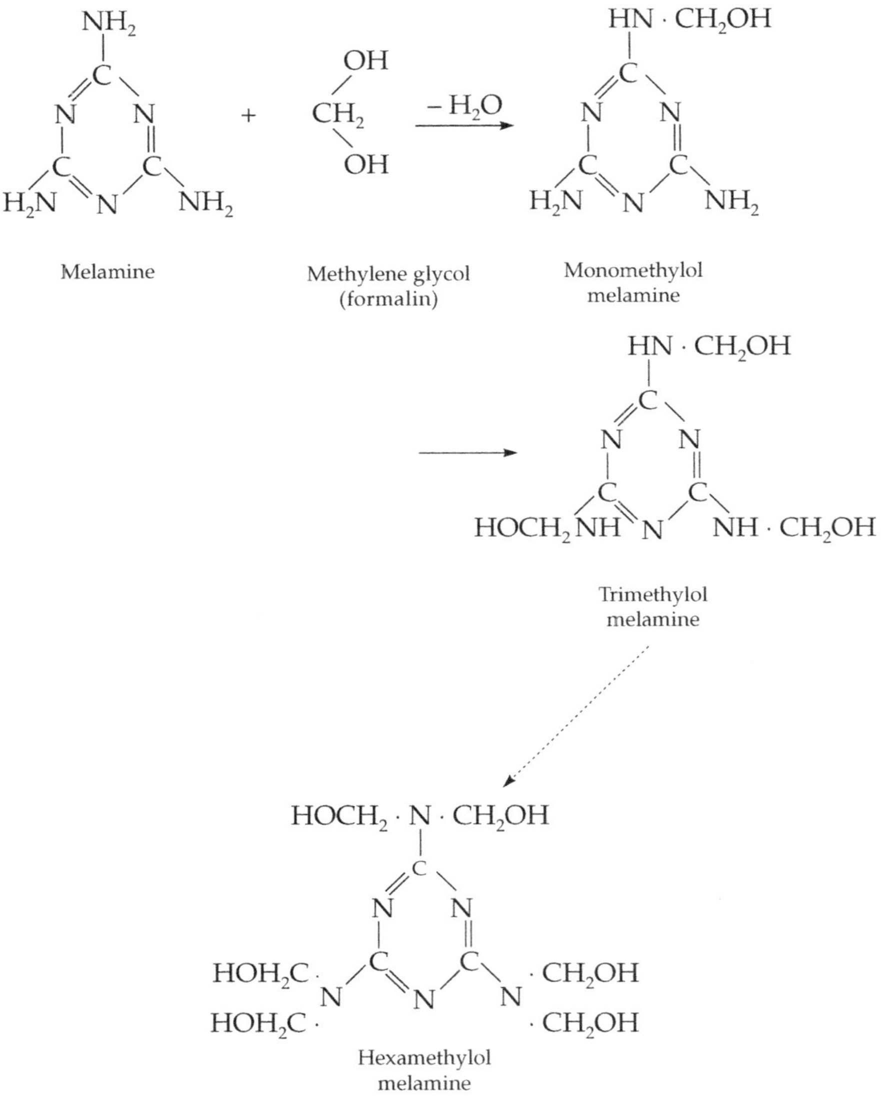

(8.48)

directly used for the finishing of fabrics, leather and paper. Spray drying of the soluble resins gives a product of higher stability.

The m–f moulding powders are made from the aqueous resin syrup produced by partial dehydration of the initial aqueous resin solution by vacuum distillation and using an initial melamine–formaldehyde molar ratio of 1 : 2. The resin syrup is intimately mixed with fillers, pigments, lubricants, stabilizers and with an accelerator if necessary in a suitable mixer, dried in an oven at 100–110°C and then ball milled into a powder. Common fillers employed are bleached wood pulp (α-cellulose), asbestos, silica or glass fibre. The m–f mouldings are superior to (u–f) mouldings with respect to hardness, scratch resistance, heat resistance, stain resistance and water resistance (the m–f mouldings having much lower moisture absorption). The m–f resin imparts better electrical insulation properties to the mouldings under humid or damp conditions and at higher temperatures. Principal applications include dinnerware, housings of home appliances, handles and knobs, bathroom fixtures, etc. They are moulded by compression, transfer or screw-injection techniques. For compression moulding, temperatures in the range 150–170°C and pressure of the order of 2000–8000 psi are commonly employed.

The high cost of the melamine resins limits their use in the production of common, general purpose laminates where p–f and u–f resins are more competitive. However, the exceptional hardness, scratch and stain resistance, solvent resistance and heat stability qualities of the m–f resins are taken advantage of in making special purpose and decorative laminates in which only the surface layers are impregnated with the m–f resins and the inner or base layers with phenolic resins. For the production of electrical grade laminates the condensation catalyst used is triethanolamine rather than sodium carbonate. The electrical grade laminates are usually made by impregnating electrical-grade glass cloth with the triethanolamine catalyzed resin.

8.23 Silicones

Polymers in which the chain backbone consists of repeat units having organic side groups linked with silicon atoms are commonly known as silicones or polyorgano-siloxanes. The polyorganosiloxanes are usually prepared by allowing chlorosilanes such as $Cl_2SiR_1R_2$ to react with water to form the corresponding hydroxy compounds which subsequently condense with each other leading to the production of polymer molecules.

$$Cl-\underset{R_2}{\overset{R_1}{\underset{|}{\overset{|}{Si}}}}-Cl \xrightarrow[-HCl]{H_2O} HO-\underset{R_2}{\overset{R_1}{\underset{|}{\overset{|}{Si}}}}-OH \xrightarrow{-H_2O} -(-\underset{R_2}{\overset{R_1}{\underset{|}{\overset{|}{Si}}}}-O-)_n- \quad (8.52)$$

Only dichlorodialkyl silanes yield linear polymer; monochloro silanes fail to produce any polymer and trichloromonoalkyl silanes yield network polymers. The intermediate silanes, R_nSiX_{4-n} (n = 0–4), where X is a hydrolyzable group or atom such as chlorine or alkoxy, may be made by direct synthesis involving reactions between alkyl/aryl halides and elementary silicon, e.g.,

$$2CH_3Cl + Si \xrightarrow[Cu]{250\text{–}280°C} Si(CH_3)_2Cl_2 \tag{8.53}$$

When methyl chloride is used, the major product under the most favourable condition is dimethyldichlorosilane (70–75%); the minor constituents in the crude product being methyltrichlorosilane (8–10%) and trimethylchlorosilane (5–6%). They are separated by fractional distillation.

Silicone fluids are linear (low) polymers or cyclic condensation products of silanols obtained by hydrolysis of a predetermined mixture of chlorosilanes in excess water under vigorous agitation. Initial products give an approximately equal mixture of cyclic compounds (mainly tetramer) and linear low polymers. To stabilize the viscosity and to obtain higher proportions of linear polymers, it is required to equilibrate the fluid by heating with dilute H_2SO_4. Chain length is controlled by using the desired proportions of the monofunctional trimethylchlorosilane as the "blocking agent". The equilibration may take several hours at 100–150°C. The oil is separated from the aqueous layer and neutralized by washing with sodium carbonate solution. Silicone fluids of different viscosity grades are all colourless, odourless, water repellent and non-toxic products of high thermal stability showing very narrow deviation in physical properties, e.g. viscosity, over a wide range of temperature (–70 to 200°C). They are used as polish additives, as release agents during the moulding of certain plastics and rubber goods and in the die casting of metals, and as greases. They are useful as antifoam additives, in drug formulations, as additives in paints and other surface coatings and as water repellent finishes in the textiles, paper and leather industry.

Silicone rubbers or *elastomers* are very high molecular weight linear siloxane polymers prepared from the difunctional monomer, dichlorodimethyl silane (or from its cyclic tetramer) often mixed with minor proportions of dichloromethyl phenyl silane. The elastomer may be vulcanized with peroxides such as benzoyl peroxide. Filler incorporation is necessary to produce vulcanizates of some strength (up to 2000 psi), since unfilled polymers are soft and they have negligible strength. Silica fillers are preferably used for reinforcement. Silicone rubbers with vinyl side groups (0.02–0.5 mole % of other organic side groups—methyl and phenyl) may be more readily cured using less reactive peroxides such as dicumyl peroxide giving more elastic vulcanizates. The silicone rubbers are much more expensive than other synthetic rubbers and they exhibit room temperature physical and mechanical properties inferior to those exhibited by other synthetic rubbers. The silicone rubbers are characterized by good resistance to weathering and lubricating oils, high compression set (20–50% after 24 h at 150°C), very high air permeability and excellent low temperature (–80°C) flexibility. Special purpose rubbers as they are, they are used in wire and cable insulation demanding fire resistance and in gaskets and seals. The physiological inertness of the silicones is taken advantage of in such applications as sterilizable blood transfusion tubing, closures of antibiotic containers, etc. The cold curing elastomers are suitable for sealing, caulking and encapsulation.

Silicone resins are commercially prepared by batch hydrolysis of a suitable blend of chlorosilanes giving an average R/Si ratio ranging between 1.2 : 1 to 1.6 : 1. The blend of chlorosilanes is dissolved in a suitable solvent and then hydrolyzed with water under controlled conditions avoiding overheating. After the required hydrolysis, the solvent layer containing the polymer is separated from the aqueous layer and neutralized with mild alkali wash. Molecular weight of the polymer is usually allowed to advance further without effective cross-linking by heating or "bodying" the solution in presence of a metallic soap to a desired viscosity for a given solids content. Cross-linking of the resin is achieved *in situ* in the finished product. The silicone resins find special applications in electrical insulation in the form of glass cloth laminates and insulating varnishes. The resin is applied by brushing the solution on the substrate and allowing solvent evaporation or by dipping the substrate in resin solution and draining or scrapping the excess resin and allowing the solvent to evaporate; cross-linking is accomplished by heating. The resins also find use in special purpose industrial paints and as encapsulating agents.

8.24 Additives for Plastics

A major aim and objective relating to polymer processing is to "get a shape and set the shape". In general, polymers of any class, particularly those meant for plastics and rubber applications are seldom used in the pure form. In most cases, they are mixed or compounded with a host of additives called mixing or compounding ingredients before moulding, forming, coating, etc. The processing itself and its controlling features are dependent in a large measure on the rheological behaviour of the mix or the compound.

The additives are incorporated in the polymer keeping one or more of the following advantages in view: (a) reduction of material cost, (b) improvements in properties, performance and service life of the end product, (c) achieving ease of processing and reduction of processing cost, and (d) minimizing degradation during processing and use. Functionally, the additives may be categorized under the following heads and the figures in the parentheses indicate part by weight of the ingredient used per 100 parts of polymer in general:

(a) Fillers (30–100)

(b) Plasticizers (30–100)
(c) Stabilizers (0.02–10)
(d) Colouring matters (2–5)
(e) Lubricants (1–2) and flow promoters (2–15)
(f) Cross-linking agents (1–5)
(g) Others (1–5)

The most important constituent of the mix or the compound is the polymer itself. The polymer plays the role of a binding material that forms the matrix which finally cements the various ingredients of the mix into a cohesive mass and holds the shape or form given to the finished article.

8.24.1 Fillers

Fillers are usually solid additives. They may simply be used as loading agents (inert fillers) with the objective of lowering the material cost; but very often filler incorporation brings about much desired improvements in physical properties such as tensile strength, modulus, abrasion resistance and tear strength (reinforcing fillers). Different types of fillers are in use, viz. particulate, rubbery, resinous or fibrous fillers. For particulate fillers such as carbon black and precipitated silica, the average particle size and particle shape have profound influence on the properties of the end product. Mineral fillers are often treated with specific chemicals to improve wetting or attachment of polymers to fillers; stearic acid-treated calcium carbonate or amine- or glycol-treated clays may be cited as examples of treated fillers. Use of coupling agents, such as specific silanes, is of more recent development whereby more effective wetting or affinity of the filler to the main polymer is achieved via the coupling agent. Rubbery fillers are incorporated to improve toughness and impact resistance. Fibrous fillers such as wood flour, wood pulp, cotton flock, glass fibre and synthetic organic fibres (rayons, nylons and polyesters) are used as reinforcing agents. For improved heat resistance, glass fibre, mica and asbestos are useful as fillers.

8.24.2 Plasticizers

Plasticizers are usually high boiling nonvolatile liquids. They are used in a variety of polymers such as the cellulosics, acrylics and poly(vinyl chloride). Plasticization is manifested essentially through solvent action. To be effective, a plasticizer is required to have a solubility parameter close to that of the substrate polymer, such that two would be largely compatible. Both polar interaction and interactions through hydrogen bonding between the plasticizer molecules and polymer segments are considered to explain the plasticization effect. Common plasticizers used are high boiling organic esters of some organic acids and phosphoric acid such as dioctyl phthalate (DOP), diisooctyl phthalate (DIOP), dibutyl phthalate (DBP), dioctyl sebacate (DOS) and tricresyl phosphate (TCP). Substances such as hydrocarbon oils and resins which are used to make the polymer soft, and easy processing only by spacing effect without any meaningful interactions with the polymer are commonly known as *softeners*. Sometimes, chlorinated paraffins and refinery oils of lower solubility parameter than the true plasticizers are used to partly replace the plasticizer giving a comparable plasticization effect. Such replacement additives used with plasticizers are known as *extenders*. Extenders are effective only in the presence of a *true plasticizer*, and they fail to impart any plasticization in the absence of a true plasticizer.

8.24.3 Stabilizers

Different kinds of stabilizers are used as additives in polymers to protect them from degradation during storage and processing and under performing conditions. The most common stabilizers are: (i) antioxidants, which protect the polymer against degradation by oxygen attack, and (ii) UV absorbers, which protect the polymer against degradation by ultra violet light.

Oxygen attack or oxidative degradation becomes very critical under processing conditions when oxygen is absorbed at a high rate resulting in peroxidation or hydroperoxidation and subsequent degradation of the polymer by a radical chain mechanism follows. Antioxidants are primarily of two types: (1) *preventive antioxidants* which prevent generation of radicals, and (2) *chain breaking antioxidants* which function effectively by interrupting the propagation of the oxidative chain degradation. Most common chain breaking antioxidants are phenols, such as styrenated phenols, and amines, such as phenyl naphthylamines. Phenols are usually preferred because of their being less prone to staining, discolouration, blooming and bleeding. Amines are, however, sometimes more effective than phenols and are advantageous on cost considerations too. Among the preventive antioxidants, those more commonly used are: (i) peroxide decomposers such as dilauryl thiopropionate and *p*-substituted phenyl phosphites, and (ii) metal deactivators in the form of selected chelating agents which form complexes with metal ion contaminants that may otherwise initiate oxidation. Antiozonants are more relevant to diene rubbers and related polymers having unsaturations. Wax additives that form an impervious layer of wax by blooming to the surface protect such polymers under conditions of static applications, but they turn out to be much less effective to prevent ozone attack under dynamic conditions. Selected *p*-phenylene diamine derivatives are more effective and useful in this regard.

Chlorine-containing polymers such as poly(vinyl chloride) and related copolymers are thermally unstable and require stabilization and protection against thermal degradation, the stabilizers commonly used being basic lead salts and selected cadmium and barium compounds and organo-tin compounds (*See* Sec. 8.10.4).

Stabilization of polymers prone to UV degradation is usually achieved by incorporation of special additives commonly known as UV-absorbers. UV-absorbers commonly in use are benzophenone and its derivatives, phenyl salicylates and benzotriazoles. Carbon black (1–5%) also gives protection against UV degradation.

8.24.4 Colouring Matters

Various types and shades of pigments and dyes and also carbon black are used as colourants in polymer compounding. The colourants show different degrees of solubility in the polymer matrix. Low solubility may lead to blooming. To be efficient and effective, the colourants should have high covering power and they should be stable under processing conditions. It is often convenient to use masterbatches or colour concentrates in appropriate proportions with the white/colourless granules to produce coloured end-products.

8.24.5 Lubricants and Flow Promoters

Lubricants used to the extent of 1–2 wt % of the polymer allow quick mould release and prevent the moulded object from sticking to the mould. There are lubricants such as silicone fluids, graphite, etc., which when rubbed against the mould surface reduce friction at the interface and hence permit easy release of the moulded articles. Stearic acid and stearates of Ca, Zn, Pb, Cd and Ba blended into the polymer mix tend to exude from the composition to the polymer-metal interface during processing at high temperatures and in this manner prevent finished object from sticking to the mould. Incorporation of certain low molecular weight materials such as glyceryl stearates and cetyl palmitate impart better flow properties to the polymer mix under processing (melt) conditions without really affecting the bulk properties under service conditions. They essentially function as flow promoters. Polymeric flow promoters are also known. Polymers exhibiting very low melt viscosity at processing temperatures and having solubility parameters close to the solubility parameters of the base polymer are suitable as flow promoters. Polymeric flow promoters such as low molecular weight PVC, chlorinated polyethylenes, acrylate copolymers and ethylene–vinyl acetate (EVA) copolymers are often used in the compounding of unplasticized and transparent PVC compositions.

8.24.6 Cross-linking Agents

Cross-linking or curing agents are important and essential ingredients for all cross-linkable or curable compositions leading to the production of thermoset plastics and vulcanized rubbers. Nowadays, almost any polymer—linear or branched and having reactive sites in its structure or otherwise—can be cross-linked using selected chemical cross-linking agents under heat and pressure or using high energy radiations such as γ-rays, X-rays and high-speed electrons. In many curing systems, suitable catalysts or accelerators in presence or absence of an activator system are used to speed up the curing process and to achieve optimum cross-link density. Peroxides, such as benzoyl peroxide and dicumyl peroxide are widely used as the curing agents for many of thermoplastics and rubbers. Other curing agents in widespread use are sulphur, selenium and sulphur monochloride for diene rubbers, diisocyanates for polymers having reactive functional groups, amines and polyamines for expoxides, formaldehyde for novolacs and polyvinyl alcohols, etc.

8.24.7 Other Additives

Among the additives other than those mentioned above, and deserving special mention are flame retardants, antistatic agents and blowing agents (for cellular or expanded products). Plastics used in building industry, transport vehicles, mining and related industries, etc., should have prominent flame retardant characteristics. Antimony compounds such as antimony oxide, antimony oxychloride, etc., are commonly used to impart enhanced flame retardancy in plastics. Chlorinated paraffins, selected bromo compounds such as tribromotoluene and phosphate plasticizers such as tritolyl phosphate and trixylyl phosphate are useful additives which enhance the fire retarding properties of plastics. The chloro- or bromo-compounds derive their effectiveness through the diluting effects of HCl, HBr or bromine formed during their thermal cracking. Flame retardants may otherwise decompose endothermically absorbing much of the heat, form impervious fire resistant layer on the surface thus preventing access of oxygen to the inner polymer layers or chemically interfere with the flame propagation process.

Selected amines and quaternary ammonium compounds are used as antistatic agents. Blowing agents of different types are in use. One type known as the chemical blowing agents, incorporated in the polymer in the compounding stage, decomposes to yield in part a gas or a volatile substance at an appropriate stage of the processing, which gets entrapped in the polymer matrix to give the expanded or blown product. Sometimes, low boiling liquids are blended and allowed to volatilize out at the most appropriate stage of processing producing the expanded product. Gases, such as air, CO_2 or N_2, are also used to produce foams or expanded items. Chemical blowing agents which are commonly used are $NaHCO_3$, azonitriles, azodicarbonamide, dinitroso pentamethylene tetramine, etc.

8.25 Plastic Processing Technology

The plastic processing begins with the mixing or compounding process. Equipment and machinery widely used in the compounding stage are open mills or two-roll mills, internal or Banbury mixers, sigma blender, ribbon blender, low-shear extruders, pulverizers and granulators. High speed equipment, such as Henschel mixer, Buss-ko-kneader or the ZSK corotating intermeshing twin screw extruder are advantageous particularly in mass production units.

For the plastic materials, the 'plastic' deformable state, attained by thermoplastic polymers at appropriate temperatures and possessed by thermosetting polymers before being 'set' or 'cured' through chemical reactions, allows them to be shaped into a spectrum of finished products. The usefulness and often the essentiality of polymers and plastic products with all their profound impact on our life-style are due as much to the versatility, flexibility and sophistication of the shaping methods as to their inherent properties.

The shaping of plastics may be accomplished by the following general approaches and techniques:

(i) Shaping in the molten state (melt processing) employing compression, transfer and injection moulding, melt spinning, calendering, blow moulding, roto-moulding and extrusion techniques,

(ii) Shaping in the leathery or rubbery state employing thermoforming techniques and calendering,

(iii) Wet processing of polymers solutions employing wet spinning, spraying, spreading, brushing and dipping techniques,

(iv) Shaping of objects using polymer suspensions or emulsions (paste and latex technology), low molecular weight resin syrup (cast phenolics) or monomer polymer syrup (casting of acrylic sheets or laying up of fibre reinforced laminates and composites), and

(v) Finishing operations.

Each of the techniques other than the finishing operations involves a setting operation (freezing, cross-linking, gelling, extended polymerization, drying or precipitation and drawing) after the desired deformation has been achieved, so as to stabilize the imparted shape. The whole spectrum of processing as outlined above may also be viewed under three major categories as detailed below:

(a) Moulding—(i) Compression moulding, (ii) Transfer moulding, (iii) Injection/jet moulding and (iv) Blow moulding.
(b) Forming—(i) Extrusion, (ii) Spinning, (iii) Calendering, (iv) Thermoforming and (v) Casting, slush moulding and rotomoulding.
(c) Others—Lamination, reinforcement and coating.

8.25.1 Moulding Techniques

Moulding is usually defined as the process of shaping of an article from a plastic material usually by the simultaneous application of heat and pressure or pressure alone in a closed chamber called the mould.

(a) *Compression Moulding* This process is widely used for thermosetting polymers, even though in principle, it is also applicable to thermoplastic polymers. Compression moulding involves forcing a measured dose of polymer granules, moulding powder or the preform, preheated or otherwise, into the shape of the mould. In general, the charge is placed in the heated mould cavity and then the mould is closed and a desired degree of compression is effected by manipulating the hydraulic system attached to the compression press, resulting in intimate contact of the polymer charge with all parts of the mould surface aided by the heat applied and pressure developed and maintained. Both heat and pressure ensure good flow of the resin or the polymer, filling of all the intricate parts and corners of the cavity and good surface finish. For thermosetting systems, pressure is maintained till cross-linking to optimum level is attained. Thereafter, the mould is opened and the moulded object is ejected from the hot mould cavity. For thermoplastics, the mould has to be cooled well below the glass transition temperature or the heat distortion temperature of the polymer before the mould is opened and the moulded object is ejected.

A moulding cycle is constituted of all the steps beginning with charging the mould cavity and ending with ejection of the moulded object. The time interval covering the moulding cycle is commonly known as the cycle time. A slight excess of material is normally placed in the mould only to ensure that it is completely filled. The excess material is squeezed out between the mating surfaces of the mould in the form of a thin film known as flash which can be easily removed. For many thermosetting systems, a short breathing spell is normally allowed seconds after the mould closing by reopening for a brief period to enable entrapped air, liberated steam or other volatiles to escape and subsequently closing it again. Use of preforms against moulding powders gives advantages in handling, ensuring low bulk factor or compression ratio. Breathing is helpful in minimizing or eliminating voids in the form of porosity, blisters and bubbles, and thereby improving the quality of the moulded part. The sketch of a compression moulding unit is shown in Fig 8.2

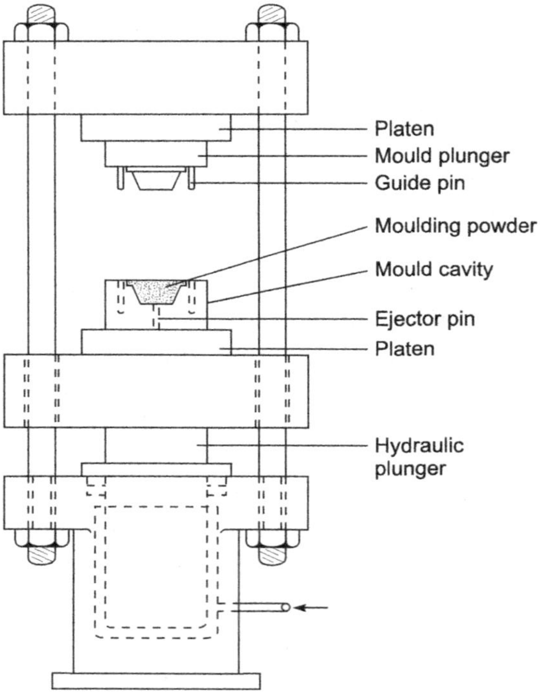

Fig. 8.2 *Sketch showing a compression moulding press and the mould parts*

(b) ***Transfer Moulding*** Poor heat transmission limits the compression moulding to a low rate process, and as a consequence it also suffers from limitations with respect to the parts that can be moulded. Transfer moulding was developed in an effort to overcome these limitations partly or substantially. In the transfer process, the charge is melted in a separate chamber or in a barrel called the "plasticizing pot" or the "transfer pot" which sometimes constitutes a part of the heated mould. The melt is forced into the mould cavity through narrow, heated runner/gate flow channels by ram pressure or a reciprocating screw plunger action (Fig. 8.3). The mould is kept at a temperature high enough to cause rapid heat curing of the resin system.

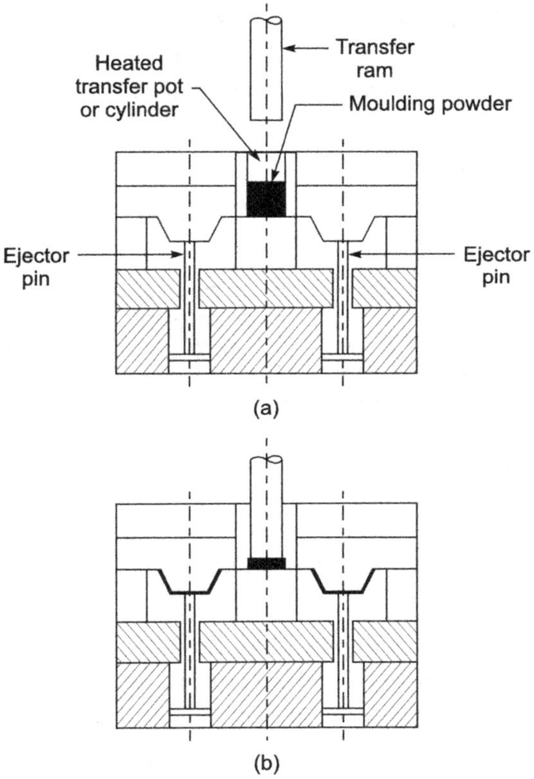

Fig. 8.3 *Sketch of a transfer moulding unit: (a) before moulding begins, (b) after moulding is just completed*

Transfer moulding cycles are usually shorter than those of compression moulding as moulding is done at higher temperatures and pressures in the transfer process. The premelted polymer charge flows more readily into the mould permitting moulding of large and intricate parts with sophisticated inserts which are not as readily attainable by compression moulding. Even thick pieces cure almost uniformly keeping the shrinkage at a minimum. The transfer moulding also produces moulded pieces of higher density and mechanical strength.

(c) *Injection Moulding and Jet Moulding* The technique of injection moulding of polymers was developed to achieve high-speed moulding of thermoplastics. In this process, a molten thermoplastic is injected or shot into a steel mould cavity under high pressure with the help of plunger system.

An injection moulding equipment (Fig. 8.4), consists of two units, the injection unit and the clamping unit. The polymer is melted into the injection unit from where it is injected into the mould by the plunger action.

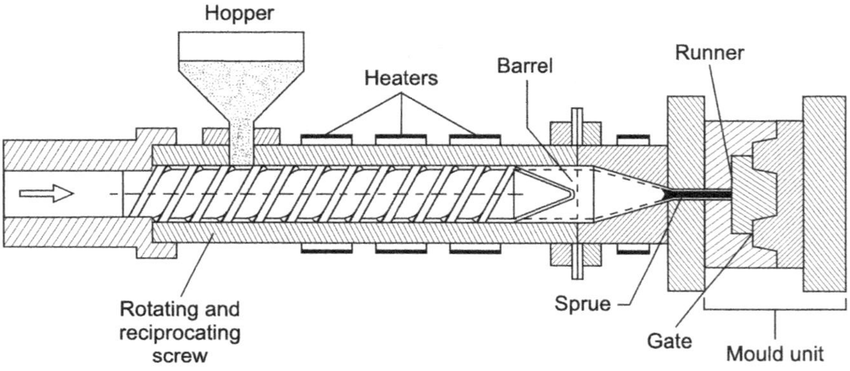

Fig. 8.4 *Sketch of an injection moulding unit*

The clamping unit is where the mould is attached. The mould can be closed or opened automatically or otherwise and ejector pins are used to eject the moulded object out of the mould. The injection unit consists of a barrel of appropriate length to allow melting of the polymer or the plastic compound as it runs down its length through successive zones of higher temperatures under the pressure applied through a ram, plunger or a screw. Present injection units are almost exclusively of the in-line reciprocating screw type. As the screw rotates, it pumps the molten polymer forward, the screw itself moving backwards allowing accumulation of enough material to fill the mould. The screw stops rotating as the mould is closed and then as it moves forward axially, it plays the role of an injection plunger. The two kinds of motion of the screw are controlled by a hydraulic system. The filling of the mould is accomplished by the flow of the polymer melt under high injection pressure along the nozzle, runners and gates into the cavities.

The rapidity of moulding and other advantages of injection moulding process prompted efforts toward developing similar techniques for the processing of thermosetting compounds. The process must be so designed and controlled as to prevent premature hardening of the thermosetting mass in the injection unit or the plasticating chamber as it is preheated there, but at the same time to heat it sufficiently to enable it flow rapidly and fill the mould under positive pressure. The jet moulding process solves this problem nicely by heating the moulding powder in the plasticating cylinder just enough to permit only flow under high pressure but allowing no measurable chemical reactions to take place in the plasticating cylinder. The plasticated mass at a relatively low temperature is then forced or injected at a high speed through the jet or nozzle into the mould which is kept at a much higher temperature leading to fast curing inside the mould.

Reaction injection moulding (RIM) technique is of a more recent development. It involves injection of low viscosity monomers or low polymers into the mould. The process is specially suitable for the production of polyurethane moulding by rapid injection of metered liquid streams of a diol and a diisocyanate into an impingement chamber where the two reactants get mixed intimately and wherefrom they are led into a mould at a relatively low pressure. Fast polymerization occurs inside the mould cavity under the influence of catalyst, heat and pressure employed. The speciality about RIM is that it allows production of solid moulded object from liquid monomers and low-viscous, low-polymer (reactive) intermediates.

(d) *Blow Moulding* Blow moulding is an important polymer processing technique for manufacturing hollow articles such as bottles and the like, and the technique is borrowed from the glass industry. The process involves in the first step the forming of a molten "parison" usually by extrusion. The parison is engaged between the mould halves and compressed air is used to blow the molten plastic into the configuration of the mould. The plastic mass quickly solidifies coming in contact with the cold mould. The process is schematically shown in Fig. 8.5. There are different types of blow moulding, processes such as extrusion, injection and stretch blow moulding, differing in the technique of forming the parison and also in the process of imparting orientation on the product. Stretch blow moulding is usually done in two steps. First, a test tube shaped preform is moulded which is then reheated, stretched and blown into a bottle. Stretching produces orientation and improves strength of the blow moulded object.

8.25.2 Forming Techniques

(a) *Extrusion* In this process, a polymer is melted and the melt is pushed ahead by a rotating screw continuously along the extruder barrel and across a metal die that shapes the melt into the desired form. End-use items of any desired length are manufactured by the extrusion process. The product range covers coated wires,

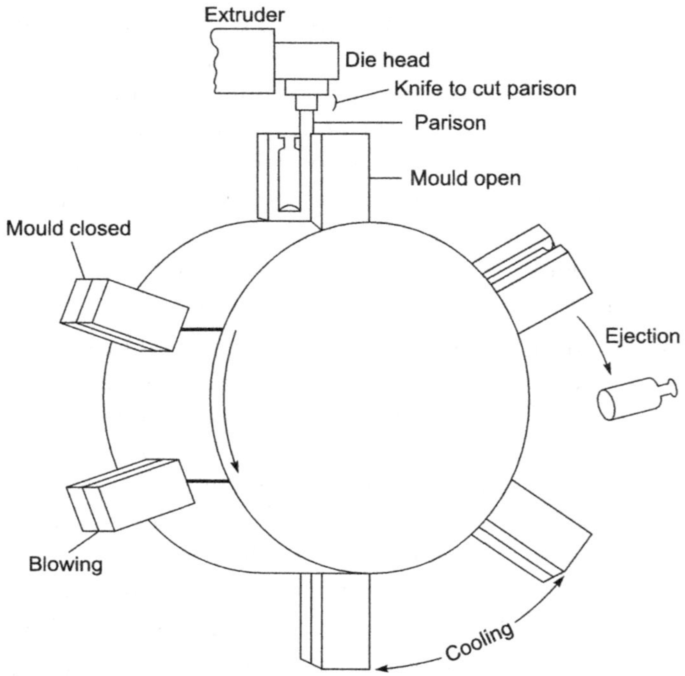

Fig. 8.5 *Diagram showing a continuous extrusion blow moulding operation–a vertical rotating wheel carrying the mould sets on the periphery*

cables and sheaths, rods, pipes and tubings, hoses blown films and a variety of other profiles including sheets and cast films. A sketch of an extruder with its components and sections is shown in Fig. 8.6.

The screw is divided into several zones each with a specific purpose. As the screw rotates, the feed zone picks up the polymer or the compounded granules from the hopper and pushes them forward into the plasticating zone where the granules melt under the influence of shearing action and external heat. The molten mass then passes through a compression zone to the metering zone. The metering zone helps in maintaining a uniform flow rate at the die point and build up enough pressure to force the molten mass out of the die through the screen pack and breaker plate assembly. The barrel is usually heated to different temperatures at different zones with the help of electrical heaters fixed around it or it may be cooled by passing cooling water through the cores, if necessary. The design variables of the extruder are the length to diameter (L/D) ratio, the screw geometry including its pitch, flight and their variations along its length, and the die.

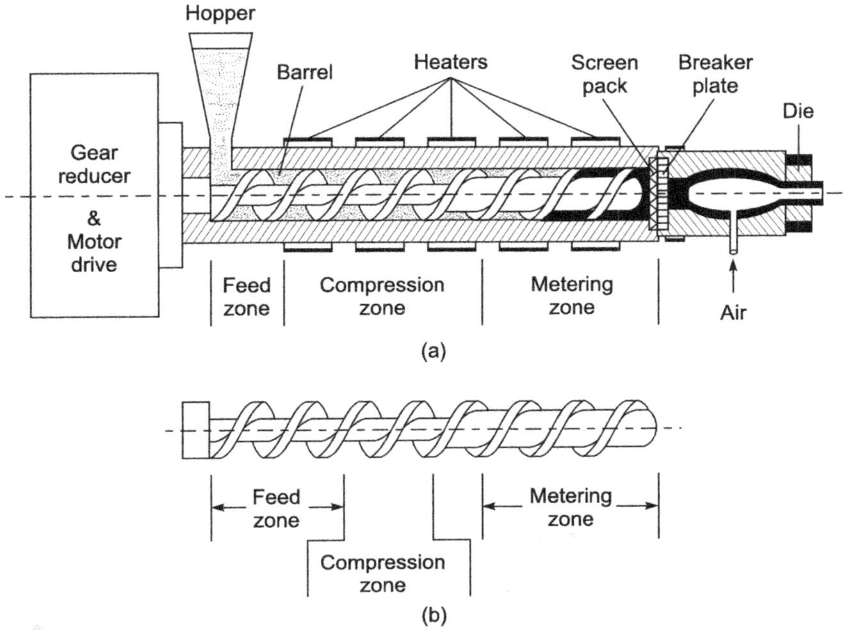

Fig. 8.6 *Sketch showing: (a) an extruder with major components fitted with a screw for polyethylene extrusion (L/D = 20 : 1 to 30 : 1; compression ratio = 3 : 1 to 4 : 1), and (b) type of a screw for nylon extrusion (L/D = 16 : 1 ; compression ratio = 4 : 1)*

In addition to single screw extruders, there are also twin screw (or multiple screw) extruders in which two or more neighbouring screws are made to work side by side rotating in opposite directions. There are also vented extruders having an opening or vent at some point along the screw or the barrel. The vent is attached to a vacuum line and is used to let off volatiles from the polymer melt. The finished product of the extrusion process is determined by the die which shapes the product and by the sizing equipment, i.e. the blowing, cooling, drawing and cutting system. Two or more extruders may be used to produce co-extruded films or other objects. Various coated and laminated objects are formed making use of the extrusion process. A typical blown film extrusion set-up and a cross head assembly for wire coating are shown in Fig. 8.7.

(b) ***Spinning*** The process of converting a suitable synthetic polymer to the fibre form is done by spinning usually from its melt or from its solution in a selected solvent. Of the different spinning techniques, melt spinning is the simplest, requiring no handling of solvents or non-solvents and hence requiring no post-spinning

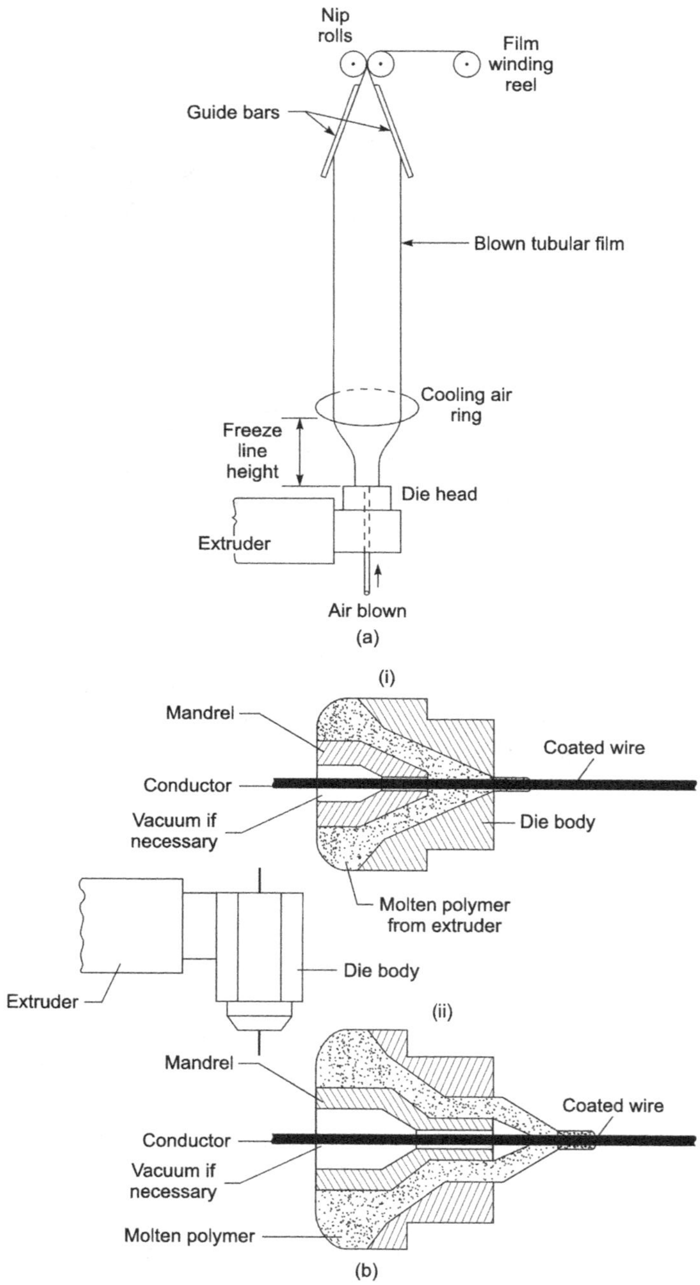

Fig. 8.7 *(a) A blown film extrusion set-up and (b) a cross-head and die showing the operational principle of wire coating*

solvent evaporation, precipitation or coagulation and drying. Solution spinning is practised when melt spinning cannot be carried out. Solution spinning may be of two kinds, (a) dry spinning, where the polymer solution is spun in a heated vertical chamber and the solvent is removed by evaporation, and (b) wet spinning, where the solvent is leached out by another liquid medium acting as a non-solvent for the fibre forming polymer or its derivative which is coagulated as soon as the spun solution impinges into the coagulating bath (non-solvent liquid medium); the chemical reaction to convert a derivative to the final polymer may take place in the coagulating bath or even later. Wet spinning is essentially done for viscose rayon and polyacrylonitrile spun from salt solution. Dry spinning is practised by spinning cellulose acetate (acetate rayon) from acetone solution and for such vinyl fibres as poly(vinyl chloride) and its copolymers and polyacrylonitrile (acrylic fibres) spun from solution in selected organic solvents. Melt spinning is common for polyamide, polyester and polyolefin (polypropylene) fibres.

As the spun polymer filaments or fibres solidify (by simply cooling, if melt spun, by solvent evaporation and cooling if dry spun and by coagulation and leaching action if wet spun), they are brought together to form a thread and subjected to drawing immediately after spinning in a continuous operation (usually for solution spinning) or in a subsequent step (usually for melt spinning).

(c) *Calendering* Calendering is a process most convenient for fast and continuous formation of sheet or film. The calender usually consists of four highly polished rolls commonly arranged in 'Z', 'I' or inverted 'L' shape (Fig 8.8.) The soft, plasticized or dough-like plastic mass is worked and metered through the nips between the heated rolls to give a product of uniform thickness. The first nip controls the feed rate while the second and the third control the thickness of the

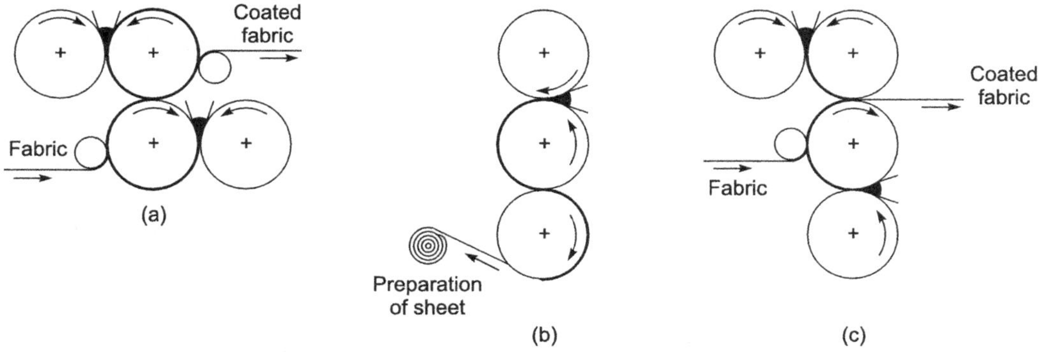

Fig. 8.8 *Typical roll assemblies for calendering*

product. An embossing effect or surface design may be produced using an appropriately engraved calender roll. One may also have a three-roll assembly.

(d) *Thermoforming* Thermoforming is a process for converting a piece of a thermoplastic sheet into a fairly deep drawn container, e.g. tray for packaging meat, hot or cold drink cups, ice cream cups, refrigerator door liners, etc., employing vacuum or air pressure from a compressor. When vacuum is used, the process is called vacuum forming. The sheet is clamped in a frame and brought close to a radiant electric heater. As it softens to a deformable state, it is placed down over a mould and the soft sagging sheet is sucked against the mould by applying vacuum and then cooled.

For deep moulds with big surface area, stretching the heat-softened web to fill the mould employing only vacuum or air pressure is often difficult or inadequately accomplished. However, the stretching of the heat-softened sheet in part may be effected using a simple mechanical aid or assist, such as a plug assist, prior to application of vacuum or air pressure to allow forming to the desired shape by contact with the mould. One such method is shown schematically in Fig. 8.9. The main advantage of the thermoforming processes is low cost for the equipment, moulds and the processing itself.

(e) *Casting, Slush Moulding and Rotomoulding* High viscosity melts or fluids cannot be conveniently cast in a mould, but they are moulded into a shape by employing injection, compression or transfer moulding techniques. But we can otherwise readily process a monomer system, a prepolymer (monomer–polymer syrup) or a polymer dispersed in excess plasticizer. All these are low viscosity systems, convenient for casting operations. Thus, polyester and epoxy resins are sometimes cast into moulds and heated to cross-linked products. Acrylic prepolymers are frequently processed into finished products (acrylic sheets) by casting. A common method of processing largely plasticized PVC is casting of plastisols which are low viscosity dispersions of PVC in a liquid plasticizer. On heating, the PVC sol swells and gels into a rubbery product.

A significant variation of straight casting of plastisols is "slush moulding". The material is poured inside the hollow heated mould which is rotated to ensure formation of a thick and uniform casting along the inner walls and the excess material is poured out. The product is gelled on heating before taking out of the mould.

Rotational moulding is a unique technique of making large, hollow objects, even of complex shapes, usually from polyethylene and related polymers in the form of fine powder. A measured quantity of powder is placed inside an aluminium/sheet

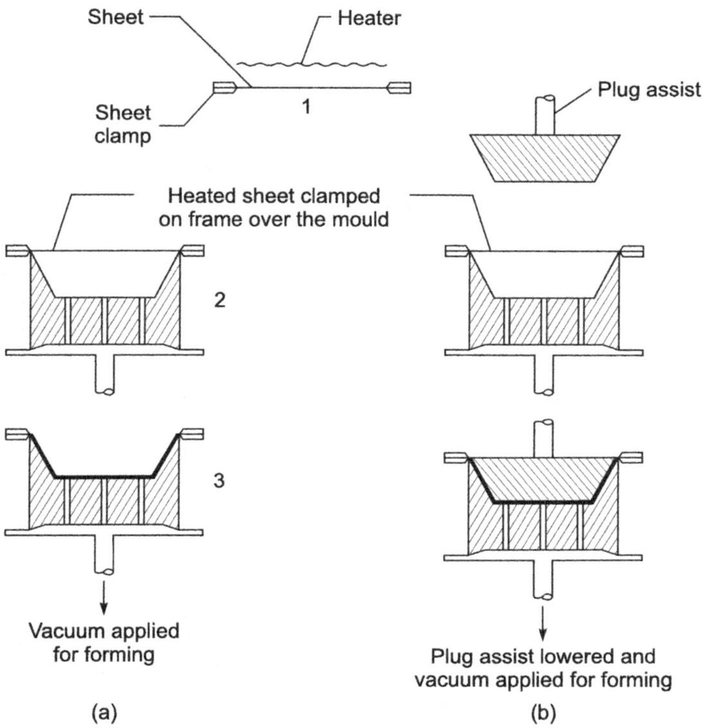

Fig. 8.9 *Sketch showing thermoforming operations: (a) simple vacuum forming; (b) vacuum forming with a plug assist*

metal mould which is then heated and rotated at a low speed. The polymer melts and coats the inside of the mould. When the melting is thoroughly achieved and the melt uniformly spread by rotation, the mould is cooled and the polymer coat solidifies duplicating the inside of the mould. Common roto-moulded objects are drums, large storage tanks, big containers, toys and dolls, etc. Crystalline sharp-melting polymers such as nylon or 'Terylene' or PET polyester are suited for the production of film by melt-casting techniques as outlined in Figs 8.10 and 8.11.

8.25.3 Other Techniques

(a) ***Lamination and Reinforcement*** Laminates, as the term suggests, consist of layers or laminae, held together by a suitable binder. The base of the layers is made of such materials as paper, woven fabric from cellulose, glass fibre, etc. which act as the reinforcing agent. The binders are usually synthetic resins used in solution or emulsion form for impregnation into the base laminae. The resins most commonly

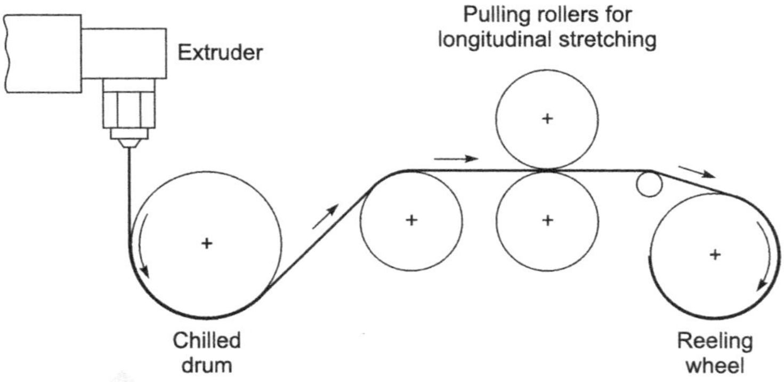

Fig. 8.10 *Melt-casting of a nylon (polyamide) film*

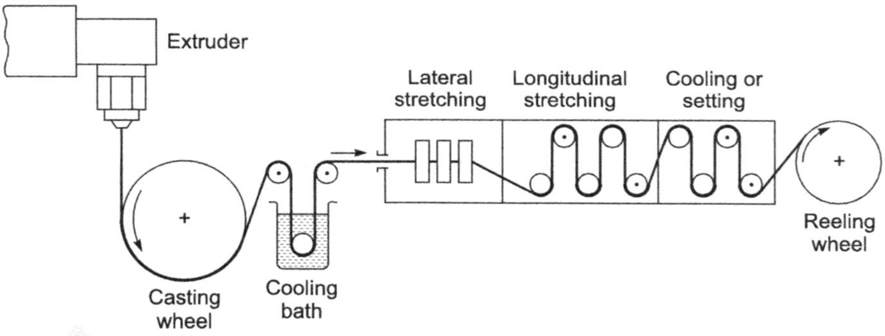

Fig. 8.11 *Melt-casting and biaxial orientation of polyester (PET) film*

used are phenolic and amino resins, epoxy resins, polyesters, silicones, etc. The essential steps in the lamination process as outlined in Fig. 8.12 are: (a) impregnation of the base with the resin solution or emulsion, (b) drying of the impregnated web and (c) finally, cutting into size, plying them up as required and then pressing under heat to a permanently bonded (infusible) unitary mass of the desired finish. Major applications of the thermosetting laminates include components of electrical and electronic equipment, parts and components used in mechanical engineering and for storage and handling of chemicals. Decorative laminates are used in building and furniture industries. Fibre reinforced plastics are made by hand lay-up, die moulding and filament winding techniques.

(b) *Coating* The application of polymers or plastics as coatings is quite important. The coating item or formulation based on a resin or polymer may be used in the form of a solution, emulsion, powder, melt or film. Coating by impregnation in a

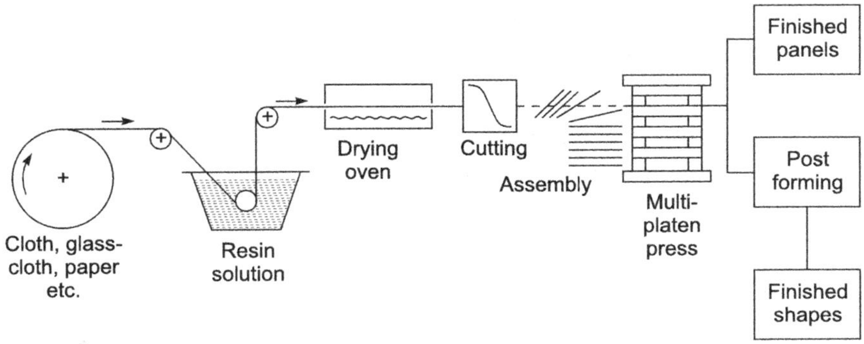

Fig. 8.12 *Sketch showing layout of a high pressure lamination unit*

solution or emulsion for lamination has already been discussed. Solution or emulsion formulations are often formed into a protective coating by brushing or spraying, and then drying into a hard finish.

Dip coating is often accomplished by dipping hot or cold metal parts into an emulsion, a plastisol or a polymer powder bed (fluidized or otherwise). Powder coating is practised using polyolefins, PVC, polyesters, nylons, epoxies and acrylics. A good and efficient technique is the electrostatic coating and spraying where the polymer particles are given an electric charge of one kind and the surface to be coated is given the opposite charge. Dielectric coating is also gaining prominence because of efficiency and uniformity.

Extrusion-coating is another coating technique of widespread use. Coated sheets and laminates produced by this technique find extensive use in various fields including packaging of food, pharmaceuticals and chemicals. Coating using polymer melts or solutions is also extensively done by what is commonly known as the spreading or roller-coating technique. "Skim coating" or "topping" and "frictioning" techniques are employed for coating of fabrics. In the topping method, the compound is fed around a calender roll from a nip and the sheet formed is applied to the fabric at a second nip. The sheeting usually but not necessarily travels at the same speed as the fabric at the point where it is laid onto the fabric. When the operation demands coating on both sides of the fabric, a four-roll calender is used. Sheeting is produced in the two terminal nips (first and the third nips) and they are brought together and laid on both sides of the fabric simultaneously in the middle nip (second nip) (Fig. 8.8). The second and the third rolls which lay the sheets on the fabric must have even speed while the first and the fourth may have even or odd speed. In case there is an appreciable difference in the speed of the fabric and the compound (sheeting) at the nip of the calender, the compound is forced into the fabric weave. If the compound is tacky so that a part sticks to the bowl and returns to the first nip, the process is commonly called "frictioning". On the other hand, if all the compound is laid on the fabric, it is known as "friction coating" or "layering".

(c) *Finishing* Finishing operations consist of numerous processes after moulding or forming, and they may include cutting, trimming, polishing, mechanical fastening, bonding, sealing, welding, dyeing, painting, electroplating, vacuum metallizing, printing, engraving, etc. Each of these processes involves use of a specific type of equipment and operational approach or technique.

(d) *Microencapsulation* Microencapsulation is the process of coating any solid or liquid material to form tiny particles of size ranging from several tenths of a micron to several thousands of micron. Depending on the nature of the core and coating materials, the thickness of the coating and the size of the microcapsules formed, microencapsulation allows: a) formation of solid entities from oil, b) protection of drugs from moisture, light and oxidation, c) delayed or retarded escape of core substrates by rapid volatilization, d) masking of disagreeable taste and odour of the core substrate, thus making drug administration easy and comfortable, and e) control and sustained release of fertilizer, pesticides and herbicides and also similar release of drugs and pharmaceuticals, making continuous chemical and drug delivery at controlled, desired rates into the environment or within body of patients possible and to achieve prolonged therapeutic effect after administration of a single, measured dose, thus making the endeavor in each case not only much more cost-effective and less dose sensitive but also in many cases free from possible adverse side effects.

The coating (polymeric) materials may be water soluble, such as gelatin, gum arabic, poly(vinyl pyrrolidone), carboxymethyl cellulose (CMC), methyl cellulose, poly(vinyl alcohol), poly(acrylic acid), or water insoluble, e.g. ethyl cellulose, polyamides (nylons), EVA copolymers and silicones; natural resins such as shellac or such items as paraffin, beewax, stearic acid, stearyl alcohol or glyceryl stearates may also be used for coating of powders of clay, $CaCO_3$ etc. used as activated fillers in polymer compounding. Biodegradable polymers such as microbial poly (3-hydroxy alkanoates), poly (lactic acid), gelatin, starch etc. are useful in microencapsulation of food and medicines.

The primary objective for uses of polymers in medicines administered in tablet form are to formulate them appropriately such that the polymers serve as the binder for shape retention and strength and regulate drug dissolution rates. Microencapsulation is widely employed for dispensation of analgesics, steroids, vitamins and for administration of antidotal drugs as a "one way trap" for toxins such as for patients suffering from kidney failure and hepatic coma. Health and beauty aids are other areas where microencapsulation has helped relevant industries; such applications include many modern cosmetic items such as skin creams or lotion (base for oil-in-water emulsion) with much improved softness and smoothness, wet-look lip-stick with oil-containing micro-capsules, encapsulated perfumes and microcapsules of flavouring oils for use in chewable gums or tablets or of dyes for making coating on paper to produce carbonless copying paper.

In a preparative method, solid, particulate core materials are dispersed in a supporting heated (if required) air stream and are coated by spraying emulsion or solution of the polymer encapsulant. Another method of microencapsulation utilizes coacervation-phase separation technique; the core material is dispersed in a solution of the coating polymer in a solvent of its own and then phase separation coacervation is effected by lowering of temperature, addition of salt or a non-solvent. Sometimes crosslinking of the polymer is effected to impart extra rigidity to the micro-capsules.

8.26 Chart of Properties

Properties of a number of selected polymers finding applications as plastics are listed in Table 8.1, mainly with the objective of enabling the reader evaluate them on a comparative basis. Table 8.2 lists some of the relevant properties of some metals and alloys. Table 8.3 gives a list of major applications of the selected polymers.

8.27 Polymer Molecular Society: A Grand Model for Human Society

8.27.1 Social Relevance and Perspective

Polymers constitute a diverse range of light and energy efficient materials discovered and developed in a large measure out of the sheer necessity of meeting the growing needs and demands of our high profile society. The present level of growth, sophistication and advancements of our expanding society have been made possible in a large measure through sustained, imaginative and creative efforts of scientists, engineers and technologists in the polymer field. The base material constituting all plastics, structural resins, rubbers, adhesives, coatings and related products is usually a polymer or a group of polymers. As a class of materials, polymers are so much distinctive that polymer education covering origin, synthesis,

Table 8.1 *Properties of plastics*

Property	*ASTM*	*Acrylics*	*Cellulosic Polymers*		
			Cellulose acetate	*Cellulose acetate butyrate*	*Cellulose propionate*
Burning rate (inflammability)	D635	Slow	Slow	Slow	Slow
Clarity/transparency	—	Transparent (92% transmission)	Transparent	Transparent	Transparent
Compressive strength (psi)	D695	12000 – 20000	2200 – 36000	2100 – 22000	3100 – 22000
Dissipation (power) factor (10^3 cycles/s)	D150	0.02 – 0.04	0.01 – 0.10	0.01 – 0.04	0.01 – 0.04
Dielectric constant (10^3 cycles/s)	D149	2.2 – 3.2	3.2 – 7.0	3.2 – 6.2	3.4 – 3.6
Dielectric strength (short time. 1/8″ thick. volts/mil)	D149	450 – 550	260 – 365	250 – 400	300 – 450
Effect of sunlight	—	Very slight	Slight	Slight	Slight
Elongation in tensile (%)	D638	3 – 10	6 – 70	40 – 90	30 – 100
Flexural strength (psi)	D790	13000 – 17000	2000 – 16000	1800 – 9300	3900 – 11000
Hardness (Rockwell)	D785	M85 – M105	R35 – R125	R31 – R116	R20 – R120
Heat distortion temperature (°F)	D648	160 – 195	110 – 209 (66 psi)	115 – 227 (66 psi)	110 – 250 (66 psi)
Impact strength (ft lb/ in notch, lzod)	D256	0.3 – 0.5	0.4 – 5.2	0.8 – 6.3	0.5 – 10.0
Low temperature resistance	—	Good	Good	Good	Good
Machining qualities	—	Good to excellent	Excellent	Excellent	Excellent
Modulus of elasticity (psi × 10^{-5})	D638	4.5	0.65 – 4.0	0.5 – 2.0	0.6 – 2.15
Mould shrinkage (in/in)					
(a) Injection	—	0.002 – 0.008	0.003 – 0.007	0.002 – 0.005	0.001 – 0.005
(b) Compression	—	0.001 – 0.004	0.003 – 0.007	0.003 – 0.009	0.003 – 0.005
Moulding qualities	—	Excellent	Excellent	Excellent	Excellent
Odour	—	None	Slight	Slight odour of rancid butter	Slight
Refractive index (η_D)	D542	1.49	1.46 – 1.50	1.46 – 1.49	1.46 – 1.49
Resistance, chemical					

Contd.

Table 8.1 *(Contd.)*

Property	*ASTM*	*Acrylics*	*Cellulosic Polymers*		
			Cellulose acetate	*Cellulose acetate butyrate*	*Cellulose propionate*
(a) Strong acids	D543	Attacked by oxidizing acids and H_2SO_4	Decomposes	Decomposes	Decomposes
(b) Weak acids	D543	Hardly affected	Slight	Slight	Slight
(c) Strong alkalis	D543	Attacked	Decomposes	Decomposes	Decomposes
(d) Weak alkalis	D543	Hardly affected	Slight	Slight	Slight
Resistance, solvent	D543	Soluble in ketones, esters and chlorinated hydrocarbons. Resistant to alcohols at room temperatures	Soluble in ketones and esters, softened or slightly soluble in alcohols, little affected by hydrocarbons		
Softening point [(Vicat)°C]	—	80 – 98	70	70	—
Specific gravity	D792	1.17 – 1.20	1.23 – 1.24	1.15 – 1.22	1.18 – 1.24
Specific heat (cal/ °C/g)	—	0.35	0.3 – 0.42	0.3 – 0.40	0.3 – 0.40
Specific volume (cu. in/lb)	D792	23.7 – 23.1	22.5 – 20.6	24.0 – 22.7	23.4 – 22.5
Tensile strength (psi)	D638 D651	7000 – 11000	1900 – 8500	2600 – 6900	2000 – 7300
Thermal conductivity (cal/s/sq.cm/°C/cm × 10^4)	C177	4 – 6	4 – 8	4 – 8	4 – 8
Thermal expansion (cm/cm/°C × 10^4)	D696	5 – 9	8 – 16	11 – 17	11 – 17
Volume resistivity (ohm cm at 23°C and 50% R.H.)	D257	$> 10^{14}$	$10^{10} – 10^{13}$	$10^{10} – 10^{12}$	$10^{12} – 10^{15}$
Water absorption (% 24 h, 1/8″ thickness)	D570	0.3 – 0.4	1.9 – 6.5	0.9 – 2.2	1.2 – 2.8

Contd.

Table 8.1 (Contd.)

Property	*ASTM*	*Fluorinated Polymer*			*Melamine Formaldehyde*	
		Fluorinated ethylene propylene	*Polytetrafluoro ethylene*	*Polytrifluoro chlorethylene*	*Unfilled*	*α-cellulose filled*
Burning rate (inflammability)	D635	None	None	None	Self extinguishing	None
Clarity/transparency	—	Translucent	Opaque	Translucent or opaque	Opalescent	Translucent
Compressive strength (psi)	D695	1400 – 2850	700 – 1800	2000	40000 – 45000	25000 – 43000
Dissipation (power) factor (10^3 cycles/s)	D150	0.003	0.005	0.01	—	0.027 – 0.045
Dielectric constant (10^3 cycles/s)	D149	2.1	2.0	2.5	—	7.2 – 8.2
Dielectric strength (short-time, 1/8″ thick, volts/mill)	D149	500 – 600	400 – 500	530 – 600	—	300 – 400
Effect of sunlight	—	None	None	None	Slight colour change	Slight colour change
Elongation in tensile (%)	D638	300 – 900	250 – 350	125 – 175	—	0.6 – 0.9
Flexural strength (psi)	D790	95000	1600	3500	11000 – 14000	10000 – 16000
Hardness (Rockwell)	D785	D55	J75 – 95	R110 – 115	—	M110 – 115
Heat distortion temperature (°F)	D648	—	—	110 – 209 (66 psi)	298	400
Impact strength (ft, lb/in, notch, Izod)	D256	No break	2.5 – 4.0	2.5 – 3.6	—	0.24 – 0.36
Low temperature resistance	—	Good	Good	Good	—	Good
Machining qualities	—	Good	Good	Good	—	Fair
Modulus of elasticity (psi × 10^{-5})	D638	0.5 – 0.7	0.38 – 0.65	1.9 – 3.0	—	13
Mould shrinkage (in/in)						
(a) Injection	—	0.001 – 0.005	0.03 – 0.06	—	—	—
(b) Compression	—	0.003 – 0.005	0.03 – 0.06	—	0.005 – 0.01	0.011 – 0.012
Moulding qualities	—	Good	Fair	Fair	Good	Excellent
Odour	—	None	None	None	None	None
Refractive index (η_D)	D542	1.34	1.35	1.43	—	—

Contd.

Table 8.1 (Contd.)

Property	*ASTM*	*Fluorinated Polymer*			*Melamine Formaldehyde*	
		Fluorinated ethylene propylene	*Polytetrafluoro ethylene*	*Polytrifluoro chlorethylene*	*Unfilled*	*α-cellulose filled*
Resistance, chemical						
(a) Strong acids	D543	Unaffected	Unaffected	Unaffected	Decomposes	Decomposes
(b) Weak acids	D543	Unaffected	Unaffected	Unaffected	Unaffected	Unaffected
(c) Strong alkalis	D543	Unaffected	Unaffected	Unaffected	Attacked	Attacked
(d) Weak alkalis	D543	Unaffected	Unaffected	Unaffected	Unaffected	Unaffected
Resistance, solvent	D543	Unaffected	Unaffected	Unaffected	Unaffected	Unaffected
	–					
Softening point [(Vicat) °C]	D789	MP 285 – 295	Decomp. pt	Decomp. pt	—	—
	53		>300	>400		
Specific gravity	D792	2.14 – 2.17	2.1 – 2.3	2.10 – 2.15	1.48	1.47 – 1.52
specific heat (cal/°C/g)	—	0.28	0.25	0.22	—	—
Specific volume (cu. in/lb)	D792	12.85 – 12.66	13.09 – 11.95	13.09 – 12.78	18.7	18.8 – 18.2
Tensile strength (psi)	D638	2500 – 3500	2500 – 3500	4600 – 5700	—	7000 – 13000
	D651					
Thermal conductivity (cal/s/sq. cm/°C/cm × 10^4)	C177	7	6	—	—	7 – 10
Thermal expansion (cm/cm/°C × 10^4)	D696	1.3 – 1.9	1.0	0.7	—	4.0
Volume resistivity (ohm cm at 23°C and 50% R.H)	D257	10^{19}	19^{19}	10^{18}	—	$10^{12} – 10^{14}$
Water absorption (% 24h, 1/8″ thickness)	D570	0.00	0.00	0.00	0.3–0.5	0.1–0.6

Table 8.1 (Contd.)

Properties	*ASTM*	*Nylons*		*Phenol Formaldehyde*	
		Type 6	*Type 66*	*Unfilled*	*Woodflour/cotton flock filled*
Burning rate (inflammability)	D635	Self-extinguishing	Self-extinguishing	Very low	Very low
Clarity/transparency	—	Translucent to opaque	Translucent to opaque	Translucent to translucent	Opaque
Compressive strength (psi)	D695	7000 – 14000	7000 – 16000	10000 – 30000	22000 – 40000
Dissipation (power) factor (10^3 cycles/s)	D510	0.02 – 0.13	0.02 – 0.06	0.015 – 0.03	4.5 – 5.0
Dielectric constant (10^3 cycles/s)	D149	3.0 – 5.0	3.6 – 6.0	4.5 – 5.0	4.0 – 7.0
Dielectric strength (short-time, 1/8″ thick, volts/mil)	D149	440 – 510	385 – 470	300 – 400	200 – 425
Effect of sunlight	—	Discolours slightly	Discolours slightly	Slight surface darkening	General darkening
Elongation in tensile (%)	D638	90 – 320	90	1.0 – 1.5	0.4 – 0.8
Flexural strength (psi)	D790	8000 – 16000	8000 – 13800	12000 – 15000	8000 – 12000
Hardness (Rockwell)	D785	R103 – 118	R111 – 118	M124 – 128	M100 – 120
Heat distortion temperature (°F)	D648	260 – 340 (66 psi)	300 – 360 (66 psi)	240 – 260	260 – 340
Impact strength (ft lb/in notch, Izod)	D256	1.0 – 3.6	1.0	0.20 – 0.36	0.24 – 0.60
Low temperature resistance	—	Good	Good	—	Good
Machining qualities	—	Excellent	Excellent	Fair	Excellent
Modulus of elasticity (psi × 10^{-5})	D638	1.5 – 3.6	2.6 – 4.0	7.5 – 10	8 – 12
Mould shrinkage (in/in)					
(a) Injection	—	0.009	0.015	—	—
(b) Compression	—	—	—	0.01 – 0.02	0.004 – 0.009
Moulding qualities	—	Excellent	Excellent	Fair to good	Fair to good
Odour	—	None	None	None	None
Refractive index (η_D)	D542	—	1.53	1.5 – 1.7	—
Resistance, chemical					
(a) Strong acids	D543	Attacked	Attacked	Decomposed by oxidizing acids	

Contd.

Table 8.1 (Contd.)

Properties	*ASTM*	*Nylons*		*Phenol Formaldehyde*	
		Type 6	*Type 66*	*Unfilled*	*Woodflour/cotton flock filled*
(b) Weak acids	D543	Resistant	Resistant	Unaffected to slight	Unaffected to slight
(c) Strong alkalis	D543	Unaffected	Unaffected	Decomposes	Attacked
(d) Weak alkalis	D543	Unaffected	Unaffected	Unaffected to slight	Slight to marked
Resistance, solvent	D543	Resistant to common solvents	Resistant to common solvents	No effect	No effect
Softening point [(Vicat) °C]	D789–53	M.P. 215 – 220	M.P. 260 – 265	—	—
Specific gravity	D792	1.13	1.09 – 1.14	1.25 – 1.3	1.32 – 1.45
Specific heat (cal/°C/g)	—	0.4	0.4	0.38 – 0.42	0.35 – 0.40
Specific volume (cu.in/lb)	D792	24.5 – 24.2	25.5 – 24.2	22.2 – 21.3	20.9 – 17.8
Tensile strength (psi)	D638				
	D651	10000 – 12000	7000 – 10900	7000 – 8000	6500 – 8500
Thermal conductivity (cal/s/sq. cm/°C/cm × 10^4)	D177	—	5.2 – 5.8	3 – 6	4.7
Thermal expansion (cm/cm/°C × 10^4)	D696	8 – 13	10 – 15	2.5 – 6	3.0 – 4.5
Volume resistivity (ohm cm at 23° and 50% R.H.)	D257	$10^{12} – 10^{15}$	$(0.45 – 4) \times 10^{14}$	10×10^{12}	$10^9 – 10^{12}$
Water absorption (% 24 h, 1/8″ thickness)	D570	1.9 – 3.3	0.4 – 1.5	0.1 – 0.2	0.3 – 1.0

Contd.

Table 8.1 (Contd.)

Properties	*ASTM*	*Polyacetals*	*Polycarbonates*	*Polyethylenes*		
				Low density	*Medium density*	*High density*
Burning rate (inflammability)	D 635	Slow	Self-extinguishing	Very slow	Very slow	Very slow
Clarity/transparency	—	Translucent to opaque	Transparent	Translucent to opaque	Translucent to opaque	Translucent to opaque
Compressive strength (psi)	D695	18000	11000	Excess cold flow	—	2400
Dissipation (power) factor (10^3 cycles/s)	D150	0.005	0.010	< 0.002	< 0.0005	<0.0003
Dielectric constant (10^3 cycles/s)	D149	3.7	2.6	2.25 – 2.35	2.25 – 2.35	2.25 – 2.35
Dielectric strength (short-time, 1/2″ thick, volts/mil)	D149	465	400	> 800	> 800	> 800
Effect of sunlight	—	Chalks slightly	Yellows slightly	Good in black or specially stabilised grades		
Elongation in tensile (%)	D638	15 injection 75 extrusion	60 – 100	400 – 600	50 – 600	50 – 600
Flexural strength (psi)	D790	14100	11000 – 13000	No fracture	4800 – 7000	2000 – 3000
Hardness (Rockwell)	D785	M94–R118	M70 – R118	D41 – 46 (Shore)	D50 – 70 (Shore)	D60 – 70 (Shore)
Heat distortion temperature (°F)	D648	338 (66 psi)	280 – 290	105 – 221 (66 psi)	120 – 150 (66 psi)	140 – 180 (66 psi)
Impact strength (ft lb/ in notch, Izod.)	D256	2.3 extrusion 1.4 injection	12 – 16	> 16	—	1.5 – 12
Low temperature resistance	—	Good	Excellent	Good	Good	Good
Machining qualities	—	Excellent	Excellent	Fair	Good	Excellent
Modulus of elasticity (psi × 10^{-5})	D638	4.1	3.2	0.17 – 0.35	0.25 – 0.55	0.8 – 1.5
Mould shrinkage (in/in)						
(a) Injection	—	0.025 av.	0.005 – 0.007	0.02 – 0.035	0.02 – 0.035	0.02 – 0.035
(b) Compression	—	—	—	—	—	—
Moulding qualities	—	Fair	Fair	Excellent	Good	Good

Contd.

Table 8.1 (Contd.)

Properties	*ASTM*	*Polyacetals*	*Polycarbonates*	*Polyethylenes*		
				Low density	*Medium density*	*High density*
Odour	—	None	None	None	None	None
Refractive index (η_D)	D542	1.48	1.584	1.51	—	1.54
Resistance chemical						
(a) Strong acids	D543	Attacked	Attacked slowly by oxidizing acids	Attacked by oxidizing acids	Attacked slowly by oxidizing acids	Attacked slowly by oxidizing acids
(b) Weak acids	D543	Resistant to some	Unaffected	Resistant	Resistant	Resistant
(c) Strong alkalis	D543	Attacked	Attacked slowly	Resistant	Resistant	Resistant
(d) Weak alkalis	D543	Resistant to some	Resistant	Resistant	Resistant	Resistant
Resistance, solvent	—	Excellent resistance to practically all solvents	Soluble in aromatics and chlorinated hydrocarbons Resistant to paraffinics	Soluble in aromatic solvents above 60°C	Soluble in aromatic solvents above 60°C	Resistant below 80°C
Softening point [(Vicat) °C]	D789–53	175	215 – 226	85 – 87	91 – 114	127
Specific gravity	D792	1.425	1.2	0.913 – 0.925	0.926 – 0.940	0.941 – 0.965
Specific heat (cal/°C/g)	—	0.35	0.30	0.53 – 0.55	0.53 – 0.55	0.53 – 0.55
Specific volume (cu. in/lb)	D792	19.3	23	30.2 – 29.9	29.8 – 29.5	29.4 – 28.7
Tensile strength (psi)	D638 D651	10000	8500 – 9500	1000 – 2300	1200 – 3500	3100 – 5500
Thermal conductivity cal/s/sq. (cm/°C/cm × 10^4)	D177	5.5	4.6	8.0	—	11 – 12.4
Thermal expansion (cm/cm/°C × 10^4)	D696	8.1	7.0	16 – 18	—	11 – 13
Volume resistivity (ohm cm at 23 and 50% R.H)	D257	6×10^{14}	2.1×10^{16}	$>10^{16}$	$>10^{16}$	$>10^{16}$
Water absorption (% 24h, 1/8″ thickness)	D270	0.12	0.3	< 0.015	< 0.01	< 0.01

Contd.

Table 8.1 (*Contd.*)

Properties	*ASTM*	*Polypropylene*	*Polystyrenes*		*Styrene Copolymers*	
			Conventional	*Toughened*	*Styrene/ acrylonitrile*	*Acrylonitrile/ butadiene/ styrene*
Burning rate (inflammability)	D635	Slow	Slow	Slow	Slow	Slow
Clarity/transparency	—	Transparent to translucent	Transparent	Translucent	Transparent	Translucent
Compressive strength (psi)	D695	8500 – 10000	11500 – 16000	4000 – 9000	14000 – 17000	2500 – 11000
Dissipation (power) factor (10^8 cycles/s)	D150	<0.0002 – 0.0003	0.0001 – 0.005	0.0004 – 0.002	0.007 – 0.010	0.007 – 0.026
Dielectric constant (10^3 cycles/s)	D149	2.0 – 2.1	2.4 – 3.1	2.5 – 3.8	2.75 – 3.1	2.7 – 4.75
Dielectric strength (short-time, 1/8″ thick, volts/mil)	D149	>800	500 – 600	300 – 600	400 – 500	310 – 410
Effect of sunlight	—	Good in black or specially stabilized grades	Unsuitable for outdoor use	Unsuitable for outdoor use	Yellow slightly	Yellow slightly
Elongation in tensile (%)	D638	50 – 600	1.0 – 2.5	7 – 60	1.5 – 3.5	10 – 140
Flexural strength (psi)	D790	—	12000 – 17000	5000 – 10000	14000 – 19000	3600 – 13500
Hardness (Rockwell)	D785	R85 – 110	M65 – 90 M35 – 70	R50 – 100	M80 – 83	R30 – 118
Heat distortion temperature (°F)	D648	210 – 230 (66 psi)	150 – 235	148 – 200	150 – 235	165 – 225
Impact strength (ft lb/in notch, Izod)	D256	0.6 – 6.0	0.2 – 0.5	0.5 – 3.0	0.35 – 0.5	3.0 – 12.0
Low temperature resistance	—	Fair	Poor but impact strength rises as temperature falls	Poor/fair	Poor	Poor

Contd.

Table 8.1 (Contd.)

Properties	*ASTM*	*Polypropylene*	*Polystyrenes*		*Styrene Copolymers*	
			Conventional	*Toughened*	*Styrene/ acrylonitrile*	*Acrylonitrile/ butadiene/ styrene*
Machining qualities	—	Excellent	Fair to good	Good	Good	Good to excellent
Modulus to elasticity (psi × 10^{-5})	D638	1.3 – 2.0	4 – 6	3 – 4.5	4 – 5.6	1.0 – 4.1
Mould shrinkage (in/ in)						
(a) Injection	—	0.015 – 0.025	0.002 – 0.006	0.002 – 0.008	0.002 – 0.005	0.003 – 0.008
(b) Compression	—	—	0.001 – 0.008	—	—	—
Moulding qualities	—	Excellent	Excellent	Excellent	Good	Good
Odour	—	None	None	None	Slight	None
Refractive index (η_D)	D542	1.49	1.59 – 1.6	—	1.57	—
Resistance, chemical						
(a) Strong acids	D543	Attacked slowly by oxidizing acids	Attacked by oxidizing acids		Attacked by oxidizing acids	
(b) Weak acids	D543	Resistant	Unaffected	Unaffected	Unaffected	Unaffected
(c) Strong alkalis	D543	Resistant	Unaffected	Unaffected	Unaffected	Unaffected
(d) Weak alkalis	D543	Resistant	Unaffected	Unaffected	Unaffected	Unaffected
Resistance, solvent	D543	Resistant below 80°C	Soluble in aromatic and chlorinated hydrocarbons, esters, ketones and higher alcohols		Soluble in ketones, esters, and some chlorinated hydrocarbons	
Softening point [(Vicat) °C]	—	150	82 – 103	78 – 100	85 – 103	85
Specific gravity	D792	0.90 – 0.91	1.04 – 1.11	0.98 – 1.10	1.075 – 1.10	0.99 – 1.10
Specific heat (cal/°C/gm)	—	0.46	0.32 – 0.35	0.32 – 0.35	0.32 – 0.34	0.33 – 0.40
Specific volume (cu. in./lb)	D792	31 – 29	26.4 – 24.8	28.1 – 25.2	25.8 – 25.2	28 – 25.2
Tensile strength (psi)	D638 D651	4300 – 5700	5000 – 12000	3500 – 6500	9500 – 12000	2500 – 9000

Contd.

Table 8.1 (Contd.)

Properties	*ASTM*	*Polypropylene*	*Polystyrenes*		*Styrene Copolymers*	
			Conventional	*Toughened*	*Styrene/ acrylonitrile*	*Acrylonitrile/ butadiene/ styrene*
Thermal conductivity (cal/s/sq. cm/°C/cm × 10^4)	D177	3.3	2.4 – 3.3	1.0 – 3.0	2.4	1.48 – 8.6
Thermal expansion (cm/cm/°C × 10^4)	D696	11	6 – 8	3.4 – 21	6 – 8	6 – 13
Volume resistivity (ohm cm at 23°C and 50% R.H.)	D257	$>10^{16}$	$>10^{13}$	$>10^{13}$	$>10^{13}$	0.5×10^{13}- 2.7×10^{16}
Water absorption (% 24 h, 1/8″ thickness)	D570	<0.01	0.03 – 0.4	0.1 – 0.3	0.2 – 0.3	0.1 – 0.3

Contd.

Table 8.1 (Contd.)

Properties	ASTM	Urea-Formaldehyde α-cellulose filled	Vinyl Polymers			
			Reigid poly(vinyl chloride)	(Rigid) Vinyl Chloride/vinyl acetate	Rubber modified PVC	Vinylidene chloride
Burning rate (inflammability)	D635	Very low	Self-extinguishing	Self-extinguishing	Self-extinguishing	Self-extinguishing
Clarity/transparency	—	Transparent to opaque	Transparent to opaque	Transparent to opaque	Transparent to opaque	Transparent to opaque
Compressive strength (psi)	D695	25000 – 35000	8000 ⟷		13000	2000 – 2700
Dissipation (power) factor (10^3 cycles/s)	D150	0.028 – 0.032	0.006 ⟷		0.14	0.05 – 0.08
Dielectric constant (10^3 cycles/s)	D149	6.4 – 6.9	3.0	3.0 – 3.5	—	3.0 – 4.0
Dielectric strength (short-time, 1/8″ thick, volts/mil)	D149	300 – 400	425	1400	1100	400 – 600
Effect of sunlight	—	None	Good in suitability stabilized grades		Rapid fall in physical strength	Slight
Elongation in tensile (%)	D638	0.5 – 1.0	2 – 40	200 – 450	—	Up to 250
Flexural strength (psi)	D790	10000 – 16000	13500	12000	12000	4200 – 6200
Hardness (Rockwell)	D785	M115 – 120	R110	—	R105	M60 – 65
Heat distortion temperature (°F)	D648	270 – 280	165 (66 psi)	145 – 170 (66 psi)	160 (66 psi)	130 – 150
Impact strength (ft lb/in notch, Izod)	D256	0.25 – 0.35	0.8	0.5 – 1.0	15 (unaged)	0.3 – 1.0
Low temperature resistance	—	Good	Good	Good	Good	Good

Contd.

Table 8.1 (Contd.)

Properties	ASTM	Urea-Formaldehyde α-cellulose filled	Vinyl Polymers			
			Reigid poly(vinyl chloride)	(Rigid) Vinyl Chloride/vinyl acetate	Rubber modified PVC	Vinylidene chloride
Machining qualities	—	Fair	Excellent	—	—	Good
Modulus of elasticity (psi × 10^{-5})	D638	15	3.5 ⟷		6	0.5 – 0.8
Mould shrinkage (in/in)						
(a) Injection	—	—	0.001 ⟷		0.05	0.005 – 0.025
(b) Compression		0.006 – 0.014	—	—	—	—
Moulding qualities	—	Excellent	Fair to good	Fair to good	Fair to good	Excellent
Odour	—	None	None	None	None	None
Refractive index (η_D)	D542	1.54 – 1.56	1.52	1.55	—	1.60 – 1.63
Resistance, chemical						
(a) Strong acids	D543	Decomposes	Unaffected	Unaffected	Unaffected	Good but attacked by strong H_2SO_4 or CH_3 COOH
(b) Weak acids	D543	Unaffected to slight	Unaffected	Unaffected	Unaffected	Unaffected
(c) Strong alkalis	D543	Decomposes	Unaffected	Unaffected	Unaffected	Resistant
(d) Weak alkalis	D543	Slight to marked	Unaffected	Unaffected	Unaffected	Resistant
Resistance, solvent	D543	None	Resists alcohols, aliphatic hydrocarbons, and oils; soluble in ketones and esters; swells in aromatic hydrocarbons			None to slight
Softening point [(Vicat) °C]	—	—	82	—	78	71 and above
Specific gravity	D792	1.47 – 1.52	1.38 – 1.4	1.37 – 1.45	1.35	1.65 – 1.72
Specific heat (cal/°C/g)	—	0.4	0.2 – 0.5	0.2 – 0.5	0.2 – 0.5	0.32

Contd.

Table 8.1 (Contd.)

Properties	ASTM	Urea-Formaldehyde α-cellulose filled	Vinyl Polymers: Reigid poly(vinyl chloride)	(Rigid) Vinyl Chloride/vinyl acetate	Rubber modified PVC	Vinylidene chloride
Specific volume (cu. in/lb)	D792	18.8 – 18.2	20.1	20.4 – 19.1	20.5	16.8 – 16.1
Specific strength (psi)	D638 D651	6000 – 13000	8500	7500 – 8500	6000	3000 – 5000
Thermal conductivity (cal/s/sq. cm/°C/cm × 10^4)	C177	7 – 10	3.5	3.7 4.5	3.0	
Thermal expansion (cm/cm/°C × 10^4)	D696	2.7 5	7	10 19		
Volume resistivity (ohm cm at 23°C and 50% R.H.)	D257	$10^{12} - 10^{13}$	10^{13} ←	→	10^{16}	$10^{14} - 10^{16}$
Water absorption (% 24 h, 1/8″ thickness)	D570	0.4 – 0.8	0.05	0.08 0.10	0.10	

Table 8.2 *Metals and alloys—applicable properties*

Properties		ASTM	Aluminium (99% pure)	Brass (Cu–Zn, 70/30)	Carbon steel (0.4% Carbon)
Compressive strength	(psi)	D695	—	—	200000
Elongation in tensile	(%)	D638	50	65–70	17
Hardness (Brinell)		—	22	100	180
Impact strength (ft in/in notch, Izod)		D256	20	66	13
Melting point	(°F)	—	1200	1715	2725
Modulus of elasticity	(psi × 10^{-5})	D638	0.21	0.59	5.27
Specific gravity		D792	2.7	8.5	7.85
Specific heat	(Btu/lb/°F)	—	0.23	0.09	0.11
Specific volume	(cu.in./lb)	D792	10.3	3.3	3.53
Tensile strength	(psi)	D638 D651	10100	46300	89600
Thermal conductivity (cal/s/sq. cm/°C/cm)	C177	0.52	0.29	0.12	
Thermal expansion	(cm/cm/°C × 10^4)	D696	0.240	0.199	0.112
		ASTM	Duralumin (Al–Si–Cu–Mg)	Mild steel (0.06% carbon)	Phosphor bronze (Cu–Sn–P)
Compressive strength	(psi)	D695	—	110000–130000	—
Elongation in tensile	(%)	D638	5	28	15
Hardness	(Brinell)	—	82	140	110
Impact strength (ft lb/in. notch, Izod)	D256	1	20	—	
Melting point	(°F)	—	1100	2760	1710
Modulus of elasticity	(psi × 10^{-5})	D638	7.16	0.61	3.88
Specific gravity		D792	2.8	7.87	8.9
Specific heat	(Btu/lb/°F)	—	0.23	0.11	0.09
Specific volume	(cu. in./lb)	D792	9.9	3.52	3.12
Tensile strength	(psi)	D638 D651	35800	44800	58200
Thermal conductivity (cal/s/sq. cm/°C/cm)	C177	0.35	0.15	0.2	
Thermal expansion	(cm/cm/°C × 10^4)	D696	0.225	0.126	0.173

Table 8.3 *Major applications of some polymers*

Polymers	*Applications*
Acrylics	Automotive parts, control knobs, dials and handles, pens and pencils, brush-backs, hospital equipment, display material, signs, lighting fittings, inspection panel covers, skylights
Cellulosic plastics	
Cellulose acetate	Toys, beads, cutlery handles, electrical parts, knobs, steering wheels, shoe heels, packaging, sheeting, toothbrushes, cosmetics
Cellulose acetate butyrate	Moulded or extruded parts for metallization (reflectors, etc.), outdoor signs, automobile tail light covers, tool handles, tooth-brushes, pipe inspection traps, piping
Cellulose propionate	Automotive parts, pens and pencils, telephone housings, shoe heels, toys, radio/TV parts, tooth-brushes, sunglass frames
Fluorinated polymers	
Fluorinated ethylene propylene (copolymer)	Coil formers, wire insulation, electronic components, encapsulations and fluidized bed coatings, non-stick valves
Polytetrafluorethylene	Gaskets, packings, valves, anti-adhesive coatings, rigid and flexible pipes, membranes, wire insulation, electronic engineering applications
Polytrifluorochlorethylene	Extruded sheet, profile and film, electronic parts, gaskets, pump sealants, dispersion coatings, liquid level indicators of particular use where resistance to aggressive chemicals is needed
Melamine formaldehyde	
Unfilled	Noted for durability, hardness and good electrical properties, suitable for appliance housings, dinnerware, closures, writing equipment, clock housings, knobs, handles, lighting fixtures, appliances, instruction panels
α-cellulose filled	As for unfilled material
Nylons	
Type 6	Moulded mechanical parts, gear wheels, bushings, elevator gibs, sliding parts for storm windows, automobile and refrigeration door closures, mixer valves, switch housings, grommets, cable clamps, tubing, filaments, aerosol bottles
Type 66	As for Type 6
Phenol formaldehydes	
Unfilled	Bottle caps, electrical parts, impact resistant, heat resistant and special purpose mouldings requiring rigid durable properties. Restricted colours, close tolerance in moulding articles
Woodflour filled	As for unfilled material
Polyacetals	Instrument panel mouldings, aerosol packs, moulded zip-fasteners, wheels for lawn mowers and push carts, impeller blades, cosmetic cases, gear wheels, conveyors

Contd.

Table 8.3 *(Contd.)*

Polymers	*Applications*
Polycarbonates	Business machine parts, electrical apparatus, sterilizable ware, draughtsman's instruments, lamp covers, safety helmets, tail lights, de-luxe housewares, engineering and industrial components
Polyethylenes	
Low density	Housewares (bottles, bowls, buckets, containers) closures for squeeze tubes, spouts for detergent cans, carboys, shoe parts, toys, packaging film, garment bags, sheet, piping for domestic, industrial and agricultural use
Medium density	As for low density polyethylene
High density	As for low density polyethylene, materials have greater rigidity, specially suitable for large carrying cases, housings, closures, appliance parts, packaging film, sheet, piping, carboys, containers and bottles
Polypropylene	Domestic, hospital and laboratory ware, textile, motor, electrical and chemical industrial usages, containers and closures, shoe heels, combs and toys, blown containers, extruded sheet, pipe, film, filament and fibre
Polystyrenes	Conventional packaging (disposable and others), dishes and utensils, refrigerator parts, emblems, signs, displays, toys, novelties, combs, brush-backs, radio and TV cabinets, lighting fixtures, rigid containers, housewares, reels/spools for film/tape, appliance panels, handles and switches
Toughened	Wheels, helmets, valve parts, refrigerator parts, bobbins, radio cabinets, electric fan blades, toys, housewares, containers, battery cases, sheeting for refrigerator linings and trays
Styrene copolymers	
Styrene/acrylonitrile (SAN copolymer)	Housewares (dishes, tumblers, cutlery handles), hair curlers, cosmetic packaging, fountain pen barrels, radio cabinets, battery cases, oil dispensing devices, laboratory utensils electric razor parts, vacuum cleaner parts and casings
Acrylonitrile/ butadiene/ styrene (ABS terpolymer)	Calendered sheet, electrical housings, office machinery components, industrial mouldings, luggage, automobile parts, domestic appliances
Urea formaldehyde (α–cellulose filled)	As for Melamine Formaldehyde. In laminate form surfacings for tables, etc.
Vinyl polymers	
Rigid poly(vinyl chloride) PVC	Extrusion of piping, profiles and sheet in applications requiring chemical inertness and scuff-resistance combined with light weight
'(Rigid) vinyl chloride/ vinyl acetate copolymer	Similar applications to those for PVC but widely used in the manufacture of calendered sheet used for toys, novelties, wall coverings, displays, templates etc.
Rubber-modified PVC	Same as for PVC
Vinylidene chloride polymers	Specialized applications including weather resistant screening: rigid pipes and fittings; in film form, uses include a variety of heat-sealing packaging applications

processing and fabrication, properties and uses and applications of polymers fast grew as a separate branch of science and technology.

In our social relevance, polymers symbolise beauty, balance, stability, creativity and growth. Hazardous volatile and highly flammable petroleum and petroleum products are turned into nonvolatile, relatively dense and stable polymer products. The first lesson in polymer education relating to polymer synthesis by polymerization of monomers or mixtures of monomers makes one realise the high if not unlimited scope for achieving strength and balance through unification. One may appreciate better the meanings of unity in diversity or diversity in unity through meaningful polymer education. A polymer material is an embodiment of a molecular society where big chain molecules of specified repeat units and of different sizes, shapes, molecular weight and molecular complexity exist in peace and harmony. As in human society, so also in the polymer society, variety is the essence of existence. Polymers come into being by different routes following different mechanistics, and taking help of different catalysts, they develop into systems showing varieties in structural features ranging from linearity and branching to cross-linking or networking with variations in molecular weight and molecular weight distributions, monomer sequence distributions (in a co-polymer), cross-link density (in a cross-linked system), tacticity, nature of end groups etc., in their growth process, environments prevailing impart a great influence. Also depending on these structural variations, polymers offer different ranges of properties under different sets of static and dynamic conditions in respect of mechanical strength, chemical and solvent resistance, photostability or photo-degradability, thermal resistance, clarity or opacity, rigidity, stiffness, resilience, toughness and flexibility, fatigue resistance, rheology and morphology.

8.27.2 Polymer Society vis-a-vis Human Society

Peaceful polymer society may be viewed as an ideal model for the strife-torn human society. In our democratic society we try to treat individuals as equals with one man one vote principle, though disallowing the voting rights to under-18 age group people or to non-members or defaulters, thus deviating from universal franchise. In the polymer system, all molecular species, big or small, like or unlike, including impurities or foreign elements contribute equally and at par when it comes to assessment of their colligative properties in an effort to have a measure of molecular weight of the system. Unlike what may happen as violation of voting rights through perpetration of violence or intimidation or snatching of ballot papers etc. in human society, the "democracy" prevailing in the polymer system shows no such aberration and it is practically foolproof; only chain molecular species, incapacitated by insolubility refrain, of their own, from casting their votes, i.e. making thermodynamic contributions in influencing the colligative property being measured, just as unwilling or incapacitated persons may of their own refrain from casting their votes in an election. The all important unifying quality of love is manifested in the polymer molecular society in the form of molar cohesion, solution tack and melt tack. It is through the property of tack that a polymer readily accommodates a host of additives including other polymers in its body matrix.

Human body, like a state or a society, remains within its bounds and does not commonly open its system spontaneously to accommodate foreign elements. We frame our own social rules and have our own likes and dislikes; we may selectively accommodate outsiders with affection, courtesy or distrust; we may try to retain them or may even throw them out sooner or later depending on proven acceptability or on stresses and strains that may develop due to economic constraints or cultural and spiritual incompatibilities. A polymer society also tends to suffer from an inclination of nonacceptance of unlike polymers or other species. In polymer blends, incompatibility and phase separation is more a rule than the exception. Incompatibility resulting in phase separation can be lowered, minimised or even eliminated by use of a right kind of compatibilizer.

8.27.3 Stabilization Role of Polymerization and Polymer Stability

Polymerization is a stabilization process as polymers are much more stable than the respective monomers; the stable polymers are, however, not stable enough and we need to practice a post polymerization stabilization process for the polymers so as to make them attain much higher durability, easy pocessability and stable performance. This necessitates addition of a host of additives for deriving advantages in respect of production cost and improving (1) aging resistance and photostability, (2) solvent and chemical resistance, (3) thermal stability, and (4) mechanical property balance. Relevant additives provide shields and weapons to the polymer system to enable them protect themselves and fight the adverse effects from the attack of hostile agencies of degradation such as thermal, electrical or mechanical loading under static or dynamic conditions, exposure to light and high energy radiations, solvents and chemicals and micro-organisms. Additives may be natural or synthetic, solid or liquid, polymeric or non polymeric particulate or fibrous and they may be used in minor or major proportions, sometimes taking the help of a compatibilizer or a coupling agent for achieving uniform dispersion with stable morphology and associated property advantages. The role of compatibilizers or coupling agents in improving mutual contacts, wetting and hence in uniform stable dispersion is similar to the role of a broker, match makers or third party arbiters, who may play vital roles in negotiated deals or contracts.

Education plays an important role in character building and in improving thinking power and overall ability of a man; likewise, simple thermo-mechanical processing imparts an educative role on polymer material systems and makes them mechanically much stronger and capable of bearing substantially higher load. Equipping the polymer system with selected reinforcing agents further allows spectacular enhancements in their nerve, strength and load bearing capabilities.

8.27.4 Helping Hand of Polymers

Man has been successful in developing modern petrochemical industries and deploying the polymer material system as substitutes for many other materials including forest wood for the welfare of human society that has finally set the stage right for much publicized globalization and establishment of one world by breaking through geographical, political and cultural barriers. The electronic mass-media and modern communication system and developments in information technology owe a lot to all-round developments in the polymer material system. More purposeful polymer education supported by R&D efforts is all the more necessary for development of technologies based on growing utilization of renewable resources and waste materials from industry, forestry and agriculture for environmental preservation, energy conservation and for striking an improved ecological balance.

Polymers are variously useful during their life-time. However, some of them when used in minor proportions may also have a lasting, formidable and favourable impact on being destroyed *in-situ* in certain high heat-resistant refractory material system. Most modern technology of making high performance magnesia carbon or alumina carbon refractory bricks or plates for furnaces of metal extraction and preparation of alloys utilizes the high adhesive or binding power of phenolic and related resins for development of high green strength and cure strength. However, during exposure to high firing temperatures (1100°–1400°C) under inert or reducing atmosphere in the final stage of refractory making, the organic resin pyrolyses to leave a network carbon skeleton with rearrangement of valence bonds when the oxides of metals (Mg, Al, Si, Zr etc.) present find chance to react *in-situ* with the network carbon skeleton being formed to randomly form the respective metal carbides, thus significantly enhancing the durability and performance level of the refractory product. Through self sacrifice, the selected polymer, a very minor component in the formulation, allows the refractory system to become higher performing and much more durable.

8.27.5 Polymer Education and its All-Round Influence

Polymer education permits one to appreciate the phenomenal role of polymer materials in arts, sports and entertainment, education and literacy drive, mass-media and information dissemination, green revolution, and operation food, space exploration, health care and surgery, aids for the disabled, preservation of fresh and processed food, drinks, drugs and pharmaceuticals, water treatment for municipalities and in chemical and power plants, desalination of sea water for drinking and irrigation purposes and in water management in a comprehensive sense, thereby minimising or eliminating loss or wastages of precious water and ensuring its efficient use.

Polymers as they are naturally born or synthesized are mostly inert, non-toxic and harmless. Polymers the base materials for plastics, rubbers and related products may, however, exhibit toxicities that find their origin in injudicious selection of toxic/hazardous additives and incorporation thereof into the end-use items. All-round awareness about property demands, safety requirements and environmental preservation can only ensure right selection of additives for safe formulation and compounding of plastics, rubbers, coatings, adhesives and composites and further ensure their disposal or recycle all the more safe and cost effective. Technology innovation and helping hand of biotechnology would have major roles to play in future in rendering polymer disposal and recycling still safer and cost effective.

8.27.6 Polymers and Environment

Growing industrialization for meeting growing social demands and needs, linked to generation of job opportunities and for achieving a better quality of life and more efficient functioning of the society, imposes some constraints and additional responsibilities on the social sector for enacting legislation and enforcing rules and laws to prevent, check or minimize environmental denigration. Packaging, particularly plastic packaging is an easily identifiable portion of solid municipal wastes. Unsystematic waste disposal and indiscriminate littering of waste items largely disturb the environmental equilibrium. There is a genuine need or imperative for evolving balanced guidelines on environmentally viable and socially acceptable approaches for dealing with the growing problem of environmental pollution as linked with use of polymers and polymer waste disposal. Harmonious guidelines are required to be evolved for national and global acceptance for comprehensive management of plastics and related wastes in order to prevent their adverse effect on the environment covering urban, rural and forest landscapes, natural and artificial drainage systems, surface and ground water reserves and air, while ensuring their steady or growing production, uses and applications; in this context it is imperative to further ensure minimum waste generation leaving maximum scope for reuse and reprocessing, and easy collection, return or recovery with efficient recycling prospects. Some progress has already been made and a lot more remains to be done to achieve this goal. Turning waste to wealth or odds to opportunities is a big challenge for modern technology.

Plastics and packaging are closely interlinked, more so, in the modern age. Plastics and related wastes constitute between 5 and 10 percent of municipal waste, but they draw much higher share of public attention as waste items because of their resistance to biodegradation and hence long life, notable lightness, being lighter than metals and alloys, minerals, glass and ceramics, ashes and clays and even water in some cases, and because of multitude of colours in which they appear in the consumer market and from there finally find refuge in dustbins and garbage heaps.

One must not be unduly obsessed with the slogan of rendering plastics and related products biodegradable. No one would like to see the plastic furniture, file covers, luggage items, pipe lines, containers and storage bins, casings and components of calculators, TV or telephone sets and computers and other electrical or electronic gadgets and innumerable essential or useful products made of polymers to perish in no time or in a short span of time as a result of biodegradation. In fact, the rapid phenomenal growth in use and popularity of polymer materials is primarily due to their non-corrosive nature and durability, exceptional lightness and significant resistance to weathering and microbial attack. Biodegradability is desired primarily for those polymer objects which are meant for packaging and related applications and are most likely to reach the garbage heaps or waste-bins soon after one time or short-time use. In the context of technology trends for development of biodegradable polymers (plastics), attention is drawn to relevant points described in Sec. 8.16.4.

Presence of chlorinated organic wastes including chlorine-containing polymers such as PVC and related polymer additives in the waste-bins or garbage heaps make their collective disposal by incineration hazardous, because of unavoidable *in-situ* generation of highly toxic dioxines (polychlorinated oxygenated aromatics or phenols) in the outcoming volatile products or relevant condensates in harmful concentrations, commonly for incineration at relatively low temperatures ($\leq$ 800°C). However, the dioxine concentration level can be substantially lowered to less than hazard levels or even practically brought to zero level for incineration at significantly higher temperatures, preferably >1000°C.

References

1. Joshi, R.M. and B.J. Zwolinsky, *Vinyl Polymerization*, G.E. Ham Ed., Marcel Dekker Inc. New York, 1967, Vol. 1, Pt. 1, Ch. 1.
2. Greenley, R.Z., J.C. Stauffer and J.E. Kurz, *Macromolecules*, **2** (1969), 561.
3. Lawrence, C.L., *Recent Advances in Liquid Crystalline Polymers*, Applied Science, London, 1985.
4. Morgan, P.W., *Macromolecules*, **10** (1977), 1381.
5. Bair, T.I., P.W. Morgan, J.R. Schaetgan and L.W. Gulrich, *Macromolecules*, **10** (1977), 1390.
6. Findlay, R.H. and D.C. White, *Appl. Environ. Microbiol.* **45** (1983), 71.
7. Capon, R. J., R. W. Dunlop, E. L. Ghisalberti and P. R. Jefferies, *Phytochemistry*, **22** (1983), 1181
8. Odham, G., A Tunlid, G. Westerdahl and P. Marden, *Appl. Environ. Microbiol.* **52** (1986), 905.
9. Abellar, M. and E. Martuscelli, *Polymer*, **29** (1988), 173.
10. Greco, P. and E. Martusulli, *Polymer* **30** (1989), 1475.
11. Pramanik, D. and T.T. Ray, *J. Appl. Polym. Sci.*, **40** (1990) 1551.
12. Pramanik, D., T.T. Ray and M.A. Bakr, *J. Polym. Mater.*, **13** (1996) 173.
13. Pramanik, D., D. Biswas and G. Betal, *J. Polym. Mater.*, **17** (2000) 69.

CHAPTER 9

Rubbers—Materials and Processing Technology

9.1 Rubber Materials—Introduction

The technology of rubber began with the natural product known as natural rubber (NR). Historically, rubber (NR) as a material was known to and used by man as early as the sixth century, as excavations subsequent to the discovery of America have revealed. The early reported uses of NR were limited to such items as playing balls and waterproof fabrics or garments. People of Europe became familiar with this natural product and its cohesive/binding properties by the end of the eighteenth century. From its popular application as eraser of pencil and ink marks developed in Europe in the middle of the eighteenth century, the name "rubber" was coined to it. Earlier, the natural product was known by the term "Caoutchouc" which, however, is now reserved in the English language to denote the pure rubber hydrocarbon.

The process technology of making waterproof objects based on rubber-coated fabrics passed into an advanced phase with the discovery of coal tar naphtha as a good solvent for rubber by Charles Macintosh (1823). This led to the development of the "sandwich" process for the so-called double texture fabric, imparting much improvements in the life and performance of the waterproof garments, but the inherent drawback in the susceptibility of rubber to changes of temperature (becoming soft and sticky in warm weather, and hard and stiff in cold weather) still remained unsolved, thus limiting expansion and diversification of its use. The difficulty was finally overcome through the discovery of vulcanization of rubber using sulphur by Charles Goodyear in 1839 in USA. The process was also developed in London by Thomas Hancock at about the same time and he applied for the first patent on vulcanization or curing of rubber in 1843. The discovery of vulcanization, which in effect is a cross-linking process, literally infused a revolution in the rubber industry and a whole range of consumer and industrial rubber products soon became available in the market.

The era of natural rubber continued till about the time of the Second World War (1939–40) without substantial introduction of synthetic rubbers in the market.[1] The empirical formula of natural rubber was recognized to be C_5H_8 by 1826 and by 1860 it was recognized to be a (linear) polymer of isoprene, $—[—CH_2—C(CH_3) = CH—CH_2—]_n—$. The establishment of the structure of natural rubber led to sustained efforts for the development of synthetic rubbers using diene monomers, such as butadiene, isoprene and chloroprene. The early synthetic rubbers of commercial importance were polybutadiene (1910–14) (butadiene polymerized by sodium, commonly named as BuNa in Germany) and polychloropene (1932–33).

9.2 Natural Rubber (NR)

9.2.1 Natural Rubber Plantation

There was a time till the middle of the nineteenth century when raw rubbers (NR) came almost entirely from the equatorial forest in the Amazon valley in South America. Rubber randomly tapped and collected from natural forests is variable in quality and is commonly known as wild rubber. Under economic, technical and other compulsions, and for a regular, uninterrupted supply of rubber, comprehensive plans for cultivation of rubber trees from seeds of *Hevea brasiliensis* in the equatorial climatic zones of South and South-East Asia began in the later part of the nineteenth century. Present-day NR or the Hevea rubber is almost entirely obtained from the plantation industries.

9.2.2 Tapping of Rubber Latex

The rubber plant produces a milk-white latex that contains the natural rubber hydrocarbon in a fine emulsion form in an aqueous serum.[2] After a thin shaving of bark of the Hevea tree has been cut, the latex that comes out is allowed to flow into a cup through a spout that is stuck into the bark below the bottom end of the cut. A little of sodium sulphite solution put into the empty cup before tapping helps prevent some darkening or discolouration of the latex which may otherwise develop as a consequence of an enzymatic reaction in the latex involving its phenolic constituents producing the dark coloured pigment melanin.

The composition of the Hevea latex varies with climatic conditions, season, age of tree, soil condition, etc. An average composition is shown in Table 9.1

Table 9.1 *Hevea latex composition*

Ingredients	*(%)*
Rubber hydrocarbon	30–40
Water	55–65
Proteins	1.90–2.5
Fats and related compounds	0.90–1.0
Sugars (carbohydrates)	1.00–1.5
Ash	0.4–0.6

9.2.3 Preservation and Coagulation of Latex

After tapping, a small amount of a preservative is mixed with the rubber latex. The latex is unstable, it undergoes putrefaction and tends to coagulate at different rates on standing, depending on condition of storage. It is preserved over longer periods of time using ammonia with or without sodium pentachlorophenate as the preservative. The Hevea rubber latex globules are about 0.2 μ in diameter and they are usually constituted of three characteristic layers—the core consisting of lower molecular weight, sol rubber with an immediate outer layer or skin of higher molecular weight, and gel rubber. The surface or outermost layer is a sorbed layer consisting mainly of proteins and soaps. On ionization, they impart a negative electrical charge to the globules.

On arrival at the factory, the Hevea latex is somewhat diluted and strained through a 60–80 mesh sieve to remove dirt and other foreign particles. Most Hevea rubber reaches the user industry in the form of pale crepe or smoked sheet. The dilution level of the latex before coagulation and creaming depends upon the type of rubber to be made. The coagulation is done by adding acetic acid or formic acid.

The coagulum contains nearly 80% of water. Much of the serum is removed from the rubber hydrocarbon by water washing and squeezing through rollers. For crepe rubber manufacture, the wet strip is heavily milled and the sheets formed are finally dried and packed. If smoked sheet is to be produced, light milling of the wet strips is allowed on the rollers and final drying of the ribbed sheets is accomplished in a smoke house. Spray dried "whole-latex" rubber may be produced and it contains all the solid impurities present in the latex. The average composition of raw Hevea rubbers is given in Table 9.2.

Table 9.2 *Composition of raw Hevea rubbers*

Ingredients	*Crepe*		*Smoked Sheet*		*Spray-dried Latex*
	Average (%)	*Limits (%)*	*Average (%)*	*Limits (%)*	*Average (%)*
Moisture	0.4	0.2–0.9	0.6	0.3–1.0	2.0
Acetone extract	2.9	2.2–3.5	3.0	1.5–3.5	5.0
Proteins	2.8	2.3–3.7	2.8	2.2–3.5	4.5
Ash	0.3	0.2–0.9	0.4	0.2–0.9	1.5
Rubber hydrocarbon (by difference)	93.6		93.2		87.0

The various non-rubber constituents in the raw rubber exhibit prominent antioxidant action which is synergestic. Acetone extracted rubber ages very rapicly. The fatty acids present act as accelerator activators and assist vulcanization.

9.2.4 Chemical Nature of Natural Rubber Hydrocarbon

The 1,4-polyisoprene structure of NR hydrocarbon was confirmed by identifying the products derived from its ozonolysis. The overwhelming products isolated on ozonolysis are levulinic aldehyde ($O{=}C(CH_3){-}CH_2{-}CH_2{-}CHO$) and the corresponding acid, i.e. levulinic acid.

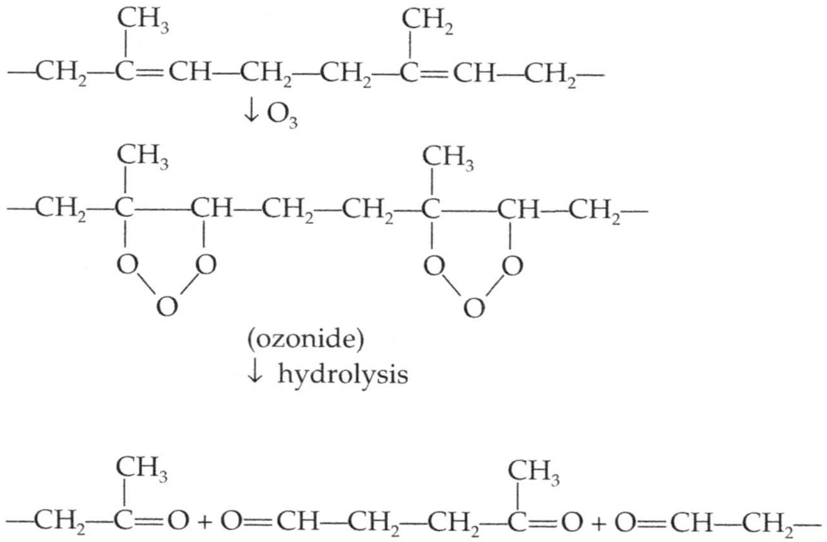

(formation of levulinic aldehyde) (9.1)

The hydrocarbons of natural rubber and *gutta percha* or balata are the same in elementary formula and in their products of ozonolysis, even though they greatly differ in physical properties. *Gutta percha* is usually lower in molecular weight than NR, but the major differences in their physical properties are appreciated from the difference in their spatial configuration. The typical elastomeric properties of natural rubber are attributed to its almost complete *cis* configuration and the prominent thermoplastic nature of *gutta percha* is attributable to its *trans* structure (*See* Sec. 5.9). Rubber hydrocarbon (specific gravity, 0.924 at 20°C) is less dense than *gutta percha* (specific gravity, 0.957 at 20°C).

9.2.5 Hydrogenated Rubber

Rubber hydrocarbon (NR) is chemically reactive and the reactive sites are the points of unsaturation and the α-methylene groups in the repeating units of the chain molecules (*See* Sec. 6.4.5). Natural rubber is readily hydrogenated into a colourless, near-transparent, wax-like product. The hydrogenation may be conveniently done taking solution of rubber (2–3%, in cyclohexane) using hydrogen at a pressure of 15–20 atm 180–200°C and employing nickel catalyst.

9.2.6 Cyclized Rubber

Cyclization as applied to natural rubber is a process whereby the linear chains of 1,4-polyisoprene rearrange to give rise to cyclic or ring structures. Cyclization is readily achieved by treating or mildly heating the rubber in the presence of mineral acid or Lewis acid catalysts, such as H_2SO_4, $SnCl_4$, BF_3, $TiCl_4$, etc.

The cyclized rubber has much lower degree of unsaturation and the product shows resinous characteristics. For strictly monocyclic structure formation at random, the unsaturation would theoretically drop to 57% of that of the original rubber. But many extensive cyclization reactions lead to drop in unsaturation to 20–30% of the original, thus indicating formation of bi- or even polycyclic structures. Formation of a typical bicyclic structure may be shown as in Eq. (9.2).

The current practice is to make cyclized rubber directly from latex. In presence of suitable stabilizers, latex may be made acidic without coagulation. Cyclized rubber masterbatch is prepared by heating such specially stabilized latex with strong H_2SO_4, mixing with untreated latex containing comparable quantity of rubber and coagulating. The coagulum is washed and dried, and the masterbatch is ready for commercial use, particularly as an ingredient for making stiff vulcanizates.

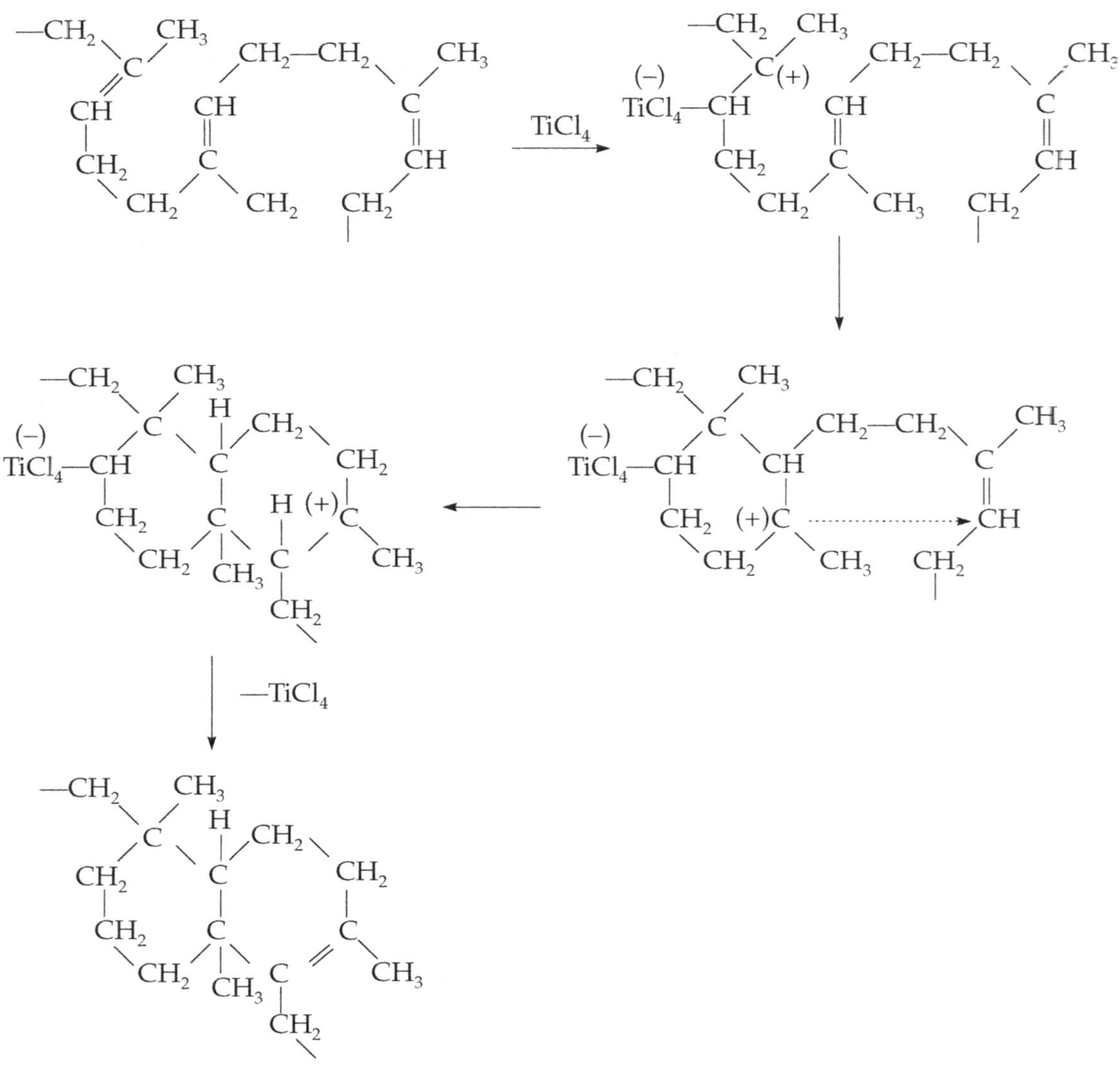

Cyclized rubber (bicychic structure)

(9.2)

9.2.7 Chlorinated Rubber

Natural rubber can be readily halogenated. Only chlorination has been commercially developed. Direct chlorination with chlorine results in both addition and substitution reactions and HCl is evolved as a byproduct. For only additive chlorination, maximum attainable chlorine content would be about 51%, while for commercial products of good stability, the range of chlorine content is 60–68%, and they are resinous in character. It is generally believed that good degree of cyclization also takes place on chlorination. Rubber, cut into small pieces is dissolved in carbon tetrachloride in presence of a small amount of benzoyl peroxide which acts as a depolymerizing agent and lowers the solution viscosity. A relatively uniform chlorination is achieved by spraying the solution at the top of a chlorinating tower in which a stream of chlorine, let in at the bottom, is allowed to ascend. The droplets of the chlorinated product collect at the bottom. The collected solution from the bottom is degassed to remove excess chlorine and then sprayed into a steam chamber where the admitted steam causes rapid volatilization of CCl_4. CCl_4 is then recovered and reused; the chlorinated rubber collected as a wet mass from the bottom is washed, dried, milled if necessary and stored. Chlorinated rubber is resistant to many chemicals. It even resists concentrated nitric acid. It is however soluble in a wide range of solvents and it is used in the formulation of many paints, lacquers, adhesives and printing inks. Because of high chlorine content, chlorinated rubber has prominent flame retardant characteristics.

Rubber can also be modified into a resinous product by hydrochlorination. The rubber hydrochloride may be prepared directly from stabilized latex using hydrochloric acid gas. It is, however, better obtained by hydrochlorination with gaseous HCl using rubber in benzene solution. Films made from the solution of the hydrochloride to which plasticizers and stabilizers have been added are used for making laminates with paper or films of other plastics such as cellulose acetate, etc., for use as heat sealable packages for dry food, cosmetics, shampoo, etc. The hydrochloride also finds use in the formulation of adhesives and bonding agents.

9.3 Synthetic Rubbers

The synthetic rubbers developed initially were based on dienes. The polymers that have been derived have unsaturated structural units in their molecules, similar to those present in the repeating units of NR. They are mostly synthesized by redox emulsion polymerization of appropriate monomer or monomer combinations, even though, in recent years, anionic (mass) polymerization has come into some prominence. The major synthetic diene-based rubbers are: polybutadiene rubber (BR), styrene-butadiene rubber (SBR) which is a copolymer of styrene and butadiene, polychloroprene rubber (CR), nitrile rubber (NBR) which is a copolymer of acrylonitrile and butadiene, polyisoprene rubber (IR) which is also known as coral rubber, and butyl rubber (IIR) which is a copolymer of isobutylene and small proportions of isoprene.

The major polyolefin rubbers are saturated binary copolymers of ethylene and propylene (EPR) and the unsaturated terpolymers of ethylene, propylene and small proportions of a diene monomer (EPDM). Besides, chlorinated polyethylene (CPE) with nearly 40% chlorine and chlorosulphonated polyethylene (CSP) are also saturated rubbers and the latter are more commonly known in the trade as Hypalon. Other major synthetic rubbers of special applicability are polysulphide rubbers—commonly known as the Thiokols, silicone rubbers, polyacrylic rubbers, polyurethane rubbers and fluorocarbon rubbers. The preparative and other aspects of EPR, CPE, CSP, silicone rubbers and polyurethane rubbers have been discussed in Chapter 8.

9.3.1 Rubbers from Stereo-regular Polymerization of Isoprene and Butadiene

Quest for synthetic polyisoprenes (IR) resembling natural rubber to a large measure in chemical structure and in physical properties proved elusive for a long time. The endeavour finally met with success with the advent of stereospecific catalysts of the Ziegler–Natta type. Both straight and oil extended grades are currently being produced using such catalysts as $TiCl_4$-$Al(i\text{-}C_4H_9)_3$ combination producing nearly 96%, 1,4 *cis* structure or butyl lithium producing about 92% 1,4 *cis* structure compared to 97–98% 1,4 *cis* structure for NR. The coordination catalyst produces polyisoprene rheologically comparable with masticated natural rubber and thus it needs no mastication; on the other hand, IR produced by using butyl lithium catalyst are much higher in molecular weight and require longer mastication than NR.

In a typical synthesis using the coordination catalyst, the monomer isoprene is allowed to polymerize in a solvent medium such as *n*-pentane or *n*-hexane containing the catalyst dispersion in a pressure reactor vessel at 50°C. At the end of polymerization, a catalyst decomposer and an antioxidant are added, the solvent is removed and the polymer is dried.

High *cis* 1,4 polybutadiene, BR (>92% *cis*) is commercially obtained by polymerization of butadiene in solution following similar approaches as for the production of 1,4-*cis*-polyisoprene from isoprene, using Ni, Co or Ti based coordination catalysts. Using butyl lithium catalyst, polybutadienes of low *cis* content (<40%) are obtained. Polybutadiene obtained by emulsion polymerization using redox catalysts is still lower in *cis* content (<10%). Its manufacture is very much similar to that of emulsion SBR. The use of emulsion BR is limited to the manufacture of rubber-modified high impact grade polystyrenes.

The high *cis* polybutadienes are also commercially available in oil-extended grades. Even when extended with 20 parts oil, the physical properties of the vulcanizate remain substantially unchanged without carbon black adjustment. However, poor tack and difficult processing characteristics of BR are limiting factors for its sole use. It is used mostly in tyres blended with NR and SBR. High *cis* BR, when blended with polychloroprene in a weight ratio of 1 : 4, imparts much improvements in low temperature flexibility of the vulcanizates without much lowering in solvent resistance and thus making such blends useful for cable sheathing. Compared to NR, BR has relatively high elasticity, high resilience exhibited by low heat build-up on repeated flexing and hence low hysteresis and somewhat improved resistance to oxidation. Used in blends to the extent of 20–40% along with NR and SBR, BR imparts excellent abrasion resistance to truck and car tyres. Use of higher proportions of BR in the blends is not recommended and practised in view of the skid resistance of the tyre then becoming poor.

9.3.2 Styrene-Butadiene Rubber (SBR)

SBR is a general purpose synthetic rubber and it surpasses all other synthetic rubbers in volume of production. Its production began during World War II using the emulsion polymerization technique. A recipe much the same as given in Table 4.7 may be used where the monomer component is a mixture of 70–75 parts of butadiene and 30–25 parts of styrene. Polymerization may be done at 40–50°C giving what is known as hot SBR or it may be carried out at 5°C or even at lower temperature (–10 to –15°C) using appropriate redox initiators (hydroperoxides with metal ion or amine activators) and adding antifreeze ingredients to the system. Polymerization at low temperatures is a later development and products thus obtained are known as cold (SBR) rubber.

Polymerization is arrested at the stage of 70–75% conversion using a shortstop or inhibitor. The unreacted monomers are then removed by blowing in steam. A requisite dose of an antioxidant is then dispersed in the latex. For making oil-extended SBR, an emulsion of the required oil is added. Appropriate grade and amount of carbon black may be dispersed into the latex at this stage to make black masterbatch or oil-black masterbatch grades of SBR. The latex is finally coagulated preferably in the form of small porous granules which are easy to wash, using solution of common salt and dilute H_2SO_4 in succession as the coagulants or simply by using acid alone after the addition of a solution of animal glue into the latex.

Alum solution may also be used as the coagulant to give SBR of lower ash content and improved heat stability.

The coagulated crumb is then separated from the serum, washed profusely with water and the water content is initially brought down to about 55–60% by draining on a vacuum filter. The amount of water is further reduced by roller pressure and then by drying in a large drier. The crumbs are then compressed into bales.

A solution copolymerization process was developed much later using coordination catalysis or anionic catalysts based on lithium metal (butyl lithium) to produce what is commonly known as solution SBR. The approach and the method are much the same as those relating to the production of IR and BR using butyl lithium catalyst. The product is usually characterized by a narrower molecular weight distribution and higher 1,4 *cis* configuration for the butadiene units in the copolymer structure in comparison with the emulsion SBR. The solution process holds prospect for wider acceptance and further development in future at the expense of the emulsion process because of higher structural uniformity, much lower branching and higher economy through increased oil extension and carbon black loading.

High-styrene SBRs containing 30–50% bound styrene are useful as rubbers in the design of tread compounds for tyres with improved road grip. These rubbers are seldom used alone; they are blended in appropriate proportions with normal SBR or NR for the production of hard vulcanizates. Copolymers with still higher bound styrene contents (up to 90%), better known as high-styrene resins serve as useful compounding ingredients imparting improved hardness and stiffness to NR or SBR based vulcanizates.

SBR is usually much superior to NR with respect to aging and ozone resistance. However, cuts or cracks grow faster in SBR than in NR. SBR is also characterized by relatively high hysteresis or heat build-up and poor resilience. The abrasion resistance of SBR is as good as that of natural rubber or slightly better.

The original development of application of SBR was in the production of tyres. Subsequent developments cover such items as moulded goods, floorings, belting and hose, extruded objects shoe soles, coated fabrics, etc. A good balance of desirable properties, favourable production economics of the raw rubber grades and easy processing characteristics contributed to the growth in popularity of SBR.

9.3.3 Polychloroprene Rubber (CR)

Polychloroprene is the world's first commercial synthetic rubber made available in the market in 1935. The commercial elastomeric polymers and copolymers of chloroprene, i.e., 2-chloro-1, 3-butadiene, are known in the trade by the generic term neoprene. The monomer chloroprene is conveniently prepared from butadiene in two steps: the first step is chlorination of butadiene to 3,4-dichlorobutene-1 which is dehydrochlorinated to chloroprene by heating with aqueous alkali in a subsequent step.

Polychloroprene is commercially produced by emulsion polymerization of chloroprene. Chloroprene is emulsified using a soap such as rosin soap and the polymerization in aqueous emulsion is done using potassium persulphate as the initiator and the temperature range commonly employed is 40–60°C. Certain grades are produced in presence of very small quantities of sulphur which acts as a modifier and a cross-linking agent; in these cases, on completion of polymerization, the latex is allowed to stand and react with tetraethylthiuram sulphide whereby the desired plasticity of the polymer is restored. The latex is coagulated by addition of acid and freezing.

The polymer has an overwhelmingly 1,4-*trans* structure and this structural feature is primarily responsible for the crystallizable nature of the neoprenes. The structural feature also makes the chlorine atom reactive, being both tertiary and allylic and providing curing sites on the polymer.

Crystallizability and rate of crystallization of CR are reduced by modification of the molecular structure of the polymer and incorporation of a second monomer in the polymerization recipe. The nature of the second monomer used is not disclosed and it may be 2-3-dichlorobutadiene known to be formed as a byproduct during manufacture of chloroprene. The major classification of neoprene is based on whether or not the product is modified with sulphur and sulphur donor compounds during manufacture, and in each class, different grades prepared differ in the extent of the second monomer incorporation (10% in type GRT, 5% in type WRT and 2.5% in type WX).

The vulcanization of CR as commercially practised is much different from that of the other diene rubbers. Vulcanization is conveniently accomplished with zinc oxide and magnesium oxide on heating. Five parts ZnO mixed with 4 parts MgO per 100 parts neoprene are used as the average curing mixture. Sulphur causes very slow curing and the common rubber accelerators are ineffective in the curing of CR. Use of small proportions (0.5–1.0 parts) of certain other chemicals such as ethylene thiourea and antimony sulphide causes faster and effective press-cure at 150°C.

In contrast to most synthetic rubbers, unfilled neoprene vulcanizates exhibit high tensile strength (3500–4000 psi). The resilience of pure gum vulcanizates of CR is lower than that of similar gum vulcanizates from NR. With increasing filler loading in each case, however, the resilience drops, though to a lesser extent for CR system. As a consequence, most filled technical CR compounds show better resilience than NR of similar loading. For optimum processability, hardness and strength, fillers are used.

CR vulcanizates are notably resistant to oxidative aging when protected with suitable antioxidants. Their ozone resistance is also quite adequate under normal conditions. They also possess very good resistance to flex cracking. CR has some inherent flame resistance because of the presence of chlorine in the molecule. The crystallizing tendencies of raw CR, particularly at low temperatures is much reduced on vulcanization and the phenomenon of crystallization can further be controlled by incorporating selected resins, plasticizers and oils in the compound.

Polychloroprene is somewhat inferior to the nitrile rubbers in oil resistance but it is substantially better in this respect than other diene rubbers. Mostly, the dynamic properties of polychloroprene are better than those of other synthetic rubbers and only marginally inferior to those of NR.

Mixing and processing of CR are done using conventional methods and equipments. CR bands readily on a cool mill. Large scale general purpose use of CR is not permitted on cost considerations. The major applications and uses include heat resistant and flame resistant beltings, industrial hoses, wires and cables, gaskets and O rings, rubber rollers, coated fabrics, adhesives, etc.

9.3.4 Nitrile Rubber (NBR)

Nitrile rubber (NBR) is a product of copolymerization of acrylonitrile and butadiene developed almost parallel to SBR in the 1930s. The commercial product developed in Germany came to be initially known as Buna N (1934) and later with the beginning of full scale production (1937), it came into the market with the trade-mark 'Perbunan'.

Copolymerization is exclusively done by the emulsion technique. In the earlier years, polymerization was done at around 25–50°C producing what is known as the "hot" nitrile rubbers with characteristic toughness derived as a consequence of the presence of 'gels' and branch units in the polymer chains. Later, easy processing grades with predominantly linear structures and having little gel content came into production by doing the copolymerization at a low temperature, say 5°C and the products are better known as "cold" nitriles. Acrylonitrile content of commercial NBR products mostly range between 25% and 40%. Relatively high nitrile grades are more costly, more oil and solvent resistant, more resistant to gas permeation and have inferior low temperature flexibility. The design of polymerization recipe including monomer ratio, nature and amount of emulsifiers and modifiers and temperature of polymerization are important factors in NBR production.

Nitrile rubbers resist mastication to some extent and cannot be plasticized by milling in the same manner as for natural rubber and hence, the plasticity of NBR as produced and usually expressed in Mooney units has a profound effect on its processing characteristics. Low Mooney NBR absorb fillers and plasticizers more readily than high Mooney grades. The hot nitriles require mastication on cold mills using a light nip between the rolls. The cold nitriles require little premastication and develop relatively less heat during mixing. Cooling of mills and internal mixers are necessarily done to prevent scroching during mixing of the vulcanizing agents.

Dispersion of ground sulphur is far from satisfactory. Coated sulphur (e.g., magnesium carbonate coated sulphur) may be used to minimize or overcome this problem. Ester and polymeric plasticizers are quite commonly used in NBR to improve processability and to modify such physical properties as low temperature flexibility, resilience, hardness, etc. In open mill, mixing is best accomplished with cold premastication followed by a sequence of addition of zinc oxide, stearic acid, sulphur, filler and plasticizer. The addition of the last two ingredients should be so controlled as to maintain a continuous band on the mill. When internal mixers are used, both sulphur and accelerator are better added in a second mixing cycle or on an open mill in the subsequent step because of considerable heat development.

Both carbon black and non-black fillers are used. NBR is compatible with PVC and many important properties such as good ozone and weathering resistance, higher abrasion and oil resistance, improved flame resistance (if suitably plasticized) may be derived from NBR–PVC blends on vulcanization. NBR is extensively used in making gasoline hoses, lining for fuel storage tanks, and in wires, cables and sheaths (particularly NBR–PVC blends) for use in oils-fields and oil installations, etc. NBR also finds special use in adhesive formulations and in the latex form as impregnating agents for paper, textiles and leather. PVC-nitrile blends may be produced in a wide range of bright permanent colours and are generally used where coloured rubber-like materials of good weathering resistance are required. Good abrasion resistance has also led to the use of blends in footwear soling.

9.3.5 Butyl Rubber (IIR)

Isobutylene can be commercially polymerized to high molecular weight poly-isobutylene using cationic catalysts such as boron trifluoride at a temperature of –80°C. Polyisobutylene is rubbery in nature but it cannot be vulcanized like NR using conventional curing systems (sulphur/accelerator) because of absence of unsaturation in the chain molecules. However, the problem of vulcanizability was solved by allowing small amount of a diene monomer, e.g. isoprene to copolymerize with isobutylene to produce what is now known as the isobutylene–isoprene rubber (IIR) or Butyl Rubber. The industrial copolymerization is also conveniently done using a cationic initiator system such as BF_3 or $AlCl_3$ in a temperature range of –65 to –80°C and taking a monomer mixture of isobutylene with 2–5% isoprene in methyl chloride diluent. The polymerization reaction is fast and exothermic and, therefore, cooling and maintenance of the low reaction temperature is of utmost importance. The rubber formed in the reactor in the form of fine particles is collected, mixed and agitated vigorously with water to remove solvent and unreacted monomers, further mixed with an antioxidant in the wet condition, dried and then baled. Isoprene has a retarding effect on the copolymerization and it leads to lowering of molecular weight of polyisobutylene. There is little scope for using high proportions of isoprene (> 5%), as in that event polymers of relatively low molecular weight and inferior qualities are obtained. Low unsaturation content in butyl rubbers makes them relatively slow curing.

Polyisobutylenes of low degrees of polymerization are liquids. Much in contrast with NR, polyisobutylene (PIB) and butyl rubber (IIR) do not crystallize with lowering of temperature and they retain flexibility even below –40°C. The T_g of PIB is around –65°C. Butyl rubbers have notably high pure-gum tensile strength (220–250 kg/cm^2), apparently showing self-reinforcing characteristics. The methyl side groups appreciably hinder elastic movements, thus giving rise to high internal viscosity and consequently, high heat build-up and poor resilience. Butyl rubber, having very low unsaturation and abundance of rigid methyl side groups in the molecular chains, is much less sensitive to oxidative degradation and ozone attack than other hydrocarbon rubbers and at the same time, it shows good resistance to heat, abrasion and chemicals. The densely packed structure of the linear hydrocarbon chains makes butyl rubber exhibit unusually low gas permeability.

Butyl rubbers show crystallizing tendencies only on stretching to quite high elongations. They have poor compatibility with most diene and related rubbers and their low unsaturation stands in the way of satisfactory covulcanization.

A distinctive feature of butyl rubber compounding relates to the favourable heat treatment of the rubber-filler mixture using carbon blacks or non-black reinforcing fillers. Such treatment of the butyl-filler mixture is done at a temperature of about 200°C in presence of a promoting chemical (0.5 phr) such as *p*-nitrosobenzene (polyac), *p*-nitrosophenol (butylac) or N-nitroso-N-methyl-*p*-nitrosoaniline (Elastopar) in an internal mixer. The reactions involved lead to very limited curing of the rubber chains and they may even persist during subsequent processing and vulcanization giving rise to improved tensile and dynamic properties. Vulcanizations of butyl rubber are more satisfactorily done using 2–6 phr of a dioxime (quinone dioxime or dibenzoyl dioxime) along with nearly 6 phr of red lead, than by using conventional sulphur-accelerator system. Vulcanization with phenol-formaldehyde resins has been found to produce vulcanizates with improved heat resistance and aging characteristics.

A major part of butyl rubber produced is consumed in the making of inner tubes for tyres and it far excels over natural rubber in this application, being much less permeable to air. Even though, with the advent of tubeless tyres, fewer tubes are manufactured, but butyl rubber consumption has not suffered as it has largely replaced natural rubber in this application. Butyl rubber vulcanizates, however, become hard at low temperatures prevailing in the winter of many countries. In tyre and inner tube applications this poses a problem which has been overcome by incorporating requisite doses of mineral oil (5–25 phr) in the compound. Its excellent heat and aging resistance make butyl rubber particularly suitable for heating bags of all types. It is also used in the manufacture of selected mechanical goods, cables, pressure-sensitive adhesive tapes, caulking compounds, etc.

Chlorobutyl or bromobutyl rubbers are structurally similar to butyl rubbers with the difference that a hydrogen atom in the isoprene units has been substituted by a halogen atom (chlorine or bromine respectively) giving, on an average, 1.2–1.3% of chlorine in chlorobutyl and 2.7–2.8% of bromine in bromobutyl rubber with practically little loss of unsaturation. Curing of these halobutyl rubbers can be conveniently done following conventional sulphur cure or using quinonoid cure (quinone dioxime), resin cure (phenolic resins), amine cure (polyamines) or simply metal oxide cure (zinc oxide, red lead) as applicable to butyl rubber.

The halobutyls are readily blended, adhered and covulcanized with other diene rubbers. Chlorobutyl rubber finds use for making air bags, steam hose, heat resistant conveyor belts, lining of tubeless tyres and for white side walls. Brombutyl rubber is faster curing; it is added to butyl rubber mixes to achieve rapid cures and it is used to make engineering objects required to resist higher temperatures.

9.3.6 Ethylene–Propylene–Diene Terpolymer (EPDM)

The ethylene–propylene rubbers (EPR) are binary copolymers of ethylene and propylene (*See* Chapter 8) and as saturated polyolefin rubbers, they are not suited for conventional sulphur/accelerator cure. To overcome this problem, EPDM was developed. Rubbers of the EPDM type are terpolymers formed by copolymerization of a nonconjugated diene as a third monomer in small proportions (up to 5%) along with ethylene and propylene. The dienes commonly used are dicyclopentadiene (DCPD) and ethylidene norbornene (ENB). ENB, yielding terpolymers of faster rates of cure, is preferred for making the terpolymers.

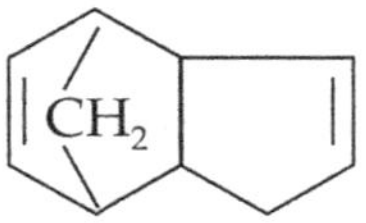
DCPD

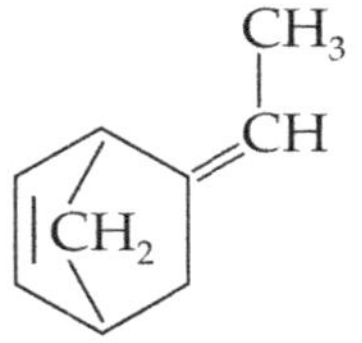
ENB

The EPDM rubbers, like other copolymer diene rubbers, show little crystallization on stretching and they require compounding with appropriate (high) dose of reinforcing fillers without which they appear poor in physical properties. Considerable oil-extension is possible and this may lead to low-cost compounds. EPDM rubbers are readily vulcanized by conventional sulphur/accelerator cure systems.

Advantages of EPDM rubbers are low density (0.85 g/cm^3), outstanding storage stability, moderate tear strength and good heat stability. Resistance to oxygen or ozone attack is good. EPDM also exhibits outstanding resistance to high pressure steam. The processing principles developed for butyl rubbers, viz. heat treatment during compounding in an internal mixer with or without the aid of promoters are also suited for EPDM. EPDM finds limited application in the manufacture of cables, tyres, coated fabrics and special moulded articles. EPDM, like EPR, is curable by peroxides.

9.3.7 Polysulphide Rubber (PSR)

The polysulphide rubbers (PSR), commonly known in the trade as "Thiokols" and developed as special purpose rubbers are generally produced by the step-growth polymerization of selected dichloroalkanes and sodium polysulphide.

$$nRCl_2 + nNa_2S_x \longrightarrow -\!\!(R - S_x)\!\!-_n + 2\ nNaCl \qquad (9.3)$$

The important variables are the nature of R and the values of x and n. S content may vary from 40 to 84%. The polymer from ethylene dichloride $(CH_2)_2 \cdot Cl_2$ and sodium tetrasulphide may be represented by the structure:

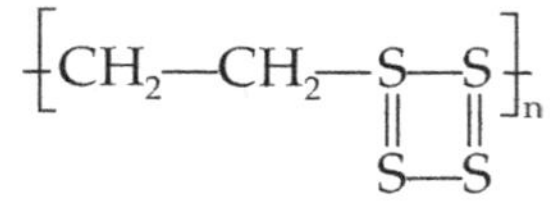

With four sulphur atoms in the repeating units, polymers of good rubbery characteristics are generally obtained; for only two sulphur atoms constituting the interunit links, dihalides having a minimum of four methylene links are required to give products of good rubbery properties. Ethylene dichloride and dichloroethyl formal are commonly used in thiokol rubber preparation.

The polysulphide rubbers have good resistance to oxygen and ozone attack. Their gas impermeability and oil or solvent resistance can be considered to be excellent. The solid polymers are used almost exclusively in areas demanding good solvent/oil resistance. Processing characteristics and mechanical properties of thiokol rubbers are very poor and they are characterized by disagreeable odours. Major applications of thiokols include tanks/hoses for gasoline and gaskets and diaphragms.

Different curing agents are used for curing thiokols. Combination of zinc oxide and stearic acid with organic accelerators such as MBTS, TMTD, etc., or with *p*-quinone dioxime are effective curatives without sulphur. Organic or metal peroxides, other oxidizing agents and metal oxides are also effective. Liquid polymers bearing terminal thiol groups are conveniently cured with metal peroxides such as lead peroxide (PbO_2) at room temperature, though over a long time period (1 day at 30°C).

$$\begin{matrix} R\text{—}SH \\ R\text{—}SH \end{matrix} + PbO_2 \rightarrow \begin{matrix} R\text{—}S \\ | \\ R\text{—}S \end{matrix} + PbO + H_2O \qquad (9.4)$$

They are good in sealing and impregnating applications. Carbon black loading of fairly high degree (40–60 phr) is essential for improvement in tensile strength. The liquid thiokols burn intensely generating large volumes of gas when mixed with strong oxidizing agents.

9.3.8 Polyacrylic Rubber or Acrylate Rubber (ACR)

Poly(ethyl acrylate) is soft and rubbery in nature. Copolymers of ethyl acrylate (95%) and 2-chloroethyl acrylate or 2-chloroethyl vinyl ether (cure site monomer) have been commercially developed and the products are known as polyacrylic rubber or acrylate rubber. These rubbers are suitably cured using aliphatic linear diamines and polyamines. Cross-linking apparently occurs by HCl elimination and intermolecular link-up through the diamines or via ester hydrolysis and establishment of intermolecular amide linkages. Small amount of sulphur is used as an anti-aging additive. The cured rubber is particularly useful as hoses, seals and gaskets. Reinforcing carbon blacks are used as fillers. For pale shades, siliceous fillers are used. The rubber has good resistance to oils and to ozone and it may be used over a wide temperature zone (–40 to nearly +200°C). The uncured polyacrylic rubbers are soluble in ketones, esters and alcohol-ester mixtures. The cured polyacrylic rubbers are better than nitrile rubbers in heat and oil resistance. Their resistances to ozone attack and to sunlight and weathering are good. A terpolymer (AEM) of methyl acrylate, ethylene and a cure site monomer, known in the trade by the name Vamac, is of more recent development (*See* Table 9.3).

9.3.9 Fluorocarbon Rubber (FKM)

Fluorocarbon rubbers or elastomers are copolymers of vinylidene fluoride and chlorotrifluoroethylene (50 : 50 or 30 : 70 ratio). Better products are obtained by copolymerization of vinylidene fluoride and hexafluoropropylene ("Viton" elastomers from Du Pont). They are usually cured with amine type curatives in presence of a metallic oxide (litharge or calcined magnesia). Curing is apparently effected by the elimination of hydrogen fluoride.

The fluorocarbon (copolymer) elastomers are prepared by batch or continuous process following the emulsion polymerization technique. The latex obtained is coagulated by hydrochloric acid and the polymer is washed and dried. The Vitons are normally soluble in lower ketones and the doughs formed are suitable for spreading over glass cloth to produce coated fabrics useful as oil seals and gaskets with a long service life at high temperatures (≥ 100 h at nearly 400°C, > 5000 h at 200°C).

The fluorocarbon rubbers exhibit excellent resistance to oils, lubricants, hydrocarbon solvents, mineral acids and chemicals and to heat; in these respects they are superior to almost all other commercial rubbers. The fluorocarbon rubbers can be suitably compounded to give vulcanizates of tensile strength of about 200 kg/cm^2 and elongation at break in the range of 200–300%. They are flame resistant and they exhibit outstanding resistance to oxygen and ozone attack. Their good low temperature flexibility makes them advantageously useful at low temperatures up to –30°C. The application of fluorocarbon rubbers is limited to only special or unusual service conditions where other rubbers are altogether unsuitable and where their high cost is not a hindrance. The applications include seals, gaskets and diaphragms, fire-resistant and protective clothing from coated fabrics, and wire and cable insulation.

A class of rubber that is still more resistant to oils, solvents, chemicals, acids, bases, strong oxidizing agents (concentrated hydrogen peroxide) and to oxygen and ozone attack is the so-called *nitroso rubber*. A typical product, an alternating copolymer of trifluoronitrosomethane and tetrafluoroethylene of high molecular weight, has the following chemical structure:

$$\left[\underset{\substack{| \\ CF_3}}{N} - O - CF_2 - CF_2 \right]_n$$

It has a high specific gravity (1.9). Just like the fluorocarbon rubbers, the nitroso rubbers are also vulcanized by diamines or polyamines or their derivatives. The nitroso rubbers are so inert and resistant that they do not ignite even in pure oxygen.

9.4 Thermoplastic Elastomers (TPE)

Development of thermoplastic elastomers (TPE) has narrowed the basic difference between the processing of thermoplastics and elastomers or rubbers for many non-tyre products. The TPEs make useful products for which the tensile and set properties are not much critical, such as in many automotive parts, footwear, cables, sealants and adhesives, hoses, coated fabrics, tubings and sheetings. Their light weight, flexibility, impact resistance and weathering resistance make them useful in many of these applications.

The thermoplastic elastomers are processed like reusable and reprocessable thermoplastics and under service conditions, they behave like vulcanized rubbers. The useful mechanical and elastic properties of conventional vulcanized rubbers are attributed to the chemical cross-links established between the rubber chain molecules during vulcanization to produce a space network structure. In the TPEs, the network is basically formed through thermally labile physical cohesive forces between specific segments of different polymer chains but not really through intermolecular chemical linkages. At elevated and processing temperatures, the thermolabile physical bonds weaken and finally break up, permitting flow under shear and thus enabling them to be moulded or formed like a conventional thermoplastic material. At and around ambient temperatures, the TPEs exist in two phases; the soft rubbery phase forms the continuous matrix in which the hard resinous phase remains dispersed in discrete domains. The volume fraction of the rubbery matrix is usually higher than that of the hard thermoplastic domains which materially act as cross-links and as stiffening fillers or points of reinforcements.

The early developed thermoplastic elastomers have been the thermoplastic polyurethanes[3,4] based on reactions between a hydroxy-terminated polyether or polyester (molecular weight: 1000–3000) a diisocyanate and a low molecular weight diol in appropriate molar proportions to yield block copolymers. The polyether or polyester blocks constitute the soft or rubbery domains, and linked with them through urethane linkages are the hard segments or blocks formed by reaction between the isocyanate (diphenyl-methane 4,4′-diisocyanate) and the small diol (1,4-butanediol) molecules. The copolyester from repeat units of tetramethylene terephthalate and poly(tetramethylene ether) terephthalate, commonly known as Hytrel in the trade, is also an early developed TPE (*See* Table 9.3).

However, the most widespread developments in TPEs have taken place with polystyrene–poly(1,3-diene) triblock copolymers such as the one having a middle block of 1,4-*cis*-polybutadiene (T_g = –85°C) forming the continuous matrix and two end blocks of polystyrene (T_g = +85°C), the polystyrene end blocks cluster to form the discrete hard domains. They are commonly known as the S-B-S thermoplastic elastomers.[5] They are elegantly prepared by anionic polymerization using such catalysts and conditions as would produce living polymers. A three-stage or two-stage sequential process may be used depending on the functionality of the initiator, that is on the number of reactive sites on the chain propagating species. For one-end chain carrier system the monomers are polymerized in three stages in the sequence: styrene, followed by butadiene, followed by styrene again. For both-end chain carrier system, the polymerization is completed in two stages using the monomers in the sequence: butadiene followed by styrene. Isoprene may be used in place of butadiene to give a middle block of polyisoprene to produce S-I-S triblock copolymers. A schematic representation of the triblock copolymers at ambient temperatures is shown in Fig. 9.1

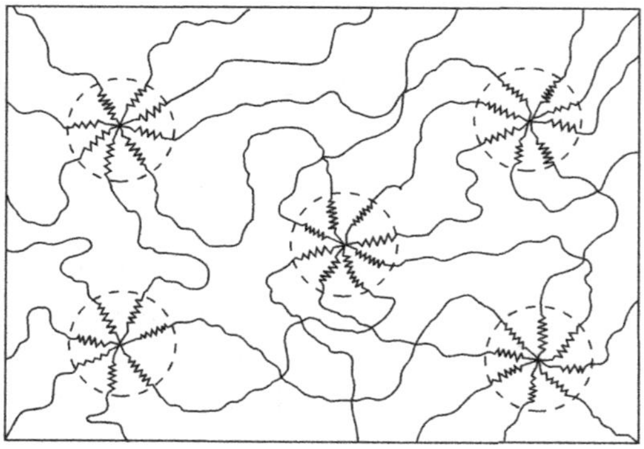

Fig. 9.1 ***Schematic representation of triblock copolymers (SBS type) showing two end blocks of styrene units and a middle block of butadiene units in each macromolecule***

A developing trend in TPEs gaining in popularity on consideration of cost and other advantages relates to products based on polyolefins—homopolymers, copolymers and blends. S–B–S or S–I–S type triblock copolymer TPEs remain susceptible to oxygen and ozone attack due to the presence of the unsaturations in the polydiene middle blocks. Polyisoprene and polybutadiene middle blocks, however, age differently, the former undergoing prominent chain-scission degradation while the latter tends to change into a cross-linked structure. The introduction of ethylene–propylene or ethylene–butylene copolymers of appropriate composition as the middle blocks improves the aging resistance and heat resistance of the styrenic triblock TPEs quite significantly.

Thermoplastic elastomers can be prepared from suitable blends of an elastomer (EPDM, EPR, NR) and a thermoplastic (PP, HDPE) by melt mixing in an internal mixer under high shearing action.[6,7] Crystallinity of the polyolefin (homopolymer) part of the blend effectively act as cross-links and points of reinforcements. Carbon black and other fillers, which impart reinforcing effects in conventional rubber compounds on vulcanization, do not really play a prominent role in improving the strength of the thermoplastic elastomer blends beyond a general stiffening of the compound and reducing the moulding shrinkage. Olefinic TPEs exhibit higher melt-viscosity than the other TPEs, necessitating the use of higher temperatures and pressures during their processing and fabrication. They also exhibit prominent non-Newtonian flow behaviour, the melt viscosities being highly sensitive to rate of shear but not as sensitive to variations of temperature beyond the melting point of the polyolefin. The hardness values of different thermoplastic elastomers range between those of typical rubbers and typical plastics (Fig. 9.2). Structure of repeat units and some physical properties of several useful rubbers are given in Table 9.3. The properties indicated should be viewed as approximate.

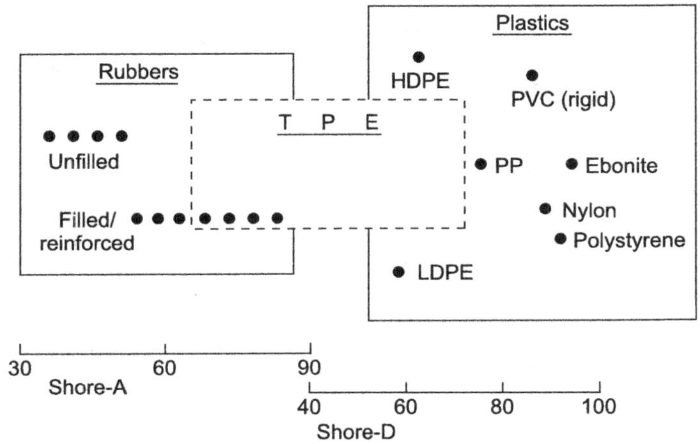

Fig. 9.2 *Hardness range of rubbers, thermoplastic elastomers (TPE) and plastics*

Table 9.3 ***Structure of repeat units and some physical properties of several useful rubbers***

S. No.	Specific Gravity	T_g °C	Continuous Use Temperature, °C	Solubility Parameter $(cal/cm^3)^{1/2}$	Tensile Strength psi	Elongation %	Modulus at 300% Elongation, psi	Tear Resistance	Abrasion Resistance	Oxidation and Ozone Resistance	Compression Set
(1)	(2)	(3)	(4)	(5)	(6)	(7)	(8)	(9)	(10)	(11)	(12)
1.	Natural Rubber (NR) $—(CH_2—C(CH_3)=CH—CH_2)_n—$ (98% 1,4 *cis*-polyisoprene)										
	0.93	–73	85	8.25	4500	600–800	2500	Good	Fair	Poor	Fair-good
2.	Synthetic 1,4-*cis*-polyisoprene (IR), (92–96%, 1,4 *cis*-structure) $—(CH_2—C(CH_3)=CH—CH_2)_n—$										
	0.93	–70	85	8.25	4500	600	2500	Good	Fair	Fair	Fair
3.	Polybutadiene rubber (BR), $—(CH_2—CH—CH—CH_2)_n—$ (> 92% 1,4-*cis*-polybutadiene)										
	0.93	–80	70	8.20	Poor-fair	Poor-fair	Poor	Excellent	Excellent	Fair	Good
4.	Styrene-butadiene rubber (SBR), $—(CH_2—CH(C_6H_5))_x—(CH_2—CH=CH—CH_2)_y—$ (25%) (75%)										
	0.94	–60	80	8.30	3000–4000	500–600	2000	Low–poor	Good	Fair	Poor–fair
5.	Butyl rubber (IIR), $—(CH_2—C(CH_3)_2)_x—(CH_2—C(CH_3)=CH—CH_2)_y—$ (97%) (3%)										
	0.92	–80	100	7.60	3000	400	1000	Excellent	Fair	Good	Fair
6.	Polychloroprene (CR) $—(CH_2—C(Cl)=CH—CH_2)_n—$ (overwhelmingly 1,4-*trans* structure)										
	1.23	–50	95	9.26	4000	800	1000	Good	Good	Good	Fair

Contd.

Table 9.3 (Contd.)

S. No.	Specific Gravity	T_g °C	Continuous Use Temperature, °C	Solubility Parameter $(cal/cm^3)^{1/2}$	Tensile Strength psi	Elongation %	Modulus at 300% Elongation, psi	Tear Resistance	Abrasion Resistance	Oxidation and Ozone Resistance	Compression Set
(1)	(2)	(3)	(4)	(5)	(6)	(7)	(8)	(9)	(10)	(11)	(12)
7.	Butadiene-acrylonitrile rubber (NBR), $-(CH_2-\underset{\mid}{CH}(Cl))_x-(CH_2-CH{=}CH-CH_2)_y-$ 30% 70%										
	1.00	–30	100	9.8	2500	550	1500	Poor-fair	Excellent	Good	Good
8.	Chlorosulphonated polyethylene (CSP) $-(CH_2-CH_2)_x-(CH_2-CH(Cl))_y-(CH_2-CH(SO_2Cl))_z-$ [35% chlorine]										
	1.20	–25	130	9.10	3000	300–500	—	Fair	Excellent	Excellent	Fair
9.	Ethylene–propylene–diene (EPDM) 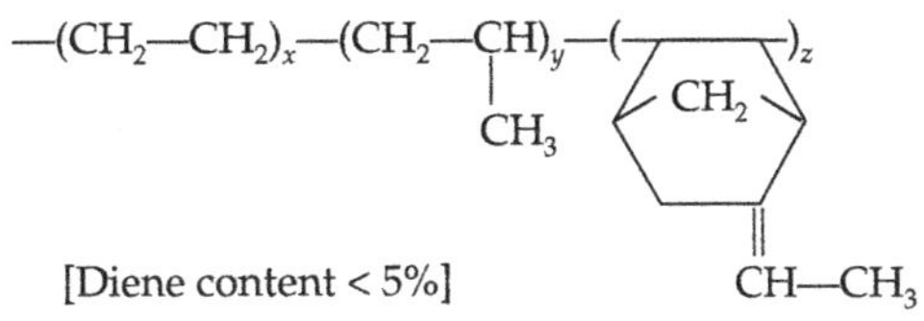 [Diene content < 5%]										
	0.86	–60	140	8.0	3000	500–600	—	Good–excellent	Excellent	Excellent	Good
10.	Fluorocarbon copolymer (FKM) $-(CH_2-CF_2)_x-(CF_2-CF(CF_3))_y-$ [65% fluorine]										
	1.85	–25	200	8.6	1500–2000	600	300	Fair–good	Poor	Excellent	Good-Excellent

Contd.

Table 9.3 (Contd.)

S. No.	Specific Gravity	T_g °C	Continuous Use Temperature, °C	Solubility Parameter $(cal/cm^3)^{1/2}$	Tensile Strength psi	Elongation %	Modulus at 300% Elongation, psi	Tear Resistance	Abrasion Resistance	Oxidation and Ozone Resistance	Compression Set
(1)	(2)	(3)	(4)	(5)	(6)	(7)	(8)	(9)	(10)	(11)	(12)
11.	Chlorinated polyethylene (CPE), 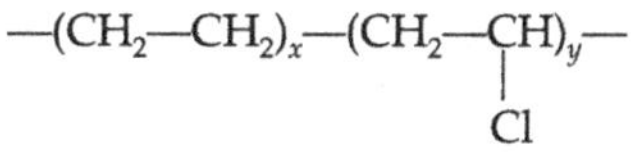 [up to 40% chlorine)										
	1.23	—	100	9.1	3000	500	1000	—	—	Good–excellent	Fair
12.	Silicone rubber 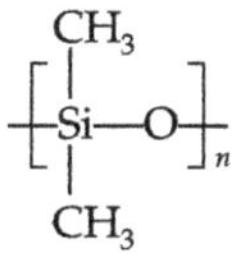										
	0.98	120	>150	7.3	1500	600	—	Poor	Poor–fair	Excellent	Excellent
13.	Polyester–ether rubber, (Hytrel) (thermoplastic elastomer) 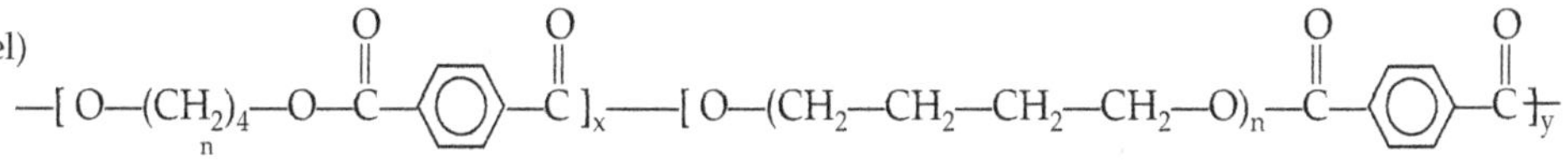 Stiff segment Soft segment										
	1.20	—	110	9.5	4000–6000	—	—	Excellent	Excellent	Excellent	Poor–fair
14.	Polyacrylic rubber (AEM), 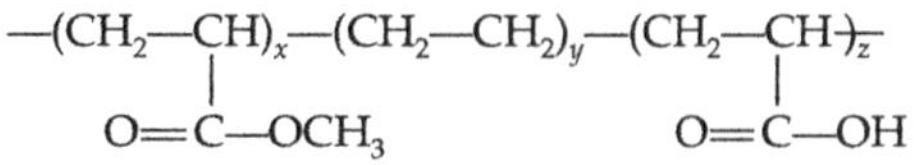 Acrylate Ethylene Cure site monomer										
	1.10	—	170	8.5	2500	400–500	—	Good	Good	Excellent	Good

9.5 Rubber Compounding and Processing Technology

9.5.1 Introduction

No rubber is considered technically useful if its molecules are not cross-linked by a process known as curing or vulcanization. The process of vulcanization is usually associated with two chemical processes taking place simultaneously, cross-linking and chain degradation, though at widely different rates. For natural rubber and many synthetic rubbers, particularly the diene rubbers, the curing agent most commonly used is sulphur. But sulphur curing takes place at technically viable rates only at a high temperature (≥140°C), and heating with sulphur alone leads to optimum curing after nearly 8 h at 140°C using a fairly high dose of sulphur (8–10 phr). Use of metal oxides, such as those of zinc, calcium, magnesium, lead, etc., brings about some advantages with respect to time of curing and improvements in physical properties without much reduction in the sulphur dose. Aniline is considered as the first organic accelerator tried in an attempt to quicken the curing process. Some of the more efficient and less toxic modern organic accelerators of rubber vulcanization are aniline derivatives. Sulphur dose has been substantially lowered with the advent of organic accelerators.

Extensive studies and experimentation have revealed that neither sulphur nor application of heat is indispensable for effecting cross-linking and associated changes in physical properties of rubbers. Peroxides, metal oxides, amines, amine derivatives and oximes have been found to bring about curing of selected rubbers quite effectively. High-energy radiations can bring about effective curing; but high-energy radiation curing has not been developed into a commercial process of even limited acceptability. Selenium and tellurium can substitute sulphur either totally or partly to effect satisfactory curing of diene rubbers. Sulphur monochloride can bring about room temperature or cold curing of diene rubbers.

Improvements in physical and mechanical properties that can be achieved through vulcanization using sulphur/accelerator systems only are rather limited. The needs of imparting colour, stability, resistance to tearing and abrasion, flexibility etc., and of improving processability and mechanical properties necessitate incorporation of a host of additives or compounding ingredients in the rubber by what is commonly known as the mixing or compounding process.

The various compounding ingredients that are intimately mixed with rubber to make what is known as the rubber compound are classified as:

i) Process aids—peptizers, plasticizers, softeners and extenders, tackifiers, etc.

ii) Accelerators
iii) Accelerator activators
iv) Curing agents
v) Antidegradants
vi) Fillers
vii) Colouring matters
viii) Others (retarders, blowing agents)

9.5.2 Mastication and Mixing

Prior to the mixing of the different ingredients, it is helpful, and for natural rubber, absolutely necessary to subject the rubber to a process of breakdown of the molecular chains. If the degradation is effected by subjecting the rubber to high mechanical work (shearing action), the process is commonly known as mastication. Oxygen of air plays a critical role during mastication (*See also* Chapter 6). Mastication and mixing are conveniently done using two-roll mills or internal mixers.

Rubber that has been masticated is more soft and it flows more readily than the unmasticated material. Mastication also allows preparation of solutions of high solid content because of much lower solution viscosity of the degraded rubber. Another consequential effect is that rubber is rendered tacky which means that the uncured rubber sticks to itself readily so that layers of masticated rubber or rubberized fabric can be used to build up articles of suitable thickness without the use of a solvent.

9.5.3 Open Mill

Two-roll open mills like the one outlined in Fig. 9.3, are extensively used in rubber industry. The roll axes are parallel and horizontal and the nip or the gap between the rolls is adjustable. The speeds of the two rolls are usually different, the front roll having a slower speed. For natural rubber mixing, a friction ratio of 1 : 1.2 for the front to back roll may be used. For some synthetic rubbers or highly filled NR mixes, friction ratios close to 1.0 produce good results.

The nip is so adjusted that when the pieces of rubber are placed between the rolls, they are squeezed through the nip to which they are returned by the operator. On repeated passage through the nip under shear, the separate pieces of rubber soon join to form a single mass and band round the front roll and a moving "bank" above the nip. Uniform treatment and mixing is ensured as the operator uses a knife to cut through the band on the front roll, removes the mass in parts from time to time and places it in a new position as the rolls keep on rotating

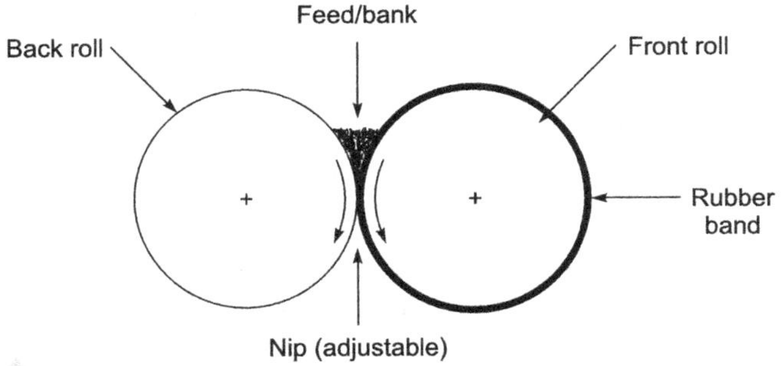

Fig. 9.3 *Section showing the features of a two-roll open mill*

With most rubbers other than natural rubber, mixing of different compounding ingredients may be normally started soon after a uniform band is formed and a bank is obtained. For natural rubber, milling is usually continued to masticate the rubber until the desired plasticity is attained; after the adequate mastication follows the mixing of the compounding ingredients. Various safety devices are attached to the mill to protect the operator and also the equipment. Mixing is effected by charging the different ingredients on to the bank. They are gradually dispersed into and absorbed by the rubber which is cut at intervals rolled over and passed through the nip of the moving rolls as described above. Time and temperature of mastication and mixing have to be controlled or kept uniform for successive compounding to get the desired uniform products from different batches.

Peptizers are added to rubber at the beginning of mastication. They act chemically and hasten the rate of breakdown of rubber chains and are used to increase the efficiency of mastication. They act fast and effectively at temperatures ≥ 65°C. Common peptizers are zinc thiobenzoate, zinc-2-benzamidothiophenate, thio-β-naphthol, etc. Compounding ingredients other than the peptizer are added after the rubber attains the desired plasticity on mastication, as stated earlier. Common process aids are pine tar, mineral oil, wax, factice, coumarone–indene resins, petroleum resins, high-styrene (styrene–butadiene) resins, phenolic resins, etc. Their main effect is to make the rubber soft and tacky so as to make the mixing easy and uniform, particularly when high loading of carbon black or other fillers is necessary. Ester plasticizers (phthalates and phosphates) are used as process aids particularly with NBR and CR. Polymerizable plasticizers such as ethylene glycol dimethacrylate are particularly useful with peroxide curing system. They act as plasticizers or tackifiers during processing and are induced to polymerization by peroxide during cure. Factice is manufactured by treating drying or semidrying vegetable oils with sulphur monochloride (cold or white factice) or by heating the oils with sulphur at 140–160°C (hot or brown factice). They are commonly used as rubber substitute in the same way as rubber reclaim. Use of factice (5–30 phr) gives a better rubber mix for the purpose of extrusion, and allows efficient mixing and dispersion of powdery ingredients.

9.5.4 Internal Mixers

An internal mixer is made up of two horizontal rotors with protrusions placed in a casing. It is provided with a feed door and hopper at the top and a discharge door at the bottom. An early developed internal mixer is the Banbury mixer (Fig. 9.4), an important and novel feature of which is a vertical ram to press the mass into contact with the two rotors. Other major machines in use are the Shaw intermix and the Baker–Perkins shearmix. Both the Banbury type and the intermix mixers have passed through subtle refinements and modifications at different points of time. Mechanical feeding and direct oil injection in measured doses into the mixing chamber through a separate oil injection port are important features of modern internal mixers. Higher rotor speeds, and in Banbury, also the higher ram pressures, are used for speedy output. In Banbury, the rotors run at different speeds while in the intermix the rotor speeds are equal but the kneading action between the thicker portion of one rotor and the thinner portion of the other produces a frictional effect.

The rubber or the mix is worked between the two rotors and between each rotor and the body of the casing. Thus, mastication takes place over a wide area, unlike in an open mill, where it is restricted only in the area of the nip between the two rolls. Personal elements are reduced to a bare minimum in the internal mixers through automatic loading and maintaining a calculated, predetermined schedule. The operation of internal mixers is power-intensive and a given job is performed at a much higher speed and over a much shorter time than on a two-roll open mill. However, the major vulcanizing agent such as sulphur is usually added later on a two-roll mill to eliminate scorching difficulties. Even if this is not practised, the mix is, of necessity, passed through a two-roll mill after being discharged from the internal mixer in order to convert it from irregular lumps to a sheet form for convenience in subsequent processing.

9.5.5 Reclaimed Rubber

Use of reclaimed rubber in a fresh rubber mix amounts to waste utilization. Reclaimed or regenerated rubbers are recognized as raw materials having some processing and economic advantages to make them highly valued in rubber compounding.[8]

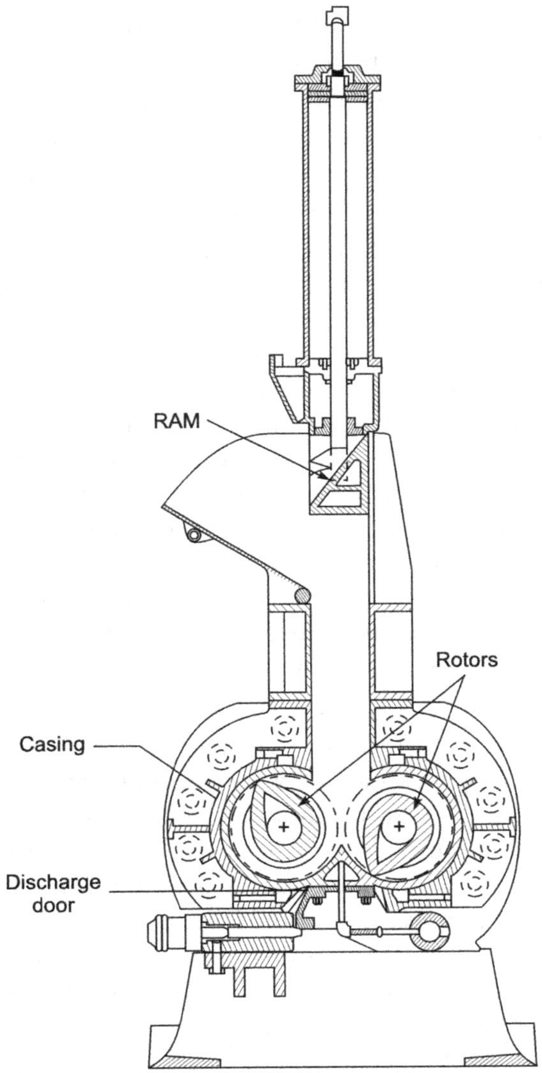

Fig. 9.4 *Section showing the features of a Banbury type internal mixer*

Waste vulcanized rubber is not normally processable. Application of heat and chemical agents to ground vulcanized waste rubber leads to substantial depolymerization whereby regeneration of the rubber compound to a soft, plastic processable state is effected. Rubber so regenerated for reuse is commonly known as reclaimed rubber or simply as 'reclaim'. The reclaim can be easily revulcanized.

A good reclaiming process must not only turn the rubber soft and plastic but must also remove reinforcing cords and fabrics that may be present.

There are a number of commercial processes for rubber regeneration: (a) alkali digestion process, (b) neutral or zinc chloride digestion process, (c) thermal process, and (d) reclaimator process. Worn out tyres and scraps and trimmings of other vulcanized products constitute the raw material.

For processes (a) and (b), debeaded tyres and scraps, cut into small pieces, are ground in grinders developed for the purpose. Mixed with a peptizer, softener and heavy naphtha, the crumbs are then charged into spherical autoclaves with requisite quantities of water containing caustic soda for process (a), or zinc chloride for process (b). Steam pressure and amount of air or oxygen in the autoclave greatly influence the period necessary for reclaiming. On completion of the process, the pressure is released, the contents in the autoclave discharged in water, centrifuged, pressed to squeeze out water and dried. They are finally passed through a two-roll mill for a refining process during which mineral fillers and oils may be added to give a product of a standard specific gravity and oil extension. In the thermal process, the debeaded tyre cut into pieces and scrap rubbers are straight away reclaimed in an autoclave using superheated steam (250–270°C) in about 6–8 h. This process is not suitable if SBR is present.

In the autoclave digestion processes, the textile is destroyed and mostly lost. The caustic soda or zinc chloride used is not recovered. Each process is a batch process and requires large volumes of water and steam. The reclaimator process is more attractive in this context. Ground scrap is mechanically treated to remove the textile material. Crumbs so obtained are mixed with mineral oil and a peptizer and the mixer is then fed into the reclaimator which acts like a high-shear extruder. Thermal depolymerization (about 180°C in presence of a peptizer) turns the rubber into reclaim that issues from the machine continuously. The whole regeneration is a dry process and may be completed in about 30 min.

Reclaimed rubber contains all the fillers present in the original scrap or waste rubber. It is characterized by less heat development during mixing and processing in comparison with mixing and processing of fresh rubber. The reclaim shows very good aging characteristics. The use of reclaim, only in part, is advantageous not on consideration of physical and mechanical properties but essentially for smooth processing and reduced cost.

9.5.6 Fillers

A major compounding ingredient is the filler. Fillers may be used from 50 phr to as high as 100–120 phr or even in higher proportions. Major fillers used in the rubber industry are classified as: (a) non-black fillers such as China clay, whiting, magnesium carbonate, hydrated alumina, anhydrous or hydrated silicas and silicates including those in the form of ground mineral such as slate powder, talc or French chalk, etc., and (b) carbon blacks.

Carbon blacks are essentially elemental carbon and are produced by thermal decomposition or partial combustion of liquid or gaseous hydrocarbons to carbon and hydrogen. Most of the carbon black used in the rubber industry is made by the furnace process. Quite a variety of grades of furnace blacks are available, e.g. SAF (super abrasion furnace) ISAF (intermediate super abrasion furnace), HAF (high abrasion furnace), GPF (general purpose furnace), HMF (high modulus furnace), SRF (semi-reinforcing furnace), FEF (fast extrusion furnace), SCF (super conducting furnace), CF (conductive furnace), etc. A few grades of channel blacks (HPC, MPC or EPC corresponding to hard, medium or easy processing channel) and thermal blacks (FT and MT referring to fine thermal and medium thermal) are also in use. Use of carbon black fillers brings about prominent reinforcements of the NR vulcanizates which then exhibit much improved modulus, tear strength, abrasion resistance and hardness. In the case of some synthetic rubbers, the use of carbon blacks or other reinforcing fillers is essential to attain physical properties of desirable levels.

Nearly 90% of the total carbon black produced goes to the rubber industry and nearly 75–80% of this is consumed in the manufacture of tyres and related products. The criterion for use of a black is its behaviour in rubber. The *p*H of carbon black has a profound influence on vulcanization. Acidic blacks (channel blacks) tend to retard the curing process while alkaline balcks (furnace blacks) produce a rate enhancing effect in relation to curing, and may even give rise to scorching effects. Another important factor is the particle size. Smaller the particle size, higher the reinforcement but poorer the processability in view of longer time needed for dispersion and greater heat developed during mixing. Blacks of the smallest particle size and most suited for use in ink formulations are unsuitable for use in rubber compounding.

For carbon black fillers, structure, particle porosity and overall physico-chemical nature of particle surface are important factors in deciding cure rate and degree of reinforcement attainable. The term "structure" is used to represent the clustering together and entanglement of fine carbon particles into long chains and three dimensional aggregates (Fig. 9.5). In 'non-structure' blacks the aggregates are almost non-existent. High-structure blacks produce high-modulus vulcanizates. High shear forces applied during mixing break the agglomerates down to many active free radical sites which bind the rubber, thereby leading to reinforcements on modulus. Particles of higher porosity have higher surface area and they tend to absorb more of other active ingredients such as accelerators and activators, resulting in measurable cure-retardation. The role of physico-chemical nature of the black particle surface in rubber reinforcement is yet to be fully understood.

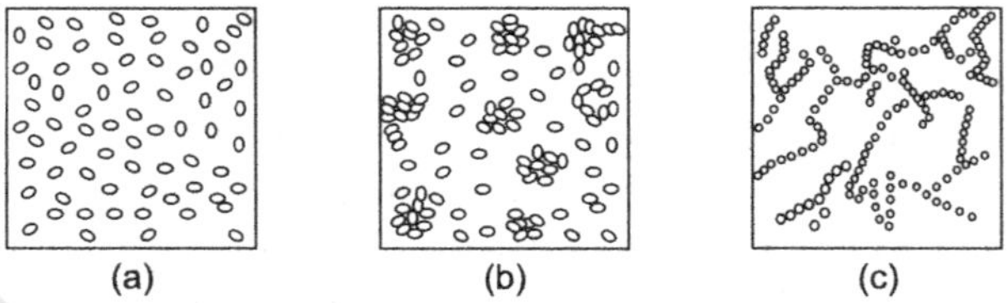

Fig. 9.5 *Schematic representation of structure of carbon blacks: (a) low structure index, (b) medium structure index and (c) high structure index*

Most of the non-black fillers are of non-reinforcing types and they are added for various objectives, the most important being cost reduction. Finely ground magnesium carbonate and aluminium silicate induce good reinforcing effects. Precipitated silica (hydrated), containing about 10–12% water with average particle size ranging 10–40 nm, produce effective reinforcements and are popular in translucent and coloured products. Precipitated calcium carbonate and those treated with a stearate to give what is known as activated calcium carbonate are used as semi-reinforcing fillers in rubber compounding. Short fibres of rayon, cotton or nylon may be used to stiffen the compound and enhance modulus and resistance to tear and abrasion of the vulcanizates. Some resins such as high styrene resins (styrene butadine copolymer) and phenolic resins (novolac type) mixed with a methylene donor such as hexamethylene tetramine may also be used as reinforcing fillers or additives.

9.5.7 Antidegradants

Major degradation of rubber is due to oxygen and ozone attack. It is therefore, mostly necessary to incorporate antioxidants and antiozonants in rubber compounds. The majority of the commercial antidegradants are of two types chemically:

(i) amine type, representing the staining variety and (ii) phenol type, representing the non-staining variety. The staining varieties cause significant discolouration of the product on exposure to light and weathering conditions. Among the staining varieties those more commonly used are phenyl (β or α) naphthylamines, 4,4′-dialkyl or dialkoxydiphenylamines and N,-N′dialkyl or diaryl-*p*-phenylenediamines. Among the non staining varieties those more commonly used are styrenated phenols, substituted phenols, *o*- or *p*-bridged substituted diphenols, etc. (*See also* Sec. 8.24.3).

9.6 Sulphur Vulcanization

The reaction between rubber (usually the diene rubbers—homopolymers or copolymers) and sulphur is very slow. Until about the first two decades of the 20th century some inorganic oxides (of Pb, Ca, Zn, Mg, etc.) were used to achieve sulphur vulcanization of NR at faster rates at high temperatures. Even then, heating for long hours using at least 8–10 parts of sulphur for 100 parts of rubber was necessary.

9.6.1 Accelerators

The technology of sulphur vulcanization as practised today is radically different and far more efficient and economical. This has been possible through incorporation of small doses of one or more organic substances, commonly known as accelerators. A variety of accelerators are currently available and they are more appropriately classified according to the speed of curing induced in their presence in NR systems. In the order of increasing speed of curing, they are classified as slow accelerator, medium accelerator, semi-ultra accelerator and ultra accelerator. The principal chemical types are guanidines, thiazoles, sulphenamides, dithiocarbamates, thiuram sulphides, xanthates, and aldehyde amines (Table 9.4).

The accelerated sulphur vulcanization is associated with many advantages over non-accelerated vulcanization. Incorporation of about 0.2–2.0 parts of accelerator allows reduction of sulphur dose from 8–10 parts to 0.5–3 parts and effective curing is achieved in a time scale of a few minutes to nearly an hour depending on temperature (100–140°C) and type of the selected accelerator. The low sulphur requirement of the accelerated sulphur vulcanization technology has eliminated bloom (migration of unreacted sulphur to the surface of the vulcanizates) which was a common feature of the earlier non-accelerated technology. Other

Table 9.4 *Accelerators for sulphur vulcanization of rubbers*

Accelerator type and formula	*Chemical name*	*Accelerator activity*
Guanidines		
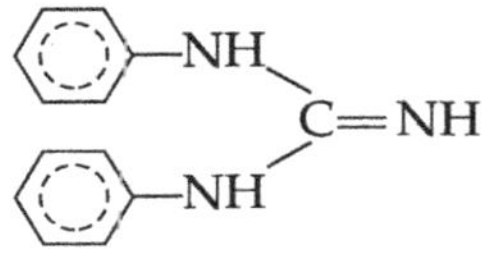	Diphenyl guanidine (DPG)	Medium accelerator
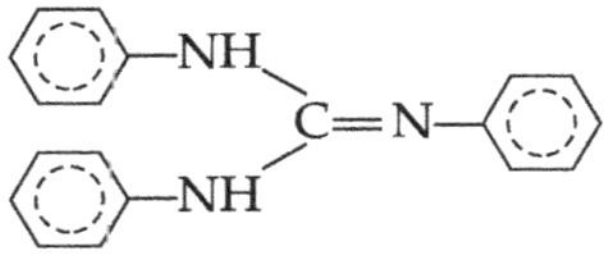	Triphenyl guanidine (TPG)	Slow accelerator
Thiazoles		
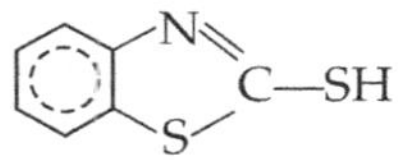	Mercaptobenzthiazole (MBT)	Semi-ultra accelerator
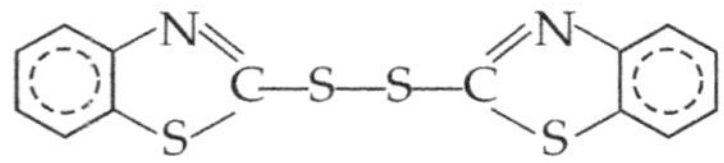	Dibenzthiazyl disulphide (MBTS)*	Semi-ultra (delayed action) accelerator
Sulphenamides		
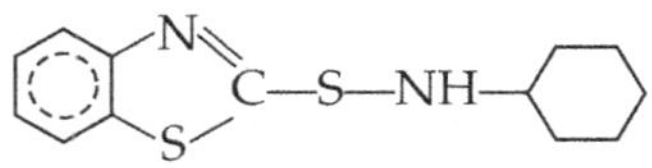	N-Cyclohexylbenzthiazyl sulphenamide (CBS)	Semi-ultra (delayed action) accelerator
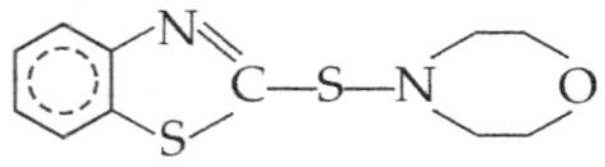	N-oxydiethylenebenzthiazyl sulphenamide (NOBS)	Semi-ultra (delayed action) accelerator
Dithiocarbamates		
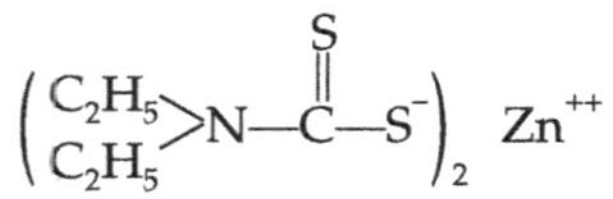	Zinc diethyl dithiocarbamate (ZDC)	Ultra accelerator
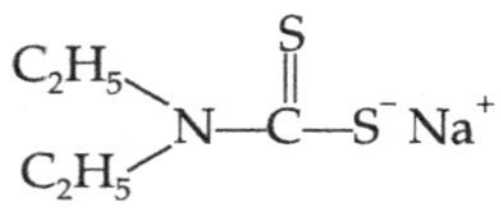	Sodium diethyl dithiocarbamate (SDC)	Ultra accelerator, water soluble (used for latex)

Contd...

Table 9.4 *(Contd.)*

Thiuram sulphides		
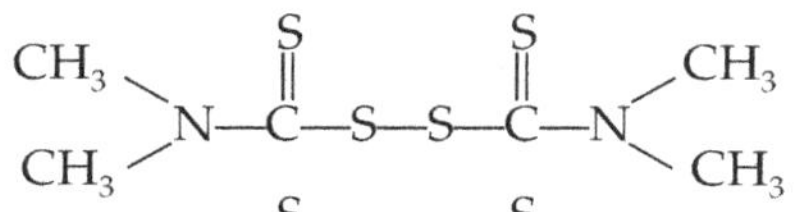	Tetramethyl thiuram disulphide (TMTD, TMT)*	Ultra accelerator
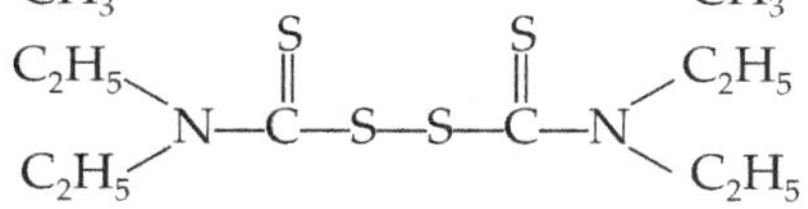	Tetraethyl thiuram disulphide (TETD, TET)*	Ultra accelerator
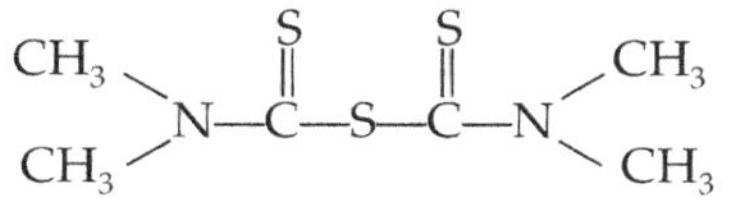	Tetramethyl thiuram monosulphide (TMTM)	Ultra accelerator
Thiophosphoryl sulphides		
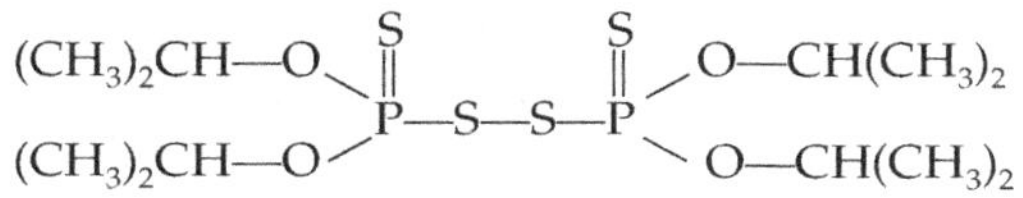	Bis(diisopropyl) thiophosphoryl disulphide (DIPDIS)	Semi ultra accelerator
Xanthates		
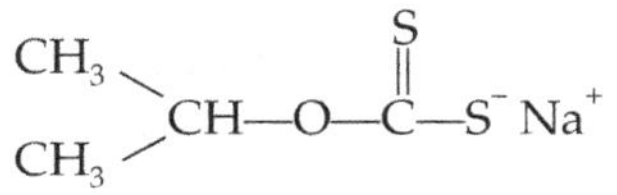	Sodium isopropyl Xanthate (SIX)	Ultra accelerator Water soluble (suited for latex)
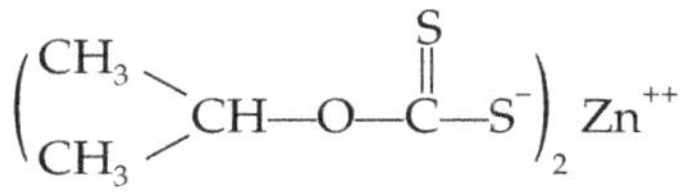	Zinc isopropyl Xanthate (ZIX)	Ultra accelerator
Morpholine sulphide		
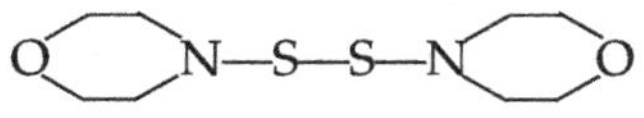	Bis-morpholine disulphide*	Semi-ultra/ultra accelerator

*Sulphur donors

consequences of low sulphur dose in accelerated vulcanization are production of vulcanizates of greatly improved physical properties and good resistance to heat and aging.

It will now be appropriate to examine and analyze the guidelines for appropriate selection of accelerator. The selection would depend on the nature of the rubber taken, the design of the product and the processing conditions.

With progress of vulcanization, the tensile strength and modulus for some systems increase very sharply up to a point, pass through a maximum and then follow a slow decreasing trend. For some other systems, the initial sharp gain in the value of the tensile parameters is followed by what is known as a long plateau region showing practically little change or may be, very slow increasing trend (Fig. 9.6). The phenomenon of decrease in the tensile parameters beyond the maximum point (completion of vulcanization or curing), (Fig. 9.6, Curve 1), usually known as reversion, is commonly observed with NR or IR. Reversion is generally not observed with SBR. The nature of change characteristic of SBR as depicted in Fig. 9.6, Curve 2, is commonly described as "marching cure".

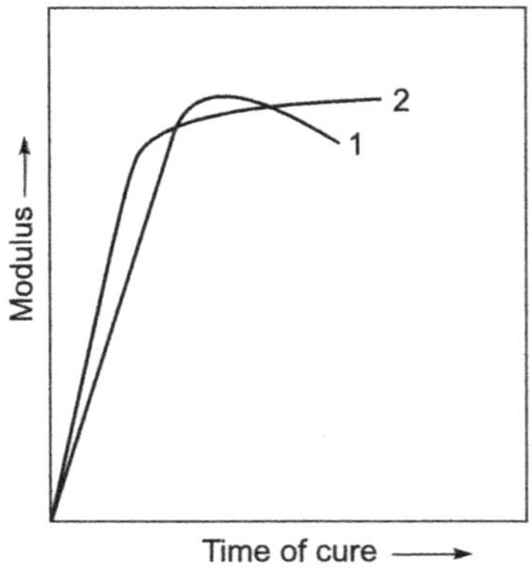

Fig. 9.6 *Plots of modulus vs time of cure showing reversion (curve 1) and marching cure (curve 2)*

A sharp drop in the physical properties such as tensile modulus for heating beyond optimum curing is undesirable; the plateau (or slow marching) effect is, however, desirable and advantageous in view of the fact that minor variations in time and temperature of vulcanization or extending the vulcanization process beyond the optimum curing time into the overcure zone would make practically little difference in or cause no deterioration of the physical properties.

Scorch or premature vulcanization during mixing or compounding and storage of the mix must be avoided. The scorch problem is very acute with ultra or fast accelerators. Rubber stocks are usually bad conductors of heat and, therefore, flow of heat to the interior of a vulcanizing stock from outside is very slow. In thick items, the outer layers, through which the heat passes to the interior of the items, may reach a state of overcuring before the core or interior layers begin to cure. For such thick items, a slow accelerator with a prominent plateau effect is most suitable.

For butyl and EPDM rubbers, having very limited unsaturations, slow accelerators are unsuitable. Fast accelerators should, therefore, be used at high temperatures for good curing at convenient rates. With butyl rubber, characterized by reversion, duration of heating at curing temperatures must be carefully controlled and prolonged heating must be avoided.

The ideal and most preferable accelerator is one that is quite stable during mixing, processing and storage of the mix, but which reacts and decomposes sharply at the high vulcanization temperature to effect rapid curing. Accelerators closely fulfilling these requirements or demands are the delayed action accelerators typical examples of which have been given in Table 9.4.

9.6.2 Retarders

Yet another approach to minimize the hazard of scorching is to incorporate an additive of the "retarder" category. Acids generally function as retarders, the more commonly used retarder acid being phthalic anhydride (up to 2.0 phr). N-nitroso diphenylamine (up to 5.0 phr) is also an effective retarder.

9.6.3 Activators

The spectacular effects of modern organic accelerators in sulphur vulcanization of rubber are observed only in the presence of some other specific additives commonly known as accelerator activators. The activators are usually two-component systems comprising combinations of a metal oxide and a long chain fatty acid. A combination of zinc oxide and stearic acid is known to produce quite satisfactory results and is almost universally used. The primary requirement for satisfactory activation or promotion of accelerator action is good dispersibility or solubility of the activators in rubber. Oxides of some other bivalent metals, such as lead, calcium, magnesium, cadmium may also act as activators in combination with stearic acid. The activators not only enhance the rate of curing but also contribute to the development of desirable properties in the vulcanizates. The importance of the activator can be appreciated from the data given in Table 9.5.

Table 9.5 *Effect of activator on vulcanization*[9]

Time of cure, min.	*Tensile strength, psi*	
	ZnO in parts	
	0.0	*5.0*
15	100	2300
30	400	2900
60	1050	2900
90	1300	2900

Base compound: NR (pale crepe): 100; Sulphur: 3;
Mercaptobenzthiazole (MBT): 0.5;
Temperature of vulcanization: 142°C.

9.7 Theory of Sulphur Vulcanization and Accelerator Action[10]

During cross-linking or curing of rubber with sulphur, the latter reacts and gets chemically bound with the rubber network in different ways. Along with establishment of sulphur cross-links (monosulphide, disulphide or polysulphide linkages between different rubber chains), pendent polysulphide or cyclic monosulphide and disulphide linkages are formed; the structural features of the vulcanizate network are shown schematically in Fig. 9.7.

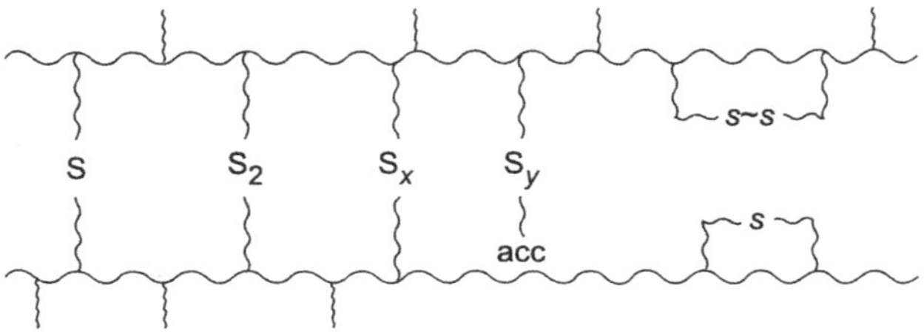

Fig. 9.7 *Schematic representation of a vulcanizate network formed during accelerated sulphur vulcanization of rubber*

Vulcanization proceeds through a sequence of complex reactions which are not yet completely identified and understood. Before the onset of the cross-linking process, zinc oxide activator, sulphur and accelerator (acc) react together to form a zinc perthio salt of the type [(acc) —S_x —Zn—S_x —(acc)]. This kind of salt is very reactive and it reacts with rubber hydrocarbon (RH) with the formation of a rubber-bound sulphur-accelerator intermediate.

$$\text{(acc)—}S_x\text{—Zn—}S_x\text{(acc)} + \text{RH} \rightarrow \text{(acc)—}S_x\text{— R} + HS_{(x-1)}\text{—(acc)} + \text{ZnS} \quad (9.5)$$

The hydrogen atom attached to the α-methylene group with respect to the unsaturation in the rubber hydrocarbon is most likely to be removed in reaction (9.5). Accelerators, which are mostly amines and related compounds, may also react through active hydrogens in their structure with sulphur to form sulphides or polysulphides which release sulphur in an active form in a subsequent step.

$$2\text{(acc)—H} + x\text{S} \rightarrow \text{(acc)—}S_{(x-1)}\text{—(acc)} + H_2S \quad (9.6)$$

The release of active sulphur may result from *in situ* reaction between the liberated H_2S and the disulphide or polysulphide accelerators or accelerator complexes.

$$R—S_x—R + H_2S \rightarrow 2R—SH + \underset{\text{(active sulphur)}}{(x-1)S} \tag{9.7}$$

It is, however, doubtful that H_2S has a major or exclusive role in the release of active sulphur. Zinc oxide and other metal oxide activators have a prominent role to play in this context. The metal oxide activator reacts with H_2S and thereby prevents undesirable and too rapid decomposition of accelerators on the one hand and contributes to reduce undue rise in acidity and thereby prevents undesirable retardation of the curing process.

It is also quite likely that radicals of the type $RS^{\bullet}$, $RN^{\bullet}$ or $R^{\bullet}$, formed by thermal dissociation of the accelerator molecules of different types, react with different relative ease with sulphur, breaking down the sulphur (S_8) rings and producing sulphur biradical chains with a variable number of sulphur atoms (active sulphur) which then link up different rubber hydrocarbon chains through sulphur cross-linkages (monosulphide, disulphide and polysulphide):

$$RS^{\bullet} + S_8 \rightarrow R—S—S_8^{\bullet} \rightarrow R—S—S_x^{\bullet} + {}^{\bullet}S_{(8-x)}^{\bullet} \tag{9.8}$$

The rubber-bound sulphur accelerator intermediate (acc)—S_x—R formed in reaction (9.5) may further react with a molecule of rubber hydrocarbon, RH, forming a sulphur cross-link and regenerating the accelerator:

$$(acc)S_x—R + RH \rightarrow R—S_{(x-1)}—R + (acc)—SH \tag{9.9}$$

or

$$R—S_x—R + (acc)—H$$

The metal oxide activators, such as ZnO, play multiple roles; besides catalyzing the reaction between accelerator, sulphur and rubber, they also contribute to the gain in physical properties of the vulcanizates through formation of strong secondary valence salt bridges of the type (rubber)-(accelerator–sulphur)$^-$Zn^{++} $^-$(sulphur–accelerator)-(rubber):

$$2H(acc)S_x—R + ZnO \xrightarrow{-H_2O} \begin{array}{c} R \\ | \\ (acc)—S_x^- \\ | \\ \vdots \\ Zn^{++} \\ \vdots \\ | \\ (acc)—S_x^- \\ | \\ R \end{array} \tag{9.10}$$

The story is still incomplete even in outline. On further heating of the vulcanizate beyond optimum curing or completion of curing, the polysulphidic cross-links, being thermolabile, progressively break-up, thus, giving rise to what is known as reversion, as commonly observed in some NR or IR systems. The net destruction of cross-links may be associated partly with formation of cyclic sulphides (Fig. 9.7) and partly with formation of some monosulphidic and disulphidic cross-links and disulphuration at different rates, depending upon the prevailing condition of heating. A sulphur cross-linked rubber network exhibiting the above features is considered to be inefficiently cross-linked. On the other hand, if the initial network is constituted of significant proportions of stable monosulphide cross-links and thermal treatment in the overcure region leads to the establishment of more cross-links with mainly monosulphidic bonds, some at the expense of the initial polysulphide and disulphide cross-linkages, it is said to be efficiently cross-linked. Curing with appropriate doses of a sulphur-donor accelerator (TMTD) or of an appropriate accelerator combination (TMTD)/(MBT or CBS) with little or limited dose of sulphur in presence of the normal doses of activators leads to efficient or semi-efficient vulcanization with better heat and aging resistance in general. But efficient vulcanization is not normally associated with the developments of optimum and most desirable physical properties combination for the vulcanizate. Vulcanizates, having primarily monosulphide cross-linkages as produced by sulphurless high TMTD efficient-curing systems, have good heat and aging resistance, but they are relatively stiff and are usually characterized by poor flex endurance or fatigue (defined as ability to withstand repeated flexing without development of severe cracking) and inferior tensile strength and resilience. Semi-efficient curing effected by combining TMTD and MBT or CBS (with a relatively high proportion of TMTD) and using a low dose of sulphur (0.6–1.2 parts) give a good balance of heat and aging resistance and physical properties including fatigue. A conventional curing system may be exemplified by a combination of CBS (0.6 parts) and sulphur (2.5 parts), and the vulcanizate exhibits good cross-link density and useful physical properties but relatively poor resistance to heat aging. Even though in most cases incorporation of a suitable antioxidant will normally improve the aging characteristics to some extent, an antioxidant additive cannot be a substitute for a good or improved curing system.

9.8 Non-Sulphur Vulcanization

A good example of non-sulphur vulcanization is provided by the sulphurless TMTD-curing, as described in Sec. 9.7. Bismorpholine disulphide may also be similarly used to effect sulphurless cure.

Sulphur monochloride was once used for what is known as cold cure of natural rubber. The use of quinone dioxime as a commercial vulcanizing agent is limited to butyl rubber and curing may be effected at fast rates when used in conjunction with red lead or MBTS. Any other diene rubber including NR may be vulcanized with the dioxime. The curing is believed to take place through oxidation of the dioxime to *p*-dinitrosobenzene which then acts as the actual cross-linking agent.

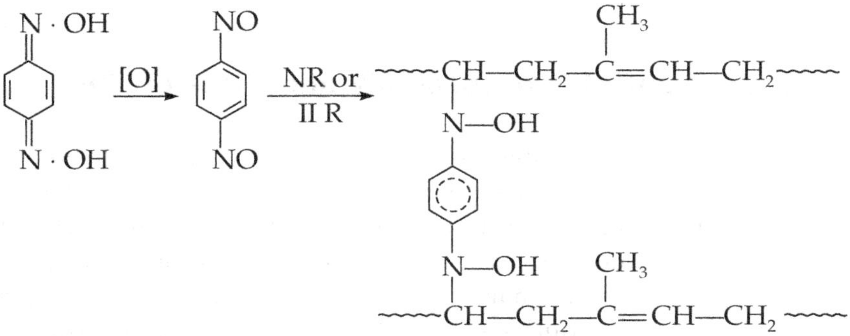

(9.11)

Peroxide curing is particularly suitable for olefin (non-diene hydrocarbon) rubbers and many thermoplastics. Peroxides may simultaneously lead to oxidation and cross-linking. With alkyl peroxides, cross-linking reaction predominates. A few commonly used vulcanizing peroxides are dicumyl peroxide, 2,5-di(t-butyl peroxide)-2,5-dimethylhexane and benzoyl peroxide. In NR vulcanization with benzoyl peroxide, vulcanizates are reported to give evidences for the formation of benzoic acid in the system indicating hydrogen atom abstraction by primary benzoyloxy radicals (formed by thermal decomposition of the peroxide) from the rubber hydrocarbon and thereby mostly establishing direct C—C cross-links. Side reactions may also lead to attachment of fragments of peroxide to the rubber chains. Peroxide curing of silicone rubbers is also accomplished by a similar mechanism.

Metal oxide cure of polychloroprene rubber is believed to be accomplished through the occasional 1,2 units in view of the fact that the double bond of each 1,4 unit is hindered by the neighbouring chlorine atom making sulphur-vulcanization difficult if not altogether impossible. The 1,2 units react in pairs with ZnO after an allylic shift of the chlorine atoms:

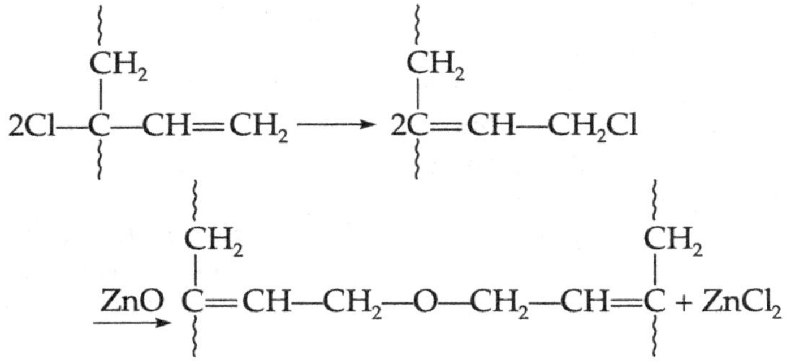

(9.12)

Halogen-bearing rubbers such as polychloroprene, polyacrylic and fluorocarbon rubbers are also conveniently cross-linked using diamines and polyamines as curatives.

Photo curing or radiation curing brings about curing through establishment of direct C—C cross-linkages.

9.9 Assessment of Processability and State of Cure

The process of vulcanization converts a predominantly plastic mass of a rubber compound into a predominantly elastic vulcanizate. The progress in the state of cure with time at a given temperature can be followed most conveniently while the rubber is curing by having a measure of the plasticity or viscosity of the compound. Such measurements are also useful in making an assessment and comparison of processability of rubber compounds. Equipment developed for the purpose and commonly in use are the Shawbury curometer, Mooney viscometer (also known as plastometer), Monsanto rheometer and Brabender plastograph (*See also* Chapter 7). The rheometers and the Brabender equipment with all the latest sophistications are very convenient and useful for continuous control of factory production. They give a plot of torque *vs* time. The minimum torque may be used for an assessment of processability of the compound; the time span over the bottom flat zone of the torque–time curve (up to 1 unit gain in torque over the minimum), is taken as a measure of scorch safety, the lower the value the more scorchy the compound, and the maximum torque attained is taken as a measure of degree of cure of a specific kind and the time required to attain 90% of maximum torque is commonly taken as the optimum curing time. Whether the curing is characterized by a plateau effect, reversion or marching cure phenomenon can be readily judged from the nature of the torque–time curve given by a rheometer. In the rheometric curve, no change in torque, or a fall in torque with time after attainment of maximum torque, is indicative of plateau effect or reversion respectively, while a slow monotonous increase in torque after the fast uptake zone is normally indicative of marching cure. Typical rheometric (torque–time) curves representative of each kind are shown in Fig. 9.8

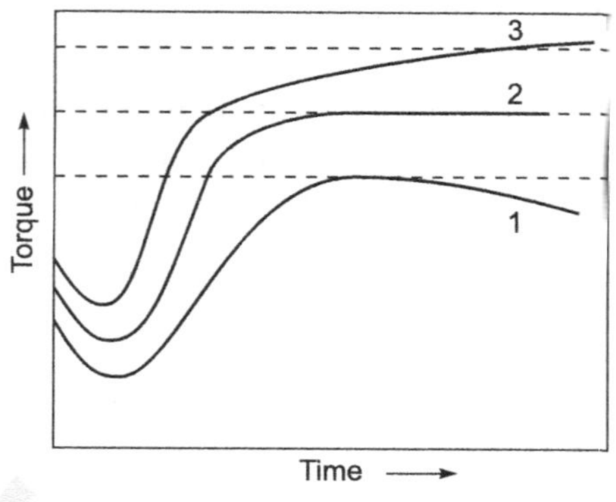

Fig. 9.8 *Rheometric curves (torque vs time) showing (1) reversion, (2) plateau effect and (3) marching cure*

9.10 Hard Rubber or Ebonite

When rubber is heated for a prolonged period with a large amount of sulphur (up to 40–45 parts), a stiff, rigid and hard product is formed and it is commonly called hard rubber or ebonite. Presence of accelerators make the ebonite compound scorch-prone. Slow or delayed action accelerators are preferred. Besides rubber, sulphur and accelerator, other ingredients commonly used are fillers in the form of factice, reclaim, bitumen, finely ground coal dust and ebonite dust.

For full saturation of NR, binding of a minimum of 32% sulphur corresponding to a product of empirical formula $(C_5H_8S)_n$ is to be expected. But a much higher proportion of combined sulphur (up to 40–45%) found in most commercial ebonites from NR points to the formation of limited polysulphide cross-links and some intramolecular cyclic di- or poly-sulphide structures. Polychloroprene is used to impart toughness to ebonite.

Ebonite has prominent thermoplastic flow characteristics. The gain in plasticity at elevated temperatures can be checked and controlled by incorporation of high proportions of fillers. The thermoplastic nature is to be viewed more as a consequence of transition from a rigid form to a viscoelastic state than due to melting. Ebonites can also be obtained from BR, NBR and SBR. Use of selected phenolic resins allow transformation of appropriate compounds into ebonite and related materials at a lower temperature over a much shorter time of cure.

Ebonite finds major application in electrical insulation and to a limited extent as mechanical parts or components. Its acid and chemical resistances make it suitable as battery boxes and as tank and vat lining in chemical plants.

9.11 Latex Technology

Natural rubber, as tapped from the Hevea tree, comes in the form of a latex. The tapping, preservation and handling of the latex and isolation of the solid rubber from the latex have been discussed in Section 9.2. Latex technology further covers all the processing steps including concentration of latex, latex compounding and fabrication into finished articles such as dipped, cast and foamed products. The processing steps using synthetic rubber lattices are also similar.

To be suitable for making commercial items, the latex requires to be compounded with major compounding ingredients such as sulphur, zinc oxide and accelerators.

Fillers, colourants and anti-oxidants may also be used. The aging characteristics of goods made from natural rubber latex are usually very good.

Normally, the compounding ingredients are first dispersed in water by grinding them in a ball mill using a proper dispersing agent. Solutions of lactic casein (10–12%) with liquor ammonia (0.5%) are used as dispersing agents. Condensation products of sulphonated naphthalene and formaldehyde are also widely used as dispersing agents. Water immiscible liquid compounding ingredients such as oils acting as softeners or plasticizers must be emulsified before directly adding to the latex. Sodium salts of organic sulphonates and sometimes alkali soluble styrene maleic ester copolymer solutions are used in measured small doses as stabilizing agents.

Latex compounding is accomplished by pouring in right quantities of selected solutions and dispersions as above into the latex under stirring conditions. Stabilizers are added first. Contact with metals and alloys harmful for rubber (such as manganese, copper, brass, etc.) must be entirely avoided. Aqueous solutions of a water soluble xanthate (SIX) or dispersions of ZDC, ZMBT are commonly used as accelerators.

After a specific latex compound has been prepared, the next step is to get the shape of the article to be made, set the shape and then vulcanize. According to the technique of getting the shape, the latex processing can be classified as follows: (i) dipping (ii) casting and moulding, (iii) spreading, (iv) spraying, (v) extrusion and (vi) electrodeposition.

The dipping process is used for the manufacture of gloves, balloons, and many such thin walled articles. It is easier, more economical and less hazardous to make such articles from rubber latex than from rubber solution. Avoidance of bubble in dipping work is important.

In the dipping process, the latex is deposited on the outer side of a former in the shape of the article to be made as it is dipped in the latex compound, whereas in the casting or (slush) moulding process, the article is built up on the inner walls of a hollow mould.

The easiest way of setting the deposited latex in the shape of the former or hollow mould after completion of the dipping or casting process is to remove the water by evaporation; otherwise a coagulation or gelling process may be followed. In contact coagulation process, the latex coagulates as soon as it comes into contact with a coagulant-coated former or mould. Calcium chloride or acetic acid solution may be used as a coagulant. In a delayed action gelling system, coagulation or gelling is effected at normal temperature using sodium silicofluoride as the most common gelling agent in presence of zinc oxide. In the heat sensitive gelling system, gelling is effected at an elevated temperature using zinc oxide in combination with ammonia or ammonium salts as the gelling agent. The latex may be prevented from gelling at room temperature through the use of a suitable nonionic stabilizer. Organic heat sensitizing agents have also been developed for easy and effective control of the gelling process.

Normally, latex rubber product is dried before it is vulcanized. A wet coagulum, showing considerable mechanical strength can be vulcanized without drying to yield a product of high elastic recovery which can be subsequently dried. This technique is advantageously employed in the production of latex foam. Vulcanization of solid coagulum in steam or water produces vulcanizates which become microporous on drying. Vulcanization is either carried out in hot air at 100–110°C or by heating latex goods in hot water or hot brine baths. For water cures, a drying operation subsequent to curing is necessary.

Latex thread is manufactured by a process of extrusion. A properly compounded latex is allowed to pass through glass nozzles under pressure into a coagulant bath. The extruded latex gels in the form of a thread. The thread is then drawn out of the bath and subjected to washing, drying and vulcanization.

Latex foam production is the most important of all latex processes. In the Dunlop process, sodium silicofluoride is used as the gelling agent. The latex compound is mechanically beaten up to a foam which is aided by blowing air below the surface of the latex. Then the requisite quantity of a dispersion of sodium silicofluoride is added, which, in the presence of zinc oxide, sets the foam into a 'gel' in a few minutes. A secondary gelling agent such as a trimene base (vulcafor EFA) in the form of 50% aqueous solution is used to reduce the gelling time so as to prevent any premature foam collapse. The trimene base additive further imparts a high modulus to the foam product, which is an added advantage. Saving on latex with the use of the trimene base additive may be as high as 10%. It also helps in the improvement of compression set of the foam. A composite aqueous dispersion of DPG and zinc oxide and some selected cationic soaps may also be used as a secondary gelling agent.

Under normal processing conditions, shrinkage of latex foam to the extent of 10–20% takes place on drying and allowance for this is made while moulds are designed and made. The dosage of gelling and curing agents and time and temperature of curing should be so adjusted as to give the best quality foamed product. 'Maturing' of the latex compound leading to its partial vulcanization is usually practised at the initial stage of deammoniation of the latex during which the ammonia content of the ammonia-preserved centrifuged latex is brought down from 0.7% to nearly 0.1–0.15%. The deammoniation minimizes or eliminates gelling problems at the stage of foam making and helps in achieving a good foam structure. The major effect of maturation is to help in the control of hardness and shrinkage and impart improved modulus and good recovery property. The initial deammoniation is usually done by the addition of formaldehyde or by blowing moist air across the surface of the latex while it is stirred.

Aluminium moulds are usually used for making foamed articles. Stainless steel may also be used. Some lubricants such as silicones are usually applied to the mould for easy removal of product. Steam at atmospheric pressure is normally used for curing. Thin layers of foam such as those used in carpet backing are conveniently cured by hot air. For moulding of thick items such as cushions and mattresses, cylindrical cores are attached to the lids of the mould to give a "honey comb" appearance to the underside of the moulded foam. This ensures adequate vulcanization in the centre of the foam, allows a great saving in material and cuts down and distributes shrinkage which would measurably distort the product if left uncontrolled. After the vulcanization, the foamed item is stripped from the mould, trimmed and then washed with water to remove salts, soaps and odouriferous residues. It is finally dried at 60–80°C in an oven allowing good circulation of air. The formulation of a typical foam compound is given below:

NR Latex (60%)	120	parts
Potassium oleate (soap) solution (20%)	3–4	"
Sulphur dispersion (50%)	3	"
ZDC dispersion (50%)	0.75	"
MBT dispersion (50%)	0.75	"
Antioxidant dispersion (50%)	0.8	"
China clay slurry (50%)	20.0	"
Zinc oxide dispersion (50%)	7.5	"
Secondary gelling agent	0.4–0.8	"
Sodium silicofluoride dispersion (20%)	4–6	"

Electrodeposition is a neat technique used in the coating industry. The negative charge on rubber particles makes it possible to deposit rubber on a conducting surface in a manner similar to the electroplating process for metals. This process, however, is of little commercial importance.

9.12 Some Major Rubber Products

9.12.1 Tyres

Most important application of rubbers relates to the transport sector. Tyre industries consume more than 60% of the rubber produced; this includes cycle and rickshaw tyres, scooter and motor cycle tyres, car, bus, truck and tractor tyres, air-plane tyres, and tubes for all pneumatic tyres.

Tyre technology is a very specialized technology and a tyre designer is faced with the magic task of trying to satisfy all the needs of the vehicle manufacturer and the user; however, the prime factors of consideration are safety and tread life. The major components of a tyre are: the bead, the carcass, the side wall and the tread.

The bead is constructed from a number of turns or coils of high tensile steel wire coated with copper or brass to ensure good adhesion of the rubber coating applied on it. The beads function as rigid, practically inextensible units that retain the inflated tyre on the rim.

The main part of the carcass is the tyre cord. The cords consist of textile threads twisted together. On an average, a tyre contains nearly 50% of its weight of actual rubber; for oil extended rubbers, it is even less. The remainder includes carbon black, textile cord and other compounding ingredients plus the beads. Rayon, nylon and polyester cords are widely used. Steel cords are also used. The practice of laying rubber-impregnated cords in position is still followed, but use of woven fabric having a very small number of weft threads is more widespread. The warp threads constitute the cords. The smooth fibre cords are compulsorily treated with an adhesive composition to achieve good bond with rubber. Water soluble resorcinol–formaldehyde resin and aqueous emulsion of a copolymer of butadiene, styrene and vinyl pyridine (70 : 15 : 15) are commonly used to promote cord-rubber adhesion. The resin-coated cord or fabric is then dried and coated with a rubber compound with calendering whereby each cord is isolated from its neighbouring cord.

For conventional tyres, the rubber coated fabric is then cut to a predetermined width and bias angle (between 45° and 90°). The bias cut plies are then joined end to end into a continuous length and batched into roll form and interleaved with a textile lining to prevent self adhesion.

The sidewall is a layer of extruded rubber compound which protects the carcass framework from weathering and damage from chafing. Overlapping the tread in the buttress region, the sidewall forms the outermost layer of the tyre together with the tread.

The tread is also formed by extrusion. Different compounds are used for sidewall and tread and they may be produced as a single unit to simplify the tyre-building operation. This may be done by simultaneous extrusion of the two compounds in a single band—the tread over the sidewall, with two extruders arranged head-to-head with a common double die. The band is then rapidly cooled only to avoid scorching and then cut to appropriate length. The characteristic pattern is imparted to the tread from the mould when the sequentially built tyre is vulcanized.

When the sidewall and the tread are made on two separate extrusion lines, it is useful to take the extruded sidewall to the second extruder which would deliver the tread to the sidewall. While building the tyre, a layer of specially compounded cushion rubber may be used so as to keep heat development on flexing to a minimum and achieve better adhesion between the tread and the carcass. One or more layers of fabric known as breakers or bracing layers may be placed below the cushion. The bracing components raise the modulus of the tread zone and level out local blows to the tread as it contacts the road.

The main feature of the tyre building appliances is a flat drum which is rotated at a controlled speed. The plies of rubber-treated cord fabric, one over the other are placed in position and rolled down as the drum rotates, the inner lining being placed next to the drum. When the building is complete, the drum is collapsed and the tyre removed. The ends of the drum are flanged to suit the bead configuration of the tyre. The overall diameter exceeds the nominal tyre bead diameter. The cylindrical shape of the raw tyre obtained from the building drum is transformed into a toroidal shape during the mould closing operation resulting in a circumferential stretch of the order of 60%.

Different methods of vulcanization are practised, but in each case the uncured tyre is pressed against the inner face of the heated mould by an internal bag or bladder made of a pre-cured heat-resistant rubber and inflated by a high pressure steam or circulating hot water. Autoclave moulding once practised is now almost obsolete because of inefficient workability. In the Bag-O-Matic or similar type of press, the uncured tyre is placed over a special type of bag or bladder and the operation of shaping, curing and ejection of the cured tyre are accomplished by an automatic sequence of machine operations. Diagrams showing construction of a standard diagonal ply tyre and a radial ply tyre are shown in Fig. 9.9.

Manufacture of inner tubes is done essentially in three steps: extrusion, cutting to length and insertion of valve and vulcanization. The ends of the cut length of the tube are joined after insertion of the valve prior to vulcanization.

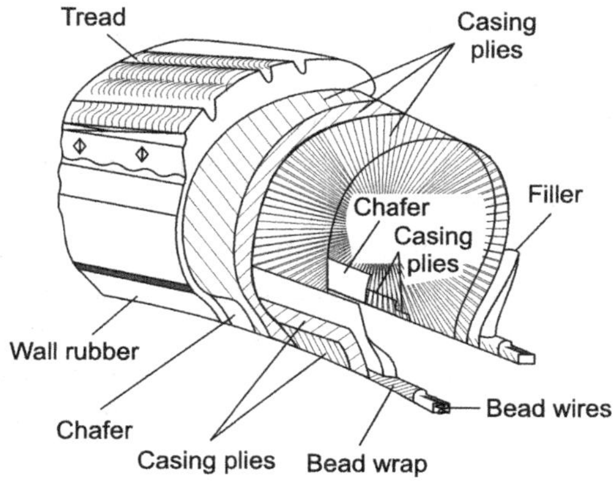

(a)

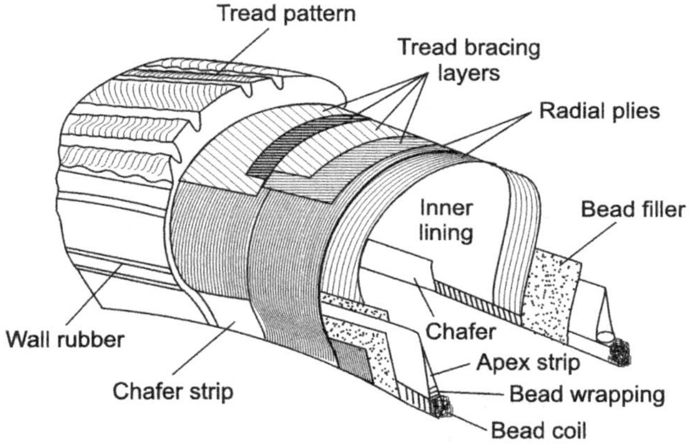

(b)

Fig. 9.9 *Diagrams showing (a) diagonal ply tyre and (b) radial ply tyre*

9.12.2 Belting and Hoses

Next in importance is the belting industry—making flat conveyor and (power) transmission belts and V-belts (for power transmission), and the hose industry—making different rubber hoses. Selected thermoplastics such as plasticized PVC are also widely used particularly for conveyor belts and hoses.

Use of conveyor and power transmission belts is to be found in almost all major industries. Textile cords or fabric and even steel cords constitute an important part of all rubber beltings and hoses. The various types of cords used in the tyre industry are also in use in the belting industry. Glass fabric may be employed at the top layer in special cases to protect the plies underneath against heat.

Essential operational features for making conveyor belts are: (a) drying the fabric, (b) frictioning of the hot fabric with a rubber compound and topping to give additional rubber between plies and the outer ply and cover, using a three- or four-roll calender, (c) belt building, and (d) vulcanization. For belt building, the frictioned and coated fabric is cut and folded to give the desired number of plies. The construction may be straight laid, folded jacket or graded ply types (Fig. 9.10). Transverse and longitudinal joints within certain limitations are permissible, the joints in neighbouring plies being staggered to eliminate weakness and failure. Finally, the cover coat is applied by calendering.

Fig. 9.10 *Construction of conveyor belts; (a) straight laid (b) folded jacket, and (c) graded ply*

Vulcanization, when done by heating in lone presses, is carried out in sections, the belt being moved between successive cures by a length less than the length of the press platens. The end or overlap zones must of necessity receive an additional cure and to minimize damage or weakness due to overcure, use of flat cure mixes and allowing water cooling at each end section of the platen are helpful and desirable. In each step of vulcanization, the corresponding section of the belt to be vulcanized is gripped and stretched hydraulically to minimize or eliminate elongation during use. The difficulties of press cure may be avoided by adopting continuous vulcanization by means of Rotocure equipment, the actual curing operation being carried out between an internally steam heated cylinder and a heated steel band. The Rotocure is also useful for the vulcanization of transmission belts and rubber sheeting.

V-belts of different types cover applications ranging from fan belts for automobiles, belts for low-power drives for domestic, laboratory and light industrial applications and high-power large industrial belts. The most common feature is their having a cross-section of the shape of a regular trapezium with the unparallel sides at an angle of 40°. The V-belts usually consist of five sections (Fig. 9.11), viz. (1) the top section known as the "tension section", (2) the bottom section, called the "compression section", (3) the "cord section" located at the natural axis, (4) the "cushion section" surrounding the cord section, and (5) the whole assembly covered by one or two layers of rubberized fabric, commonly known as the "jacket section". Relatively short length belts are built layer by layer on rotatable collapsible drum formers. The separate belts are then cut out with knives and transferred to a skiving machine that imparts the desired V-shape. Finally, the fabric jacket is applied and the belts are vulcanized in open steam using multi-cavity ring moulds for smaller belts. Long length belts are similarly built on V-groove sheave pulleys taking weftless cord fabric in place of individually wound cord. They are vulcanized endlessly by moulding in a hydraulic press under controlled tension.

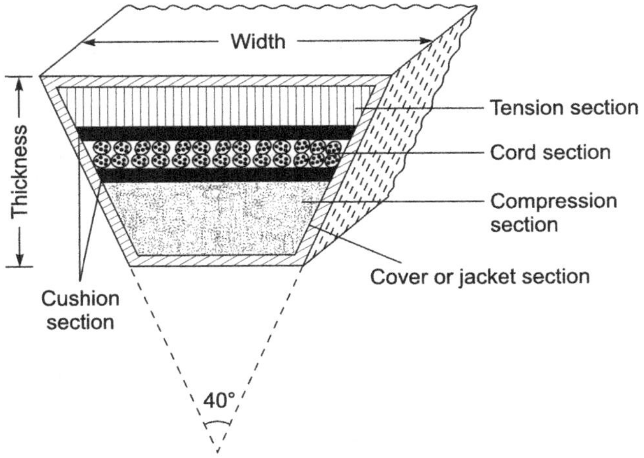

Fig. 9.11 *V-belt construction*

Three different rubber compounds are required to be used in the fabrication of a V-belt: (i) the major constituent on a weight basis is the base compound, (ii) the soft and resilient compound surrounding and protecting the reinforcing cord assembly is the cushion compound and (iii) the compound used for rubberizing the fabric casing of the belt is the friction compound.

A rubber hose has three major parallel layers along the length. The innermost part consisting of a rubber lining or tube is required to resist the action of the material that world pass through. The top layer or the cover is meant to play the role of a protective layer. Between the outer cover and the inner lining is given a layer of reinforcement in the form of a textile yarn or steel wire applied by knitting, braiding, circular loom weaving or spiralling. A cut woven fabric wrapped straight or on the bias may also be used to reinforce the inner lining or tube. More than one ply of textile may be used and in that event, it is customary to apply a layer of rubber between two layers of the reinforcing textile items just to prevent chafing during service. The cover or the outer protective rubber layer of the hose should be so compounded as to resist weathering, temperature surges over day and night and with change of seasons, soil condition, oils and chemicals, abrasion etc.

The outline of hose building consists of extruding the lining or tube, braiding or spiralling the textile around the cooled tube and applying outer cover of rubber to the reinforced hose using a cross-head extruder. In one manufacturing process, the hose is then passed through a lead press or lead extruder to give a layer of lead cover to the hose. The hose is then wound on a drum, filled with water or air and the ends are sealed. The whole assembly is then placed in large pans and heated to achieve vulcanization. The water or air inside expands and the vulcanizing hose is pressed against the lead acting as the mould. On completion of cure, the seals or clamps at the ends are cut and the lead cover is removed by slitting lengthwise in a stripping machine and the cured hose is coiled up. A few other methods are also employed for the manufacture of hoses.

9.12.3 Cellular Rubber Products

Cellular rubber may be defined as an assembly of multitude of cells distributed in a rubber matrix more or less uniformly. The cells may be interconnected as in a natural sponge or separate. Foam rubber is a cellular product made from a liquid starting material such as the latex described earlier, and here, the cellular structure is intercommunicating. Cellular products made from solid rubber are commonly called sponge (intercommunicating cells) and expanded rubber (closed cell structure). The technology of making cellular products from solid rubber is solely dependent on the incorporation of a convenient gas, usually nitrogen or a chemical blowing agent in the rubber compound. The most widely used chemical blowing agent for this purpose is dinitroso pentamethylene tetramine (DNPT). The sponge and expanded products include carpet backing, sheets, profiles and mouldings. The curing is carried out either freely using hot air or steam or in a mould which is only partially filled with the moulding compound. Synthetic rubbers, particularly SBR are widely used for precise control over level of viscosity needed for processing and for obtaining consistent product quality.

Microcellular soles The development of microcellular rubber soling and resin-rubber soling has brought a revolution in the footwear technology. With extremely fine non-communicating cell structure and very comfortable wearing properties, microcellular rubber is the lightest form of soling that can be produced. Density of the soling as low as 0.3 g/cc may be obtained with high dose of DNPT (8–10 parts) at a curing temperature of 140–150°C. For common solings, the density normally varies between 0.5 and 0.8 g/cc.

Low density, high hardness and improved abrasion resistance can be achieved at the same time using SBR, and high styrene resins along with NR in right proportions. Higher proportion of high styrene resin gives compounds of higher hardness, and abrasion resistance, and lower density. Silicious fillers such as precipitated silica and aluminium or calcium silicate are also useful in giving high hardness, resistance to abrasion and split tear strength, much desired in the microcellular soling. Microcelluar crumbs are normally used in considerable quantity along with china clay and whiting to reduce the compound cost. Higher proportions of stearic acid (5–10 parts) are commonly used in microcellular compounds in order to bring down the decomposition temperature of the DNPT type blowing agents. Post-cure heating of the microcellular sheets in an air oven stabilizes them and reduces the delayed shrinkage after cure to a minimum.

9.12.4 Miscellaneous Applications of Rubber

Other important and special applications of rubber are in the areas of adhesives, coated fabrics including packaging and luggage items and rainwear, footwear, pipes and tubings, wire insulation, cables and sheaths, tank lining for chemical plants and oil storage, gaskets and diaphragms, rubber mats, rubber rollers, sports goods, toys and balloons and wide varieties of moulded mechanical and miscellaneous products. Formulations of a few selected rubber compounds are given below:

1. Passenger Tyre

(i) *Tread Compound*		(ii) *Side Wall Compound*	
Smoked sheet	50	Smoked sheet	50
SBR 1712	70	SBR 1712	70
Zinc oxide	5	Zinc oxide	5
Stearic acid	2	Stearic acid	2

(i) *Tread Compound (Contd.)*

ISAF Black	60
Softener	2
Antioxidant (diphenylamine–acetone condensate)	1.5
Accelerator (CBS)	1
Sulphur	2.2
(Cure: 40 min. at 150°C)	

(ii) *Side Wall Compound (Contd.)*

HAF Black	45
Softener	2
Antioxidant (diphenylamine–acetone condensate)	1.5
Accelerator (CBS)	1
Sulphur	2.2
(Cure: 30 min. at 150°C)	

2. Tube Compound for Car Tyres

Butyl (polysar 301)	100
p-dinitroso benzene (polyac)	0.15
FEF black	60
Mineral oil (butyl grade)	20
Zinc oxide	5
Stearic acid	2
Accelerator:	
MBT	1
TMT	1
Sulphur	1.5
(Cure: 30 min. at 160°C)	

3. Conveyor Belts

(i) *Friction Compound*

Smoked sheet	100
Zinc oxide	5
Stearic acid	2
SRF black	10
Whiting/activated calcium carbonate	15
Tackifier/softener	5
Antioxidant	1
Accelerator (CBS)	0.6
Sulphur	2.5

(ii) *Cover Compound*

Smoked sheet	50
SBR 1500	50
Zinc oxide	5
Stearic acid	2
Tackifier/softener	5
ISAF black	40
Antioxidants	1.5
Accelerator (CBS)	1.0
Sulphur	2.0

(Cure: 20 min. at 150°C)

4. Insulation Compound for Cables

Smoked sheet	100

Zinc oxide	20
China clay	30
Precipitated calcium carbonate	45
Paraffin wax	2
Stearic acid	0.5
Antioxidants	1.0
Accelerators:	
DPG	0.5
MBTS	1.0
Sulphur	1.5
(Cure: 15 min. at 140°C)	

5. Shoe Soles

(i) *Translucent Soling*

Pale crepe	100
Zinc oxide	3
Stearic acid	1
Precipitated silica	50
Paraffin wax	1
Spindle oil	2
Diethylene glycol	2
Antioxidant (styrenated phenol)	1
Accelerators:	
TMTM	0.5
ZDC	0.75
Sulphur	2.5
(Cure: 7 min. at 150°C)	

(ii) *Microcellular Soling*

Smoked sheet	20
SBR 1500	20
SBR 1958	60
Peptizer	1
Microcellular crumb	60
Zinc oxide	5
Stearic acid	3–5
Paraffin wax	1
Mineral oil	10
Cumarone–indene resin	5
Styrenated phenol	1
Aluminium silicate	40
China clay	100
Blowing agent (DNPT)	5
Accelerators:	
DPG	0.5
MBTS	1.0
Sulphur	2.5
(Cure: 8 min. at 150°C); Oven stabilization at 100°C; 4 h).	

9.13 Equation of State for Rubber Elasticity[11]

The change in internal energy E accompanying the stretching of an elastic body may be generally expressed as:

$$dE = dQ - dW \tag{9.13}$$

where dQ is the element of heat absorbed and dW is the element of work done by the system on the surrounding environment. If P is the external pressure and f the external force of extension (equal to the force of retraction developed), then

$$dW = PdV - fdL \tag{9.14}$$

When conducted reversibly, $dQ = TdS$, S being the entropy of the elastic body and dW signifying the element of reversible work, then one may write

$$dE = TdS - PdV + fdL \tag{9.15}$$

One may also consider the expression for Gibbs free energy defined by

$$F = H - TS = E + PV - TS \tag{9.16}$$

where $H = (E + PV)$ is the heat content; then

$$\begin{aligned} dF &= dE + PdV + VdP - TdS - SdT \\ &= VdP - SdT + fdL \end{aligned} \tag{9.17}$$

Equation (9.17) is obtained by taking help of Eq. (9.16); the last expression relates differential of the free energy with the differentials of convenient experimental variables P (pressure), T (temperature) and L (linear dimension).

It further follows from Eq. (9.17) that

$$(\partial F/\partial L)_{T,P} = f \tag{9.18}$$

or

$$f = (\partial H/\partial L)_{T,P} - T\,(\partial S/\partial L)_{T,P} \tag{9.19}$$

and also

$$(\partial F/\partial T)_{P,L} = -S \tag{9.20}$$

The second derivatives obtained by differentiating $(\partial F/\partial L)_{T,P}$ with respect to T at constants P and L and by differentiating $(\partial F/\partial T)_{P,L}$ with respect to L at constants T and P are identical; so one may write from Eqs (9.18) and (9.20)

$$-(\partial S/\partial L)_{T,P} = (\partial f/\partial T)_{P,L} \tag{9.21}$$

Equation (9.19) may then be modified as

$$f = (\partial H/\partial L)_{T,P} + T(\partial f/\partial T)_{P,L} \tag{9.22}$$

This thermodynamic equation of state for elasticity is analogous to the ordinary thermodynamic equation of state, viz.

$$P = -(\partial E/\partial V)_T + T(\partial P/\partial T)_V \tag{9.23}$$

From the definition of H, one may write,

$$(\partial H/\partial L)_{T,P} = (\partial E/\partial L)_{T,P} + P(\partial V/\partial L)_{T,P} \tag{9.24}$$

and in view of the fact that for all practical purposes, for external pressures $P \leq 1$ atm/moderate pressure, $P(\partial V/\partial L)_{T,P}$ the second term on the right hand side of Eq. (9.24) will be negligible compared with the first term. One may therefore genuinely approximate and use the following relations:

$$(\partial H/\partial L)_{T,P} \cong (\partial E/\partial L)_{T,P} \text{ such that}$$

$$f = (\partial E/\partial L)_{T,P} + T(\partial f/\partial T)_{P,L} \tag{9.25}$$

This form of the equation of state for elasticity closely resembles Eq. (9.23).

The major structural effect of elastic deformation being an orientation of the polymer chains, one may be tempted to equate $(\partial E/\partial L)_{T,P}$ and $(\partial S/\partial L)_{T,P}$ respectively to the energy and entropy changes associated with the molecular orientation effected due to application of the stretching force (f); but this way of equating may lead to some serious misinterpretations. This is because orientation is not the only consequence of elongation at constant pressure and owing to the decrease in internal pressure (not to be mixed up with the external pressure, P) that accompanies the stretching process, the volume of the rubber also increases. Orientations and associated volume increase, lead to energy and entropy changes. In mathematical terms one may write:

$$(\partial E/\partial L)_{T,P} = (\partial E/\partial L)_{T,V} + (\partial E/\partial V)_{T,L}\,(\partial E/\partial L)_{T,P} \tag{9.26}$$

In an effort to find a more appropriate relationship allowing evaluation of derivatives of entropy and energy at constant volume, it suits to introduce the work function $A = E - TS$. Proceeding as outlined above

$$dA = dE - TdS - SdT$$

$$= -PdV - SdT + fdL \tag{9.27}$$

We may then write

$$f = (\partial A/\partial L)_{T,V}$$

or

$$f = (\partial E/\partial L)_{T,V} - T(\partial S/\partial L)_{T,V} \tag{9.28}$$

and also

$$(\partial A/\partial T)_{V,L} = -S$$

One may therefore write in analogy with Eq. (9.21)

$$-(\partial S/\partial L)_{T,V} = (\partial f/\partial T)_{V,L} \quad (9.29)$$

Thus, one may transform equation (9.28) to

$$f = (\partial E/\partial L)_{T,V} + T\,(\partial f/\partial T)_{V,L} \quad (9.30)$$

Equation (9.30) is another version of the thermodynamic equation of state for elasticity. An experimental measure of $(\partial f/\partial T)_{V,L}$ would enable assessment of thermodynamic derivatives as in Eqs (9.29) and (9.30). However, this will require measurement of the force of retraction (f) at constant length and at constant volume. To fulfill the latter condition, a hydrostatic pressure would have to be applied to undo the increase in volume due to thermal expansion, and the pressures required would be very high indeed for a considerable range of temperature, and this makes such an experiment unattractive.

However, it is possible to show that

$$-(\partial S/\partial L)_{T,V} \cong (\partial f/\partial T)_{P,\alpha} \quad (9.31)$$

where α is the elongation given by $\alpha = L/L_0$, L_0 being the length at zero stress at the temperature T and pressure P. The coefficient $(\partial f/\partial T)_{P,\alpha}$ (i.e. the change in retractive force with temperature at constant pressure and at constant elongation (%) meaning that the length L is varied in the manner which will maintain a constant ratio between L and the unstressed length L_0 at each temperature) can be easily obtained experimentally.

We may therefore express the equation of state for rubber elasticity as:

$$f = (\partial E/\partial L)_{T,V} + T\,(\partial f/\partial T)_{P,\alpha} \quad (9.32)$$

A measure of f as a function of temperature at constant pressure P and elongation α would enable assessment of $(\partial E/\partial L)_{T,V}$ and $(\partial S/\partial L)_{T,V}$ using Eqs (9.32) and (9.31).

Hysteresis in the stress–strain/stress–elongation behaviour of rubber and rubber-like materials poses a serious problem in the experiment. This is due to the viscoelastic as opposed to only elastic nature of rubbery materials. The stress exerted by the sample held at a fixed length decays with time toward a limiting value, but the rate of approach to the limiting value, however, decreases with time in a manner that virtually precludes attainment of the actual limiting stress within any reasonable time period. The rate of equilibration of the stress increases fast with temperature.

The stresses observed within each set of measurements (allowing stress relaxation to proceed at fixed length for a reasonable time at the highest safe temperature to be initially employed, avoiding chemical changes and then proceeding immediately with measurements at successively lower temperatures at the same length without waiting for further relaxation at the lower temperature as the rate of relaxation is much slower at the lower temperatures) are reversibly related to one another. A state of true elastic equilibrium is hardly to be attained, however. The stress observed at the various temperatures will be consistently higher on this count than the equilibrium values for the given length of the test specimen. The degrees of relaxation attained at different lengths will not be generally equivalent. Hence measurement with equal lengths is necessary at all temperatures.

For natural rubber and for most synthetic rubbers, experimental observations indicate that the internal energy coefficient value $(\partial E/\partial L)_{T,V}$ is very low or equal to zero or slightly negative over low to moderate elongations. For elongations up to 350%, the retractive force generated is mostly or solely due to the decrease in entropy with elongation and hence, the term $-T(\partial S/\partial L)_{T,V}$ is very nearly equal to the total force, f.

The features outlined above imply that the energy of intermolecular interaction (Van der Waals interaction) is affected negligibly by deformation at constant volume—which is rarely surprising and not only that the intermolecular distance must remain unchanged, but also that contributions from deformation of valence bonds or from hindered rotations about bonds are likewise very small.

For ideally elastomeric chains characterised by bonds having lengths and angles fixed with unhindered rotation about single bonds, it may be predicted for ideal rubber that $(\partial E/\partial L)_{T,V}$ would be exactly equal to zero. The correspondence to an ideal gas for which $(\partial E/\partial V)_T = 0$, is thus, immediately apparent.

For condition of elastic ideality applying over a range of temperatures, the equation of state for elasticity [Eq. (9.32)] then implies that the force of retraction, f should be proportional to the absolute temperature, T, i.e.

$$f = T\ (\partial f/\partial T)_{P,\alpha} = -\ T\ (\partial S/\partial L)_{V,T} \tag{9.33}$$

Likewise, the condition $(\partial E/\partial V)_T = 0$ for an ideal gas implies direct proportionality between P and T and on the basis of Eq. (9.23), we may write,

$$P = T\ (\partial P/\partial T)_V \tag{9.34}$$

Further, since $(\partial P/\partial T)_V = (\partial S/\partial V)_T$ for an ideal gas,

$$P = T\ (\partial S/\partial V)_T$$

Thus, just as pressure of an ideal gas is attributed to the increase in entropy with volume at a given temperature, the retractive force in an ideal rubber is due entirely to the decrease in entropy with increase in length or deformation at a given temperature, as if each is owing to the thermal motions of its particles.

At high elongations (> 350%), The $(\partial E/\partial L)_{T,V}$ values for natural rubber (NR) tend to become progressively negative, while the $-T\ (\partial S/\partial L)_{T,V}$ values increase sharply and the region of abrupt changes in each case precede the region of sharp rise in the stress-elongation curve obtained by isothermal stretching, Fig. 9.12(b), from which stress-induced crystallization effectively commences. The pronounced drop in the $(\partial E/\partial L)_{T,V}$ values at higher elongation and concomitant enhancement in the $-\ (\partial S/\partial L)$ parameter are pointed indications of an associated ordering or orienting process. These are the essential thermodynamic requirements for a crystallization process.

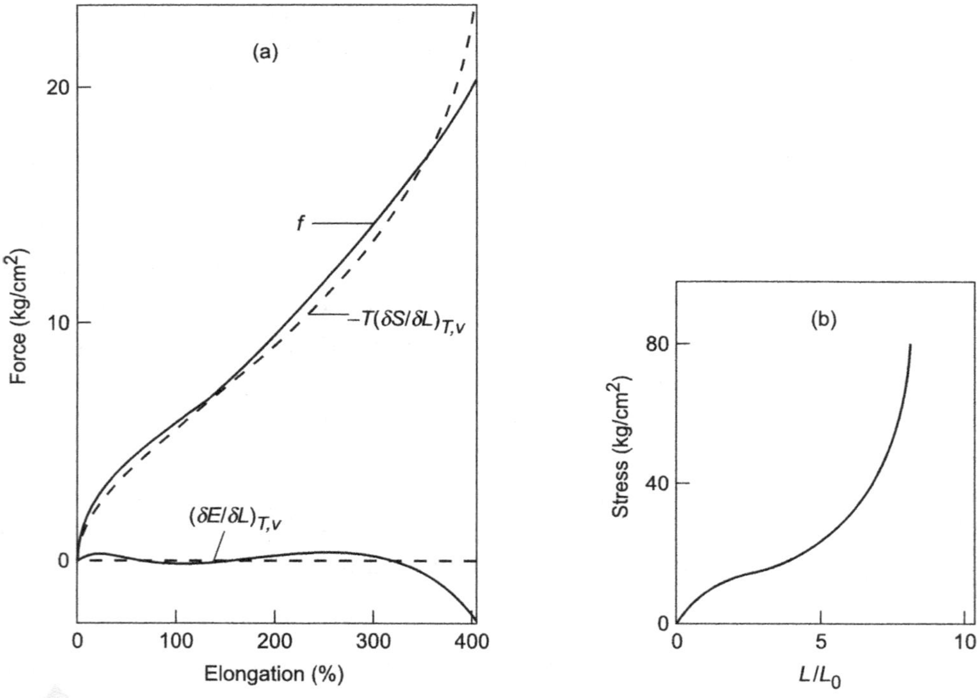

Fig. 9.12 *(a) Variation of force of retraction, f and its components vs elongation % [Eqs (9.28) and (9.29)] derived from force–temperature plots at constant elongation, and (b) Stress–strain curve for natural rubber gum vulcanizate*

Rubbers which do not crystallize on stretching give no evidence for significant lowering in $(\partial E/\partial L)_{T,V}$ or $(\partial S/\partial L)_{T,V}$ at higher elongation. The so-called non-crystallizing rubbers inherently lack in capacity to crystallize on stretching, thus causing them to fail to withstand a large tensile stress.

References

1. Schidrowitz, P. and T.R. Dawson, Eds., *History of the Rubber Industry*, IRI, London, 1952.
2. Flint, C.F., *Chemistry and Technology of Rubber Latex*, Chapman and Hall, London, 1938.
3. Schollenberger, C.S., H. Scott and G.R. Moore, *Rubber World*, 137 (1958), 549.
4. Wells, S.C., *Elastoplastics*, 5 (1973), 102.
5. Porter, L.M., U.S. Pat. 3,149,182 (1964).
6. Fisher W.K., U.S. Pat. 3,806,558; 3,835,201 (1974).
7. Morris H.L., *J. Elastomer Plast.*, **6** (1974), 1.
8. Nourry, A. Ed., *Reclaimed Rubber—Its Developments, Applications and Future*, Maclaren, London. 1962.
9. Garvey, Jr. B.S., and G. Thompson, *Ind. Eng. Chem.*, **25** (1933), 1295.
10. Hoffmann, W., *Vulcanization and Vulcanizing Agents*, Maclaren, London, 1967.
11. Flory, P.J., *Principles* of Polymer Chemistry, Chapter IX, Cornell University Press, Ithaca, New York, 1953.

CHAPTER 10

Identification, Testing and Evaluation of Plastics and Rubbers

10.1 Identification of Common Plastics and Rubbers[1–6]

The first step in the identification of polymers is visual examination. Just the appearance of the sample for testing may indicate whether it is essentially a virgin or raw polymer or a compounded or processed item. The first hand idea that may be gained about its size, shape or form, feel, odour, colour, bounce, softness, stiffness or rigidity, transparency or opacity, surface texture and appearance without the use of any chemical or physical testing equipment is very important in the process of identification of the polymer or resin of which the sample is made. After a critical visual examination, the following preliminary tests and chemical analysis are performed one by one.

10.1.1 Heating Tests

A small piece or amount (0.1 g) of the test sample is placed on a well cleaned glass or stainless steel spatula, previously heated to remove any traces of combustible or volatile materials. It is then gently warmed over a small colourless bunsen flame avoiding ignition till it begins to fume. The decomposing sample is then removed from the flame and the nature of the fume or gas is examined with respect to colour, odour, chemical identity including acidity, alkalinity or otherwise.

The sample is next moved to the hottest zone of the small bunsen flame and the following observations are made:

i) if the material burns and if so, how readily
ii) the nature and colour of the flame as the material burns
iii) whether or not the material continues to burn after removal from the flame, and
iv) the nature of the residue.

Observations made with some common polymers on heating are listed in Table 10.1.

10.1.2 Tests for Readily Detectable Elements

The results of tests for readily detectable elements such as N, S and halogens may serve to roughly indicate the nature of unknown material including the nature of the base polymer or resin and the additives, if present. These results would become more meaningful when viewed against other preliminary observations, viz., thermal response (fusibility or otherwise, heat distortion temperature, etc.), solubility or extractability in water or different organic solvents, flame tests and tests for thermal degradation, ash content, etc. The different polymers may be classified into several groups according to the elements present (Table 10.2). Compounding ingredients, may in certain instances, interfere with the identification of the polymer; so, the compounded polymer should be subjected to certain other tests and analysis including solvent extraction, solubility and swelling in selected solvents, determination of ash content and its chemical composition including tests for different metal ions (Ca^{2+}, Cu^{2+}, Mg^{2+}, Al^{3+}, Pb^{2+}, Zn^{2+}, etc.) and acid radials (Cl^- Br^-, SO_4^{2-}, CO_3^{2-}, PO_4^{3-}, etc.) along with elemental analysis of the resin or polymer component following established analytical approaches and procedures.

10.1.3 Some Specific Tests for Ready Identification of Specific Polymers

(a) ***Tests for Natural Rubber (NR) and Synthetic Rubbers***

(a) Odour Test About 0.1 g of the sample is gently heated in a test-tube in presence of chromic acid solution (5 ml) prepared by dissolving chromium trioxide (2 g) in water (5 ml) and adding concentrated H_2SO_4 (1.5 ml). 1,4 Polyisoprenes yield acetic acid which can be readily identified by its odour.

(b) Colour Test—(i) Weber Test If the above test is positive for an unknown product, it may be further subjected to modified Weber colour test for polyisoprenes. The acetone-extracted polymer (0.05 g) is dissolved or suspended in CCl_4, treated with a little bromine solution in CCl_4 and heated in a water bath to remove excess

Table 10.1 ***Heating tests of some common polymers***

Polymer	*Flame Colour*	*Odour*	*Other Notable Points*
(1)	*(2)*	*(3)*	*(4)*
A. The material burns but self-extinguishes on removal from flame			
Poly(vinyl chloride)	Yellow with green base	Acrid	Strongly acidic fumes, liberation of HCl, black residue
Poly(vinylidene chloride)	"	"	"
Polychloroprene	Yellow, smoky	"	"
Phenol–formaldehyde resin	Yellow	Phenol and formaldehyde	Very difficult to ignite
Melamine–formaldehyde resin	Pale yellow, light blue–green edge	Fish-like, formaldehyde	Alkaline, very difficult to ignite
Urea–formaldehyde resin	"	"	"
Nylons	Blue, yellow tip	Resembling burning vegetables (celery), burnt hair	Melts sharply to clear, flowing liquid; melt can be drawn into a fibre
Chlorinated rubber	Yellow green	Acrid	Strongly acidic fumes, liberation of HCl; swollen, black residue
B. The material burns and continues burning on removal from flame			
Polybutadiene (BR)	Yellow, blue base, smoky	Disagreeable, sweet	Chars readily
Polyisoprene (NR and gutta percha, synthetic)	Yellow, smoky	Pungent, disagreeable	"
SBR	Yellow, highly smoky	Pungent, fruity, smell of styrene	"
Nitrile rubber (NBR)	Yellow, smoky	Burnt, resembling protein	"
Butyl rubber (IIR)	Practically smoke free candle-like	Slightly sweetish	Melt does not char readily
Polysulphide rubber (polymer itself emits unpleasant, mercaptan like odour)	Smoke-free, bluish	Pungent; smell of H_2S	Yellow, acidic (SO_2) fumes
Cellulose (cotton, cellophane, viscose rayon, etc.)	Yellow	Burnt wood	Chars, burns without melting
Cellulose acetate	Yellow	Acetic acid	Acidic fumes

Contd...

Table 10.1 (Contd.)

Polymer	*Flame Colour*	*Odour*	*Other Notable Points*
(1)	(2)	(3)	(4)
Cellulose acetate butyrate	Yellow	Acetic acid/butyric acid	Acidic fumes
Cellulose nitrate (plasticized with camphor)	Yellow	Champhor	Burns very fast, often with explosion
Ethyl cellulose	Pale yellow with blue-green base	Burning wood	Drips on ignition
Polyacrylonitrile	Yellow	Resembling burnt hair	Dark residue, volatiles having an alkaline reaction
Poly(vinyl acetate)	Yellow, smoky	Vinyl acetate, acetic acid	Sticky residue, acidic volatile
Poly(vinyl alcohol)	Yellow, candle-like, limited smoky	Unpleasant, charry smell	Black residue
Polyethylene	Yellow with blue base	Resembling burning paraffin wax	Clear melt
Poly(α-olefins) (PP, EPR, etc.)	"	"	"
Polyesters (alkyds, polyester fibre)	Yellow, smoky	Unpleasant, sweety small	Melts or chars giving acidic distillate and black residue

Table 10.2 *Classification of most common polymers according to elements present*

Group	*Element found*	*Rubbers*	*Plastics/Fibres*
Group 1	N	Nitrile (NBR), Polyurethane (ester/ether urethanes), nitroso rubber, etc.	Cellulose nitrate, cyanoethyl cellulose, silk, polyamides, polyimides, polyurethanes, Polyacrylonitrile, SAN and ABS resins, unrea formaldehyde resins, melamine formaldehyde resins, etc.
Group 2	S	S-vulcanized diene rubbers (NR, IR, SBR, BR, CR, IIR, EPDM, NBR), polysulphide rubbers and chlorosulphonated polyethylenes	Wool, polysulphones, ebonite, etc.
Group 3	Cl	Polychloroprene, chlorosulphonated polyethylenes, etc.	Poly(vinyl chloride), poly(vinylidiene chloride) and related copolymers polychlorotrifluoroethylene, chlorinated or hydrochlorinated rubber, etc.
Group 4	Absence of N, S, Cl, etc.	Peroxide cross-linked or unvulcanized hydrocarbon rubbers (NR, IR, SBR, BR, IIR, EPDM, EPR, etc.)	Petroleum resins, coumarone–indene resins, cellulose and cellulosics other than cellulose nitrate, polyesters, polyethers or acetal resins, polyolefins, polystyrene, poly(methyl methacrylate), poly(vinyl acetate/alcohol), etc.

bromine. The residue is then warmed with a little phenol. In the presence of polyisoprenes the solid or the solution turns violet or purple and gives a purple solution with chloroform. The above test is positive for NR, IR and butyl rubber (IIR) but not for alkali reclaim.

(ii) Trichloroacetic Acid Test Acetone-extracted NR, isomerized rubber and rubber hydrochloride and *gutta-percha*, when warmed with trichloroacetic acid give mauve-pink melts which on boiling turn red or red–brown and on dilution with water result in grey-mauve precipitates that turn yellow on cooling and standing. Chlorinated rubber does not respond to this test. The synthetic rubbers are different in the sense that they either produce no effect or produce yellow-colouration turning to pale yellow/grey precipitates on dilution. Polysulphide rubbers, however, impart pink colour on treatment with trichloroacetic acid which turns into ruby-red solution on boiling, giving rose-pink precipitate on dilution.

(iii) Other tests Rubbers may also be identified by using the pyrolytic vapours from test samples. A strip of filter paper soaked in an ethereal solution containing *p*-dimethylamino benzaldehyde (3%) and hydroquinone (0.05%) and then moistened with a 30% solution of trichloroacetic acid in isopropanol produces different colour reactions in presence of pyrolytic vapours of different rubbers: those from NR produce a deep blue or blue-violet colour and vapours from SBR pyrolysis turns the paper green or blue with a distinct green tinge. Butyl rubber and polyisobutylenes resemble NR and silicone rubbers resemble SBR in this colour reaction. Pyrolytic vapours from NBR or nitrile rubbers give a brown or brown–yellow colour and those from polychloroprenes (CR) turn the test paper grey with a yellow tinge.

Pyrolytic vapours from polyisobutylenes or butyl rubbers (IIR) produce a bright yellow colour on filter paper freshly soaked in a solution obtained by dissolving yellow mercuric oxide (5 g) in concentrated H_2SO_4 (15 ml) and water (85 ml) on boiling. Other rubbers may either produce little change or give an uncharacteristic brown colour.

(b) *Test for Cellulose and Cellulosics* Cellulose and regenerated cellulose (viscose rayon) are insoluble in water and organic solvents. They readily respond to the Molisch test for carbohydrates carried out as follows: a small sample soaked in 1 ml water is shaken well with a few drops of a 10% solution of α-naphthol in chloroform and then a volume of 2–2.5 ml of concentrated H_2SO_4 is so added as to form a lower layer. A violet ring is produced at the interface and on uniform mixing a purple solution results which on dilution with water produces a violet precipitate. Esters and ethers of cellulose respond to this test feebly or much less readily and the purple colour generally appears on heating the liquid mixture; a notable exception is cellulose nitrate which does not respond to Molisch test at all. Cellulose nitrate on the other hand gives an intense blue colour reaction when a drop of a solution of diphenylamine in concentrated H_2SO_4 (5% w/v) is added to the sample in absence of other oxidizing agents.

(c) ***Colour Test for Formaldehyde*** Formaldehyde enters into the composition of several resins and polymers; it is also evolved when some of them are thermally treated or boiled with water particularly in presence of an acid (H_2SO_4). Common resins/polymers of this type are poly(vinyl formal), polyoxymethylenes, urea–, melamine–, and phenol–formaldehyde resins. The aqueous extracts or acid distillates of these polymers or resins containing or liberating formaldehyde and taken in clear aqueous solution are tested with chomotropic acid (1,8-dihydroxynaphthalene–3, 6-disulphonic acid). A few drops of 5% aqueous chromotropic acid solution are added to the aqueous test solution and then an excess of concentrated sulphuric acid is added and the mixture preferably warmed to nearly 100°C for a few minutes. In presence of formaldehyde, the solution turns violet/dark violet. Urea, melamine and phenol are also obtained as intermediates on acid hydrolysis of the corresponding formaldehyde condensate resins and appropriate tests for urea, melamine and phenol may be employed for identification purposes.

(d) ***Test for Polyacrylonitrile and Acrylonitrile Copolymers*** When a strip of cupric acetate paper, freshly moistened with a dilute solution of benzidine in dilute acetic acid is held in the pyrolytic vapours of polyacrylonitrile or its copolymers, a bright blue colouration is readily developed. On the other hand, if the condensed pyrolyzate of polyacrylonitrile or a related copolymer is made alkaline, boiled with a trace of ferrous sulphate and then acidified, Prussian Blue precipitate is readily obtained.

(e) ***Test for Polyamides*** When a strip of filter paper soaked in a fresh saturated solution of *o*-nitrobenzaldehyde in dilute NaOH is held over the pyrolytic vapours of polyamides containing adipic acid, a mauve-black colour is readily developed while vapours of polyamides from diacids other than adipic acid produce a grey colour when tested similarly.

(f) ***Test for Poly(Vinyl Chloride)*** On addition of a few drops of alcoholic NaOH/ KOH solution (10%) to 2–5 ml of 5% solution of poly(vinyl chloride) the mixture changes from colourless through light yellow, yellow–brown, dark brown to black with time. Poly(vinyl chloride) and vinyl chloride copolymers readily form brown colouration and eventually dark brown precipitates when their pyridine solutions are heated, specially in presence of a few drops of alcoholic NaOH solution.

(g) ***Test for Poly(Vinylidene Chloride)*** Poly(vinylidene chloride), when immersed in morpholine, develops a brown colour both in the liquid and the partially swollen polymer; the change takes place quicker if the mixture is warmed in a water bath.

(h) Tests for Poly(Vinyl Acetate) and Poly(Vinyl Alcohol) When a small volume of iodine solution (0.1 g iodine and 1.0 g KI dissolved in 20 ml of a 1 : 1 alcohol–water mixture and diluted to 100 ml using 2 N hydrochloric acid solution) is added to an equal small volume of 0.25% neutral or acidic solution of polyvinyl alcohol, a blue colour develops almost immediately or on addition of a pinch of borax into the solution.

Poly(vinyl acetate), on the other hand, turns deep brown on contact with the dilute iodine solution.

(i) ***Tests for Polystyrene and Styrene Copolymers*** The test depends on the aromatic rings of the styrene units in the polymer chain. The polymer sample (0.1 g) is warmed under reflux in concentrated HNO_3 and then the clear mixture is poured into water (25 ml) to yield a pale yellow precipitate which is then extracted with ether (2 × 5 ml). The ethereal extracts are combined and washed thoroughly with water (2 × 5 ml), extracted with dilute NaOH solution (2 × 5 ml) and finally with water (5 ml). The alkaline extracts are combined with the final aqueous extract and the nitro compounds present in the mixture are reduced using concentrated hydrochloric acid (5 ml) and granulated zinc (1 g) and gently warming. The solution is cooled subsequently, filtered and diazotized with a little dilute solution of $NaNO_2$ under ice cooled condition and finally, the solution is poured into excess of alkaline β-naphthol solution producing a deep red colour.

Polystyrene, the general purpose polymer, produces a characteristic metallic tinkle when objects moulded from it are dropped or struck. Its transparency and brittle nature are also helpful in its identification.

10.2 Physical Testing[4–6]

10.2.1 Stress–Strain Measurements: Mechanical Properties

Stress–strain test is probably the most important and very commonly employed test method for evaluation of plastics, rubbers and related materials, but it is yet to be fully understood from the theoretical standpoint and its practical significance is often overrated. The inadequacy of stress-strain data measured at a single temperature and at a single speed of testing stems from the tremendous variations in the properties of plastics and related materials as a function of time, temperature and thermal history compared to other kinds of materials such as metals and alloys.

Despite the theoretical inadequacies and some practical inconsistencies, stress-strain measurements made over a wide range of temperatures and speeds of testing are very useful for technical evaluation of polymeric materials. Such measurements are among the few that reveal meaningful information about their strength or the conditions under which they will break or fail to perform.

Stress–strain measurements are generally made in tension in standard testing equipment using dumbbell shaped specimens of specified dimensions under specified conditions by stretching the testpiece at a uniform rate and simultaneously measuring the force or stress(τ) on the specimen and the elongation or strain(γ) induced until the specimen breaks. A stress–strain diagram for a typical thermoplastic is shown in Fig. 10.1.

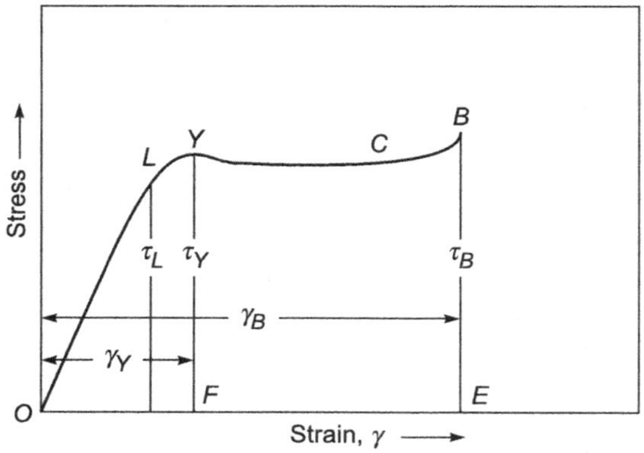

Fig. 10.1 *A schematic stress–strain diagram for a typical thermoplastic polymer*

The stress–strain curve as given by Fig. 10.1 may be examined to find out stress corresponding to the elastic limit (τ_L), yield stress (τ_Y), the ultimate strength or simply the tensile strength as given by the breaking stress (τ_B), the elongation at break or the breaking strain (γ_B) and the modulus of elasticity given by the slope of the linear zone of the curve (OL) up to the elastic limit.

Stiffness of a thermoplastic material is understood by the ability to carry stress without suffering permanent set, the determining parameter being the modulus of elasticity. Resilience corresponds to the ability of the material to absorb energy without suffering permanent set and the area under the elastic portion of the stress–strain curve (OLYFO) is used to evaluate resilience. This can also be taken as the work required to deform the material to its elastic limit.

The strength of the material is its ability to carry dead load and is commonly taken as equal to the breaking or ultimate stress. The material's ability to carry stress without suffering permanent set is a measure of its elasticity which is judged by the criterion of yield point (γ_Y). Toughness of the material is judged by the criterion of ultimate energy resistance or the work done on the material to deform it up to the breaking point and is given by the total area under the stress–strain curve (OLYCBEO). For a material having no well defined yield point, resilience and toughness cannot be meaningfully distinguished. Brittleness arises due to poor resilience or toughness. Stress–strain diagrams typical of some different kinds of plastic materials are shown in Fig. 10.2.

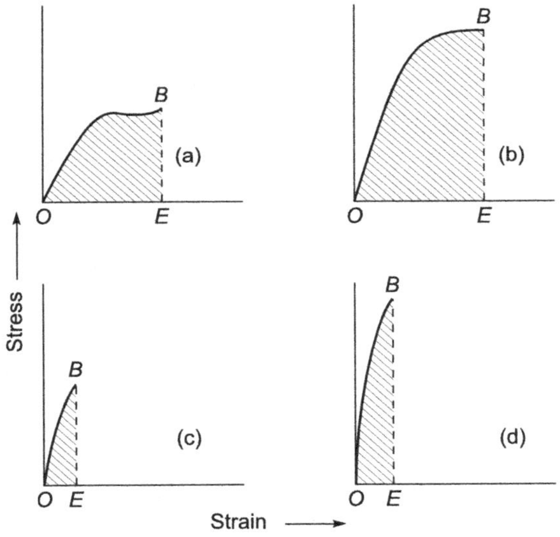

Fig. 10.2 ***Stress-strain diagrams of different types of plastic materials: (a) weak and soft, (b) strong and tough, (c) weak and brittle, and (d) strong and stiff***

The tensile test as commonly done is a *uniaxial test* in which the force is applied to the test specimen in one direction. *Biaxial tests* are those in which stresses from two directions at right angles to each other are applied to the specimen simultaneously. The test is performed using films or sheets and stretching them in two directions by applying pressure to circular diaphragms clamped at the edges or by other suitable means.

In many areas of application, plastic objects are required to withstand biaxial stresses—two good examples being pipes or tubes and packaging bags filled with a fluid or powder. Toy balloons are subjected to very large biaxial deformation when they are blown to sizable extents.

A biaxial stress–strain curve is different from one obtained by the application of a uniaxial stress. Generally, a polymer behaves more in a brittle manner when under a biaxial stress than when subjected to a uniaxial one. The initial slope of the stress–strain curve is greater for a biaxial test than for a uniaxial test, usually giving a relatively low elongation at break and correspondingly a high yield stress value for the biaxial test.

For a uniaxial test, the stress–strain relationship up to the elastic limit is given by Eq. (7.2), i.e., $\tau = E\gamma$, whereas for a biaxial test the relationship is:

$$\tau = \frac{E\gamma}{1-\nu} \tag{10.1}$$

where ν is Poisson's ratio, i.e., the ratio of the lateral contraction to elongation. In the extreme case where the density of the material remains unchanged under stress, $\nu = 0.5$ and under this condition, for a given elongation or deformation, the stress in each of the two directions of biaxial stretch is twice the value corresponding to a uniaxial test. If the material decreases in volume or gains in volume under stress, the value of ν is greater than 0.5 or less than 0.5 respectively. The distribution of the orientation of the chain molecules in uniaxially and biaxially oriented polymeric materials is schematically shown in Fig. 10.3. The tensile strengths of most plastics are lower than the tensile strengths of most common metals and alloys, but in terms of specific strength (tensile strength/density) the plastics are comparable in a large measure with the metals and alloys (Table 10.3).

The particular material for which the stress–strain curve is given in Fig. 10.1 suffers large scale elongation without the requirement of additional stress between points Y and C. Some internal process appears to be aiding the externally applied stress in producing the large scale extension. The internal process in question that follows the stress-induced plastic deformation is commonly recognized to be the stress-induced intermolecular orientation resulting in an alteration of the structure of the molecules in relation to each other which results in measurable reinforcements as clearly indicated by the rise of the stress–strain curve beyond the point C. In thermoplastics, the strain-hardening, i.e. the reinforcement achieved beyond the point Y over the zone YC is permanent and retraction to original dimension on release of stress is not possible. The degree of reinforcement attainable is however low for thermoplastics characterized by low molar cohesion while it is very high for others characterized by high molar cohesion and high potentiality for extensive intermolecular hydrogen bond formation such as for fibre forming polymers. For a

Table 10.3 *Specific strength of different materials*

Material	*Mean Tensile Strength* (τ_B) kg/cm^2	*Density* (ρ) g/cm^3	*Mean Specific Strength* (τ_B/ρ) *cm*
Aluminium	2200	2.7	815,000
Brass	4000	8.5	470,000
Steel	6000	7.8	769,000
Copper	5000	7.3	640,000
Zinc	1700	7.2	236,000
Tin	350	7.3	48,000
U–F resin	700	1.5	466,000
P–F resin	500	1.4	357,000
Poly(methyl methacrylate)	550	1.2	420,000
Polystyrene	500	1.05	476,000
Cellulose acetate	350	1.3	270,000
Polycarbonates	650	1.2	541,000
Polypropylene	320	0.9	355,000
Poly(vinyl chloride)	490	1.4	350,000
Nylon 66	800	1.14	700,000

typical rubber which has been vulcanized, the stress–strain curve, however, does not exhibit any distinct Hookian zone; it assumes a shape shown by the loading curve in Fig. 7.14. Reinforcement on extension is also quite prominent for vulcanized rubber, but it is of a temporary nature, since on release of the applied stress from a stage of high extension, the material snaps back to or near the zero-stress dimension though mostly by a different route as shown by the unloading curve (Fig. 7.14).

Stress–strain tests may also be done in compression. Bulk modulus may be obtained from the initial slope of the compressive stress–strain curve. Rigid materials (plastics) appear significantly less brittle under compression than under tension. Normally hard polymers, brittle under tension, may appear to assume some order of ductility under compression. The breaking stress or the ultimate strength under tension is much less than the ultimate strength under compression, often by a factor of nearly 0.5.

For a vulcanized rubber, evaluation of permanent set is technically useful as it gives a measure of the extent to which the material falls short of ideal elasticity. The parameter is also a helpful guide in assessing degree of vulcanization. When done

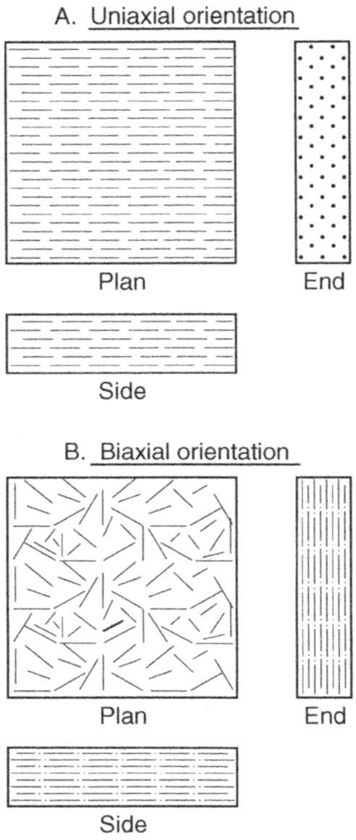

Fig. 10.3 *Schematic representation of the distribution of orientation of chain molecules in (A) uniaxially oriented and (B) biaxially oriented polymers*

in tension, a strip of the test vulcanizate is marked with reference lines at a known distance apart and then stretched and kept in the stretched position for a specified time period. After a specified time period following release of the applied stress, the distance between the reference lines is measured again.

Compression set may be determined by applying a constant load on the specimen in the form of a flat disc with the help of a calibrated spring. In the determination of compression set at constant strain, the test specimen of the same kind is compressed to a fixed percentage of its original thickness by clamping it between rigid parallel plates using different spacers or gaskets so as to control the degree of compression. Technically, the set value is the difference between the original length (or thickness) of the test specimen and that after recovery following the specified procedure and is expressed as a percentage of the initially attained elongation (or compression).

10.2.2 Effect of Temperature and Plasticizer on Stress–Strain Relationship

The stress–strain properties of polymers are temperature sensitive. The modulus, yield strength and tensile strength usually follow decreasing trends as the temperature increases. An increase in temperature brings about a change in the nature of the stress–strain curve quite significantly, say from hard–brittle type to a hard–tough or even a soft–tough type. Below T_g, the modulus is rather insensitive to temperature changes for amorphous polymers. The modulus or the slope of the stress–strain curve, however, undergoes a sharp lowering, may be by a factor of reciprocal of a thousand, by increasing the temperature by a few degrees around T_g. For a crystalline polymer, the drop in modulus usually takes place by a factor of 0.1 through the glass transition region (*See also* Chapter 1). In this respect, heavily cross-linked thermosetting resins such as the phenolic (p–f) and amino (m–f) resins are much less sensitive to variation of temperature than the linear thermoplastics—amorphous or crystalline.

Incorporation of a plasticizer in a rigid polymer produces a change comparable with that effected by increase in temperature. Heat, in fact, is a physical plasticizing agent. Plasticizer incorporation usually brings about a sharper lowering in T_g than in T_m of a polymer system resulting in a widening of the separation of the two transition zones (*See also* Chapter 7, Fig. 7.18) on increasing plasticizer incorporation.

10.2.3 Dynamic Mechanical Behaviour

All polymers are known to be characterized by some degree of viscoelasticity. When a strain is developed in or a stress is applied to such as material, part of the energy is stored elastically and the rest is lost in the form of heat in the deformation process (*See also* Secs 1.6.6 and 7.18). The ratio of the energy dissipated to that stored gives a measure of the damping characteristics of the material. The delayed or time-dependent reaction of a polymer to strain or stress influences its dynamic properties.

Application of time variable sinusoidal strain (γ) to a specimen at an angular frequency ω and with a small amplitude produces a sinusoidal stress (τ) the magnitude of which varies with time (t), but it is out of phase with the applied strain by a phase angle δ (Fig. 10.4), such that one may write:

$$\gamma = \gamma_0 \sin \omega t \tag{10.2}$$

and

$$\tau = \tau_0 \sin (\omega t + \delta) \tag{10.3}$$

where γ_0 and τ_0 are the peak strain and peak stress respectively.

The tangent of the phase angle (tan δ) measures the internal friction or the dissipa-tion factor $\Delta E/E$, where ΔE is the energy dissipated or the damping energy in taking the specimen through a stress or strain cycle and E is the energy stored in the sample when the stress developed is maximum. In other words, the value of tan δ gives a measure of the relative importance of the viscous and elastic behaviour of the material system of the test specimen.

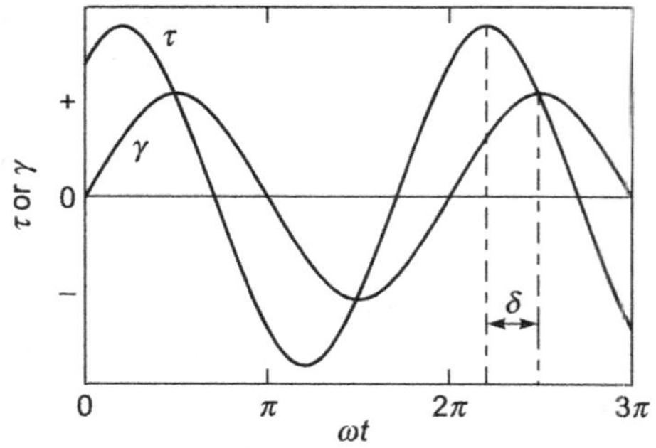

Fig. 10.4 *Strain (γ) and stress (τ) as a function of time (t) on application of a sinusoidal strain to a viscoelastic specimen at a frequency, ω*

The peak stress can be resolved into two components, one having a value of τ_0 cos δ that is in phase with the (applied) strain which is characteristic of the stored elastic energy, and the other having a value of τ_0 sin δ that is 90° out of phase with the strain which is characteristic of the loss or dissipated energy. Rheovibron (dynamic) viscometer is widely used for measurements of dynamic mechanical properties such as loss modulus, storage modulus and dissipation factor, each as a function of temperature. In this instrument, the test specimen is clamped between strain gauges and subjected to low orders of sinusoidal (tensile) strain at a specified frequency. The value of tan δ is directly read and the storage and loss moduli are calculated using sample dimensions and instrument readings.

In order to compare data given by dynamic tests following different approaches and methods or techniques, it is necessary to convert them to a common basis. The dynamic mechanical properties are interrelated through use of a complex modulus.[7] The complex shear modulus G^* is expressed as $G^* = G' + iG''$. Similarly, the complex elastic (Young's) modulus E^* is expressed as $E^* = E' + iE''$, where $i = (-1)^{1/2}$. G' and E' (each considered equal to (τ_0/γ_0) cos δ) constitute real parts of shear modulus and of elastic modulus respectively, while G'' and E'' (each considered equal to (τ_0/γ_0) sin δ) correspondingly constitute imaginary parts of the moduli. For very small damping, $G' = G$ and $E' = E$. In general, the absolute value of shear modulus (G) and of elastic modulus (E) are expressed as $G = |G^*| = [(G')^2 + (G'')^2]^{1/2}$ and $E = |E^*| = [(E')^2 + (E'')^2]^{1/2}$.

The imaginary parts of the moduli are the damping terms that give a measure of the dissipation of energy into heat as the material is deformed. The useful damping term called the dissipation factor or loss tangent is given by (G''/G') or (E''/E').

The dynamic modulus and damping of a Maxwell unit as a function of frequency are schematically shown in Fig. 10.5. At low frequencies, the dynamic modulus is very low as most of the deformation or motion comes from the dashpot (viscous component). At very high frequencies, enough time is not available to allow appreciable flow in the dashpot in a given cycle, and hence, the deformation or motion is mostly or overwhelmingly due to the stretching of the spring such that the dynamic modulus nearly equals the modulus of the spring (elastic component). At intermediate frequencies, the deformations or motions of both the spring and dashpot take place measurably and in this frequency range, the dynamic modulus (E') increases rapidly with frequency. The loss modulus (E'') approaches zero at low as well as high frequencies. At intermediate frequencies, both shear rate and the motion of the dashpot are appreciable and hence the loss or dissipation energy is large. The loss modulus passes through a maximum when the frequency $\omega = 1/\lambda$, where λ is the relaxation time. The absolute value of the modulus, E is equal to E'' at low and to E' at high frequencies.

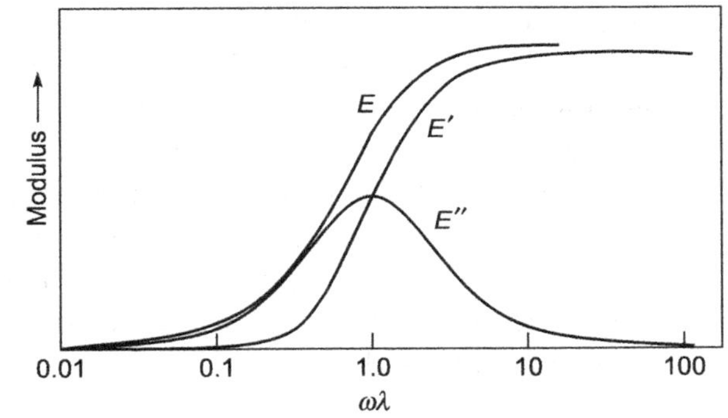

Fig. 10.5 *Dynamic behaviour of a Maxwell unit as a function of frequency, ω*

10.2.4 Stress Cracking

Brittle fractures occurring in such polymers as polyethylene may arise due to: (1) environmental stress cracking, (2) thermal stress cracking, and (3) fatigue failure. Environmental stress cracking is due to chemical or solvent attack on the polymer; thermal stress cracking is a consequence of (repeated) heating and cooling and fatigue failures arise as a consequence of subjecting the material to repeated stress–strain cycles.

A very sharp incidence of environmental stress cracking may be observed when the outer curved surface of a bent strip of poly(methyl methacrylate) is held over acetone vapour arising from an open container of acetone. The strip will shatter into pieces almost instantaneously, often with explosive violence. This phenomenon may be encountered with many other polymers and resins in specific critical environmental conditions. Factors playing important roles in environmental stress cracking are: minor defects in the surface of the material, a sensitizing agent and built-in stresses. A crack will appear if only the accumulated stresses at the defect points exceed the cohesive strength of the material. The crack propagates along the intercrystalline boundaries bridged by fewer polymer chains.

Average molecular weight and molecular weight distribution are also important factors to control stress cracking behaviour. The stress cracking increases as average molecular weight decreases. With more tie molecules constituting the intercrystalline regions, as for systems of higher average molecular weight, the crack propagation becomes more difficult. For a polymer of a given structure, higher order of crystallinity and bigger crystalline size make the polymer more prone to stress cracking. Built-in stresses consequent to freezing of orientation of chain molecules in moulded, extruded or calendered objects give rise to failures due to stress cracking. More active solvents or contaminants causing more swelling or wetting produce more severe stress cracking. Improved formulation, part design and moulding methods may go a long way in reducing the stress cracking. A simple way of measuring stress cracking is to bend strips of a plastic material of specific dimensions to a specific angle and keep or immerse them in the desired environment, solvent or reagent while holding them in the bent condition and the time to stress crack is measured. The observed time may vary from seconds or minutes to years depending on factors relating to nature of polymer, processing and test conditions and the environment used.

10.2.5 Hardness

Hardness of a material is generally measured by its resistance to local deformation or indentation by a rigid body of a specified dimension under a specified condition. Hardness is a complex property and it is related to such mechanical properties of materials as modulus, strength, elasticity and plasticity, even though the relationship is not quite straightforward. In a given class of plastics and rubber materials, higher modulus and strength usually result in higher hardness. Hardness is a relative concept and it cannot be absolutely defined. It can be measured and expressed following standardized method of testing using a standard equipment.

Hardness evaluation is very commonly used as a simple and rapid means of production and quality control in the plastics and rubber industries in the processing sector. Hardness testing is particularly useful in getting an indication of degree of cure in thermosetting resin moulding and rubber vulcanization.

Rockwell Hardness The standard test for hardness of plastics of different degrees of rigidity comprises procedures designated by ASTM number D785. The Rockwell hardness testing is commonly done by using different scales of measurements, such as R, L, M, E, K all coloured red and scale X coloured black, each scale having a characteristic combination of a minor load, a major load and a specific diameter for the indentor steel sphere. The Rockwell hardness numbers are not measures of total indentation but of the non-recoverable indentation after the major load applied for 15 s is reduced to the minor load for 15 s.

Durometer Hardness Test (Shore A and Shore D) This test is described in two similar ASTM methods D 1706 and D 2240, each employing two types of Durometers—type A and type D, for measuring the indentation hardness of materials ranging from soft rubbers to hard rubbers and plastics. These equipments measure the depth of indentation under load when a hardened steel indenter is forced into a specimen by means of a calibrated spring. The scale has a graduation from 0 to 100 hardness numbers and has an inverse relationship to indentation giving an indentation of 0.1 in at zero and 0.000 in at 100.

For hardness tests on vulcanized rubbers, the depth of indentation is converted to international rubber hardness degrees (IRHD) the scale of which also ranges from 0 (infinitely soft, zero elastic modulus) to 100 (infinitely hard, infinite elastic modulus) following standards and procedures as per ISO recommendations (ISO R–48). The readings of IRHD are approximately the same as those of the shore Durometer type A.

10.2.6 Tear Strength or Tear Resistance

For flexible polymers, such as some soft thermoplastics, plasticized PVC, vulcanized rubbers, etc. tear strength or resistance to tear is a property of special consideration; the testing commonly practised is a special form of tensile testing, making use of the same equipments and several form of standard tear test-pieces. A small cut is given in the test piece, particularly for rubbers, leaving a standard initial untorn width, w, in view of the fact that rubbers tear more readily once the process has been initiated. The tear strength is taken as the force per unit thickness of test-piece required to cause a nick cut in the test-piece to extend until the piece is broken or torn and is expressed in the unit of kg/cm or lb/in.

10.2.7 Resilience

Resilience is technically evaluated as rebound resilience by allowing a metal ball to fall on the flat surface of the test material or a ball of the test material to fall on a flat metal surface and noting the extent of rebound. If h_0 is the height of fall and h the height of rebound, the rebound resilience is expressed as $(h/h_0) \times 100\%$. It is, however, difficult to measure resilience by the falling ball method to a good degree of precision or accuracy.

Most standard tests measure with convenience the rebound of a pendulum from the test material. They, however, differ considerably with one another in the design of the pendulum leading to variations in the corresponding test results. Instruments widely in use are the Lupke pendulum, the Schob pendulum, the Goodyear–Healey instrument and the Dunlop Tripsometer.

10.2.8 Flex Cracking Resistance

Failure of a flexible material such as a flexible plastic item or more particularly a sample of rubber may appear due to development of cracks on the surface in the region where the material has been repeatedly flexed. The surface cracks developed are small at first but gradually become deeper and may grow through the entire thickness of the substrate on extended flexing. Resistance to flex cracking, also commonly referred to as flex endurance, assumes importance for rubbers and rubber-like materials, with special reference to such applications as tyre treads and side walls, conveyor and power transmission belts, different parts of footwears particularly the shoe sole, etc. The most widely used machine is the De Mattia flex testing machine which utilizes moulded flat test strips of specific dimensions that are flexed at a specified rate, say 300 cycles per minute.

The task of assessing flex cracking resistance is far from simple. The number of flexing cycles required to initiate a crack or to result in a specified degree of cracking is to be measured. The cycle frequency is important, though not critical, but it is necessary to operate in an ozone-free atmosphere, as ozone, hastens the initiation of cracks quite markedly and critically.

10.2.9 Abrasion Resistance

Abrasion is a mechanical property not easy to define or to measure properly. Methods or instruments developed are many, but very few of them measure the same thing in reality. Abrasion loss or simply abrasion is closely related to friction even though friction in itself does not necessarily lead to material wear. Abrasion may result in physical damage of surfaces, and build-up of high temperatures locally at points of abrasion or friction is often accompanied by chemical attack such as oxidation in some cases which often results in embrittlement and wear that is commonly taken as abrasion loss.

Abrasion phenomenon assumes special importance for rubbers in view of their use in tyres. For rigid polymers, we refer to a property called scratch resistance which, however, is closely related to abrasion resistance. Abrasion loss is generally the volume abraded in a given time or over a given length of abrasion path. Most machines to evalute abrasion loss allow moulded or punched out test specimens (usually disc shape) of specified dimensions pressed against a rotating abrasive surface by a constant force or by a force producing a constant frictional drag.

10.2.10 Impact Resistance

The capacity of a material to absorb energy of a high speed blow without failure or rupture is a property of great technological significance. Impact tests are in fact made to get a measure of resistance to breakage of materials under high velocity impact conditions. There are many empirical methods of measuring impact resistance or impact strength in current use in the plastics industry, though with little correlation among them. Methods that are more commonly used are the Izod impact test, the Charpy impact test, the falling weight impact test and the high-speed stress–strain tests.

Both the Izod and Charpy instruments are pendulum instruments. In each case a specially designed hammer or pendulum released from a specific height strikes the test specimen placed or clamped at the base of the instrument with enough kinetic energy to fracture the specimen. In Izod method, a notched cantilever beam type test specimen, firmly clamped in a vertical position in a vise fixed at the base of the apparatus, is struck horizontally by the pendulum on the top (free) and above the notch. The pendulum continues on its arc after the specimen has been fractured and the remaining energy is measured by the extent of the excess swing. The Charpy test on the other hand makes use of a horizontal rectangular beam specimen supported at two ends, which is struck at the middle by the pendulum. Here also, the drop in the extent of backward swing is taken as the measure of the energy absorbed consequent to impact on the specimen. In addition to the true energy required to initiate a crack and propagate it through the test specimen, energies required to deform it and to throw the broken pieces apart are also measured in the process. The energy used up in throwing the broken pieces, unrelated to the actual impact energy may constitute a major part of the total energy measured by the pendulum tests.

Falling weight type impact tests utilize a dart, a ball of a standard weight dropped from a specific height on to a bar or sheet made from the test material. The standard striker is attached to a load carrying device to which selected weights can be attached. The striker assembly slides freely in vertical guides when released from a height to strike centrally on a specimen supported at the base of the equipment. The impact strength of a specimen corresponds to the experiment where the falling weight barely has enough energy to crack or break the test specimen. A "staircase" method is often followed in which the load is changed either in the increasing order or decreasing order on successive standard specimens, each struck once, until the specimen just fails or just resists failure. In view of the fact that in falling weight method no energy is wasted in knocking out the broken pieces, its results are more relevant in a practical sense, even more so because the test can measure just the energy to form a crack rather than that required to completely fracture the specimen.

High speed stress-strain tests appear to be the most meaningful for evaluating the impact strength of material—particularly for those which are relatively tough and ductile—but not as meaningful for brittle materials such as the brittle thermosetting resins. The impact strength is given by the area under the stress–strain curve obtained under the high speed test condition (> 50 cm/min) and the high speed test results are often correlated well with the falling weight impact values.

A point of critical consideration in impact measurements is a notch in the test specimen which may lower the measured impact strength drastically. A notch acts as a stress concentrator and it has apparently a greater effect on ductile materials than on brittle things. For brittle thermosetting polymers, the impact strength varies little with temperature over a wide range such as between –50 and 200°C. The impact strength of thermoplastic materials is usually strongly temperature dependent. Fibrous fillers generally improve the impact strength of brittle thermosetting resins and polymers. Fibres distribute the stress over a large volume at the base of the notch and they also help prevent propagation of a crack by carrying a large part of the load in the area round the crack. Crystallinity has also a role to play in controlling impact strength—higher the degree of crystallinity or the size of the crystallites, lower is the impact strength. In fact, crystallites act as stress concentrators in most crystallizable polymers.

10.3 Electrical Properties

Till about the first few decades of the twentieth century, insulator items consisted primarily of glass, wood, paper, rubber, asphalt, mica, amber and related materials mostly of natural origin. The availability of a large spectrum of synthetic polymers later has changed our options in this regard quite remarkably. For electrical application, the selection is required to be made on the basis of evaluation of electrical property parameters.

10.3.1 Resistivity

A good insulating material is one that offers a high electrical resistance under a variety of conditions. Volume resistivity of a material is the resistance between opposite faces of a unit cube when the current flow is confined to the volume of the test specimen, and it is commonly expressed in ohm cm. The reciprocal of volume resistivity is conductivity.

Most high polymeric materials are very good to excellent insulating materials and they are less subject to conduction by imperfection and structural irregularities than other types of insulating materials such as the ceramics in view of the polymeric materials being non-porous or of very low porosity in comparison. The ductility of many of polymeric insulators and the greater control in keeping them free from ionic and metallic impurities during synthesis and processing are also added factors in their favour. The *surface resistivity* is given by the surface resistance between two electrodes forming the opposite sides of a unit square of the test material and is expressed in ohm. In some cases the volume and surface resistivities of polymers are much higher than ordinarily required. The resistivity measurements are often used to check uniformity in synthesis and processing. Humidity affects volume resistance of different insulators to markedly different extents. Non-polar polymers such as polystyrene and polyethylene are unaffected but polar organic polymers are measurably affected by moisture. The extent to which they are affected depends on the degree of their moisture absorption and consequent solvation. Porosity favours moisture absorption and lowers volume resistance. Polar inorganic polymers such as quartz and glass, however, remain practically unaffected by moisture if they are non-porous because they undergo little solvation. Resistance of polymeric insulators suffers appreciably with increase in temperature.

10.3.2 Dielectric Constant

If a potential of V volts is applied to the plates of an air filled parallel plate capacitor, a certain amount of charge, say, Q, will be induced on the two plates such that $Q = CV$, where the proportionality constant C is called the capacitance. If a slab of an insulating material is interposed between the two plates, a larger value of capacitance, C_s will be measured. The ratio C_s/C is defined as the dielectric constant of the slab material or the insulator. The dielectric constant is characterized solely by the inherent nature of the material and is independent of geometry of the test specimen and the applied voltage.

10.3.3 Electrical Breakdown

Failure of electrical insulation occurs under high voltages. The maximum potential gradient that an insulating material can withstand without breakdown and passage of discharge is known as the *breakdown voltage*. The potential gradient or the electric intensity at a point in the material may be considered as the electrical stress. When acting on the material, it produces an electric displacement proportional to the employed electric stress as long as the latter is within a critical value. Exceeding this critical value, the electric displacement increases sharply till a failure and collapse occurs due to the structure of material becoming unstable. The pattern of change is analogous to what is experienced when a solid is stressed mechanically within and beyond its yield point.

The voltage required to produce breakdown is dependent on many factors such as duration and rate of voltage application, thickness of test specimen, frequency of the applied voltage, temperature, dimensions and geometry of the electrodes and the nature of the prevailing environment. All these factors necessitate laying down of a specific test condition. The determination of breakdown voltage allows calculation of dielectric strength with a measure of the thickness of the specimen at the point of rapture. The dielectric strength is expressed in volts per mil (0.001 in) or in volts per millimetre. Flexible plastics characterized by high dielectric and mechanical strength in thin sections are useful as insulating tapes.

10.3.4 Power Factor

Power factor is an electrical property parameter that represents, on a simple consideration, the amount of electrical energy absorbed by an insulator subjected to an alternating current field. For a perfect insulator, there will be no power loss, i.e., no power would be absorbed by the material.

Let us consider that a dielectric is quite simple, manifesting only dipole and electronic polarizations. At a low frequency range, as the direction of the applied field changes or reverses at a relatively slow rate, both the dipoles and electrons always have sufficient time to orient and remain in phase with the field and little energy is absorbed by the dielectric neglecting d.c. conductance. Moreover, the dielectric constant remains essentially unchanged, both the dipole and electronic polarizations contributing to it.

At sufficiently high frequencies, the dipoles, owing to the internal viscosity of the material, are totally incapable of following the reversals of the applied field even though the electrons can still do so, causing stoppage of dipole polarization. Under such conditions too, there will be little electrical energy absorption by the dielectric since the electronic polarization is always in phase with the field.

In the range of frequencies between those stated above, the dipoles follow the reversal of field, but not fast enough to remain in phase or completely oriented with it. The dipole orientation thus lags behind the applied voltage and in each cycle the electrical energy absorbed in the process is lost as heat. The frequency range where this occurs is known as the region of anomalous dispersion and the feature is also referred to as dielectric absorption. In this frequency range, the dielectric constant is no longer a constant but changes with the frequency. Owing to this dielectric loss, the current in the practical capacitor leads the voltage by some angle ϕ which is slightly less than 90°C as shown in Fig. 10.6. This arises due to the delay in reversing the molecular dipoles in the polymer. Power factor is expressed as the cosine of the dielectric phase angle ϕ or sin δ, where $\delta = (90 - \phi)$ is known as the loss angle. For an ideal capacitor, $\phi = 90°$ and the loss angle δ is zero. The current through the dielectric, I_d may be regarded as composed of two components (Fig. 10.6): (a) the capacitative or displacement current, I_c which is out of phase with the voltage and does not contribute to dissipation of power, and (b) the ohmic or conductance current, I_o in phase with the applied voltage that contributes to power dissipation. I_d is the vector sum of I_c and I_o. The ratio I_o/I_c, given by tan δ is termed as the dissipation factor. Polymers without polar groups are superior to those with such groups as high frequency dielectrics.

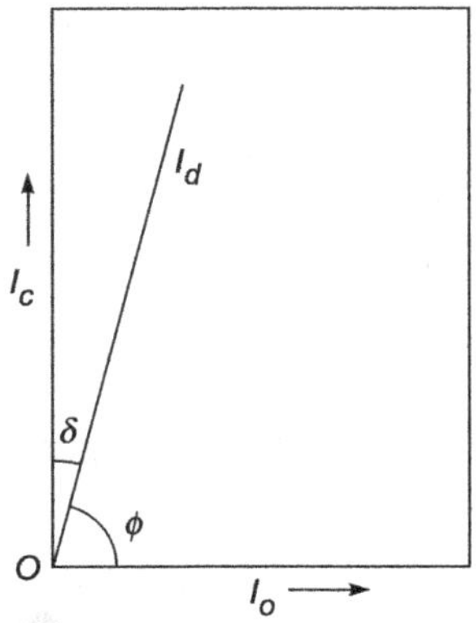

Fig. 10.6 *Current in the dielectric I_d and its components I_c and I_o*

10.4 Softening Temperature Tests

Softening temperatures of plastics are technically evaluated by distortion or penetration tests and are correspondingly referred to as heat distortion temperature or heat penetration temperature. All softening or deflection tests reveal the temperature at which deformation of the plastic material reaches some specified limit under specific set of test conditions. Test conditions usually consist of either flexural, penetration or tensile stressing of a test specimen under a specified set of conditions while increasing its temperature at a constant rate and recording the deflection, penetration or tensile strain that takes place.

In most flexural methods, the specimen is freely supported at a short distance from each end and a load is applied midway between the two supports. The temperature of the specimen is raised at a constant rate, keeping the specimen immersed in a silicone oil bath. The temperature at which a small specified deflection (say, 0.01 in or 0.0254 cm) occurs is recorded as the heat distortion (deflection) temperature. Measurement with automatic recording of specimen deflection is helpful. The gradient of the deflection-temperature curve gives a meaningful engineering design information as to how rapidly a specimen fails.

In a penetration method, an indentor is forced against the specimen and its temperature is increased at a constant rate until a specified penetration occurs. The Vicat softening point test is widely used. A needle or indentor of a specified area (1 sq. mm) is applied to the specimen surface under a specified load (5 kg). The temperature is raised at a specified rate (50°C/h). The temperature at which the indentor/needle penetrates 1 mm is called the Vicat softening point.

Both the penetration and deflection methods give nearly the same information. Such tests are useful in determining the highest application temperature of a plastic and in assessing degree of cure, plasticization or copolymerization effects, development of crystallinity and other structural modifications.

10.5 Melt Flow Index (MFI)

The apparatus used for MFI determination is essentially an extrusion plastimeter. The evaluation relates to melt flow behaviour of polyolefins (polyethylenes, polypropylene, etc.) under a specified condition. For a given polyolefin type, the MFI is inversely related to average molecular weight, higher average molecular weight giving lower MFI. The test polymer may be used in any convenient form: powder, granules or moulded pieces. The polymer sample is contained in a vertical metal cylinder, subjected to controlled heating at 190°C and then extruded through a jet by a loaded piston. MFI is given by the amount of polymer in gram extruded in 2.5 min. The variables prescribed are dead load on the piston and jet size, depending on the MFI range.

References

1. Haslam, J. and H.A. Willis, *Identification and Analysis of Plastics*, Iliffe, London. 1965.
2. Kline, G.M. Ed., *Analytical Chemistry of Polymers*, Wiley-Interscience, New York, 1959–1962 (Three Volumes).
3. Saunders, K.J., *Identifications of Plastics and Rubbers*, Chapman and Hall, London, 1966.
4. Krause, A., A. Lange and M. Ezrin, *Plastics Analysis Guide—Chemical and Instrumental Methods*, Harver Pub., New York, 1983.
5. Crompton, T.R., *The Analysis of Plastics*, Pergamon Press, Oxford, 1984.
6. Brown, R.P., *Physical Testing of Rubbers*, Applied Science, London, 1979.
7. Ferry, J.D., *Viscoelastic Properties of Polymers*, Interscience, New York, 1961.

CHAPTER 11

Polymer Blends and Composites

11.1 Polymer Blends

Mixing together of two or more different polymers or copolymers is known as blending. Blending is a process somewhat similar to compounding. In a polymer blend, the constituent polymers are usually present in significant weight or volume proportions with respect to each other, but seldom all in equal proportions. Blending of two or more different polymers is often viewed as similar though not equal to alloying of two or more metals, the objective in either case being development and production of new uniform materials having ranges of properties far different from those of the constituents. In metal systems, the alloy is a solid solution necessarily giving a homogeneous mass, while the polymer blends or alloys are seldom homogeneous in that sense.

11.2 Consideration of Compatibility

The question of compatibility between polymers is a very important consideration in making blends from them.[1] In a very general manner it may be said that polymer–polymer incompatibility is more a rule than the exception. Even if two different polymers are by and large compatible under a specified condition, they slowly develop incompatibility as they are put to use under a different set of conditions. It is at the same time questionable if complete miscibility or compatibility between the component polymers in a blend is at all the most desired feature.

A useful blend has the characteristics of a uniform dispersion—fine discrete particles of one component being uniformly dispersed through the medium or matrix of the other component. It would be advantageous if the size and size distribution of the domains or the dispersed phase remain practically unaltered over long time periods from the condition of blending to the condition under which the blend is required to perform. Use of selected compatibilizers[2,3] in the form of appropriate block copolymers or coupling agents such as the silane or titanate coupling agents may substantially help in dispersion of additives in polymers by improving adhesion or by minimising phase separation in binary or multicomponent polymer blends by playing the role of a surfactant.

11.3 Types of Polyblends

Polymer blends are also commonly termed as polyblends and they may be of the following different kinds:

(i) Mechanical polyblends,
(ii) Chemical polyblends,
(iii) Mechano-chemical polyblends,
(iv) Solution cast polyblends, and
(v) Latex polyblends.

11.3.1 Mechanical Polyblends

A mechanical polyblend is made by melt blending of the constituent polymers. In the case of binary blends from two different amorphous polymers, the processing temperature must at least be well above the glass transition temperature (T_g) of both the constituting polymers and for blends of semicrystalline polymers, it must be higher than the melting points of the two constituent polymers. Because of the high heat input necessary to maintain a low melt viscosity and also because of the high shearing rates needed for thorough mixing or blending, some degree of chain branching or chain extension leading to formation of block and graft copolymer structures, chain scission and even cross-linking may occur in the polymers making the blend. Melt mixing is used for systems in which thermal degradation does not ordinarily occur.

11.3.2 Chemical Polyblends and Mechano-chemical Polyblends

A chemical polyblend is given by polymeric systems in which long monomeric sequences of one kind are chemically linked to similar long monomeric sequences of a different kind in either the axial direction or in the cross direction giving block copolymer or graft copolymer structures respectively. Selective or random cross-linking of mechanical blends may ultimately lead to mutual grafting, cocross-linking or intercross-linking resulting in formation of what may be termed as mechano-chemical polyblends. The resultant polyblend may often appear as an interpenetrating polymer network (IPN) system of structurally different polymers. Such polyblends are commonly characterized by a relatively uniform phase morphology with remote chance for gross phase separation, improved mechanical strength, thermal stability, chemical resistance and durability.

11.3.3 Solution-cast Polyblends

In solution blending, selected diluents are used to dissolve the component polymers; the diluents also lower the temperature and shear force necessary for satisfactory and uniform mixing without having any degradative effect in the bulk properties of the blend prepared. But removal of the diluents after solution blending may ultimately lead to prominent and uncertain changes in the phase morphology, thus weakening the blend and lowering its performance reliability. Incomplete removal of solvent may further pose potential complication for lack of reliability in performance at the desired level and uncertain failures.

11.3.4 Latex Polyblends

Latex mixing is one of the most important techniques for the preparation of commercial polyblends. In a latex polyblend, polymers are present as suspended, stabilized micropheres of a specific size and size distribution. Interactions between the neighbouring particles are prevented by the stabilizers present in the medium. When two different latexes are blended together, the mixture contains a random suspension of dissimilar kinds of particles, each unaffected by the presence of the others of like or unlike kinds. As the flocculation is entirely contingent upon the emulsifier concentration and not on the polymer characteristics, the coagulation is manifested as a random process. Blends thus obtained on coagulation appear as relatively intimate mixtures of the constituent polymers.

11.4 Properties of Polyblends

The behaviour of two polymers in a mixture or blend is not necessarily the same as expected from the behaviour exhibited by them in their isolated/separate forms. The ultimate behavioural pattern of polyblends is dependent on:

(i) Extent of phase separation,
(ii) Nature of the phase provided by the matrix material,
(iii) Character of the dispersed phase,
(iv) Interaction between the component polymers, etc.

It has been commonly observed that for a miscible polyblend, the following arithmetical semiempirical rule[4, 5] is obeyed by different physical properties:

$$P = P_1\,\phi_1 + P_2\,\phi_2 + I\,\phi_1\,\phi_2 \tag{11.1}$$

where P is the property of interest, ϕ is the concentration and I is an interaction term that can be positive, negative or zero. If I is zero, the rule of mixtures (additivity principle) is observed. If it is positive, the polyblend property would be higher than the weighted average of the constituent polymers and the polymers produce a synergistic effect in the blend. If I is negative, the polyblend property would be lower than what is expected from simple average and the blend system is then referred to as antisynergistic.

For an immiscible polyblend giving a continuous phase (phase 1) and a dispersed phase (phase 2), Eq. (11.1) is not tenable for property analysis. In such cases, the semiempirical relationship to analyze the physical properties is given by Eq. (11.2):

$$\frac{P}{P_1} = \frac{1 + AB\,\phi_2}{1 - B\psi\phi_2} \tag{11.2}$$

where ϕ_2 is the concentration of the dispersed phase constituent. The value of A varies between zero and ∞ depending on the shape and orientation of the dispersed phase as well as the nature of the interface. B depends on the relative values of the properties P_1 and P_2 and A, and ψ is a reduced concentration term that is a function of the maximum packing fraction. It is a common experience that for $A \to 0$, the dispersed phase is soft and for $A \to \infty$, the dispersed phase is hard. Equation (11.2) is useful for analysis of almost all physical properties of immiscible blends excepting the failure and toughness properties.

11.5 Glass Transition of Polyblends

The glass transition phenomenon characterizes almost all polymers. At glass transition points, different polymers are said to exist in an isoviscous, isoconfigurational and iso(free volume) state. A miscible polyblend exhibits a single and sharp glass transition temperature (T_g), intermediate between those of the constituent polymers[6]. However, polyblends showing gross phase heterogeneity such as styrene-acrylonitrile (SAN) copolymer and grafted polybutadiene[7] exhibit glass transition for each constituent polymer present with some shifting and broadening, however. The T_g method of characterizing blends is limited by poor resolution particularly when the constituent polymers have T_gs differing within 20°C.

11.6 Techno-commercial Consideration for Development of Polymer Blends

Commercially available polyblends are either homogeneous or heterogeneous. Plasticization of a given rigid polymer by blending with appropriate polymeric plasticizers leads to homogeneous polyblends. In comparison with simple, low molecular weight liquid plasticizers, polymeric plasticizers are relatively immobile and show low or little volatility. Their effect on T_g is small and they yield blends with reasonable or undiminished rigidity[8] under ambient conditions and allow measurable lowering in heat distortion temperature and melting temperature and make the product flow at relatively low temperatures more readily during processing. Incorporation of polymeric plasticizer renders the polyblend product stable, free from proneness to migration or blooming of the plasticizer additive to the surface of finished products. Use of polymeric plasticizers, therefore, often yields blends endowed with energy saving processability. Thus, low molecular weight poly(methyl vinyl ether) may be used to plasticize polystyrene, and polyethylene glycols may be used to plasticize cellulose nitrate.

Heterogeneous polyblends possess unique and wide application potentials in the field of high impact plastics. If a small to moderate proportion of a suitable rubber (gum) is blended with a brittle plastic, a softer, low modulus material of improved impact resistance results and if the rubber is appropriately cross-linked as the blend is made, the resultant product would normally show much improved toughness and impact resistance with measurably improved stiffness as well if the uniform dispersion and prevalent size distribution of the dispersed rubber phase are retained or they remain practically undisturbed on cross-linking. Relevant feature is shown schematically in Fig. 11.1 for a styrene copolymer (SAN) toughened with a diene rubber (NBR).

Homogeneous, miscible polyblends from two different amorphous polymers would mostly appear clear, showing in each such blend a single refractive index that is intermediate between those of the constituent polymers. A heterogeneous blend (discrete phase of one constituent dispersed or embedded in the matrix of the other) usually scatters light according to the size of the dispersed particles and

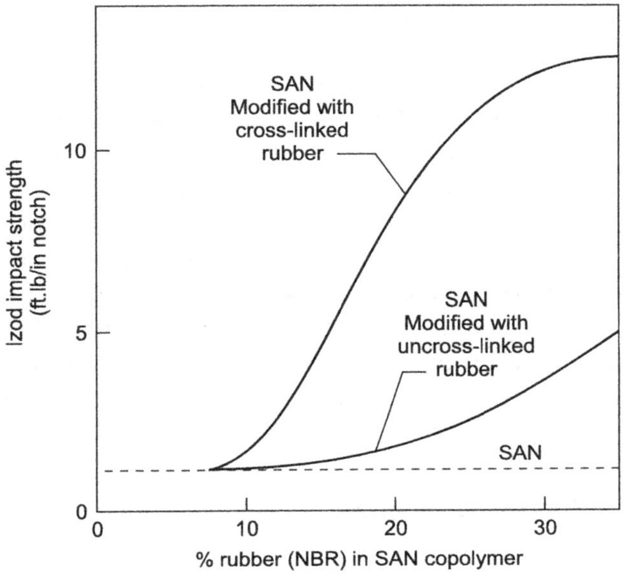

Fig. 11.1 *Toughening of SAN copolymer by rubber (NBR) and effect of rubber cross-linking in imparting enhanced toughening or impact resistance*

would tend to appear opaque if the dispersed particles are large enough and the films are thick enough to make the scattering extensive if not complete.

The imperatives that induces one to go for blending of polymers are: (a) to maintain a more favourable cost-performance ratio, and (b) to achieve reinforcement of a desired property. An expensive polymer whose property spectrum is much higher than is needed for a specific application may be blended with an inexpensive polymer having a property spectrum of a level that makes the resulting polyblend exhibit an attractive cost/performance ratio for the contemplated application. On the other hand, the demerits from a poor property level of a polymer for a particular application may sometimes be effectively or more than compensated by blending it with a second polymer which is rich in that particular respect. Thus, the high level of performance demanded is often satisfied by an appropriate blend of commercially available polymers without having the need to develop a new polymer or to invest in a new plant. Commercial polyblends are commonly of three types:

(i) Elastomer–elastomer blends,
(ii) Plastomer–plastomer blends, and
(iii) Elastomer–plastomer blends.

11.6.1 Elastomer–Elastomer Blends

Such blends are widely made and used in view of the fact that a single elastomer often fails to offer the desired balance of properties and overall cost advantage. Natural rubber is selectively blended with synthetic rubbers to improve a range of properties including tack, resilience, tear strength, heat-build-up and sometimes fatigue resistance. Likewise, selected synthetic rubbers are blended with a primarily natural rubber based compound to obtain advantages with respect to wear or cracking resistance, bounce and cost-effectiveness. Sometimes different synthetic rubbers are blended between themselves. Nitrile rubber (NBR) may be blended with SBR, EPDM, polychloroprene and polysulphide rubbers to selectively achieve advantages on cost, ozone resistance, oil and heat aging resistance and low temperature flexibility.

11.6.2 Plastomer–Plastomer Blends

In the thermoplastic range, commercial polyblends include combination of poly(methyl methacrylate), polystyrene, polycarbonates, nylons, polyethylene (PE), polypropylene (PP), polyethylene terephthate (PET), polyimides, thermoplastic polyurethanes, ionomers etc.

Poly(vinyl chloride), PVC, an inherently flame retardant, relatively low cost thermoplastic resin has the disadvantage of being thermally unstable and relatively brittle. The ABS plastics, on the other hand, are relatively ductile and thermally stable but not flame retardant. When mixed appropriately, ABS and PVC together yield polyblends with much improved heat stability, excellent impact resistance and flame retardance. Such ABS–PVC polyblends are useful in power tool handles, computer devices, communication relays, electrical terminal blocks, kitchen appliances, housing for electronic gadgets, partition items and furniture for the building and housing sector, etc. Polyblends based on ABS and polycarbonate combine high impact strength excellent thermal resistance and good processing characteristics with improved environmental stress cracking resistance and cost advantage over use of only polycarbonates.

11.6.3 Elastomer–Plastomer Blends

General immiscibility of polymers has been turned into a unique advantage in the development of rubber-toughened plastics[4, 9] such as high-impact polystyrene (HIPS). Such materials find applications in packaging, kitchen wares, toys, appliances, furniture, electronic/electrical goods, etc.

Blending of an elastomer with a selected plastomer provides a convenient means of making thermoplastic elastomers (TPE) of diverse nature. The olefinic class of thermoplastic elastomers have been growing in use and acceptability. Polyolefin thermoplastics, viz., polypropylene (PP) and polyethylene (HDPE) are blended with such elastomers as EPR or EPDM and natural rubber (NR). To be widely useful, the elastomer component should have a T_g <–50°C and the thermoplastic polymer used must have a T_m >> the service temperature of the blend. EPR and EPDM are preferable to NR and polybutadiene rubber (PBR) in this context.

The properties profile of TPEs from elastomer–plastomer blends is contingent upon (i) the critical surface tension for wetting of each component, (ii) The weight fraction of the crystallites in the hard phase providing the matrix, and (iii) the strength and modulus of the hard phase. The elastomer component may be cross-linked by use of selected curative under the influence of heating plus high shearing action during blending by what is known as dynamic vulcanization or cross-linking[10], or the vulcanization process may be accomplished statically in a heated compression press as practised commonly.

Dynamic vulcanization effectively enhances the elastomeric properties of the blend while enhancing processing advantage and recycling potential by retaining thermoplastic processability. The advantage of dynamic vulcanization can be drawn to full potential on allowing adequate cross-linking of the elastomeric phase. Even when the rubber is the major component of the blend, it gets dispersed in the form of discrete particles or domains in the thermoplastic matrix under heat and shear action in a shear mixer[10], thus conferring the thermoplastic processing character to the blend. Such dynamically cross-linked elastomer–plastomer blends are also called thermo-plastic vulcanizates (TPVs).

11.6.4 Formation of Interpenetrating Polymer Network (IPN)

Due to very small entropy of mixing and positive heat of mixing, phase separation during mixing or shortly or long after mixing is more common in two-component polymer blends even in systems where the component polymers are very close in solubility parameter. Even though limited phase separation is necessary or desirable for proper balancing of properties between the constituent polymers making the blend, uncontrolled or large-scale phase separation is altogether undesirable. Domain size (of the dispersed phase, elastomer or plastomer) is viewed as important in determining the physical and mechanical behaviours of elastomer–plastomer blends.

Interpenetrating polymer network (IPN)[9–13] is a good example of controlled morphology chemical polyblend. The IPNs are novel blends of network and linear polymers of two or more polymer networks that are held together with prominent resistance to gross phase separation such as a consequence of stable or permanent chain entanglements or catenation than by establishment of mutual covalent bonds, i.e. by grafting of one polymer or the other. IPNs where only one polymer component is selectively cross-linked leaving the other component uncross-linked are known as semi-IPNs or pseudo-IPNs. Full-IPNs are blend networks where the constituent polymers are all selectively crosslinked either simultaneously (giving simultaneous IPNs) or sequentially (giving sequential IPNs). Control of morphology including domain size and domain distribution and prevention of gross phase separation is often better achieved due to some degree of unavoidable mutual grafting during network formation or due to controlled grafting induced by design, more so, in full IPNs than in semi IPNs. Figures 11.2 and 11.3 showing SEM micrographs, highlight the differences in

(a) (b) (c)

Fig. 11.2 *SEM micrographs of NBR–PVC (70 : 30) blends: (a) uncured blend, (b) semi IPN (only NBR component cross-linked using TMTD-Sulphur combination as the curative, and (c) full IPN (both NBR and PVC components simultaneously cross-linked using a combination of ethylene thiourea (ETU) and sulphur as the curative)*

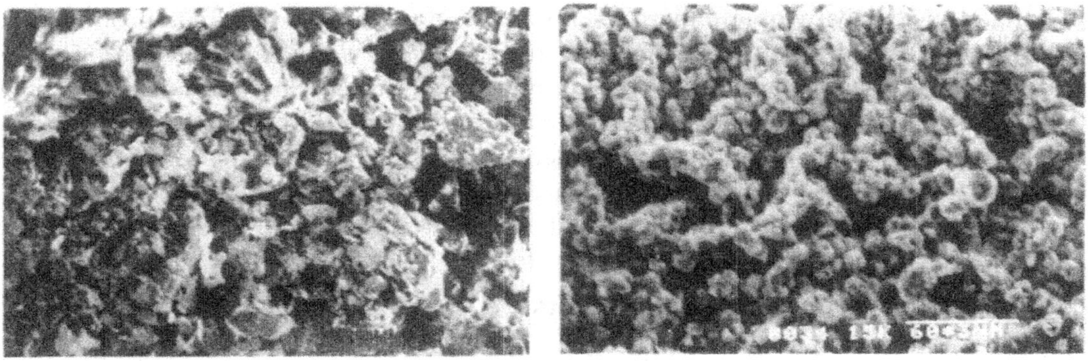

(a) (b)

Fig. 11.3 *SEM micrographs of PE–EPDM (30 : 70) blends (semi IPNs) cross-linked by TMTD–MBTS–Sulphur curative system: (a) Statically cured (showing flaky, layered phase morphology) and (b) dynamically cured (cocontinuous phase morphology), showing fine threads or fibrils of PE passing through the cores of the doughnut-like clustered beads of EPDM*

phase morphology for (i) semi and full IPNs from nitrile rubber (NBR)–poly(vinyl chloride) (PVC) blends, and for (ii) polyethylene (PE)-EPDM semi IPNs where the EPDM was vulcanized statically and dynamically using accelerator-sulphur curative system.

11.7 Polymer Composites

11.7.1 Historical Background

The course of development of composite materials runs parallel to the course of transformation of our civilized society over the ages. Complex combination of materials was used by man even in the early periods of history for housing and hunting purposes. Mud and clay are being used for building walls for thousands of years. It was realised even in the early stage of development of the mud-walls that use of chopped straw and strips or mats of bamboo, canes or the like imparted higher strength, safety and durability to the mud walls. Uses of papyrus reinforced bitumen in building fishing boats and of shellac as a stiff matrix material with reinforcements based on fibrous materials collected from forest were known to people in ancient Babylonia, Egypt and India. At a much later date, composites such as linoleum, asbestos cement items, and reinforced concrete grew into prominence and universal acceptance. More recent developments are the high performance composite materials such as glass fibre reinforced plastic (FRP) composites, carbon-black or carbon fibre reinforced polymer/resin based composites and plastic or rubber based composites reinforced with natural or synthetic fibres or fibrics, steel cords, and fibres from some other metal, nonmetal or ceramic items.

11.7.2 Definition and Basic Concepts

The word composite, as technically defined, means a macroscopic combination of two or more distinct materials having an identifiable interface between them, but more pointedly, the definition is confined to only such materials as would contain a fibrous or particulate reinforcement supported by a binder or matrix material. The discontinuous phase constituted of the fibre or particulate reinforcement is commonly stiffer and stronger than the continuous (matrix) phase. In most cases, a substantial volume fraction (> 10%) of the discontinuous phase is necessary just to achieve significant reinforcement. The difference between a composite and a filled system lies in the fact that while in a composite the discontinuous phase imparts significant mechanical reinforcement, it only plays the role of a loading agent or diluent without property enhancement or reinforcement in what is called a filled system. Fillers may, however, impart many advantages such as fire/flame retardancy, shrinkage control, improved thermal/electrical conductivity and above all, cost benefit. One may also have composites where the stiff component may constitute the major, continuous phase such as the rubber modified (toughened) plastics.

Composites, in general, and fibre reinforced plastics (FRP) or glass fibre reinforced plastics (GRP) in particular, have been developed out of sheer necessity imposed by the stringent requirements of modern technology. The demand of the automotive and particularly of the aerospace community prompted developments in reinforced composites which combine light weight and high strength of Aluminium and Al-alloys and excel them in respect of chemical resistance, corrosion/weather resistance and dent resistance (good recovery of deformation). Composites are in growing demand from architects and construction engineers in the transport, sports, aerospace, defence, building and many other sectors. Composites have been developed because no homogeneous structural material could offer all the desired attributes for a given application.

11.7.3 Prospects and Limitations

As materials of construction in automotive and aerospace sectors, plastics are undoubtedly growing in acceptance. Even then, in engineering status, simple plastics still find their place at or near the bottom; but they hold high potential for a much brighter future through material development in the form of FRP composites. In many cases, there are clear objections to use of plastics on consideration of demands on thermal and aging resistance and lack of confidence centering limitations in a desired balance of thermal and mechanical properties. The best way to overcome these limitations and constraints to enable plastics to not only compete with or even overshadow light metals and alloys in stress and safety-sensitive engineering applications is to make composites[14] through fibre reinforcement of prospective polymers coupled with judicious cross-linking of the polymer matrix.

The chemical, thermal and often the electrical performance of composites are controlled by the polymer or the matrix material while the mechanical performance within the polymer's ultimate temperature capability is dictated by effectiveness of the reinforcement. Each polymer has basic attributes that no amount of incorporation of reinforcing or other additives of different dose levels and varieties can disguise; no plastic can reliably function even briefly at a temperature above the softening or heat distortion temperature of the base polymer. However, the heat distortion temperature and mechanical (tensile) strength of selected polymers may be substantially enhanced through judicious incorporation of a reinforcing fibre such as glass fibre and by cross-linking. Data in Table 11.1 will substantiate this point. With appropriate support and fibre-reinforcement, the polymer or the matrix material enables us to partly realise the inherent high strength and stiffness of the reinforcing fibre material. The principal benefits of fibre reinforcements are: improved resistance to hot deformation, lowering or avoidance of low temperature embrittlement, higher strength, modulus, rigidity and stiffness, improved creep resistance and hence improved dimensional stability.

The discovery of light-weight, strong and stiff carbon fibres in the mid 1960s proved to be a turning point in the development of high performance plastic composites offering serious threats to the light alloys in their traditional use as materials of construction in the aerospace sector. A chain of further developments in strong stiff, light weight synthetic fibres, particularly the aromatic polyamides popularly known as aramid fibres and by the trade name Kevlar, and also developments in aromatic resins and other engineering polymers (as matrix materials) suitable for use in the design of high performance FRP composites followed in the wake of developments of carbon fibres

Table 11.1 *Effect of glass fibre reinforcement (30% w/w) of some crystalline and amorphous polymers: data listed highlight actual heat distortion temperature (HDT) and tensile strength (TS) and enhancement effected due to fibre reinforcement*

	HDT (°C)		*TS (MPa)*	
Matrix material in composite	*Actually observed*	*Enhancement on fibre reinforcement*	*Actually observed*	*Enhancement on fibre reinforcement*
A. Crystalline				
Nylon 66	248	150	180	100
PEEK	300	145	175	75
PBT	210	140	135	75
Nylon 6	210	130	135	100
PP	148	82	85	50
Acetal	164	55	140	70
B. Amorphous				
PES	215	16	145	55
PPO	140	15	125	60
PC	140	10	120	52
ABS	100	10	90	40

(short fibres and continuous filaments).[15] With the developments in high performance aromatic polymers, the old division between the functional and processing potentials of thermoplastics and thermosets is getting blurred progressively. Thermosets, however, commonly exhibit lower shrinkage, improved chemical resistance and thermal performance, while thermoplastics give advantages in higher outputs, better design of parts over and above their recycling potential.

Amorphous thermoplastics such as ABS, poly(phenylene oxide) (PPO), polycarbonate etc. are valued for low shrinkage and good impact behaviour, even though their fatigue resistance is rated low and on fibre reinforcement, they offer low or limited benefits; crystalline polymers on the other hand exhibit higher shrinkage and are dimensionally affected more on heating but they are characterized by substantial improvements in heat distortion temperature and tensile/flexural strength on fibre reinforcements. (*see* Table 11.2).

A composite is viewed as a laminate if the reinforcing fibres are arranged in layered fashion such as in the form of webs, fabrics or mats, Fig. 11.4(a); mixing of the matrix resin with randomly oriented short fibres, Fig. 11.4(b); allows the resin fibre mix or the dough to be injection moulded or compression moulded with or without

Table 11.2 *Properties of nylon 66 composites: effect of incorporation of particulate and fibrous reinforcing agents*

Nature of reinforcing additive	*Tensile strength (MPa)*	*Flexural modulus (GPa)*	*Heat distortion temperature (°C)*	*Stiffness ratio (Anisotropy Index)*
1. None	85	2.5–3.0	95–100	1.0
2. Particulate (50%)	95	5.6–6.0	130–170	1.0–1.2
3. Particulate (35%) and Fibrous (15%)	125	6.0–7.0	220–230	1.4–1.5
4. Fibrous (50%) (Glass Fibre)	200	9.0–9.5	240–250	1.7–1.8

the presence of a particulate filler and such dough-like short fibre resin mixes are commonly known as dough moulding compound. Short fibres are not necessarily randomly arranged. Maximum strength and stiffness are obtained when continuous fibres/filaments are aligned parallel to the applied load. Such unidirectionally aligned fibres in resin matrix, Fig. 11.4(c), can be most efficiently achieved by such methods as pultrusion (combining pulling with extrusion).

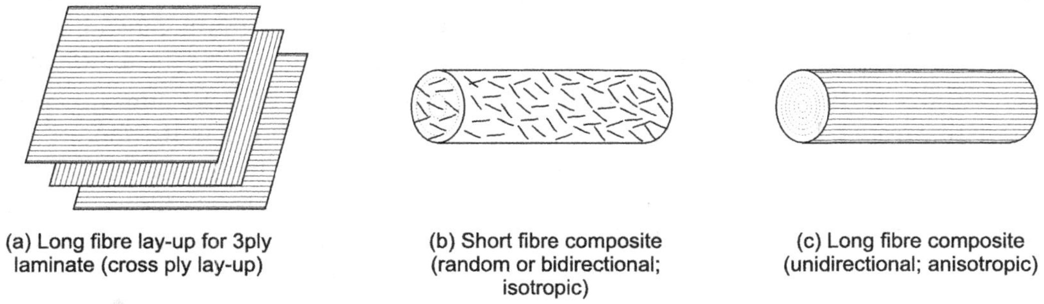

Fig. 11.4 *Diagrams showing different arrangements of short (– –) and long (—) fibres in composite making*

11.7.4 Controlling Features and Property Demands

An unfavourable competitive position of a plastic object vis-a-vis a metal component may originate from incompetent processing and from bad part design or the designer's ignorance of the plastic conversion process. Unknown features and uncontrolled factors relating to thermal or thermomechanical history and flow geometry of the tool, particularly in injection moulding may be the root cause of performance shortcomings and unexpected failures. Improved flow geometry coupled with controlled rate of cooling from moulding temperature may give substantial advantages in having a good to excellent balance of properties and performance stability to moulded parts.

High performance applications frequently call for fibre reinforcements for thermoplastic or thermosetting resins or elastomers. For thermoplastics, the flow geometry of tool is critical. Stiffness ratio (ratio of modulus in machine direction to modulus on cross direction) or anisotropy index is also an important consideration and this parameter would also largely depend on particulate or fibrous nature of the reinforcing agent, Table 11.2. In respect of anisotropy index commonly observed, the positioning of the reinforcing agents will be in the order fibrous >> particulate > none. Equally if not more important considerations are the fibre length or more intrinsically the fibre aspect (length/diameter) ratio and bond strength (inter laminar shear strength, ILSS) between the fibre and the matrix material or the contact angle between the reinforcing fibres and matrix resin indicating extent of wetting of the fibre with resin (a lower contact angle indicating better wetting). Fibre-resin or filler-resin interactions and wetting may be improved substantially by treating the fibre or filler system with what is known as a coupling or compatibilising agent.[16–18] A long range of silicones have been developed as coupling agents for this purpose. The nature of dependence of composite strength on length of the given reinforcing fibre (i.e. on aspect ratio) and on use of a coupling agent is shown schematically in Fig. 11.5. For a fine fibre of a given diameter a fibre length of 8–10 mm appears to impart the upper limiting reinforcing effect. For a given fibre length, the strength of composite is higher for use of a suitable coupling agent. Table 11.3 lists specific gravity and mechanical property levels of some fibres, metals or alloys and FRP composites. Data in Table 11.4 highlight the level of reinforcements possible for use of fibres (short or continuous length) in different weight proportions in the composite for use of a number of thermoplastic matrix resins.

11.7.5 Mixing or Impregnation

It is difficult to mix fibres with a viscous polymer. Crude mixing is of little practical advantage. The entire surface of all the fibres must be forced into close contact with the polymer. A low initial matrix viscosity is helpful for mixing and impregnation. Molecular weight must either be low initially or a solvent/diluent that would finally turn into an integral part of the matrix must be used. Use of a simple thinner poses more problems than it solves.

Table 11.3 *Properties of some materials; fibres, metals and alloys and composites*

Materials	*Specific Gravity*	*Tensile Strength* $lb/in^2 \times 10^{-5}$	*Tensile Modulus* $lb/in^2 \times 10^{-6}$	*Specific Strength* $lb/in^2 \times 10^{-5}$	*Specific Modulus* $lb/in^2 \times 10^{-6}$
S-glass Fibre	2.49	6.5	12.45	2.61	5.02
B-fibre	2.69	4.5	35.5	1.67	17.1
C-fibre (high-modulus)	2.00	2.5	55.0	1.25	27.5
C-fibre (high-strength)	1.74	4.5	35.0	2.70	20.1
Aramid fibre	1.44	4.0	19.0	2.77	13.2
Al-alloy (RR-58)	2.77	0.59	10.0	0.23	3.59
Ti-alloy (Ti-6Al-4V)	4.50	1.50	16.0	0.33	3.55
Steel (FV 520)	7.80	1.55	28.0	0.198	3.59
S-glass-epoxy composite	2.08	2.60	7.6	1.25	3.65
C-fibre (high modulus)-epoxy composite	1.67	1.50	30	0.93	18.7
Aramid fibre-epoxy composite	1.38	2.0	11	1.44	7.97
B-fibre-epoxy composite	1.97	1.95	39	0.99	19.8

Table 11.4 *Flexural strength of some FRP composites*

Polymer Matrix	*Fibre used*	*Flexural Strength, MPa*		
		No fibre	*Short fibre* (30%)	*Continuous length fibre* (65%)
None	Glass	–	–	3400*
PP	"	30	85	350
Nylon 66	"	80	180	600
PET	"	60	140	800
PES	"	90	145	500
PEEK	"	91	150	780
PEEK	C-fibre	90	250	1100

*Tensile strength

The rise in viscosity of the initial low viscous resin system on curing or chain extension, which is guided by what is known as chemorheology, is important for moulding or shaping of parts of intricate shapes. Orientational effects are important

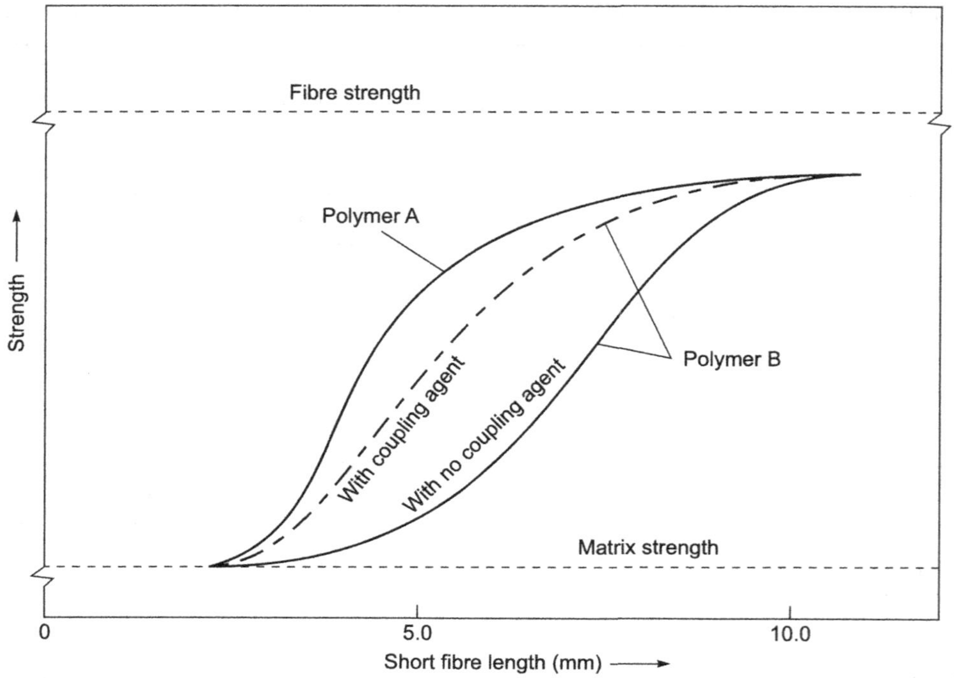

Fig. 11.5 *Effect of variation of fibre length (aspect ratio), nature of polymer (polymer A and polymer B) and of use of a coupling agent for a given polymer (polymer B) on the strength properties of FRP composites*

in reinforced plastics and this is still one of the least understood features of FRP composites. Designers are required to view loaded or reinforced parts in terms of stress–strain–time–temperature–orientation relationships.

The easiest way to achieve a composite with long continuous fibres with high axial orientation is to allow continuous impregnation followed by shaping into profiles by pultrusion technique (extrusion with pull in the machine direction). Such impregnated systems have the added advantage of providing very high reinforcing fibre content with near 100% orientation in the finished composite in the machine direction.

11.7.6 Scope for Use of Natural or Man-made Fibres in Composite Making

In many applications, use of very high-cost glass fibres, aramid fibres or carbon fibres in composite making is not necessary or essential. Relatively low-cost natural fibre substitutes such as jute and synthetic fibre substitutes such as rayon, nylon or polyester fibres are considered promising with respect to energy and property advantages. Use of rayon, nylon and polyester cords in construction of tyres, beltings and hoses is well known. Technologies based on use of steel cords in constructing high speed radial tyres have been developed and are in the process of developing into higher sophistications.

Use of jute and allied fibres as reinforcing agents in place of traditional glass fibres in different forms partly or fully in composite making stems from higher specific modulus and lower specific gravity of jute (40 GPa and 1.29 respectively) compared with those of glass fibre (30 GPa and 2.5 respectively). Besides, much lower cost, and renewable nature of jute and much lower requirement of energy for production and processing of jute (only 2% of that required for glass) make it attractive as a reinforcing fibre in resin-based composites. Properties of jute fibre reinforced plastic (JRP) composites may be substantially improved by minor chemical modification of jute by grafting of polyacrylonitrile (PAN) and related polymer and by incorporation of phenolic resins.[19] Only 10–20% acrylic polymer-grafting on jute may impart up to 30% improvements in flexural strength and flexural modulus of polyester resin based JRP composites. Likewise, phenolic resin binding (4% resin uptake) on jute fibre may impart up to 40% improvement in flexural strength of such JRP composites. Improvements in mechanical properties due to modification of jute as referred to above is a consequence of better wetting and hence improved bonding between the matrix resin and the reinforcing (jute) fibres. This has been evidenced by results of (i) measurements of contact angle between the fibres (modified and unmodified) and the matrix resin, (ii) resistance to hydrothermal aging of the respective composites, and (iii) examination of tension fractured surfaces of the different composites by scanning electron microscopy.[21,22]

Properties of FRP composites, more importantly, their dynamic mechanical properties are likely to depend as much on the nature and distribution of the matrix materials and the reinforcing fibres as on the nature of the fibre-matrix interfaces and of the interfacial regions.[20–27] Even a small change in the chemical and physical nature of the fibre for a given matrix resin and in relative weight proportions of fibre and matrix may result in notable changes in the overall properties of the composite. Subtle variations in reinforcing fibre selection or combinations of different fibres in hybrid composites and in processing conditions may lead to uncertain presence of voids, odd thermal effects due to odd differences in the expansion coefficients of the main phases, and hence odd variations in static and dynamic mechanical properties,[27] more so, for hybrid composites consequent to uncertain nature and degrees of interactions at the different fibre-matrix interfaces.

Proximity of the stiff fibre and preferential adsorption of readily diffusable constituents (usually low molecular weight curatives) on the fibre surface may impose a relatively high cross-link density and hence lead to development of a more than expected stiffness level at the fibre-matrix interface. This may, at the same time, lead to some softening of the matrix in the zone next to the interface due to notable depletion of the curative. There is a common unmistakable general trend in having a progressively lower loss modulus value with progressive replacement of glass fibre with jute fibre starting with all-glass unitary composite and finally ending with all jute unitary composite and giving different glass jute hybrid compositions in between. Fibre reinforcement in general causes lowering in the damping parameter (tan δ) value of the resin matrix.[27]

References

1. Schultz, A.R., *Polymer Preprints*, **8** (1967) 1455.
2. Coran, A.Y. and R. Patel, *Rubber Chem. Technol*, **56** (1983) 1045.
3. Asaletha, R., S. Thomas and M.G. Kumaron, *Rubber Chem. Technol*, **66** (1995) 671.
4. Nielson, L.E., *Predicting the Properties of Mixtures*, Marcel Dekker, New York, 1978, p. 13.
5. Hill, R., *J. Mech. Phys. Solids*, **13** (1965) 189.
6. Bikales, N.M, Ed., *Encyclopedia of Polymer Science and Technology*, Wiley Interscience, New York, 1969, Vol. **10**, p. 701, 703.
7. Uemura, S. and N. Takayahagi, *J. Appl. Polym. Sci.*, **10** (1966) 113.
8. Hopff, H. in *Plasticization and Plasticizer Process*, R.F. Gould Ed., Advances in Chemistry Series, Amer. Chem. Soc., Washington, D.C., 1965, No. 48, p. 87.
9. Ghosh, P. and P. Ray, *Rubber Reporter*, **14** (1989).
10. Ghosh, P., B. Chattopadhyay and A.K. Sen, *Polymer*, **35** (1994) 3958.
11. Ghosh, P. and P. Ray, *Polymer Science, Contemporary Themes*, S. Sivaram Ed., Vol. 2, pp. 871–876, Tata McGraw-Hill, New Delhi, 1991.
12. Ghosh, P. and P. Ray, *J. Mater. Sci.*, **26** (1991) 6104.
13. Ghosh, P., A.K. Sen and P. Ray, *Polymer*, **33** (1992) 744.
14. Ghosh, P., *Rubber Reporter*, **17** (1992) 25.
15. Mallick, P.K., *Fibre Reinforced Composites*, Marcel Dekker, New York, 1988.
16. Pluedemann, E.P. et al., *Modern Plastics*, **39** (1962) 135.
17. Clark, H.A. and E.P. Pluedemann, *Modern Plastics*, **40** (1963) 133.
18. Wu, S., *Polymer Interface and Adhesion*, Marcel Dekker, New York, 1982.
19. Ghosh, P. and P.K. Ganguly, *Plastics, Rubbers and Composites—Processing and Application*, **20** (1993) 171.

20. Jang, B.Z., *Composite Sci. Technol.*, **44** (1992) 333.
21. Ghosh, P. and N.R. Bose, *J. Mater. Sci.*, **20** (1991) 4759.
22. Ghosh, P. and N.R. Bose, *J. Appl. Polym. Sci.*, **58** (1995) 2117.
23. Jinen, E., *Composite Sci. Technol.*, **33** (1988) 19.
24. Ide., S., T. Takahashi and A. Shimamoto in *Controlled Interphases in Composite Materials*, H. Ishida Ed., Elsevier, London, 1990.
25. Kardos, J.I., In *Molecular Characteristics of Composite Interfaces*, H. Ishida and G. Kumer Ed., Polymer Sci. and Tech. Series, Plenum Press, New York, 1985, Vol. 27.
26. Nielsen, L.E. and R.F Landel, *Mechanical Properties of Polymers and Composites*, 2nd Ed., Marcel Dekker, New York 1994, 131.
27. Ghosh, P. et al., *J. Appl. Polym. Sci.*, **64** (1997) 2467.

CHAPTER 12

Conducting Polymers

12.1 Introduction

The interest in the field of organic semiconductors originated from the suggestion that the transfer of π electrons from molecule to molecule may play an important role in fundamental physical processes of living organisms. Very small but measurable electrical conduction was reported for anthracene in 1906. Single crystals of purified anthracene and other condensed aromatic ring systems formed interesting objects of preliminary study of the fundamental behaviour of some charge carriers in such systems, even though such molecular crystals showing some degree of electrical conductivity would be prominently brittle, low melting and easily soluble. It obviously follows that a high polymeric system obtained in a semi-conducting or conducting form would combine the interesting electrical property of molecular crystals of condensed polynuclear aromatic compounds with the useful range of mechanical (strength, modulus, rigidity, etc.) and thermal (melting/softening point or heat distortion temperature) properties and good corrosion resistance.

The common synthetic polymers are electrical insulators in general. A conducting polymer would, however, be a much desired product for use as corona shield or as a separate layer in an insulation system in film/sheet form to prevent corona discharge and for providing a conducting surface for electroplating non-conductors or for use in printed circuit boards (PCB), electronic devices, photovoltaic devices, solar cells, batteries, etc.

12.2 Different Approaches for Making Conducting Polymers

Highly conducting polymers for use as high quality conducting coatings, strong films or filaments are in the process of development. A practical and more obvious approach involves incorporation of metallic powders, flakes or whiskers or other conductive fillers such as graphite powder or conducting carbon blacks into common plastics or rubbers. However, such filled conducting compositions have their own limitations. Despite infusion of conductivity due to the incorporation of the conducting filler particles, the prominent insulation or resistance character of the matrix polymer generates heat during passage of electricity. Accumulation of heat and consequent rise in temperature and thermal expansion cause the conducting filler particles in such filled polymer systems move further apart, thereby leading to a dropping trend in conductivity with time of passage of electricity. As for synthesis of inherently conducting polymers (ICP), there are developing trends in two directions (i) synthesis of conjugated polymers and (ii) preparation of polymeric charge transfer complexes.

The first approach is based on the fact that electron delocalization favourably occurs as a molecular system becomes more conjugated. Likewise, in charge-transfer (CT) complexes, a partial electron transfer occurs from the donor site to the acceptor site of the molecule; a good example is a crystalline complex of quinoline with tetracyanoquinodimethane (TCNQ), having a room temperature conductivity of the order of only 1 s cm^{-1} (or $ohm^{-1}cm^{-1}$) and a temperature dependence corresponding to an activation energy of 0.1 eV. By complexation of this nature, a conductivity level in the upper side of the semiconducting range, Fig. 12.1, is readily achieved for an organic high polymeric system.

Polymeric CT complexes of this kind are attractive for their use in electronic engineering as they are commonly endowed with good range of mechanical and thermal properties with the advantage that they could be processed into such useful shapes as cohesive films, sheets, discs, rings and thin wires.

The electrical resistance of metals is caused by the scattering of conduction electrons when they collide with ion cores during flow under a potential gradient, thus contributing to a heating effect. The displacement of the cores from their positions arises from their thermal motion. The thermal motion of the ion cores decreases with lowering of temperature with consequent lowering in the resistance character or increase in the conductivity level.

The resistivity of the semiconducting materials on the other hand decreases with rise in temperature because of the fact that with higher thermal excitation more

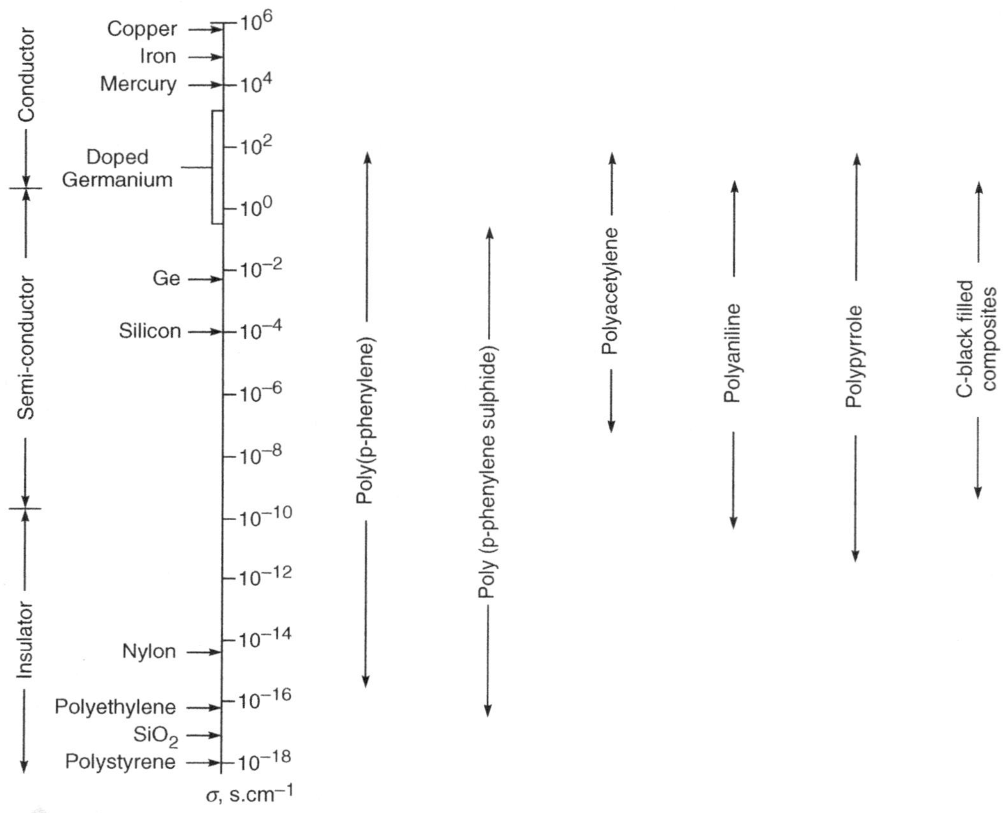

Fig. 12.1 *Electrical conductivity or conductivity range in s. cm^{-1} of some typical insulating, semi-conducting and conducting materials (for a given conductivity range, the upper values are for appropriately doped samples)*

electrons are released from interatomic bonds and the applied electric field enables the released electrons in higher numbers more through the mass giving rise to increased electronic conduction with rise in temperature.

12.3 Inherently Conducting Polymers

Inherently conducting polymers (ICP), also commonly known as synthetic metals, somehow possess the electrical, electronic, magnetic and optical properties of a metal while also exhibiting[1] the physical and mechanical properties, rheological features and processabilities associated with a conventional polymer. The applicability of the concept of "doping" is the unifying theme that distinguishes the inherently conducting polymers (ICPs) from all others. Addition of electron donors or acceptors causes doping that results in dramatic electronic and magnetic changes in the ICPs along with increase in conductivity to or approaching the metallic range.

Doping is reversible; removal of dopant or dedoping produces the original low-conducting, semiconducting or insulating (base) polymer usually without degradation of the polymer backbone.

12.3.1 Polyacetylene

A simplistic approach to an inherently conducting polymer is to remove one hydrogen atom from each carbon atom in the chain molecules of polyethylene, leaving one spare electron on each carbon atom forming the chain backbone; the electrons can be made to flow through the polymer mass so that the hypothetical polymer intermediate can be called a "metallic polymer". A material of this kind can be synthesized by polymerization of acetylene. In polyacetylene, a conjugated polymer[2, 3], the spare electrons are held by formation of alternate double bonds and single bonds in the polymer structure.

$$\underset{\text{acetylene}}{n\ HC \equiv CH} \rightarrow \underset{\text{Polyacetylene (PA)}}{-CH = CH \left(CH = CH \right)_{n-2} CH = CH-} \qquad (12.1)$$

The conjugated structure of PA makes it behave like a semiconductor and not as an insulator like polyethylene (PE). Some of the π electrons of PA can be thermally excited out of the bonds giving rise to a small electrical conductivity.

Polyacetylene (PA) may exist in the geometrical isomeric forms as shown in Fig. 12.2.

Polymerizations of acetylene catalysed by Ziegler–Natta type catalyst,[4, 5] metal carbonyls[6, 7], metal acetylacetonate complexes[8] and glow discharge[9] have been reported. The polymer as commonly obtained in powder, gel or film form is commonly cross-linked, insoluble and intractable showing different ranges of semiconducting character and it has no defined melting point. It decomposes gradually at elevated temperatures and reacts readily with oxygen to yield oxygenated products.

Doped derivatives of PA are ionic compounds and doping of PA is viewed as a redox reaction.[10] The net step in doping is the oxidation or reduction of PA molecules to polycations or polyanions:

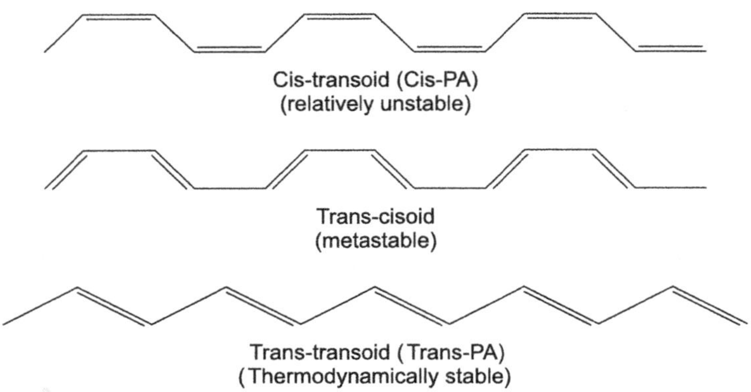

Fig. 12.2 *Geometrical structural forms of polyacetylene (PA).*

$$[\text{—(CH=CH)}_x^{(-)}\text{—} + (\text{reductant})^{(+)}] \xleftarrow{\text{reductant}} \text{—(CH=CH)}_x\text{—} \xrightarrow{\text{oxidant}} [\text{—(CH=CH)}_x^{(+)}\text{—} + (\text{oxidant})^{(-)}]$$

Appropriately doped PA may show conductivities in the metallic range ($10–10^3$ $ohm^{-1}cm^{-1}$).

Substituted acetylenes (R_1—C ≡ C—R_2, R_1—C ≡ C—H) may also be polymerized employing Ziegler–Natta catalysts, transition metal complexes and by free radical, cationic or anionic initiation. Many substituted polyacetylenes are soluble in organic solvents and are amorphous and thus, they are markedly different from PA; they also usually exhibit conductivities[11] far lower than the same exhibited by PA in undoped or doped forms. While PA is usually dark red or black, substituted polyacetylenes are generally white or yellow in colour. Unlike PA, substituted polyacetylenes have generally high oxygen stability.

12.3.2 Polydiacetylenes

Another interesting group of polymers are the polydiacetylenes (PDA). Use of soluble polydiacetylenes has enabled casting of thin film from their solution. In such solution-cast films, the parallel packing of the chain molecules in the polydiacetylene crystal lattice is lost, thereby largely impairing the electrical conductivity character of the PDA film.

Formation of a polydiacetylene (PDA) from the corresponding monomer is an example of a topochemical polymerization in which a conjugate, 1,4-addition of the 1,3 diyne units occurs in the crystalline state. The reaction is activated by heating or irradiation of the monomer crystals by visible/uv light, γ-ray or x-ray (Fig. 12.3).

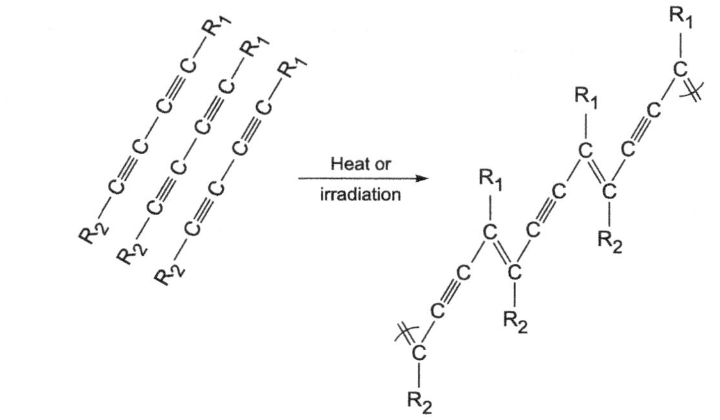

Fig. 12.3 *Formation of polydiacetylenes (PDA) from corresponding diacetylenes*

The unreacted monomer may then be extracted out by a suitable solvent leaving behind deeply coloured PDA single crystals[12]. The intense colour of PDA is attributable to fully conjugated linear carbon backbone similar to PA with the difference that the main chain of PDA comprises alternate double and triple bonds interspaced by single bonds while that of PA comprises alternate single and double bonds, i.e. a conjugated polymer structure. PDA crystals are not doped by all conventional chemical dopants. The low doping efficiency in PDA is a direct consequence of perfect close packing of the PDA crystal lattice that, in a way, restricts dopant diffusion. Solution cast PDA films having less crystalline order allows a much deeper penetration of dopant in comparison with the single crystal, whereby conductivity as high as $10^{-1}ohm^{-1}cm^{-1}/scm^{-1}$ can be attained.

Because of their unique one-dimensional unsaturated structure, the PDAs hold interest for their non-linear optical properties, potential applications in optical communications and integrated optical technology.

Other than PA, the inherently conducting polymers shown in Table 12.1, such as polyparaphenylene (PPP), poly (paraphenylene vinylene) (PPV), polythiophene (PTh), poly (paraphenylene sulphide) (PPS), and polypyrrole (PPy) also undergo redox doping.

Table 12.1 *Some inherently conducting polymers and their electrical conductivities*

Polymer	*Repeat unit (Undoped)*	*Dopant material*	*DC-Electrical conductivity ($s\ cm^{-1}$)*	
			Undoped	*Doped*
Polyacetylene (PA)	$-\!(CH{=}CH)\!-_n$	I_2, AsF_3, $AlCl_3$, $FeCl_3$, K, Li	$<10^{-7}$	$10–10^5$
Poly(p-phenylene) (PPP)	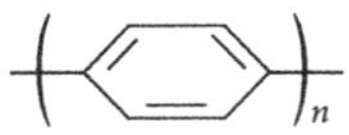	AsF_5, $AlCl_3$, $FeCl_3$, K, Li	$<10^{-7}$	$10^{-2}–10^2$
Poly(p-phenylene vinlylene) (PPV)	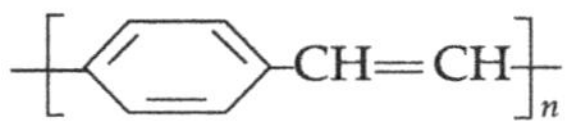	AsF_5	$<10^{-7}$	$10^{-2}–10^1$
Polythiophene (PTh)	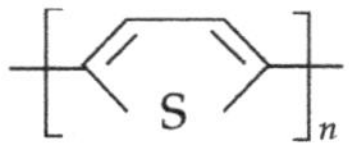	$SO_3CF_3^-$, ClO_4^-, BF_4^-, tosylate	$<10^{-7}$	$10^{-2}–10^2$
Polypyrrole (PPy)	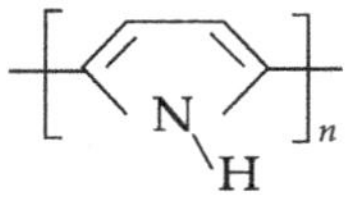	$FeCl_3$, Cu $(ClO_4)_2$, AsF_5, tosylate	$<10^{-7}$	$10–10^3$
Poly(p-phenylene sulphide) (PPS)	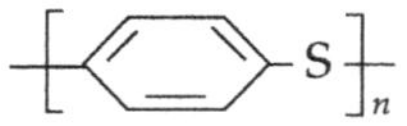	AsF_5, FSO_3H, SbF_5	$<10^{-7}$	$10^{-9}–10^1$
Polyaniline (PANI)	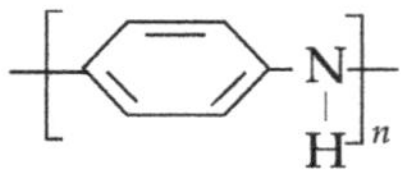	HCl, $HClO_4$, Camphor sulphonic acid	$<10^{-7}$	$10^{-1}–10^2$

12.3.3 Polyaniline

The most important inherently conducting polymer from the viewpoint of low cost and high application potential is polyaniline (PANI). It has drawn the attention of many workers in recent years in respect of synthesis by chemical and electrochemical means. Its non-redox doping character is also an added attraction. Polyaniline's relatively low cost, ease of synthesis in easily processable dispersion forms and satisfactory environmental stability are also some advantageous features.

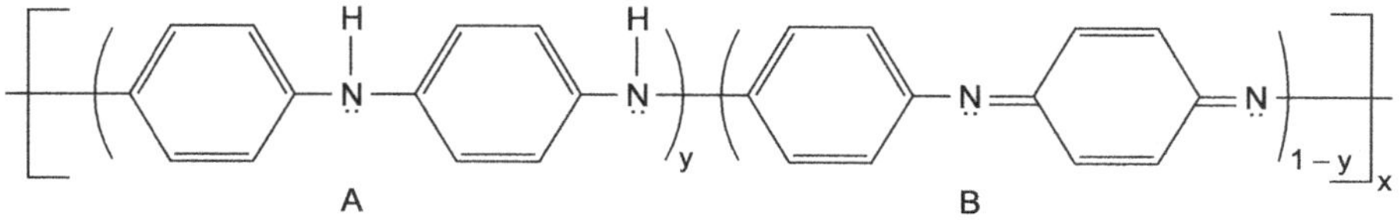

The base form of polyaniline (PANI) has the general formula consisting of reduced (A) and oxidized (B) repeat units[13–15]. The average oxidation state, $(1 - y)$ may vary from zero (to give the completely reduced polymer characterized only by the repeat unit A) to 0.5 (to give the "half oxidized" polymer characterized by 1 : 1 alternating sequences of A and B) to one (to give $y = 0$ and the completely oxidized polymer characterized by the repeat unit B only). For the presence of 1 : 1 alternating sequence of A and B repeat units in PANI, the polymer is called emeraldine base. The imine nitrogen atoms can be protonated in full or in part to give the respective salts, the degree of protonation being dependent on the oxidation state and pH of the aqueous acid used.

The partly protonated emeraldine hydrochloride salt is prepared readily in the form of a black-green precipitate by polymerizations of aniline by oxidative coupling in aqueous acid (HCl) media by such oxidising agents as $(NH_4)_2$ S_2O_8, H_2O_2, Cr^{6+} – complexes/salts etc. It can also be prepared electrochemically from aniline.

Different structures (II and III) for fully protonated emeraldine base (I) may be obtained on use of ≥ 1M HCl for doping (Fig. 12.4):

From the alternative resonance forms (II, showing bipolaron structure and III, showing separated polaron structure), it follows that the overall structure would have extensive spin and charge delocalizations[16] resulting in a half filled polaron conduction band. The conductivity of PANI increases with extent of doping and the conductivity of the doped polymer increases monotonically with molecular weight up to a molecular weight of 150,000. Soluble polyaniline can be obtained by solution polymerization of aniline in N-methyl pyrrolidone (NMP) and in aqueous acidic solution[17–19] using the support of a water soluble polymer such as poly(vinyl alcohol) (PVA), methyl cellulose, carboxy methyl cellulose, polyacrylic acid, polyacrylamide, poly(vinyl pyrrolidone) etc. PANI has application potentials in electromagnetic interference (EMI) shielding, as gas sensors, in gas separation, as electrodes in rechargeable batteries and in static charge dissipation.

12.3.4 Poly(p-phenylene Sulphide) PPS

PPS can be obtained by homopolymerization of thiophenol in the presence of H_2SO_4[20] or by oxidative condensation of thiophenol in the presence $SOCl_2$ and a Lewis acid[21].

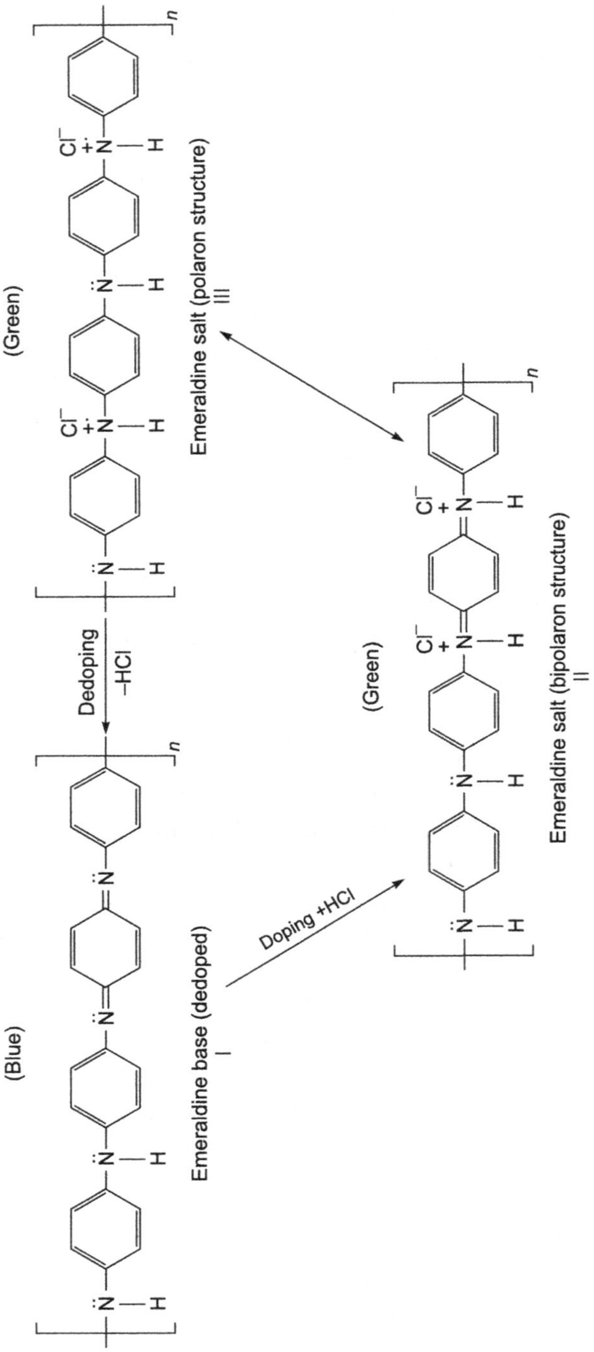

Fig. 12.4 *Dedoped and HCl - doped structures of polyaniline (PANI)*

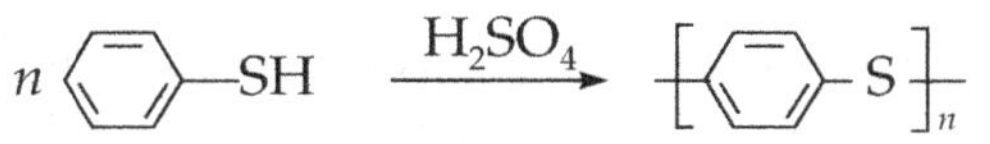 (PPS)

 (PPS)

PPS may also be synthesized by self-condensation of metal-*p*-halogeno-thiophenoxide. This method[22] uses a less drastic condition than Macallum's synthesis and it involves extensive washing of the product in order to remove the residual metal contaminants.

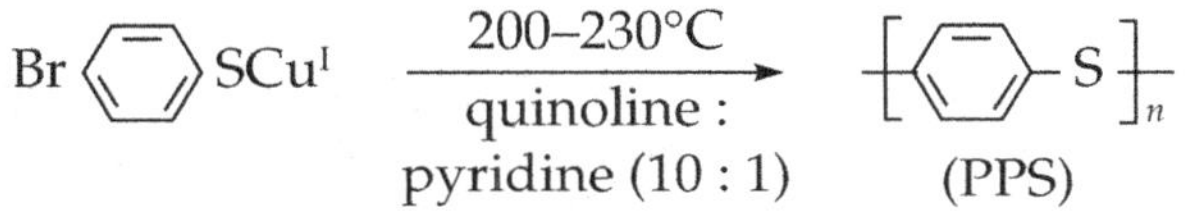

PPS can also be synthesized by electro-oxidative polymerization of thiophenol[23] in nitromethane at room temperature in the presence of an acid (trifluoro acetic acid, a sulphonic acid or stannic chloride). The polymer separates out as a white precipitate during synthesis. Its melting point is 180–190°C and it is soluble in N-methyl pyrrolidone (NMP). The polymerization of thiophenol to PPS is believed to proceed via intermediate formation of diphenyl disulfide. Oxidative polymerization of diphenyl disulphide in high yields can also be done in the presence of a small amount of dichloromethane and vanadyl acetylacetonate V $(acac)_2$. PPS is easily doped using AsF_5 and $FeCl_3$.

12.4 Photoconducting Polymers

Inducement or enhancement of electrical conductivity on exposure to light or irradiation is called photoconductivity. Photoconductive polymers are commonly insulators in the dark and they behave as semiconductors when exposed to light. Genuine photoconductors do not contain free charge carriers; in fact, the charge carriers are generated in them by the action of light. The electrical charge carriers may also be photo-generated extrinsically in an adjacent photo-conductor and then transferred to the polymer that would act as a charge transporting medium.

Poly(N-vinyl carbazole) (PNVC), I, exhibits high level of photoconductivity when sensitized with suitable electron acceptors. A photoactive polymer may be one with pendant or in-chain photoactive groups with large π-electron systems (e.g. carbazole units (I, II) or condensed aromatic ring systems (III).

A photoconducting polymer may also be provided with a highly conjugated main chain such as polyacetylene (PA) mentioned earlier in this chapter and also poly-(phynylene vinylenes), PPV (IV and V).

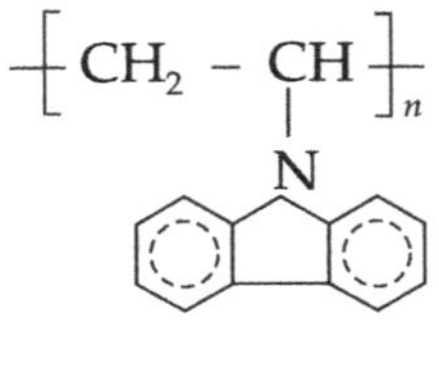

I
Poly (N-vinyl carbazole)
PNVC

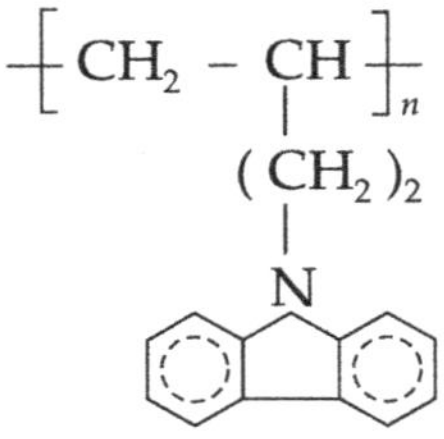

II
Poly (N-ethyl-2-vinyl carbazole)
PNEVC

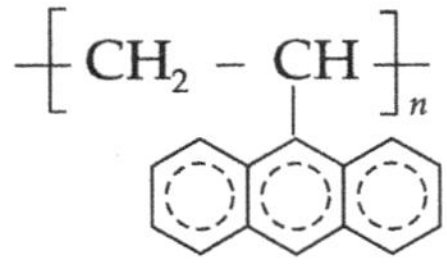

III
Poly (9-vinyl anthracene)
PVAn

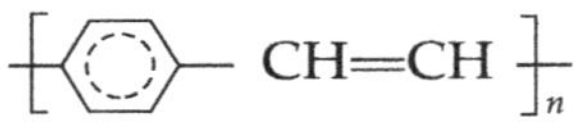

IV

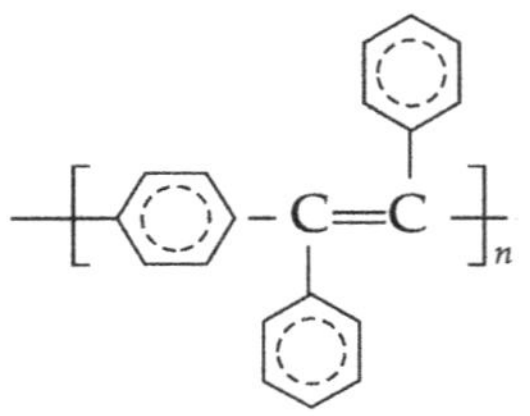

V

Poly (phenylene vinylenes)

12.5 Metal-filled Conducting Polymer Composites

Metals may be used in the form of powders, flakes, whiskers, filaments and cords, wires and wire mesh as filler or reinforcing agents in a polymer matrix for making various moulded, formed or coated objects. The advantage of metal powders over other particulate fillers is that they impart good electrical and thermal conductivity into the composite.

Uniform dispersion of metal powders etc. (particulate fillers) are achieved by shear mixing using the matrix polymer above its T_g or favourably above the T_m. For use of a liquid polymer or resin intermediate, the metallic powders are normally dispersed in the liquid resin/polymer by continuous stirring which is then further continued with addition of suitable chain extenders or curatives before transferring the mix to a moulding or forming equipment for ultimate conversion into a useful end-product. The end use items so produced exhibit a higher order of electrical conductivity if the metal powder-liquid polymer mix is subjected to an external magnetic field to orient the dispersed metallic particles.

A conductive composite can be made with any combination of matrix polymer and conductive particulate filler, provided enough filler is used to form electrically conductive network. Incompatibilities between matrix polymer and particulate filler materials can prevent formation of a well dispersed stable conductive filler network. It is also necessary that the filler and the matrix materials remain chemically inert toward one another to ensure stable electrical conductivity with little lowering in the property with time. The polymer matrix should not oxidise the surface of the metal fillers or it should prevent surface oxidation of the metal fillers by its presence. Metals of importance are copper, aluminium, iron, nickel, stainless steel and silver. Silver is the most expensive and at the same time a highly conductive metal. Iron, copper and aluminium are relatively inexpensive but they all oxidise readily. The oxide layer on the particle surface can reduce effective conductivity significantly. Nickel or stainless steel are more resistant to oxidation but they are relatively expensive compared to iron, copper or aluminium. Stainless steel is also lower in conductivity than the pure metals.

12.6 Carbon Black/Carbon Fibre Reinforced Conductive Polymer Composites

Carbon blacks as particulate reinforcing fillers span a wide range of prices and particle morphologies. They are produced by controlled pyrolysis of hydrocarbon fuels. Carbon fibres are likewise made by controlled pyrolysis of selected organic polymer fibres, e.g. polyacrylonitrile (PAN) fibre. The industrial carbon fibres are available in a variety of product forms such as continuous filament, chopped fibre and woven or nonwoven mats of varying area density.

Carbon blacks having low particle size and forming fibrous aggregates can promote formation of conductive networks. Highly structured and porous carbon black particles are even better network formers and hence higher conducting: composites based on use of these porous carbon blacks produce orders–of–magnitude higher electrical conductivity than those based on low structure, less porous carbon blacks. In view of having a high aspect (length to diameter) ratio, carbon (graphite) fibres impart a much higher electrical conductivity than particulate carbon black for comparable loading in composites based on a given polymer matrix.

The electrical conductivity of a conductive composite depends on intrinsic properties of the filler material as well as matrix-filler interactions and processing conditions. The relationship between conductive filler concentration and composite conductivity has been widely studied and reported. For conducting carbon black filled EPDM vulcanizates, physical and mechanical properties such as density, hardness, tensile strength and electrical conductivity pass through notable sharp uptake, Fig 12.5, with increase in filler concentration, commonly over the zone of 15–30 phr of conducting carbon black that is recognized as the percolation threshold.

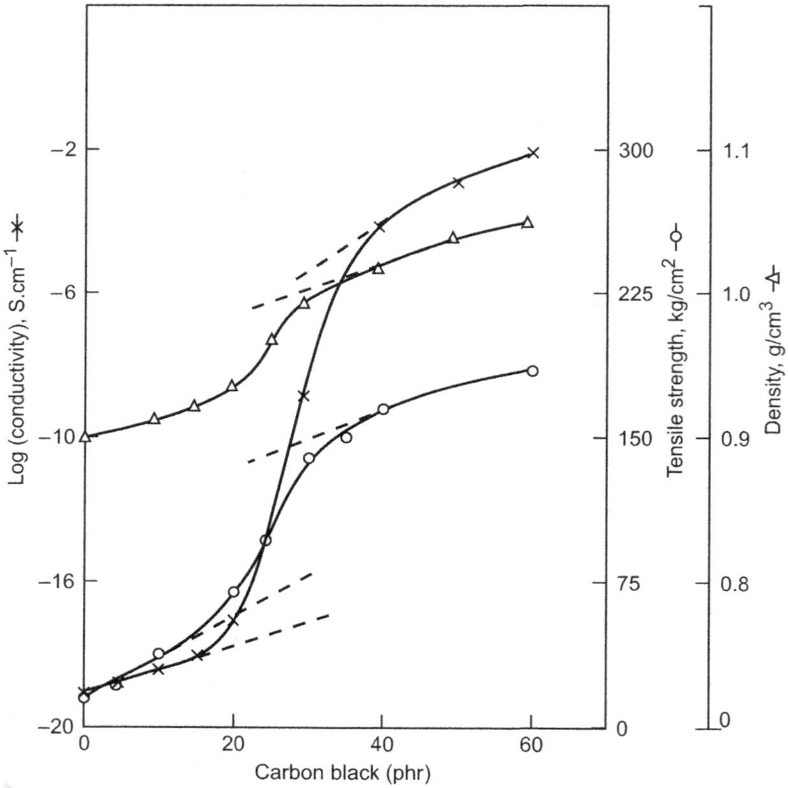

Fig. 12.5 ***Plots of tensile strength, electrical conductivity and density of conducting carbon black filled EPDM composites***

Carbon black as a particulate filler is well known[24] for its ability to reinforce rubber and related polymer systems. The reinforcement is manifested[24, 25] mainly through physical interactions or interfacial attachments via operation of surface forces between them and perhaps by some minor degree of chemical interactions too through anchorage of the rubber chains on the reactive sites of filler particles. The existence of a percolation threshold in each case of physical and mechanical properties and electrical conductivity is fundamentally related and it points to a developing trend in filler particle networking. The effect of filler networking on the property parameters can be understood on the basis of (a) percolation concept (PC)[26] and (b) a kinetical cluster–cluster aggregation (KCCA) model.[27]

The percolation concept (PC) is based on pure geometrical consideration and it is understood as a simple case of formation of chain-like clusters on a lattice resulting form random positioning of the filler particles to the adjacent/nearby lattice sites which, however, admits of some variations in the number of particles in a cluster and in the structure and size distribution of the clusters, ultimately leading to filler networking. Filler networking through the percolation concept does not really depend much on the chemical nature of the polymer matrix because pure geometrical arguments are considered. For the kinetical cluster-cluster aggregation (KCCA) model, however, the chemical structure parameters of the matrix polymer as well as the size of the filler particles or particle networks also become important. The KCCA model considers fluctuation of the colloidal filler particles or their clusters around their mean positions in the elastomer matrix that behaves much like a liquid. The fluctuating particles find greater chance to cluster together on contact when they are present in the matrix at a higher concentration; the particles thus growing in size are held in the cluster irreversibly by strong adhesive energy. In this way, continued cluster–cluster fluctuation leads to formation of larger clusters and finally to cluster networking in different directions. The network growth continues until all particles within a distance smaller than the average entanglement spacing are aggregated. Chemical anchorage and occlusion of the elastomer segments on the surface and within the pores of the filler particles may offer physical hindrances to or may even impede further cluster growth and networking under a given set of conditions of time, temperature, filler loading level and shear action.[28–30] A levelling off trend in the property parameters with increase in the conducting filler dose above what is known as the percolation threshold, may then be easily understood and interpreted. Morphological analysis by scanning electron microscopy is helpful in this regard.

For low filler loading (10 phr) in conducting carbon black filled EVA or EPDM vulcanizates, the carbon black particle clusters are shown to be distinctly and discretely accommodated and pressed between fine strands or clusters of the elastomer phase with varied degrees of discontinuity in different parts, Fig. 12.6, micrograph (a), showing a fine mosaic pattern. As the carbon black loading is increased from less than the percolation threshold range (15–30 phr) to above the same, the distribution of the filler phase becomes more uniform through break down or coming together of the strands and clusters of the elastomer matrix accepting the networking carbon black filler particles more intimately between them; consequently, an unmistakable trend of development of filler-matrix cocontinuous phase morphology becomes noticeable with increase in filler loading. For conducting carbon black loading (50 phr) much above the percolation threshold range, the phase morphology appears largely halo or isotropic and more uniform indicating mutual masking, interpenetration and overlapping of the two phases, micrograph (b), Fig. 12.6; the phase uniformity is also effectively aided by the high shear action during mixing, greater degree of filler-matrix physical interaction including occlusion of segments of the chain molecules of the elastomer matrix within the pores of the filler particles at high filler loading, besides some degree of chemical anchorage of chain segments on the reactive sites (bearing –OH, –COOH, C = O groups) of the filler particles.

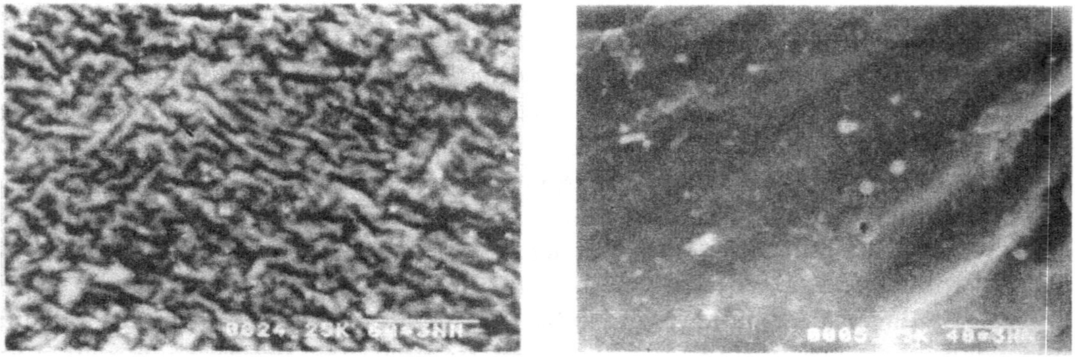

(a) (b)

Fig. 12.6 ***SEM micrographs of conducting carbon black filled EVA composites (× 800): (a) 10 phr carbon black as filler and (b) 50 phr carbon black as filler***

References

1. Skotheim, T.A Ed., *Handbook of Conducting Polymers*, Vols. 1 & 2, Marcel Dekker, New York, 1986.
2. MacDiarmid, A.G. et al., *Phys. Rev. Lett.*, **39** (1977) 1098.

3. MacDiarmid, A.G. et al., *J. Am. Chem. Soc.*, **100** (1978) 1013.
4. Franzus, B. et al., *J. Am. Chem. Soc.*, **81** (1959) 1514.
5. Halane, M., *J. Polym. Sci.*, **51** (1961) 26.
6. Hubel, W. et al., *Chem. Ber.*, **93** (1960) 103.
7. Woon, P.L. et al., *J. Polym. Sci., Polym. Chem. Ed.*, **12** (1974) 1749.
8. Cierdelli, F. et al., *Makromol. Chem.*, **103** (1967) 1.
9. Bradley, A. et al., *J. Electrochem Soc.*, **110** (1963) 15.
10. Chung, T.C. et al., *J. Chem. Phys.*, **74** (1981) 5504.
11. Chien, J.C.W., et. al., *Macromolecules*, **14** (1981) 479.
12. Wegner, G., *Disc. Faraday Soc.*, **68** (1980) 494.
13. MacDiarmid, A.G. and A.J. Epstein, *Faraday Disc., Chem. Soc.*, **88** (1989) 317.
14. Chiang, J.C. and A.G MacDiarmid, *Synth. Met.*, **13** (1986) 193.
15. Epstein, A.J. et al., *Synth. Met.*, **21** (1987) 63.
16. Wenk, G.E., *Synth. Met.*, **15** (1985) 213.
17. Ghosh, P., S.K. Sidhanta and A. Chakrabarti, *Eur. Polym. J.*, **35** (1999) 699.
18. Ghosh, P., A Chakrabarti and S.K. Sidhanta, *Eur. Polym. J.*, **35** (1999) 803.
19. Ghosh, P., and S.K. Sidhanta, *J. Polym. Sci.*, Polym. Chem. Ed., **37** (1999) 3243.
20. Viswanathan, P. and P. Vasudevan, *Angew, Makromol. Chem.*, **17** (1982) 102.
21. Wegchan-Jwdek, M., E. Rogal and A. Zuk, *Polymer*, **22** (1981) 845.
22. Port, A.B. and R. Still, *J. Appl. Polym Sci.*, **24** (1979) 1145.
23. Tsuchida, E. et al., *Macromolecules*, **20** (1987) 2315.
24. Dannenberg, E.M., *Rubber Chem. Technol.* **48** (1979) 410.
25. Krans, G. Ed. *Reinforcement of Elastomers*, Interscience, New York, 1995.
26. Stanffer, D. and A. Aharony, *Introduction to Percolation Theory*, Taylor and Francis, London, 1992.
27. Meakin, P., *Prog. Solid State Chem.* **20** (1990) 135.
28. Ghosh, P. and A. Chakrabarti, *Eur. Polym. J.*, **36** (2000) 607.
29. Ghosh, P. and A. Chakrabarti, *Eur. Polym. J.*, **36** (2000) 1043.
30. Ghosh, P. and A. Chakrabarti, *J. Polym. Mator.*, **17** (2000) 291.

CHAPTER 13

Miscellaneous Polymers and Nanotechnology

13.1 Dendritic Polymers

The synthesis of dendritic polymers is the outcome of sustained efforts to synthesize specified, well defined polymer molecular architecture. Dendritic polymers are highly branched polymers with tree-like branching in all possible directions[1-4] with potential for propagation of branches on branches on branches, and so on. In molecular shapes and dimensions, they are intrinsically and considerably different from more common linear or lightly branched polymers. The dendritic polymers are characterized by a *globular* shape in general, unlike the relatively *elongated* shapes with wide scope for molecular chain entanglements for the much more common linear or lightly branched polymer systems.

The scope for manifestation of intermolecular cohesive (attractive secondary) forces and molecular chain entanglements is far greater in linear (elongated) polymers, and this scope is much limited in comparison in dendritic (globular) polymer molecular systems. Dendritic polymers, as a consequence, exhibit easy swelling and greater solubility and miscibility with other materials not excluding other polymeric materials. Their globular architecture helps them to exhibit relatively *low hydrodynamic volume and lower solution and melt viscosities*. They are useful as *viscosity modifiers* and *flow promoters* and *as lubricants and compatibilizing agents.*

In conformity with their growth and structural features, the dendritic polymers are essentially characterized by the existence of a much greater number of well identifiable chain ends on their outer periphery. They may, in suitable cases carry many circumferential end functional groups such as $-NH_2$, $-OH$ or $-COOH$, which make them highly potential as carrier molecules for use in *catalysis, drug delivery* and *gene therapy,* and *as sensors.*

13.1.1 Random-Growth (Hyperbranched) Dendritic Polymers

A polyfunctional monomer molecular system of the type AB_f (see Sec. 2.10.1) where, $f \geq 2$ may lead to formation of a hyperbranched polymer system such as of the type (I) shown in Fig. 13.1 for $f = 2$ by a controlled, random step–growth polymerization process[5].

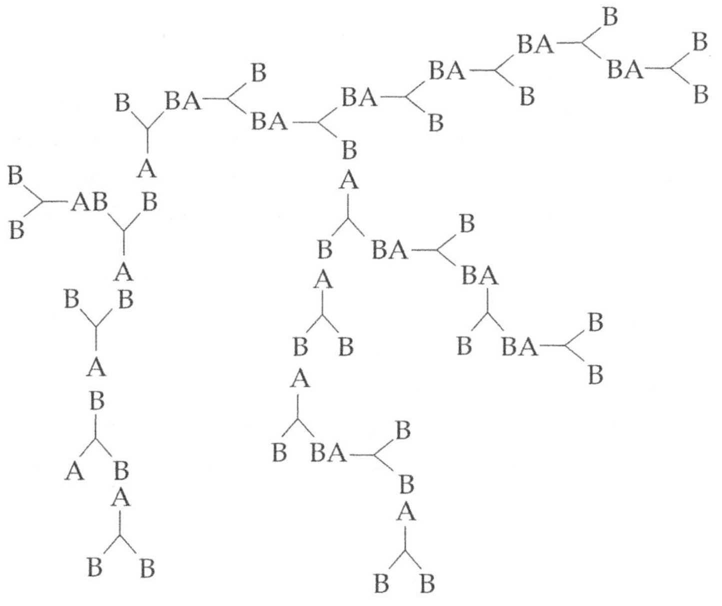

(I)

Fig. 13.1 ***A hyperbranched, partially dendritic polymer segment formed by random poly-condensation of monomer type AB_f, (divergent synthesis) where, $f = 2$ ($\bar{X}_n = 20$, unreacted B groups = 21)***

The degree of branching (DB) for such hyperbranched polymers would depend very much on the value of f and the extent of reaction. Starting with the AB_2 type of monomer systems one may identify two distinct types of chain segments or repeat units such as (i) dendritic, having two terminal B groups, and (ii) linear with only one free terminal B group attached to AB linkages forming the main chain/ branch segments.

Due to the random nature of their growth, hyperbranched polymers of this kind characteristically exhibit molecular weight polydispersity. Progressive steric crowding at the growth ends with increasing molecular size contributes to limit further growth to higher molecular weight that also results in limiting molecular shape variations. One can use systems such as (A_2 + B_3), (AB_2 + B_3) and (AB_2 + AB) combinations for obtaining hyperbranched polymers of varying degrees of structural complexity, molecular size distribution and observed property profile.

13.1.2 Dendrimers

Dendritic polymers with more compact and controlled structures and better possible alignments of the branch units are commonly termed *Dendrimers*[4]. One may also use such terms as *starburst, cauliflower, cascade* etc., to indicate dendrimers. Dendrimers are built outward from a core unit each by attaching successive layers of monomeric units sequentially one after the other. A repetitive sequence of two different reactions, viz. an amine addition to an α, β–unsaturated ester to be followed by nucleophilic substitution of the ester by an amine, as depicted in reaction scheme[1] given by Eq. (13.1); the reaction scheme starts with ammonia as the (initial) core molecule and then successively and sequentially one uses methyl acrylate (CH_2=$CHCOOCH_3$) and ethylene diamine (H_2N–CH_2–CH_2–NH_2) as the branch extension reactants, as shown hereafter by Eq. (13.1):

Finally, a globular structure that would emerge is shown in Fig. 13.2. In synthesis of dendrimers of this kind, the process turns more difficult for growing to each new branch length (new generation). The steric crowding within the molecule of a

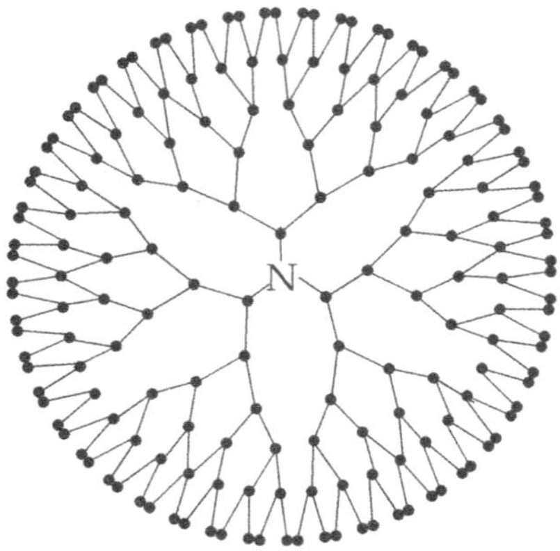

Fig. 13.2 *A typical dendritic polymer (schematic)*

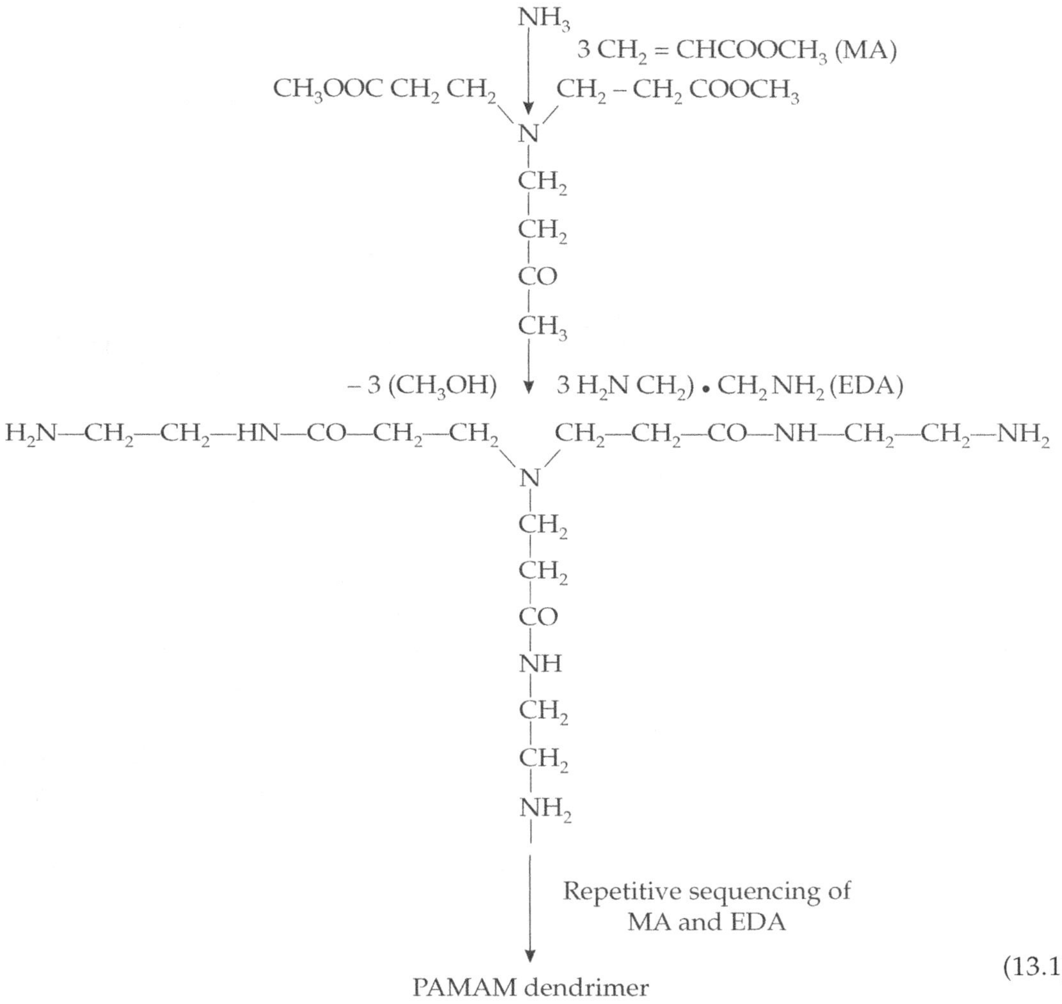

(13.1)

Synthesis of Polyamidoamine (PAMAM) dendrimer

dendrimer and also such crowding of the peripheral functional groups increases progressively, particularly for higher generation dendrimers. Steric crowding becomes significant and possibly also overwhelming after about five or six generations, thereby affecting the purity of the higher dendrimers. It is in fact nearly prohibitive to take the dendritic growth beyond 10–15 generations.

As opposed to a divergent synthesis of a dendrimer ultimately leading to a structure as shown by Eq. (13.1) and schematically depicted by Fig. 13.2, one may also conceive a convergent method as outlined in Fig. 13.3.

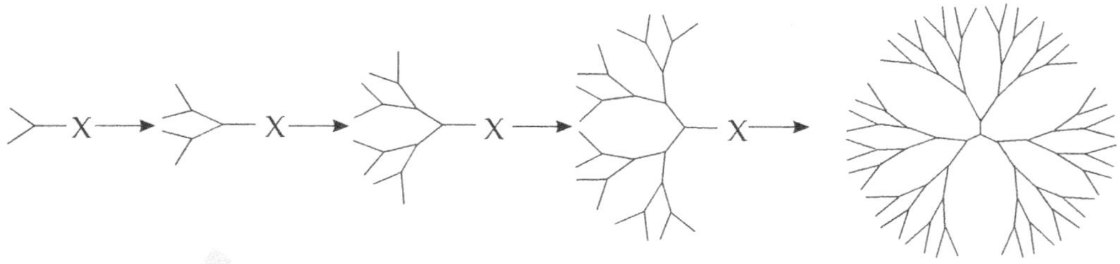

Fig. 13.3 *Convergent synthesis of a dendrimer (schematic)*

The core-forming functional group X may be a protected functional group which is made functional/reactive at the appropriate time to complete the dendrimer structure.

13.2 Inorganic Polymers

Inorganic polymers occurring in abundance in nature in the forms of sand and quartz, i.e., silica, and as ground soil, rocks and stones (mostly assorted metallic polysilicates, oxides, carbonates, etc.) find important applications as building materials, in making different kinds of refractories, glass and ceramics and in arts and crafts. Their high resistances to corrosion, fusion and solubility and their fusibility at high temperatures are notable and their exceptionally high rigidity make them unsuitable for replacement of or an alternative to organic polymers. Sand or quartz (silica, SiO_2) bearing the three-dimensional network polymeric structure (II):

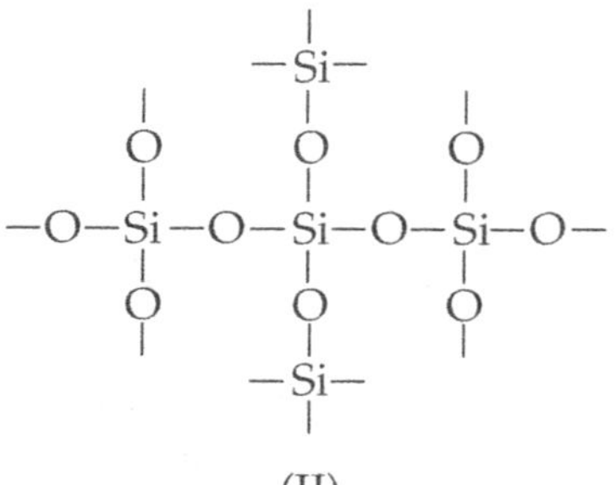

(II)

is highly rigid, infusible and insoluble. If this structural model is changed to poly (organo siloxane) systems (Sec. 8.23), where the presence of two organic groups ($-CH_3$,$-C_6H_5$ etc.) attached to each silicon atom in the chain molecular structure largely changes chain rigidity as compared to the (mother) polymer silica and imparts into the polysiloxane chains (also called silicone polymers) solubility, fusibility and ready processability to moulded and formed objects, films, coatings, etc. (see also Sec. 8.23).

13.2.1 Novel Organometallic Polymers[6]

Carboxylate or sulfide groups are known to have been used to obtain specific organometallic polymers (III or IV)

$$\underset{\text{(dialkyl stannic chloride)}}{R_2SnCl_2} + \underset{\text{(sodium salt of a diacid)}}{NaOOC{-}R'{-}COONa} \xrightarrow{-NaCl} \underset{\text{(III)}}{\left[-\overset{\displaystyle R}{\underset{\displaystyle R}{\overset{|}{\underset{|}{Sn}}}}-OCO-R'COO- \right]_n} \qquad (13.2)$$

$$\underset{\text{(dicyclopentadienyl titanium dichloride)}}{n(Cp)_2\,TiCl_2} + \underset{\text{(dithiol)}}{nHS{-}R{-}SH} \xrightarrow{-HCl} \underset{\text{(IV)}}{\left[-\overset{\displaystyle Cp}{\underset{\displaystyle Cp}{\overset{|}{\underset{|}{Ti}}}}-S-R-S- \right]_n} \qquad (13.3)$$

The product of reaction (13.3) contains the titanocene group in which two cyclopentadienyl legands are coordinated to the metal titanium. Coordinate–bonded metal atom containing organometallic polymers are commonly referred to as coordination polymers. Comparable reactants bearing a variety of other metals (e.g., Hf, Zr, Si, Ce, Pb, Sb, As, Mn, Bi, etc.) are known to have been used to synthesize similar organometallic polymers.

13.2.2 Linear Polymers from Cyclosiloxanes

Polysiloxanes or silicone polymers (VI) of high molecular weight are synthesized by ionic (anionic/cationic) ring opening polymerizations of cyclic siloxanes[7] (e.g., of octamethyl cyclotetrasiloxane, V):

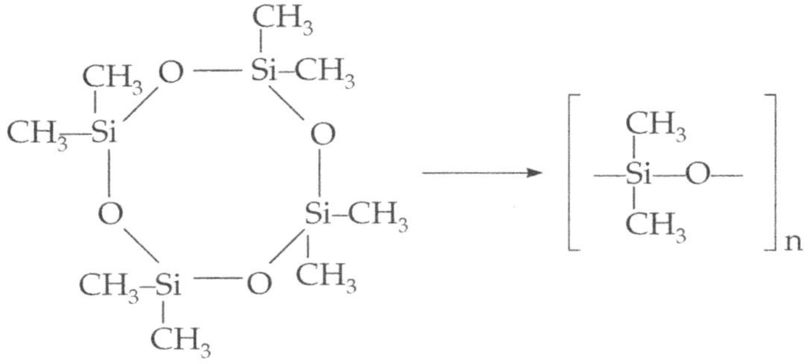

(V) (VI)

(13.4)

Anionic (A^-) chain initiation (by bases such as alkali metal alkyls, alkoxides, hydroxides, etc.) may be depicted as:

$$A^- + \overset{\frown O}{SiR_2} - (OSiR_2)_3 \longrightarrow A - (SiR_2O)_3 - SiR_2O^- \qquad (13.5)$$

and then propagation follows according to:

$$\sim\sim SiR_2O^- + \overset{\frown O}{SiR_2} - (OSiR_2)_3 \longrightarrow \sim\sim(\sim SiR_2O)_4 - SiR_2O^- \qquad (13.6)$$

Interestingly, for this polymerization, ΔH is low or nearly zero, while ΔS is significantly positive (+ 190 $J mol^{-1}K^{-1}$); in fact, the driving force here is the significant increase in entropy of polymerization[8] much like the polymerization of cyclic S and Se, each showing a characteristic floor temperature (Sec. 3, 17.1). Positive ΔS values in such cases result from the much higher flexibility of the relevant linear polymer chains. The flexibility imparts greater degrees of freedom in the linear polymer compared to the cyclic monomer.

13.2.3 Cyclotriphosphazenes and Relevant Polymers[9]

Poly (dichlorophosphazene) (VIII) is readily obtained by thermal polymerization of hexachlorocyclo triphosphazene (VII).

Hydrolytically unstable elastomeric poly (dichlorophosphazene), (VIII) can be turned to stable alkoxy (IX) and amino (X) derivatives as shown by reactions (13.7) and (13.8) respectively. The polyphosphazene chains have high order of intrinsic

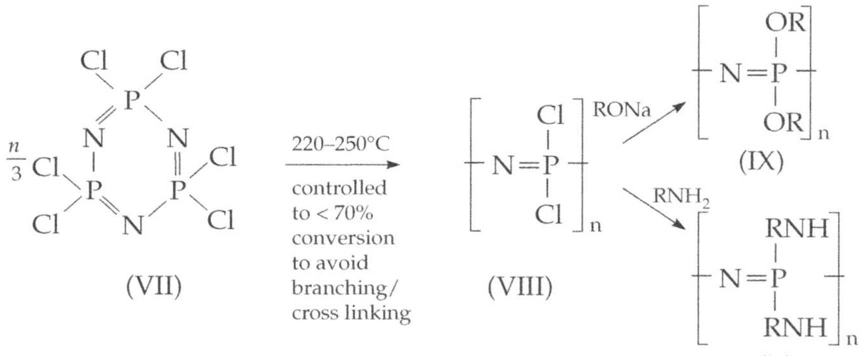

(VII) (VIII) (IX) (13.7)

(X) (13.8)

flexibility (Tg ranging between –60 and –100°C) and they exhibit good photolytic, oxidative and solvent resistance. The organic moiety attached is key to the range of property that they exhibit. With good elastomeric property range, the poly (organophosphazenes) are suited to applications as O–rings, oil–seals or sealants in general, oil–field pipelines and also as selected inner lining materials. Some derivatives are biocompatible, some are biodegradable or bioactive and are considered for biomedical applications. Some varieties have potential as solid electrolytes for light weight rechargeable batteries.

Other cyclophosphazenes that have been likewise studied and polymerized are carbophosphazene (XI), thio phosphazene (XII) and thionyl phosphazene (XIII):

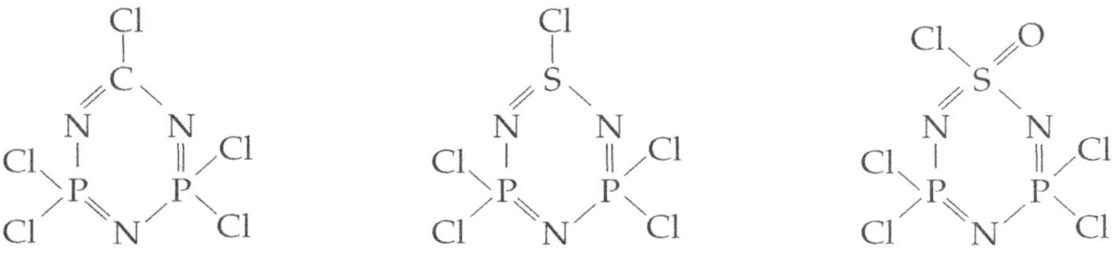

(XI) (XII) (XIII)

13.2.4 Cyclic Organophosphates and Organophosphonates[10]

Cyclic organophosphates (XIV) and organophosphonates (XV) are polymerized anionically to yield high molecular weight polymers, e.g.,

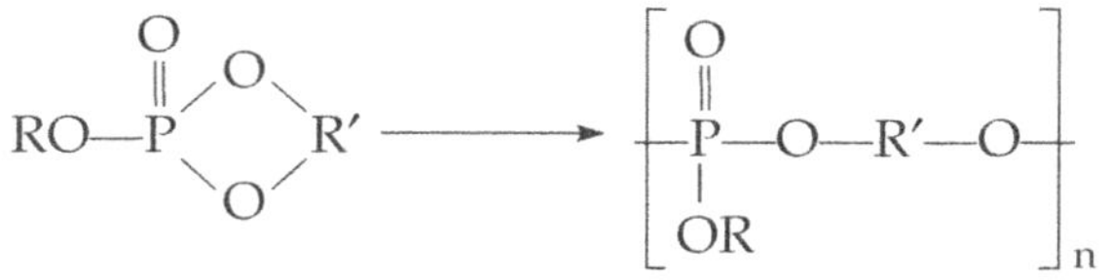

(13.9)

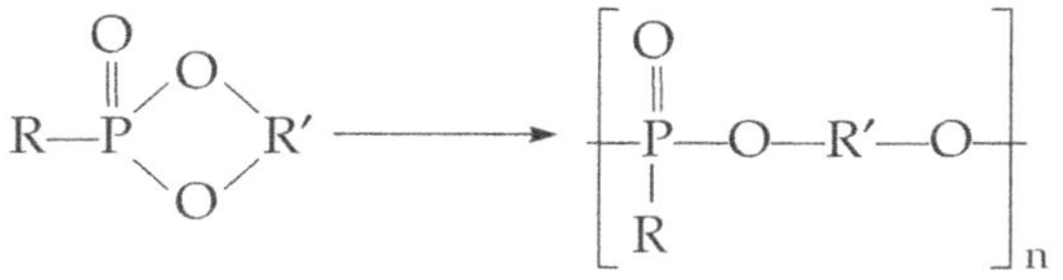

(13.10)

Cationic polymerization, however, yields low to very low molecular weight products in view of the inherent extensive chain transfer reactions.

13.3 Power Polymers[11]

A very important scientific and technological development of our time relates to harnessing and storing solar energy in a package for proper distribution and use as a source of power as and when needed. The advantages of solar power are obvious; every minute, the sun pounds the earth with more energy than the whole world consumes in a full year. So, in reality, the sun remains, as it were, a potential source of unlimited power or electricity that is free and clean.

13.3.1 Solar Cell Based on Silicon Crystals

Research in the development of solar cells gathered momentum from the day the semiconducting nature or more exactly, the photoconducting nature of silicon crystals was established. When sunlight strikes the surface of such a semiconductor, the photons transfer energy to electrons in the material, the excited electrons gather momentum and a photovoltaic effect is generated. In a workable solar cell, that energy is captured, stored and put to use. Solar cells, better known as 'Photovoltaics' take advantage of the same electronic properties that make a semiconductor such as crystalline silicon so vital to the computer industry. For solar cells, the bottom line is the ability of the relevant semiconductor to silently transfer absorbed sunlight into usable current electricity.

A film or sheet of the semiconductor material is laid in such a manner that it is sandwiched between two layers of electrode material. In a well designed photovoltaic unit, a built-in electric field draws the photoexcited electrons to the top electrode which carries them out of the cell and puts them into a well designed circuit; the bottom electrode gathers the flowing electrons from the circuit to fill the 'holes' that the excited electrons left behind.

13.3.2 Search for Low-Cost Technology

Search for low-cost technology has been driving the solar cell manufactures to leave out silicon crystals altogether. The current developing trend has been directed toward developing photovoltaics employing cheap amorphous silicon as the semiconductor. Semiconductor alloys, such as copper-indium diselenide, copper-indium-gallium diselenide, cadmium telluride, etc. which can be easily spread into a very thin film (few thousandths of a millimeter that is nearly hundred times thinner than the films of silicon crystals used in the conventional, first-generation solar cells) have grown into much higher prospect, prominence and potential. As a consequence, significant to substantial material saving has been possible and in mass production, much lighter solar cells based on this new advanced technology has turned largely cheaper. They give higher output per unit mass of semiconductor, though showing relatively poor efficiency. Cheap, thin films of amorphous silicon or of the relevant semiconducting alloys have the potential of capturing as high as 20% solar energy.

Silicon crystals may justify their cost in microprocessors, but for the purpose of generation of power, they are nearly four times, as expensive as the power generated by burning fossil fuels. The low cost and inherently flexible nature of amorphous silicon make it suitable for application as fine, thin coatings on a range of materials including glass panels and roof tiles. Switching to thin film, silicon may not really bring down the cost and make solar cells competitive in most respects. Even with five fold uprise in output and use over the past five years (2004–2009), solar power still does not account for even 1% of the worldwide power generation.

13.3.3 Prospect for Use of Photoconducting Polymers

Materials far cheaper and thinner than the thin film of amorphous silicon and relevant semiconducting alloys are needed to accelerate the penetration of solar power as a major player in the power scenario. Experts in the field are pitching hopes on technology developments based on thin films from special organic materials including pigments and fullerenes (carbon in a special form) and the so called power polymers, viz., photoconducting polymers such as poly (N–vinyl carbazole), poly (vinyl anthracenes), poly (phenylene vinylenes) etc. (Secs. 12.3 and 12.4) in order to carry forward the relevant technology to maturity.

It may further take a decade or two for thin films from such potential materials, perhaps in selected combinations, to transform solar power package from a marginalized technology to a mainstream source of energy to power most machines and devices and also as substitutes for conventional fuels/energy items for heat generation.

Over the past few decades, we have become familiar with polymers conducting electricity and emitting light (light emitting diodes) as they do in flat panel displays. Now, we are required to run the reverse process too, viz., polymers getting/ absorbing light and generating electricity in what is known as the photovoltaic device. Photo conductive polymers, thus form the basis of inexpensive solar cells. Solar cells being developed and made from highly thin films of such photoconductive polymers and relevant plastics containing selected additives have the potential to finally turn relatively decentralized solar power generation, storage and distribution commonly and widely easy and affordable, and made readily available to many in wide locations for a wide spectrum of application areas than hitherto been possible through the use of crystalline or amorphous silicon and relevant semiconductors. The science and technology behind synthesis of such polymers and relevant plastics and further turning them to ultra thin films are fast growing simpler. Some of the polymers and their prospective blends have turned a lot cheaper and they can now be easily made and processed to desired usable forms.

A promising approach in developing solar cells/batteries relates to novel synthesis of easy processable blends of two (or more) different polymers of controlled rheology where one component polymer is photoconducting while the other(s) acts as the supporting durable matrix providing strength and stability and other desired property combinations and processing ease. Polymer-based photo detectors being developed at the current stage rival the sensitivity of conventional commercial photo detectors that employ silicon and related semiconductors. Nonconducting, semiconducting, conducting and photoconducting polymers duly and selectively blended, filled, doped, or otherwise are currently being developed and employed for making wide range of solar energy packages, gas/chemical sensors and bio-sensors.

13.4 Nanotechnology[12]

Cheap microelectronics really revolutionized human activities relating to computing and information technology and infused a whole new trend in scientific research through development of microelectronic aided sophisticated automated equipments instantly processing or scanning a system and providing recorded graphic and printed data that are highly accurate, reproducible and dependable within a low coefficient of variation.

By the turn of the 20th century, we entered the era of technology of nanosize (billionths of a meter) materials, commonly referred to as nanotechnolgy. It almost began since the discovery of single walled carbon nanotubes (SWNTs) in 1993 and with the development of nanoclay/polymer nanocomposites at about the same time. The novel combination of structural, mechanical, electronic and surface properties of SWNTs and other nanosized materials (e.g. nanoclays) henceforth developed and made available in particulate, layered, wire or fibre/whisker forms have rendered them attractive candidates for nanoscale electronic devices and circuitry and for high-performance novel structural composites called nanocomposites using selected polymers or polymer blends as matrix materials and the nanoscale particulate or fibrous fillers at low doses (3–10%) as the reinforcing agents, producing novel composites of unusual performance profile.

The carbon nanotube (Fig. 13.4) is associated with impressive list of attributes: they may be 2 mm long and may be hundred times as strong as steel. The average diameter is in the range of 1.2–1.5 nm. Having a high aspect ratio, they are flexible, light, thermally stable and chemically inert. They may show metallic conductivity or semi-conducting character depending on size and geometry, stretch or twist infused. They may conduct electricity as good as or even better than copper and can transmit heat better than diamond. Most other nanoscale materials come in the common size range of 1.0–10.0 nm. Lower particle size provides higher surface area and hence offers much higher property profile and much enhanced performance prospect manifested through significantly higher efficiency and more intense particle surface-matrix interfacial interactions.

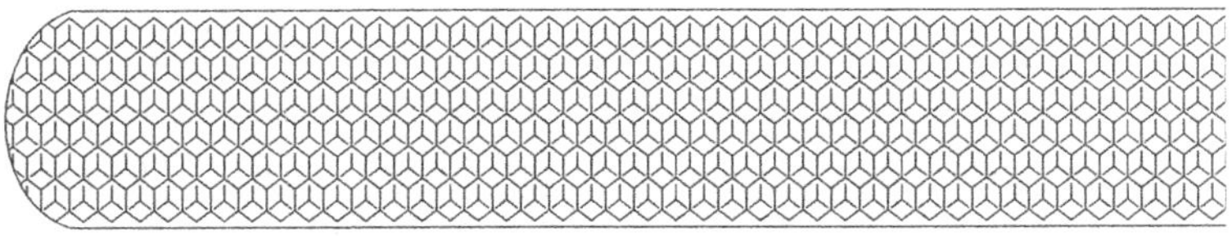

Fig. 13.4 *A carbon nanotube (schematic)*

13.4.1 Methods and Markets[13]

Nanoparticles can be made using wet chemical processes yielding dispersions in colloidal or emulsion forms of a wide variety of polymeric or non-polymeric materials. The mechanical process covers size reduction by milling and grinding (including cryogrinding) and mechanical alloying. They provide low-cost and operation-simplicity. Other techniques worthy of mention are flame pyrolysis, high-temperature drying/evaporation, laser ablation and plasma processes. Production of fullerene and carbon nanotube is based on gas phase pyrolytic synthesis.

There remains a vast market for nano materials and relevant composites. The re-engineering of much of the existing composite materials technologies by bringing them down to a state allowing use of nano–fine materials as the dispersed phase, has the high scope and potential in respect of production of relatively light weight high performance nanocomposites, permitting in most cases 1.5–3.0 fold or even higher level of enhancement in strength/modulus, hardness/rigidity and also significant enhancement in heat–distortion temperature at a dose level of only ≤ 5.0% of the nano–fine component materials (in particulate, whisker, fibrillar or fibrous forms) used during composite making. The electronic market accrues benefit from miniaturization of chips, and processors so developed enable display of short–time or instant response and exhibit high performance. There is expanding scope for relevant composite's applications in the automotive and other transport sectors, space technology sector, magnetic recording material systems and in micro–electro mechanical system (MEMS) technologies.

13.4.2 Polymer Nanocomposite Systems

Polymer Nanocomposites have blossomed into a major R&D area throughout the world over the past 15–20 years. The nanofillers that have drawn major attention are the conductive single walled carbon nanotubes and (layered) graphite nanoparticles, nanofine metal powders, nanofine silica and nanofine (natural/synthetic) layered clay materials. Others that have grown in use are nanofine TiO_2, $-SiO^-_3$, Al_2O_3, $CaCO_3$, ZnO etc. all being formed and isolated from solution or dispersion by careful precipitation and flocculation techniques.

Use of highly anisotropic polymer nano fibres (diameter 10–1000 nm) including cellulose or cellulosic nano fibres in making polymer nanocomposites is also worthy of mention. Their outstanding property range makes the polymer nano fibres optimal candidates for many important applications such as tissue engineering, wound dressing, drug delivery and related biomedical applications. A number of processing techniques such as drawing (melt or dry spinning), template synthesis, phase separation, electrospinning, etc. are employed for preparation of polymer nano fibres. Suitable polymers are electrospun into ultra fine/nanofine fibres mostly from solution and for some, from melt. The electrical forces allow the polymer jet elongate thousands/millions of times (allowing solvent evaporation or melt solidification as the case may be) thus, producing the nano fibre that collects in a spiraling shape on an electrically grounded metal sheet.

13.4.3 Clay-Polymer Nanocomposites

Clay (natural/synthetic) materials used in nanocomposite-making appear in nano scale layered (silicate) structures[14], that can be schematically shown as stacks of plate-like structures or platelets, Fig. 13.5. Each layer is 1–2 nm in thickness and 100–200 nm in length and may be in the range of 30–60 nm in width. The space between the platelets is called gallery gap. The silicate (SiO_3) moieties accommodate aluminium, magnesium, iron, calcium, sodium or some other metal ions in appropriate amounts and selective combinations, depending on nature and origin of the clay material. The hydrophilicity of the clay is a mismatch against the commonly hydrophobic nature of the matrix polymers such as polystyrene (PS) and selected PS-block (acrylate/methacrylate) copolymers, selected phenolics, nylons, polyethylene, polypropylene, polyurethanes and different grades of EVA etc. Replacing the clay cations by long-alkyl ammonium ions, the hydrophobicity of the layered clays can be significantly enhanced along the outer surfaces and within the gallery gap; the gallery gap is also measurably widened as a consequence.

Enhancement of hydrophobic nature and widening of the gallery gap allow ready accommodation of the matrix polymer chains between the clay platelets by a process known as intercalation much like sandwiching by penetration from outside. Intercalation may be attained through dispersion of matrix polymer by three different approaches, such as polymer intercalation by allowing *in situ*

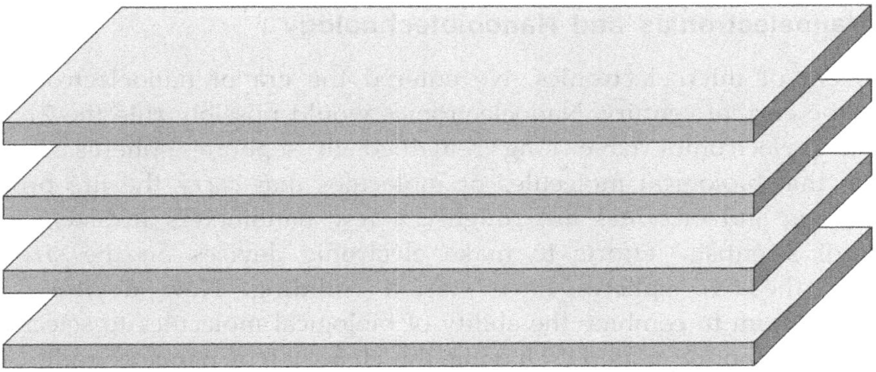

Fig. 13.5 *Stacked platelet structure of a typical nanoclay material*

polymerization of a selected liquid monomer/monomer mixture (sol–gel process), solvent (polymer solution) intercalation followed by solvent evaporation and polymer melt intercalation usually under shear followed by solidification of the melt. Excessive extent of intercalation due to polymer entry aided by external shear may sometimes push the clay layers wide apart leading to what is called exfoliation, often causing random dislocations of the clay layers or the clay platelets. The successive changes from stacked layers to intercalation and exfoliation due to polymer growth or polymer entry is schematically shown in Fig. 13.6. Infusion of long-alkyl ammonium compounds over the clay platelet surfaces renders the clay adequately compatible with the polymer matrix, thus resulting in higher performance and enhanced resistance and barrier properties, and also durability for the resulting composites. In case the polymer fails to intercalate, a micro-composite would result.

Consequent to the surface/structural modifications as detailed above for the clay–polymer nanocomposites, a wide range of engineering properties can be significantly improved at a low level (<5.0%) of filler loading. Additionally, at such a low nano scale filler dose level, the optical properties of the polymer are not significantly impaired. The thickness of individual clay layers is much smaller than the wavelength of visible light, so that in case of an optically clear polymer, well exfoliated clay polymer nanocomposites also by and large remain optically clear. Preparation, processing and handling of nanofine fillers entail the danger from spread of nanofine particles, toxic or otherwise in the surrounding environment. All due precautions and remedial measures need to be adopted to prevent spread of toxicity and environmental pollution.

13.4.3 Nanoelectronics and Nanobiotechnology

From the era of microelectronics, we entered the era of nanoelectronics by the advent of the current century. Nanoelectronics would possibly rule the 21st century. Biology and electronics have long remained in separate spheres. Having the knowledge that biological molecules or molecules that carry the life process, viz. DNA, proteins and enzymes are roughly a few nanometers in size, and in the backdrop of scientists' efforts to make electronic devices on the size scale of nanometers, the two spheres have started colliding. New devices are being developed that aim to combine the ability of biological molecules to selectively bind with specified nanofine materials having the character of nanoelectronics, enabling instant detection of electrical changes caused by such binding. Nanoelectronics so

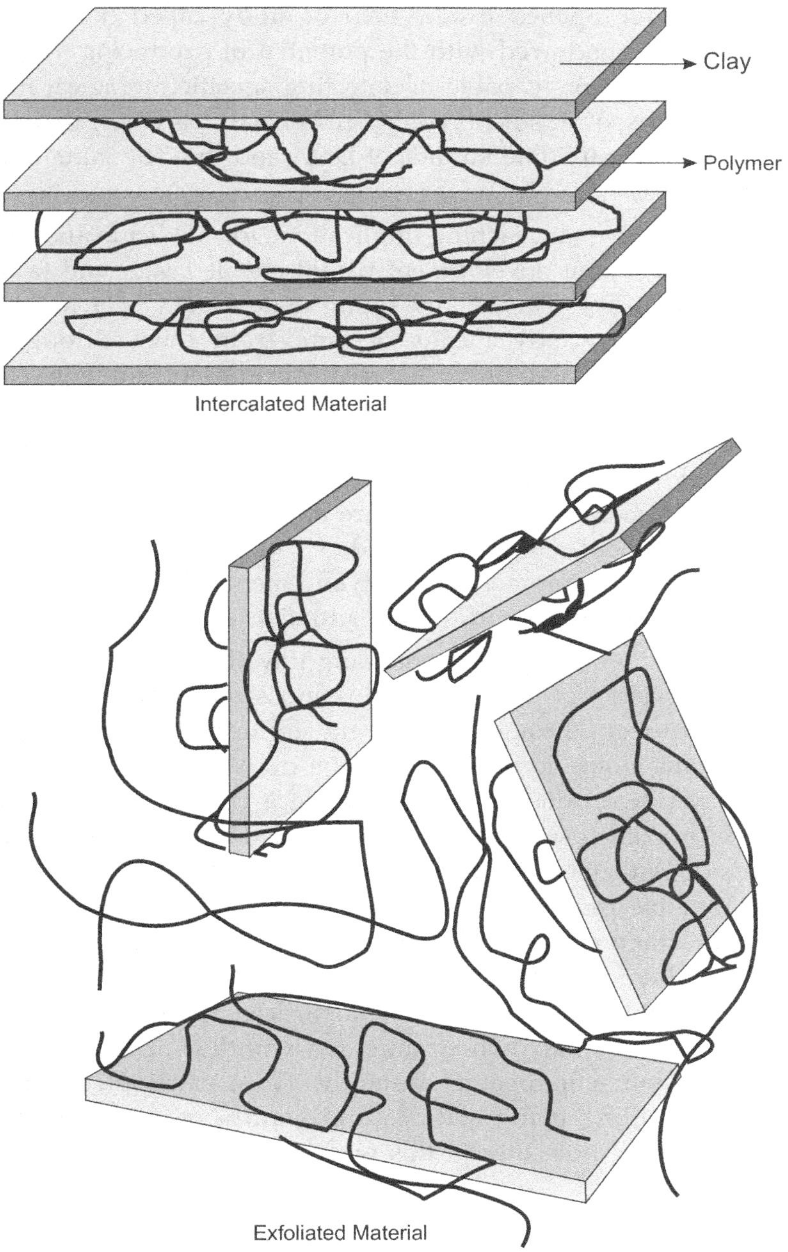

Fig. 13.6 *Schematic representation of (a) intercalation and (b) exfoliation of layered clay structural units due to limited polymer growth and higher level of polymer growth respectively*

wedded to biology have opened a new field of study called 'nanobiotechnology'. This hybrid discipline is endowed with the potential of producing a variety of tools in the form of tiny biosensors capable of detecting specific biological molecules and microscopic systems based on silicon nanofibres or nanoparticles, thereby providing a road map for reading individual strands of DNA and thus, opening a new window on biological molecules.

The new disciplines have far reaching implications for medicine and biotechnology. The devices in the process of development would enable easier and faster diagnosis of complex diseases such as cancer and immuno-deficiency related killer deseases and could even open up easy means of cure from them through ready and appropriate drug selection and provide early warnings of ailments such as heart attack or cerebral attack from a reading of developing changes in the mix of dozens of proteins in selected locations. Nanobiotechnology chips and gadgets could lead to tools that would enable rapid discovery and evaluation of potential drug candidates at once. Nanoelectronic devices are aiming at doing away with elaborate apparatus or equipment hitherto needed for ultra sensitive detection. Nanoelectronics has made commendable advancements through miniaturization of devices needed in such studies and such a situation is now a close reality.

It is possible to build a silicon chip containing tiny channels, each about 50–60 nm in width and depth. Through such a small, nano size channel, a single strand of DNA can barely squeeze through. Under the influence of an electric field, the normally coiled chain molecule of DNA may be drawn into such a channel, uncoil and thread its way down, thus, possibly creating a chance to read if it contains a specific sequence, making use of fluorescent labelled DNA probes attached to the sample beforehand. With passage of each molecule of DNA down the channel, an optical detector may be used to identify the fluorescent labels passing by. The DNA is then played like a tape player. One may also affix gold nanoparticles (1.5–2.0 nm) to a piece of DNA where each gold particle serves as an antena. Application of radio frequency magnetic fields causes the gold nanoparticles to heat up, thereby snapping the double stranded DNA into two strands. On withdrawal of the magnetic field, the strands regroup and pair together instantly. Thus, nanobiotechnology provides us powerful tools enabling manipulations and control of things at the molecular level. Probing life's basic molecules in this way looked like fantasy some years ago. It is now a distinct reality.

References

1. Tomalia, D. A., A. Naylor and W. A. Goddard, *III, Angew. Chem. Int. Engl. Ed.*, **29**, (1990), 138.
2. Tomalia, D. A., and J. M. J. Frechet, *J. Polym. Sci., Polym. Chem. Ed.*, **40** (2002), 2719.
3. Bosman, A. W., H. M. Jannsen and E. W. Meijer, *Chem. Rev.*, **99**, (1999), 1665.
4. Frechet, J. M. J. and D. A. Tomalia, *Dendrimers and Other Dendritic Polymers*, Wiley-Interscience, New York, 2002.
5. Bolton, D. H. and K. L. Wooley, *J. Polym. Sci., Polym. Chem. Ed.*, **40**, (2002), 823.
6. Carraher, C. E. Jr., J. E. Sheets and C. U. Pittman, Jr., (Eds.), *Organometallic Polymers*, Academic Press, New York, 1978.
7. Bostick, E. E., "Cyclic Siloxanes and Silazanes", Chap. 8 in Ring Opening Polymerization, K. C. Frisch and S. L. Reegan (Eds.), Marcel Dekker, New York, 1969.
8. Lee, C.L. and O.K. Johannson, *J. Polym. Sci., Polym. Chem. Ed.*, **14**, (1976), 729.
9. Majumdar, A. N., S. G. Scott, R. L. Merker and J. H. Magill, *Macromol. Chem.* **190**, (1989), 2293; Macromolecules, **23**, (1990), 14.
10. Penczek, S., P. Kubisa and R. Szymanski, *Macromol. Chem. Macromol. Symp.*, **3**, (1986), 203.
11. Ghosh, P., *Indian Sci. Cruiser*, **19** (1), (2005), 11.
12. Ghosh, P., *Indian Sci. Cruiser*, **19** (3), (2005), 61.
13. World Nanotechnology Market – An Industry Update (2005).
14. Alexandre, M. and P. Dubois, *Mater. Sci. Eng.*, **28**, (2000), pp. 1 – 63.

CHAPTER 14

Polymers in Wastes and their Environmental Impact

14.1 Introduction

Whenever one thinks about materials that one uses in one's everyday life, one cannot lose sight of the nearly all–encompassing materials called polymers finding outlets as (i) Plastics (various moulded objects, films, laminates, electrical and electronic gadgets, pipes, tubes and rods, storage tanks and bins, bottles and carboys, cables and wire coatings etc.) (ii) Rubbers (tyres and tubes, beltings, hoses, coated fabrics and sheets, foot-wears, foams and microcellular objects, etc.), (iii) fibres, yarns, cords, ropes, tows and fabrics, and also as various types of (iv) adhesives, coatings and composites[1].

Growth of production and use of all types of polymers together on worldwide basis has been following a nearly steady upward trend at about 6–7% per annum, surpassing the world average GDP growth rate of about 3–4% per annum. In India, the current consumption of major polymers as plastics/rubbers etc. is only between 4–5% of global consumption. The very high growth rate of polymer products consumption worldwide as well as in India is linked with the multiple advantages that they provide, viz., (i) helping to improve quality of life, (ii) helping to preserve land, water and forest reserves, (iii) allowing most efficient use of non–renewable energy resources, (iv) offering a more favourable cost–benefit ratio in respect of material saving through improved design and through waste minimization consequent to high potential for recovery, recycle and reuse, (v) having a versatile range of applications, and above all, (vi) playing a significant role in energy conservation. Developing technology trend of having growing use of renewable resources i.e., agro/forest residues/wastes as filler or reinforcing components in polymer compounding and processing, particularly, for making specified moulded objects, laminates and structural composites, matches well with the present day concept of green technology and clean technology, that is, a technology trend that is eco – friendly[2,3].

14.2 Natural Resources Scenario

With the progress of our civilization and with rapid growth of population, our dependence on use and exploitation of natural resources including forest wealth and mineral wealth covering reserves of natural gas, petroleum, coal and related fossil fuels and inland water resources as well, have grown by leaps and bounds, so much so, that resources conservation including energy conservation and water conservation are currently viewed as some of the most burning issues related to the survival of life on the earth.

Consumptions of fossil fuels, particularly of petroleum and to some extent of natural gas reserves for the production of polymer products are still limited to less than 6% of the overall quantities tapped or processed. Bulk of the quantity tapped is used for running transport vehicles in the aviation, automobile, shipping and railway sectors and also in the industrial sectors for production of chemicals and fertilizers and for power generation, where coal is still the major player. On the other hand, whenever one thinks of prospects of power generation from wastes, one thinks of only organic wastes that may be judiciously collected for efficient combustion/incineration for energy recovery with the ultimate objective of deriving heat for steam and power generation.

14.3 Waste Items

Household/municipal wastes, agro/farm or forest/wet land wastes and wastes from industrial activities come as solid wastes, liquid wastes (effluents) and gaseous wastes, all putting heavy burdens on the environment. The first and perhaps the best way of managing any kind of waste is not to generate or create it in the first place, or to extend efforts to minimize waste by reducing it right at the source and meaningfully prevent its accumulation only to prevent rapid spread of filth and disagreeable odour and pollution of the adjoining land, air space and water reserves, thereby endangering public health and sanitation, and adversely affecting eco-aesthetics.

Significant portions of solid organic wastes, particularly, the synthetic polymer-based materials in the forms of films and nets, fibres and textiles, moulded objects, packaging items, laminates, pipes, tubes and conduits, printed circuit boards, and a host of waste items from automobile, carpet, furniture and footwear sectors and from the electronic and electrical industry are non-biodegradable. In fact, synthetic plastics have, by now, earned for themselves the status of 'punching bag' for most environmentalists.

14.3.1 Plastic vs. Paper Bags for Packaging and Textile Nets for Prevention of Soil Erosion

Some people think of use of biodegradable paper or textiles (e.g. cotton/jute yarns, cords, nets and cloths) as better alternative as compared to use of bioresistant plastics (polyethylene, polypropylene, PVC, etc.) for packaging purposes, and for prevention of soil erosion at river banks and of landslides in landslide-prone slopes of hills and mountains. Laying of selected textiles in the form of nets of proper design and blends (e.g. blends of jute and polyolefin (PE/PP)) are getting more and more popular and technically acceptable for durable road/highway construction, reclamation of wetlands, landslide prevention and prevention of erosion of arid or irrigated agro lands and for protection of vast stretches of river banks.

As for resolving the dilemma between use of paper or natural textiles in place of films of synthetic plastics for packaging, storage and transportation, one must not lose sight of the fact that the paper or textile (cotton jute etc.) industry causes massive deforestation (use of soft forest wood, bamboo etc.) or use of vast stretches of vegetable or foodgrain growing land, needing support through application of manures/fertilizers and pesticides, proper dispensation of irrigation water and also needing deployment of educated/trained agricultural labour force. Paper as well as textile industries involve major polluting chemical transformation activities causing release of massive volumes of toxic effluents in each case thereby, putting heavy burdens on environment.

Production of plastic grocery bags requires much less total energy than production of paper/jute grocery bags. It apparently takes five to six trucks to carry the same number of paper bags as would be carried by only one truck-load of as many plastic bags of comparable holding/packaging capacity.

14.4 Classified Waste Materials

Our major waste materials may be classified and sub-classified under the following heads and categories:

(I) Solid Wastes (industrial, agro and municipal)
All solid wastes come under the Heads: (A) Inorganic or (B) Organic.

(A) Inorganic	*(B) Organic*
1. **Ash, Sand and Cinders** (from household/industrial sectors, metallurgical plants, power plants etc.)	1. **Pulp and Paper** (from household, business and industrial sectors)
2. **Glass** (glass sheets, rods/tubes, bottles, casings, novelty items etc.)	2. **Forest and Agro/Farm Wastes** (plant leaves, stalks and tree-branches, logs, wood chippings and saw dust, a range of biomass including straws, feathers, sea-weeds, water hyacinths, animal dung etc.
3. **Ceramic Items** (broken earthen wares and potteries and crockeries, ceramic-made broken toys, dolls and other images etc.)	3. **Polymer and Related Items** (rubbers and plastics, leather, fibres and textiles, waste laminates, sheets, films, cans, carboys and bottles, used packages, furniture and moulded items, worn-out bags, beddings, cushions, luggage etc. – all made from rubbers, plastics and fibres/textiles etc.)
4. **Slag, Minerals Metals and Alloys** (from mineral handling activities and metallurgical process units and foundries, broken metal utensils, used metal toys, cans, panels and other such items etc.)	4. **Food and Animal Residues** (intimately mingled with municipal wastes/forest or agro wastes)
5. **Radioactive Items/Residues**	
6. **Refuse Materials from Construction and Demolition Sectors.**	

(II) Liquid Wastes (industrial and municipal effluents etc.)

(III) Gaseous Wastes (combustion gases i.e. chimney/furnace gases and incinerator gases, green house gases arising from agro wastes or from natural vegetation decomposition, leakage gases from industries, store houses etc.)

Of the listed wastes, only the organic solid waste, natural or synthetic obtaining in garbage heaps or in accumulated waste heaps has the potential for recovery and exploitation for power generation.

14.5 Power Scenario

We may now take a look at the (I) conventional and (II) non-conventional power generation activities currently being followed in different countries as described below:

(I) Conventional:

A. Thermal Power, where the energy items in use are: (i) coal/lignite, (ii) natural gas and (iii) oil (petroleum fractions).

B. Hydro Power, where tidal and hydal energy are utilized for power generation.

C. Nuclear Power, where nuclear (atomic) energy is utilized for power generation.

(II) Non-Conventional:

A. Bio-Gas Generator: Gaseous energy items derived from bio-decomposition of bio-mass that are utilized for heating, lighting and electric power generation.

B. Windmill Generator: Wind energy utilized for power generation.

C. Solar Power Packages: Solar energy utilized to generate photovoltaic effects that are transformed into photovoltaic devices to generate power for storage (cells/batteries), and for distribution to (homes, offices and industries) localities for heating and lighting and for running machines.

D. Household Automotive and Industrial Storage Cells/Batteries: Chemical energy packaged for power supply.

E. Energy/Power Saving and Conservation: This is to be achieved through stringent energy audit with watchful attention and monitoring aimed at elimination of wasteful/avoidable consumption. Power (energy) saved is as good as or even better than power (energy) generated/produced.

Besides all options listed above, one also should focus one's attention in respect of converting waste to wealth, i.e., by waste-utilization for recovery of energy (power generation) and by conversion of wastes to value-added materials. Present-day municipal/industrial effluent treatment processes utilize small doses of polymeric flocculants based on water-soluble polymers/(graft) copolymers such as selected polysaccharides and polyacrylamide or –COONa group containing polyacrylates. Doing away with our dependence on wood in construction and furniture industry by legislation and putting emphasis on widespread use of plastics in such application areas have gone to check deforestation considerably at the current stage.

14.6 Municipal Solid Wastes (MSW)

Let us now focus our attention on the prospects of utilization or disposal of municipal solid wastes (MSW) having average composition as in Table 14.1.

Table 14.1 *Municipal solid waste composition (%)*

Constituents	*India*	*Some developed countries*
Paper	3–8	10–40
(Organic) Polymers (plastics, rubbers fibres etc.)	2–5	4–10
Metals/Alloys etc.	0.5–1.0	4–15
Glass/Ceramics	0.5–1.0	4–15
(Other) Organic Matters	10–30	15–25
Moisture	20–25	15–25
Ash, Fine Earth etc.	20–40	15–30

An unmistakable modern trend that can not escape one's attention is the increasing use of polymers (plastics, rubbers, fibres, etc.) in automobiles and other forms of transport vehicles, agriculture and farming, horticulture and sericulture, sports and entertainments, electrical and electronics, computer and radio engineering, communication network and IT sector, interior decoration, advertisement and media activities, water management, wire coating, cables and sheaths, road construction and civil engineering, geo-engineering and what not. The current status of plastics and related materials may be described as hereunder in Table 14.2.

Table 14.2 *Current status of plastics*

Status	*India*	*World*
Per capita Consumption (kg)	4–6	20–30
Recycling (%)	45–60	15–20
Plastics in Solid Waste	1–4	3–9

The energy consumption map for making some common packaging materials that may find refuge in garbage heaps stands as in Table 14.3.

Table 14.3 *Energy map of some packaging materials*

Material	*Energy (kWh/kg)*
Al	74.1
Steel	13.9
Glass	7.9
Paper	7.1
Plastics	3.4

Energy evolved or released on burning some common virgin polymers in MJ/kg varies between 150–350 while that released by burning coal is in the range of 18–30 MJ/kg. A rubber tyre releases about 30–32 MJ/kg.

14.7 Waste Management

The mantra of waste management relates to 'conserve' (materials and energy) and reduce, reuse, recover and recycle (wastes). Waste disposal primarily relates to landfill, composting, incineration for power generation or adoption of processes ending up in producing value–added alternatives.

Among the major packaging materials found in waste heaps, viz. aluminium, steel, glass, wood, paper and polymers (LDPE, HDPE, LLDPE, PP, Polystyrene, PVC and their copolymers and blends, rubbers, the nylons, PET and polycarbonates (PC) and different fibres and textile items), only polymers contribute to notable energy efficiency and they significantly retard or prevent deforestation and can even contribute to enhance activities in aforestation, agriculture, plantation and farming leading to enhanced crop/food production with much desired progress toward striking good ecological balance. Polymer items are more visible in garbage heaps as they are light and they come in bright colours.

14.7.1 Life Cycle Assessment

Figure 14.1 illustrates life cycle assessment as an appropriate model to provide an insight into environmental aspects of use of materials:

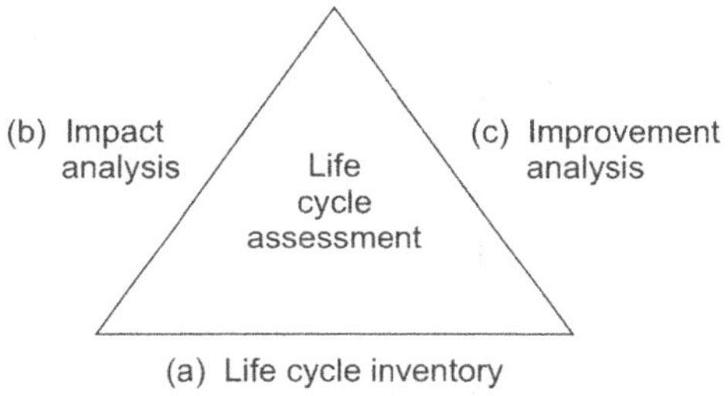

Fig. 14.1 *Life cycle assessment*

The three interrelated components of life cycle assessment are : (a) life cycle inventory, (b) life cycle impact analysis and (c) life cycle improvement analysis represented by the three sides of the model triangle. The inventory component quantifies the raw materials and energy requirements, emissions and effluents released as by-products and outcome of refuse items/solid wastes, and also, residues and other environment disturbing releases incurred during the life cycle of the product, process or activity. The impact analysis component relates to technical (qualitative or quantitative) process parameters that characterize and assess the outcome of the environmental loadings as identified in the inventory constituent. The life cycle improvement analysis relates to systematic assessment and evaluation of the needs, or of the chances of reducing the environmental burden that may be caused or identified by the inventory component. The improvement analysis component may cover the need for changes in product design and process route, raw material alternatives, consumer usage pattern or acceptance and finally, waste management.

14.8 Recovery and Recycling of Organic Wastes

Energy recovery offers an interesting, technically practicable and economically viable alternative to municipal solid waste (MSW) recycling. It reduces or eliminates over-dependence on mechanical recycling through the helping hands of rag-pickers. For this to happen, we would need to ensure abundant availability of MSW feedstock at minimal cost in big metros or clusters of townships. One also needs to remember that plastics and some other non-perishable or perishable organic wastes including biomass and waste food items add to the calorific value of the MSW; MSW may thus turn out to be a source of low-cost energy to meet part of the city community needs when fed for energy recovery for steam production for power generation using properly designed modern incinerators.

The plastic waste may be segregated and burned separately or can form a part of mixed combustibles in MSW for use as solid fuel in boilers or other heat exchanger units. Plastics are viewed as high-value, low-moisture and variable ash-containing fuels for MSW incineration. Combustion data for major components of MSW are summarized in Table 14.4.

Table 14.4 *Combustion data for components of MSW*

Component in waste	*Heat energy MJ/kg*	*Ash (%)*	*Moisture °C*
Compostibles	4–7	15–30	60–70
Cardboard/Paper	12–17	8–10	5–15
Textiles (fibres, cords, ropes, nets, yarns, cloths)	13–17	2–5	20–23
Plastics and related items	35–40	2–10 (or higher depending on inorganic filler loading)	4–10

Power generation from MSW incinerators is on the rise in different parts of the world, not excluding India. According to data available, in Tokyo City itself, more than 4 million tons per year of MSW are currently being collected and processed in about 20 incinerators generating a total of 1000 MW of power. Municipal and industrial incinerators account for 30–50% of total trash disposal in most advanced countries including USA. Incinerators should be designed to operate with excess air (110–150% of calculated dose). The temperature should be 800–1000°C, so as to keep smoke and odour formation and possible pollution releases to a minimum. High combustion volume provides adequate residence time for the burnout of all flying particulate matter. Hospital (bio-medical) wastes are not required to be recycled; they are disposed of through autoclaving and incineration.

14.9 Composting[4]

Composting is a means of reducing volume of solid waste and returning a partly/reasonably usable product to the plant or community. Composting is an accelerated biological degradation of organic wastes leaving a low–volume residual matter called compost, containing mainly nitrogen, phosphorous and potassium etc. as nutrients for vegetation. Spread of disagreeable odour associated with the degradation process during composting of solid wastes often creates a problem and turns the process to one of limited acceptability, particularly in countries of the western hemisphere. Composting plants of 10–300 t/day capacity are running in different countries for over four decades. The modern plant is a rotary cylinder much resembling a cement kiln. The selectively pretreated/sorted waste material is fed at the top, and pathogenically pure compost is delivered at the lower end. In a special pretreatment step, glass and metal components are removed from the trash. Shredding/granulating at this stage ensures homogeneity and controls moisture content. Pretreatment accelerates the decomposition/degradation rate at temperatures where pathogenic organisms are destroyed. As the pretreated waste material is admitted in the kiln, air is admitted at low pressure and at a controlled rate over the length of the kiln. An environment is created inside, whereby the action of aerobic microorganism ensures further desired degradation/decomposition of the waste. Normal operating temperature is ~ 60°C and it is spontaneously developed without additional heat input. The weight reduction from feed to final compost is 30–40%. The resulting compost is good as plant nutrient or soil conditioner.

14.10 Integrated Waste Management for Sustainable Development

Sustainable development and striking a good ecological balance need to be ensured through what may be termed as integrated waste management. Environmental preservation needs to be closely linked with industrial production activities with industries' focus on material saving and energy saving and at the same time maintenance of specified standards for processes and products. Production further needs to be controlled as per regulatory authorities prescription with focus on avoidance or minimization of wastes or emissions and of pollutants of all other kinds, ensuring recovery, recycle, return or reuse as far as practicable and prescribing means for associated waste disposal. Lastly, use, storage or consumption of the product made and its desirable property range must remain under the scrutiny of the producers and consumers with constant focus on demand–supply situation and proper distribution, avoiding inappropriate or wasteful use.

Finally, Chart–I detailed hereunder gives a comprehensive outline of integrated waste management for sustainable development.

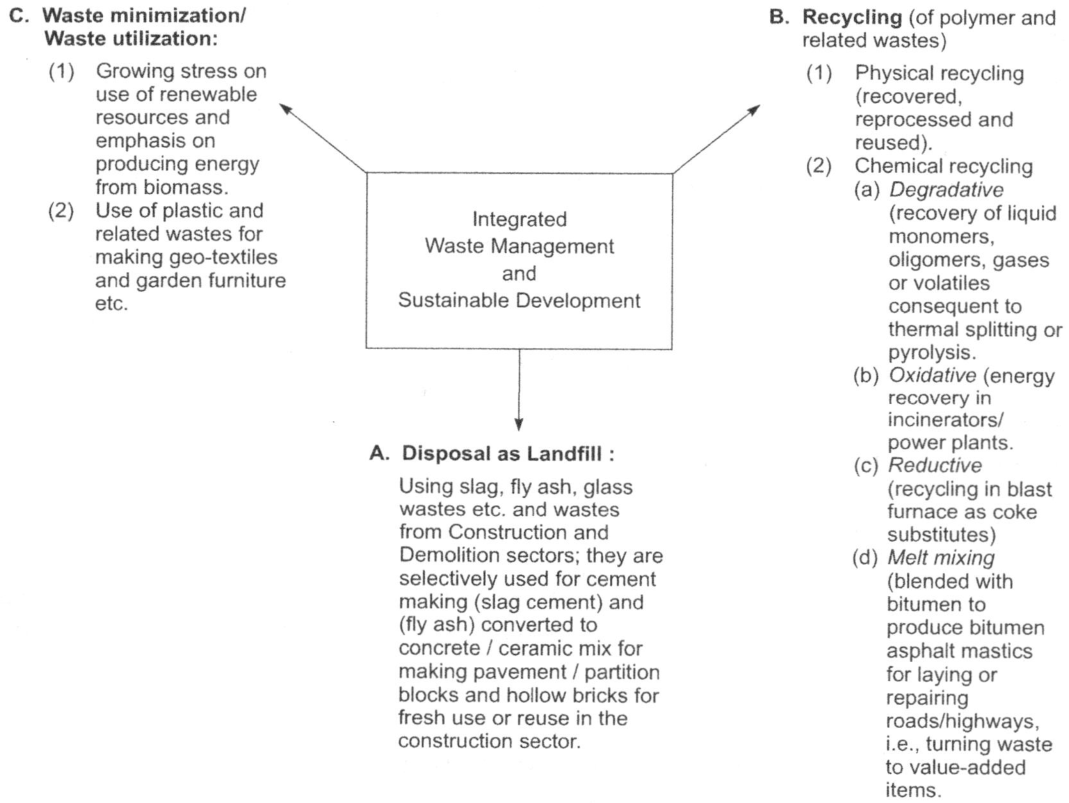

Chart–1 *Integrated waste management*

References

1. Ghosh, P., Power from Waste, Plastics India, 5 (3), (2007), pp. 24–27.
2. http: // www.icpenvison.org.
3. Plastics for Environment and Sustainable Development, ICPE, Delhi and CIPET, Chennai,
4. Encyclopedia of Polymer Sci., and Engg., H.F. Mark, N. M. Bikales, C.G. Overberges and G. Menges, Eds. Vol. 17, pp. 722 – 740 (1985).

Selected Reading References

Akelah, A and A. Moet, *Funtionalized polymers and their Applications,* Chapman and Hall, London 1990.

Baijlal, M.D. Ed., *Plastics, Polymer Science and Technology*, Wiley, New York, 1981.

Bark, L.S. and N.S. Allen, Ed., *Analysis of Polymer Systems*, Applied Science, London, 1982.

Billmeyer Jr., F.W., *Text Book of Polymer Science*, Interscience, New York, 1971.

Blow C.M., Ed., *Rubber Technology and Manufacture*, Butterworths, London, 1971.

Blow, C.M. and C. Hepburn, *Rubber Technology and Manufacture*, 2nd Ed., Butterworths, London, 1982.

Blythe, A.R., *Electrical Properties of Polymers*, Cambridge University Press, Cambridge, 1979.

Brown, R.M. Ed., *Cellulose and Other Natural Polymer Systems: Biogenesis, Structure and Degradation*, Plenum Press, New York, 1982.

Brown, R.P., *Physical Testing of Rubbers*, Applied Science, London, 1979.

Brydson, J.A., *Plastics Materials*, Butterworths, London, 1982.

Brydson, J.A., *Rubber Chemistry*, Applied Science, London, 1978.

Butt, L.T. and D.C. Wright, *Use of Polymers in Chemical Plant Construction*, Applied Science, London, 1980.

Chanda, M. and S.K. Roy, *Plastics Technology Handbook*, Marcel Dekker, New York, 1987.

Ciferri, A. and I.M. Ward, Eds, *Ultra High Modulus Polymers*, Applied Science, London, 1979.

Cassidy, P.E., *Thermally Stable Polymers*, Marcel Dekker, New York, 1980.

Dubois, J.H. and F.W. John, *Plastics*, Van Nostrand, New York, 1981.

Dyson, R.W., *Speciality Polymers*, Blackie, Glasgow and London, Chapman and Hall, New York, 1987.

Flokes, M.J., *Short Fibre Reinforced Thermoplastics*. American Chemical Society, Washington, D.C., 1982.

Flory, P.J., *Principles of Polymer Chemistry*, Cornell University, Press, Ithaca, New York, 1953.

Ghosh, P., *Fibre Science and Technology*. Tata McGraw-Hill, New Delhi, 2004.

Ghosh, P., *Adhesives and Coatings Technology*. Tata McGraw-Hill, New Delhi, 2008.

Ghosh, P. and P.K. Ganguli, *Jute, Encyclopedia of Polymeric Materials*, J.C. Salamone Ed., CRC Inc., New York, Vol 5, 3504-3513, 1996.

Hatada, K. and T. Kitayama Eds., *Macromolecular Design of Polymeric Materials*, Marcel Dekker, New York, 1999.

Henson, J.H.L. and A. Whelan Eds, *Developments in PVC Technology*, Applied Science, London, 1977.

Hiemenz, P.C., *Polymer Chemistry—The Basic Concepts*, Marcel Dekker, New York, 1984.

Huilgol, R.R. and N. Phan-Thien, *Fluid Mechanics of Viscoelasticity*, Elsevier, New York, 1997.

Karger-Koesis, J. Ed., *Polypropylene, An A–Z Reference*, Kluwer Academic, London, 1998.

Klempner, D. and K.C. Frisch, *Handbook of Polymeric Foam and Foam Technology*, Hanser, Munich, 1991.

Klenin, V.J., *Thermodynamics of Systems Containing Flexible Chain Polymers*, Elsevier, New York, 1999.

Mark, H.F. and N.G. Gaylord Eds., *Encyclopedia of Polymer Science and Technology* Wiley Interscience, New York, 1964-1972.

Minsker, K.S., S.V. Kolesov and G.E. Zaikov, *Degradation and Stabilization of Vinyl Chloride Based Polymer*, Pergamon Press, Oxford, 1988.

Monk, J.F. Ed., *Thermosetting Plastics, Practical Molding Technology*, George Godwin, London, 1981.

Morton, M. *Introduction to Rubber Technology*, Chapman and Hall, London, 1959.

Nass, L.I., *Encyclopaedia of PVC*, Marcel Dekker, New York, Vol. 1 (1976) and Vol. 2 (1977).

Naunton, W.J.S., *The Applied Science of Rubber*, Arnold, London, 1961.

Nelson, W.E., *Nylon Plastics Technology*, Newnes—Butterworths, London, 1976.

Odian, G., *Principles of Polymerization*, 4th ed., McGraw-Hill, New York, 2004.

O'Driscoll, K.F., *The Nature and Chemistry of High Polymers*, Reinhold, London, 1964.

Ray, N.H., *Inorganic Polymers*, Academic Press, London, 1978.

Richardson, M.O.W., *Polymer Engineering Composites*, Applied Science, London, 1977.

Robert, A.D. Ed., *Natural Rubber Science and Technology*, Oxford University Press, Oxford, 1988.

Roff, W.J., *Fibres, Plastics and Rubbers—A Hand Book of Common Polymers*, Butterworths, London, 1956.

Saunders, K.J., *Organic Polymer Chemistry*, Chapman and Hall, London, 1973.

Schmidt, A.X. and C.A. Marlies, *Principles of High Polymer Theory and Practice*, McGraw-Hill, New York, 1948.

Seymour R.B., *Conductive Polymers*, Plenum Press, New York, 1981.

Shonaike, G.O. and G.P. Simon Eds., *Polymer Blends and Alloys*, Mercel Dekker, New York, 1999.

Siginer, D.A., D. De kee and R.P. Chhabra Eds., *Advances in the Flow and Rheology of Non-Newtonian Fluids*, Elsevier, New York 1999.

Stern, H.J., *Rubber, Natural and Synthetic*, Maclaren, London, 1967.

Stille, J.K., *Introduction to Polymer Chemistry*, John Wiley, New York, 1962.

Tanner, R.I. and K. Walters, *Rheology: An Historical Perspective*, Elsevier, New York, 1998.

Treloar, L.G.R., *Introduction to Polymer Science*, Wykeham Publication, London, 1970.

Tsuruta, T. and K.F. O'Driscoll, Eds., *Structure and Mechanism in Vinyl Polymerization*, Marcel Dekker, New York, 1969.

Ulrich, H., *Introduction to Industrial Polymers*, Henser Publisher, Munchen, 1982.

Whelan, A. and J.A. Brydson Eds., *Developments with Thermosetting Plastics*, Applied Science, London, 1974.

Whelan, A. and J.L. Craft, Eds., *Developments in PVC Production and Processing I*, Applied Science, London, 1977.

Wilson, A.S., *Plasticizers: Principles and Practice*, The Institute of Materials, London, 1995.

Wypych, J., *Polyvinyl Chloride Stabilization*, Polymer Science Library 4, A.D. Jenkins Ed., Elsevier, Amsterdam, 1986.

Young, R.J., *Introduction to Polymers*, Chapman and Hall, London, 1981.

Index

Author's Profile

Dr Premamoy Ghosh, Ph.D., is former Professor and Head, Department of Polymer Science and Technology, Calcutta University. He is a Fellow of the Institution of Engineers, (India) and the Institute of Materials (London)—formerly, Plastics and Rubber Institute (London). He is a life member of the Indian Institute of Chemical Engineers and the Indian Association for the Cultivation of Science, Kolkata. Additionally, he is the current president of the Institute of Science, Education and Culture (ISEC), Kolkata, and chairman of the Editorial Board of its journal *Indian Science Cruiser*. Dr Ghosh is also a member of the editorial board of the *Journal of Polymer Materials*, M.D. Publications, New Delhi, and a former member of the editorial board of the *Indian Journal of Technology*, CSIR, New Delhi.

Dr Ghosh was born in 1936. He acquired his B.Sc. (Applied Chemistry) in 1958, and went on to do his Ph.D. (Polymer—Chemistry) in 1962—both from Calcutta University. He researches on polymer synthesis/modification, photopolymerization, polymerization kinetics and mechanistics, polymer microstructure, random, block and graft copolymerization, rubber compounding and vulcanization, reactive processing of polymers, polymer finishes on fibres and textiles, conducting polymers, polymer blends and composites, and polymer rheology and morphology.

Dr Ghosh has so far published more than 320 research publications and review articles. He has authored two books titled *Fibre Science and Technology* (2004) and *Adhesives and Coatings Technology* (2008), both published by Tata McGraw Hill. He has also written books on Society Science and Spiritualism in Bengali language, including one in an autobiographical style and another as a collection of poems.

Made in the USA
Monee, IL
08 November 2026

67812132R00328